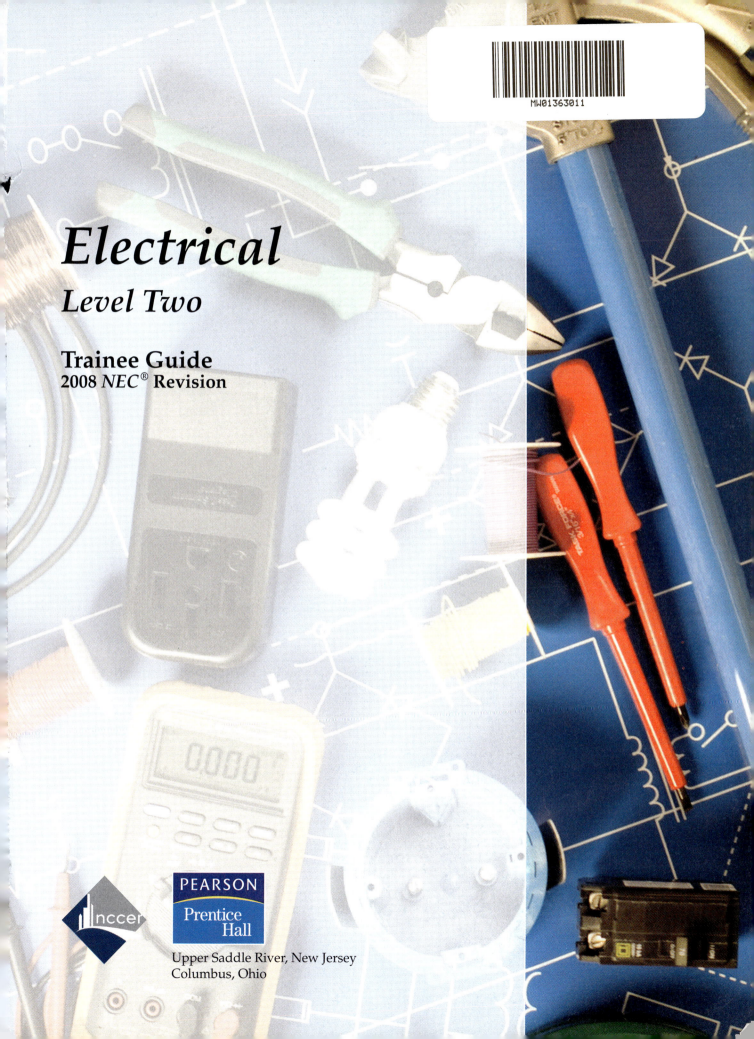

National Center for Construction Education and Research
President: Don Whyte
Director of Product Development: Daniele Stacey
Electrical Project Manager: Daniele Stacey
Production Manager: Tim Davis
Quality Assurance Coordinator: Debie Ness
Desktop Publishing Coordinator: James McKay
Editors: Rob Richardson, Matt Tischler, Brendan Coote

Writing and development services provided by Topaz Publications, Liverpool, NY
Lead Writer/Project Manager: Veronica Westfall
Desktop Publisher: Joanne Hart
Art Director: Megan Paye
Permissions Editors: Andrea LaBarge, Jackie Vidler
Writers: Tom Burke, Gerald Shannon, Nancy Brown, Charles Rogers

Pearson Education, Inc.
Product Manager: Lori Cowen
Product Development Editor: Janet Ryerson
Project Managers: Stephen C. Robb, Christina M. Taylor
AV Project Manager: Janet Portisch
Senior Operations Supervisor: Pat Tonneman
Art Director: Diane Y. Ernsberger
Text Designer: Kristina D. Holmes
Cover Designer: Kristina D. Holmes
Cover Photo: Tim Davis
Director of Marketing: David Gesell
Executive Marketing Manager: Derril Trakalo
Senior Marketing Coordinator: Alicia Dysert
Copyeditor: Sheryl Rose
Proofreader: Bret Workman

This book was set in Palatino and Helvetica by S4Carlisle Publishing Services and was printed and bound by Courier Kendallville, Inc. The cover was printed by Phoenix Color Corp.

Copyright © 2008, 2005, 2002, 1999, 1997, 1993 by the National Center for Construction Education and Research (NCCER), Gainesville, FL 32606, and published by Pearson Education, Inc., Upper Saddle River, NJ 07458. All rights reserved. Printed in the United States of America. This publication is protected by Copyright and permission should be obtained from NCCER prior to any prohibited reproduction, storage in a retrieval system, or transmission in any form or by any means, electronic, mechanical, photocopying, recording, or likewise. For information regarding permission(s), write to: NCCER Product Development, 3600 NW 43rd St., Bldg. G, Gainesville, FL 32606.

Pearson Prentice Hall™ is a trademark of Pearson Education, Inc.
Pearson® is a registered trademark of Pearson plc
Prentice Hall® is a registered trademark of Pearson Education, Inc.

Pearson Education Ltd., London
Pearson Education Singapore Pte. Ltd.
Pearson Education Canada, Inc.
Pearson Education—Japan

Pearson Education Australia Pty. Limited
Pearson Education North Asia Ltd., Hong Kong
Pearson Educación de Mexico, S.A. de C.V.
Pearson Education Malaysia Pte. Ltd.

10 9 8 7 6 5 4 3 2 1

Perfect bound	ISBN-13:	978-0-13-604466-6
	ISBN-10:	0-13-604466-2
Case bound	ISBN-13:	978-0-13-604465-9
	ISBN-10:	0-13-604465-4
Loose leaf	ISBN-13:	978-0-13-604467-3
	ISBN-10:	0-13-604467-0

Preface

Electricity powers the applications that make our daily lives more productive and efficient. Thirst for electricity has led to vast job opportunities in the electrical field. Electricians constitute one of the largest construction occupations in the United States, and they are among the highest-paid workers in the construction industry. According to the U.S. Bureau of Labor Statistics, job opportunities for electricians are expected to be very good as the demand for skilled craftspeople is projected to outpace the supply of trained electricians.

Electricians install electrical systems in structures. They install wiring and other electrical components, such as circuit breaker panels, switches, and light fixtures. Electricians follow blueprints, the *National Electrical Code*®, and state and local codes. They use specialized tools and testing equipment, such as ammeters, ohmmeters, and voltmeters. Electricians learn their trade through craft and apprenticeship programs. These programs provide classroom instruction and on-the-job training with experienced electricians.

We wish you success as you embark on your second year of training in the electrical craft and hope that you will continue your training beyond this textbook. There are more than 700,000 people employed in electrical work in the United States, and, as most of them can tell you, there are many opportunities awaiting those with the skills and desire to move forward in the construction industry.

NEW WITH *ELECTRICAL LEVEL TWO*

NCCER and Pearson Education, Inc. are pleased to present the sixth edition of *Electrical Level Two*. This edition has been updated to meet the 2008 *National Electrical Code*® and includes revisions to "Electric Lighting" and "Boxes and Fittings" (now "Pull and Junction Boxes") modules. "Electric Lighting" was revised to include lamps, ballasts, and components, while "Pull and Junction Boxes" was updated to cover boxes over 100 cubic inches and include *NEC*® requirements for handholes.

The opening page of each of the eleven modules in this textbook features photos of award-winning construction projects from Associated Builders and Contractors, Inc. and The Associated General Contractors of America. Check out these awesome projects and see where a career as an electrician could take you.

We invite you to visit the NCCER website at **www.nccer.org** for the latest releases, training information, newsletter, and much more. You can also reference the Contren® product catalog online at that site. Your feedback is welcome. Email your comments to **curriculum@nccer.org** or send general comments and inquiries to **info@nccer.org**.

CONTREN® LEARNING SERIES

The National Center for Construction Education and Research (NCCER) is a not-for-profit 501(c)(3) education foundation established in 1996 by the world's largest and most progressive construction companies and national construction associations. It was founded to address the severe workforce shortage facing the industry and to develop a standardized training process and curricula. Today, NCCER is supported by hundreds of leading construction and maintenance companies, manufacturers, and national associations. The Contren® Learning Series was developed by NCCER in partnership with Pearson Education, Inc., the world's largest educational publisher.

Some features of NCCER's Contren® Learning Series are as follows:

- An industry-proven record of success
- Curricula developed by the industry for the industry
- National standardization providing portability of learned job skills and educational credits
- Compliance with the Office of Apprenticeship requirements for related classroom training (CFR 29:29)
- Well-illustrated, up-to-date, and practical information

NCCER also maintains a National Registry that provides transcripts, certificates, and wallet cards to individuals who have successfully completed modules of NCCER's Contren® Learning Series. *Training programs must be delivered by an NCCER Accredited Training Sponsor in order to receive these credentials.*

Special Features of This Book

In an effort to provide a comprehensive, user-friendly training resource, we have incorporated many different features for your use. Whether you are a visual or hands-on learner, this book will provide you with the proper tools to get started in the electrical industry.

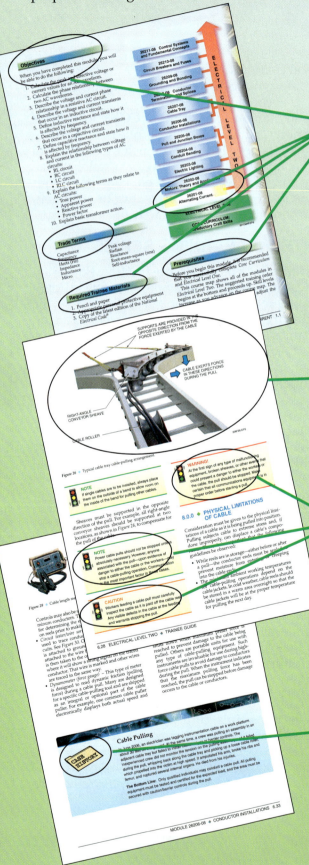

Introduction Page

This page is found at the beginning of each module and lists the Objectives, Trade Terms, Required Trainee Materials, Prerequisites, and Course Map for that module. The Objectives list the skills and knowledge you will need in order to complete the module successfully. The list of Trade Terms identifies important terms you will need to know by the end of the module. Required Trainee Materials list the materials and supplies needed for the module. The Prerequisites for the module are listed and illustrated in the Course Map. The Course Map also gives a visual overview of the entire course and a suggested learning sequence for you to follow.

Color Illustrations and Photographs

Full-color illustrations and photographs are used throughout each module to provide vivid detail. These figures highlight important concepts from the text and provide clarity for complex instructions. Each figure reference is denoted in the text in *italic type* for easy reference.

Notes, Cautions, and Warnings

Safety features are set off from the main text in highlighted boxes and are organized into three categories based on the potential danger of the issue being addressed. Notes simply provide additional information on the topic area. Cautions alert you of a danger that does not present potential injury but may cause damage to equipment. Warnings stress a potentially dangerous situation that may cause injury to you or a co-worker.

Case History

Case History features emphasize the importance of safety by citing examples of the costly (and often devastating) consequences of ignoring *National Electrical Code®* or OSHA regulations.

Inside Track

Inside Track features provide a head start for those entering the electrical field by presenting technical tips and professional practices from master electricians in a variety of disciplines. Inside Tracks often include real-life scenarios similar to those you might encounter on the job site.

What's wrong with this picture?

What's wrong with this picture? features include photos of actual code violations for identification and encourage you to approach each installation with a critical eye.

Think About It

Think About It features use "What if?" questions to help you apply theory to real-world experiences and put your ideas into action.

Going Green

Going Green looks at ways to preserve the environment, save energy, and make good choices regarding the health of the planet. Through the introduction of new construction practices and products, you will see how the "greening of America" has already taken root.

Step-by-Step Instructions

Step-by-step instructions are used throughout to guide you through technical procedures and tasks from start to finish. These steps show you not only how to perform a task but also how to do it safely and efficiently.

Trade Terms

Each module presents a list of Trade Terms that are discussed within the text and defined in the Glossary at the end of the module. These terms are denoted in the text with **blue bold type** upon their first occurrence. To make searches for key information easier, a comprehensive Glossary of Trade Terms from all modules is located at the back of this book.

Review Questions

Review Questions are provided to reinforce the knowledge you have gained. This makes them a useful tool for measuring what you have learned.

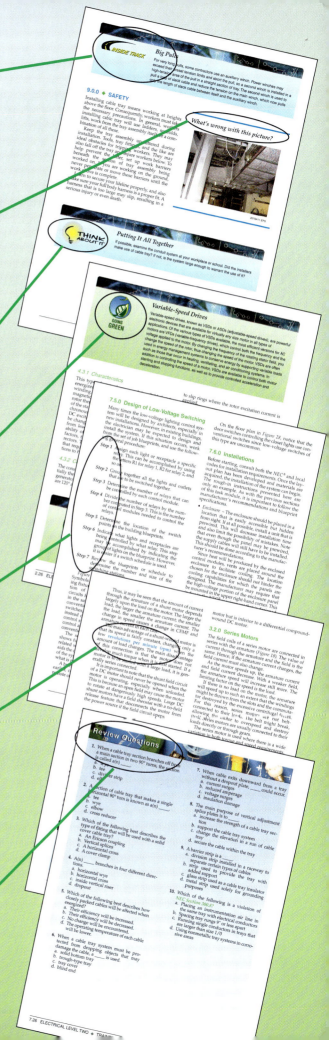

Contren® Curricula

NCCER's training programs comprise over 50 construction, maintenance, and pipeline areas and include skills assessments, safety training, and management education.

Boilermaking
Cabinetmaking
Carpentry
Concrete Finishing
Construction Craft Laborer
Construction Technology
Core Curriculum: Introductory Craft Skills
Drywall
Electrical
Electronic Systems Technician
Heating, Ventilating, and Air Conditioning
Heavy Equipment Operations
Highway/Heavy Construction
Hydroblasting
Industrial Coating and Lining Application Specialist
Industrial Maintenance Electrical and Instrumentation Technician
Industrial Maintenance Mechanic
Instrumentation
Insulating
Ironworking
Masonry
Millwright
Mobile Crane Operations
Painting
Painting, Industrial
Pipefitting
Pipelayer
Plumbing
Reinforcing Ironwork
Rigging
Scaffolding
Sheet Metal
Site Layout
Sprinkler Fitting
Welding

Pipeline
Control Center Operations, Liquid
Corrosion Control
Electrical and Instrumentation
Field Operations, Liquid
Field Operations, Gas
Maintenance
Mechanical

Safety
Field Safety
Safety Orientation
Safety Technology

Management
Introductory Skills for the Crew Leader
Project Management
Project Supervision

Spanish Translations
Andamios
Currículo Básico: Habilidades Introductorias del Oficio
Instalación de Rociadores Nivel Uno
Introducción a la Carpinteria
Orientación de Seguridad
Seguridad de Campo

Supplemental Titles
Applied Construction Math
Careers in Construction
Your Role in the Green Environment

Acknowledgments

This curriculum was revised as a result of the farsightedness and leadership of the following sponsors:

ABC of Iowa
ABC of New Mexico
ABC Pelican Chapter Southwest, Westlake, LA
Baker Electric
Beacon Electric Company
Cuyahoga Valley Career Center
M.C. Dean Inc.
Duck Creek Engineering
Hamilton Electric Construction Company
IMTI of New York and Connecticut
Lamphear Electric
Madison Comprehensive High School/Central Ohio ABC
Pumba Electric LLC
Putnam Career & Technical Center
Rust Constructors Inc.
TIC Industrial
Tri-City Electrical Contractors, Inc.
Trident Technical College
Vector Electric and Controls Inc.

This curriculum would not exist were it not for the dedication and unselfish energy of those volunteers who served on the Authoring Team. A sincere thanks is extended to the following:

John S. Autrey
Clarence "Ed" Cockrell
Scott Davis
Tim Dean
Gary Edgington
Tim Ely
Al Hamilton
William "Billy" Hussey
E. L. Jarrell
Dan Lamphear
Leonard R. "Skip" Layne
L. J. LeBlanc
David Lewis
Neil Matthes
Jim Mitchem
Christine Porter
Michael J. Powers
Wayne Stratton
Marcel Veronneau
Irene Ward

A final note: This book is the result of a collaborative effort involving the production, editorial, and development staff at Pearson Education, Inc., and the National Center for Construction Education and Research. Thanks to all of the dedicated people involved in the many stages of this project.

PARTNERING ASSOCIATIONS

ABC Texas Gulf Coast Chapter
American Fire Sprinkler Association
Associated Builders & Contractors, Inc.
Associated General Contractors of America
Association for Career and Technical Education
Association for Skilled & Technical Sciences
Carolinas AGC, Inc.
Carolinas Electrical Contractors Association
Center for the Improvement of Construction Management and Processes
Construction Industry Institute
Construction Users Roundtable
Design Build Institute of America
Electronic Systems Industry Consortium
Merit Contractors Association of Canada
Metal Building Manufacturers Association
NACE International
National Association of Minority Contractors
National Association of Women in Construction
National Insulation Association
National Ready Mixed Concrete Association
National Systems Contractors Association
National Technical Honor Society
National Utility Contractors Association
NAWIC Education Foundation
North American Crane Bureau
North American Technician Excellence
Painting & Decorating Contractors of America
Portland Cement Association
SkillsUSA
Steel Erectors Association of America
U.S. Army Corps of Engineers
University of Florida
Women Construction Owners & Executives, USA

Contents

26201-08 *Alternating Current* 1.i
Focuses on forces that are characteristic of alternating-current systems and the application of Ohm's law to AC circuits. **(17.5 Hours)**

26202-08 *Motors: Theory and Application* 2.i
Covers AC and DC motors, including the main components, circuits, and connections. **(20 Hours)**

26203-08 *Electric Lighting* 3.i
Introduces the basic principles of human vision and the characteristics of light. Focuses on the handling and installation of various types of lamps and lighting fixtures. **(15 Hours)**

26204-08 *Conduit Bending* 4.i
Covers all types of bends in all sizes of conduit up to 6 inches. Focuses on mechanical, hydraulic, and electrical benders. **(15 Hours)**

26205-08 *Pull and Junction Boxes* 5.i
Driven by the *NEC*®. Explains how to select and size pull boxes, junction boxes, and handholes. **(12.5 Hours)**

26206-08 *Conductor Installations* 6.i
Covers the transportation, storage, and setup of cable reels; methods of rigging; and procedures for complete cable pulls in raceways and cable trays. **(10 Hours)**

26207-08 *Cable Tray* 7.i
Focuses on *NEC*® installation requirements for cable tray, including cable installations. **(7.5 Hours)**

26208-08 Conductor Terminations
and Splices 8.i

Describes methods of terminating and splicing conductors of all types and sizes, including preparing and taping conductors. **(7.5 Hours)**

26209-08 Grounding and Bonding 9.i

Focuses on the purpose of grounding and bonding electrical systems. Thoroughly covers *NEC®* regulations. **(15 Hours)**

26210-08 Circuit Breakers and Fuses 10.i

Describes fuses and circuit breakers along with their practical applications. Also covers sizing. **(12.5 Hours)**

26211-08 Control Systems and Fundamental
Concepts 11.i

Gives basic descriptions of various types of contactors and relays along with their practical applications. **(12.5 Hours)**

Glossary of Key Terms G.1

Figure Credits FC.1

Index I.1

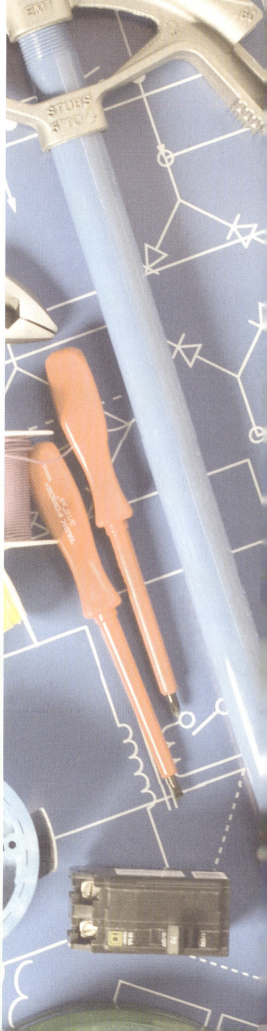

Alternating Current

Pawnee Health Center

W.W. Enterprises won an Excellence in Construction award from ABC in the Commercial-Electrical division for its work on the Pawnee Health Center in Pawnee, Oklahoma, which serves the Native American Pawnee Nation.

26201-08

26201-08
Alternating Current

Topics to be presented in this module include:

1.0.0	Introduction	1.2
2.0.0	Sine Wave Generation	1.2
3.0.0	Sine Wave Terminology	1.4
4.0.0	AC Phase Relationships	1.6
5.0.0	Nonsinusoidal Waveforms	1.9
6.0.0	Resistance in AC Circuits	1.11
7.0.0	Inductance in AC Circuits	1.12
8.0.0	Capacitance	1.15
9.0.0	LC and RLC Circuits	1.21
10.0.0	Power in AC Circuits	1.32
11.0.0	Transformers	1.35

Overview

The vast majority of your work as an electrician will be with alternating current (AC) circuits, because AC provides the power to operate the appliances and machines in our homes and businesses. Although direct current (DC) was used in the original electrical systems developed by Thomas Edison, the dynamic nature of AC makes it much easier to generate and distribute than DC. On the other hand, circuit analysis is more difficult because, unlike DC, AC does not act the same way in components such as coils and capacitors as it does in resistive components. When circuits contain these AC components, Ohm's law cannot be directly applied. Special formulas must be used to determine the circuit resistance.

Because alternating current is in a constant state of change with regard to polarities and values, its dynamics make it very useful. Induction-dependent components such as transformers, motors, and coils rely on the changing polarity and voltage values of alternating current in order to function.

Note: NFPF 70®, *National Electrical Code*®, and *NEC*® are registered trademarks of the National Fire Protection Association, Inc., Quincy, MA 02269. All *National Electrical Code*® and *NEC*® references in this module refer to the 2008 edition of the *National Electrical Code*®.

Objectives

When you have completed this module, you will be able to do the following:

1. Calculate the peak and effective voltage or current values for an AC waveform.
2. Calculate the phase relationship between two AC waveforms.
3. Describe the voltage and current phase relationship in a resistive AC circuit.
4. Describe the voltage and current transients that occur in an inductive circuit.
5. Define inductive reactance and state how it is affected by frequency.
6. Describe the voltage and current transients that occur in a capacitive circuit.
7. Define capacitive reactance and state how it is affected by frequency.
8. Explain the relationship between voltage and current in the following types of AC circuits:
   - RL circuit
   - RC circuit
   - LC circuit
   - RLC circuit
9. Explain the following terms as they relate to AC circuits:
   - True power
   - Apparent power
   - Reactive power
   - Power factor
10. Explain basic transformer action.

Trade Terms

Capacitance
Frequency
Hertz (Hz)
Impedance
Inductance
Micro
Peak voltage
Radian
Reactance
Root-mean-square (rms)
Self-inductance

Required Trainee Materials

1. Pencil and paper
2. Appropriate personal protective equipment
3. Copy of the latest edition of the *National Electrical Code®*

Prerequisites

Before you begin this module, it is recommended that you successfully complete *Core Curriculum* and *Electrical Level One*.

This course map shows all of the modules in *Electrical Level Two*. The suggested training order begins at the bottom and proceeds up. Skill levels increase as you advance on the course map. The local Training Program Sponsor may adjust the training order.

ELECTRICAL LEVEL TWO

- 26211-08 Control Systems and Fundamental Concepts
- 26210-08 Circuit Breakers and Fuses
- 26209-08 Grounding and Bonding
- 26208-08 Conductor Terminations and Splices
- 26207-08 Cable Tray
- 26206-08 Conductor Installations
- 26205-08 Pull and Junction Boxes
- 26204-08 Conduit Bending
- 26203-08 Electric Lighting
- 26202-08 Motors: Theory and Application
- 26201-08 Alternating Current

ELECTRICAL LEVEL ONE

CORE CURRICULUM: Introductory Craft Skills

1.0.0 ◆ INTRODUCTION

Alternating current (AC) and its associated voltage reverses between positive and negative polarities and varies in amplitude with time. One complete waveform or cycle includes a complete set of variations, with two alternations in polarity. Many sources of voltage change direction with time and produce a resultant waveform. The most common AC waveform is the sine wave.

2.0.0 ◆ SINE WAVE GENERATION

To understand how the alternating current sine wave is generated, some of the basic principles learned in magnetism should be reviewed. Two principles form the basis of all electromagnetic phenomena:

- An electric current in a conductor creates a magnetic field that surrounds the conductor.
- Relative motion between a conductor and a magnetic field, when at least one component of that relative motion is in a direction that is perpendicular to the direction of the field, creates a voltage in the conductor.

Figure 1 shows how these principles are applied to generate an AC waveform in a simple one-loop rotary generator. The conductor loop rotates through the magnetic field to generate the induced AC voltage across its open terminals. The magnetic flux shown here is vertical.

There are several factors affecting the magnitude of voltage developed by a conductor through a magnetic field. They are the strength of the mag-

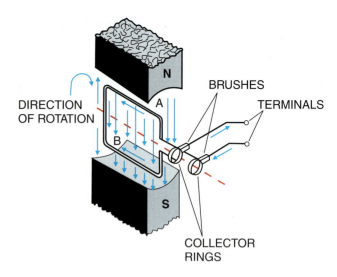

Figure 1 ◆ Conductor moving across a magnetic field.

netic field, the length of the conductor, and the rate at which the conductor cuts directly across or perpendicular to the magnetic field.

Assuming that the strength of the magnetic field and the length of the conductor making the loop are both constant, the voltage produced will vary depending on the rate at which the loop cuts directly across the magnetic field.

The rate at which the conductor cuts the magnetic field depends on two things: the speed of the generator in revolutions per minute (rpm) and the angle at which the conductor is traveling through the field. If the generator is operated at a constant rpm, the voltage produced at any moment will

INSIDE TRACK

Why Do Power Companies Generate and Distribute AC Power Instead of DC Power?

The transformer is the key. Power plants generate and distribute AC power because it permits the use of transformers, which makes power delivery more economical. Transformers used at generation plants step the AC voltage up, which decreases the current. Decreased current allows smaller-sized wires to be used for the power transmission lines. Smaller wire is less expensive and easier to support over the long distances that the power must travel from the generation plant to remotely located substations. At the substations, transformers are again used to step AC voltages back down to a level suitable for distribution to homes and businesses.

There is no such thing as a DC transformer. This means DC power would have to be transmitted at low voltages and high currents over very large-sized wires, making the process very uneconomical. When DC is required for special applications, the AC voltage may be converted to DC voltage by using rectifiers, which make the change electrically, or by using AC motor–DC generator sets, which make the change mechanically.

depend on the angle at which the conductor is cutting the field at that instant.

In *Figure 2*, the magnetic field is shown as parallel lines called lines of flux. These lines always go from the north to south poles in a generator. The motion of the conductor is shown by the large arrow.

Assuming the speed of the conductor is constant, as the angle between the flux and the conductor motion increases, the number of flux lines cut in a given time (the rate) increases. When the conductor is moving parallel to the lines of flux (angle of 0°), it is not cutting any of them, and the voltage will be zero.

The angle between the lines of flux and the motion of the conductor is called θ (theta). The magnitude of the voltage produced will be proportional to the sine of the angle. Sine is a trigonometric function. Each angle has a sine value that never changes.

The sine of 0° is 0. It increases to a maximum of 1 at 90°. From 90° to 180°, the sine decreases back to 0. From 180° to 270°, the sine decreases to −1. Then from 270° to 360° (back to 0°), the sine increases to its original 0.

Because voltage is proportional to the sine of the angle, as the loop goes 360° around the circle the voltage will increase from 0 to its maximum at 90°, back to 0 at 180°, down to its maximum negative value at 270°, and back up to 0 at 360°, as shown in *Figure 3*.

Notice that at 180° the polarity reverses. This is because the conductor has turned completely around and is now cutting the lines of flux in the opposite direction. This can be shown using the left-hand rule for generators. The curve shown in *Figure 3* is called a sine wave because its shape is generated by the trigonometric function sine. The value of voltage at any point along the sine wave can be calculated if the angle and the maximum obtainable voltage (E_{max}) are known.

The formula used is:

$$E = E_{max} \sin \theta$$

Where:

E = voltage induced
E_{max} = maximum induced voltage
θ = angle at which the voltage is induced

Using the above formula, the values of voltage anywhere along the sine wave in *Figure 3* can be calculated. Sine values can be found using either a scientific calculator or trigonometric tables. With an E_{max} of 10 volts (V), the following values are calculated as examples:

$\theta = 0°$, sine = 0
$E = E_{max} \sin \theta$
$E = (10V)(0)$
$E = 0V$

$\theta = 90°$, sine = 1.0
$E = E_{max} \sin \theta$
$E = (10V)(1.0)$
$E = 10V$

$\theta = 180°$, sine = 0
$E = E_{max} \sin \theta$
$E = (10V)(0)$
$E = 0V$

$\theta = 270°$, sine = −1.0
$E = E_{max} \sin \theta$
$E = (10V)(-1.0)$
$E = -10V$

$\theta = 45°$, sine = 0.707
$E = E_{max} \sin \theta$
$E = (10V)(0.707)$
$E = 7.07V$

$\theta = 135°$, sine = 0.707
$E = E_{max} \sin \theta$
$E = (10V)(0.707)$
$E = 7.07V$

$\theta = 225°$, sine = −0.707
$E = E_{max} \sin \theta$
$E = (10V)(-0.707)$
$E = -7.07V$

$\theta = 315°$, sine = −0.707
$E = E_{max} \sin \theta$
$E = (10V)(-0.707)$
$E = -7.07V$

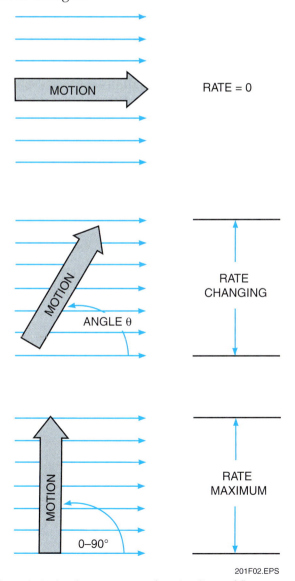

Figure 2 ♦ Angle versus rate of cutting lines of flux.

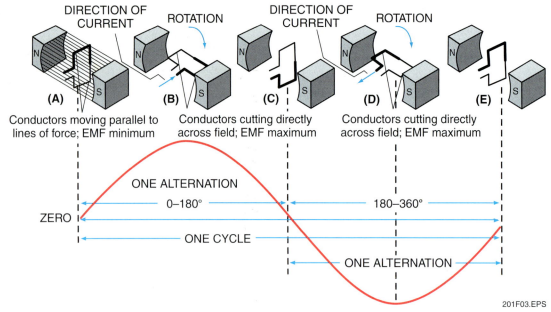

Figure 3 ◆ One cycle of alternating voltage.

3.0.0 ◆ SINE WAVE TERMINOLOGY

There are a number of AC voltage terms that are specific to sine waves. The following sections discuss some of these terms, including frequency, wavelength, peak value, average value, and effective value.

3.1.0 Frequency

The frequency of a waveform is the number of times per second an identical pattern repeats itself. Each time the waveform changes from zero to a peak value and back to zero is called an alternation. Two alternations form one cycle. The number of cycles per second is the frequency. The unit of frequency is hertz (Hz). One hertz equals one cycle per second (cps).

For example, let us determine the frequency of the waveform shown in *Figure 4*.

In one-half second, the basic sine wave is repeated five times. Therefore, the frequency (f) is:

$$f = \frac{5 \text{ cycles}}{0.5 \text{ second}} = 10 \text{ cycles per second (Hz)}$$

3.1.1 Period

The period of a waveform is the time (t) required to complete one cycle. The period is the inverse of frequency:

$$t = \frac{1}{f}$$

Where:

t = period (seconds)
f = frequency (Hz or cps)

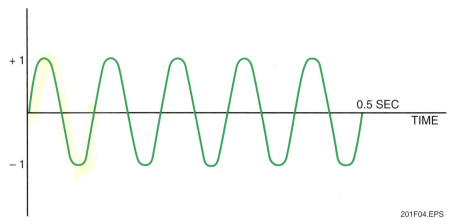

Figure 4 ◆ Frequency measurement.

1.4 ELECTRICAL LEVEL TWO ◆ TRAINEE GUIDE

For example, let us determine the period of the waveform in *Figure 4*. If there are five cycles in one-half second, then the frequency for one cycle is 10 cps (0.5 ÷ 5 = 10). Therefore, the period is:

$$t = \frac{1}{cps}$$

$$t = \frac{1}{10} = 0.1 \text{ second}$$

3.2.0 Wavelength

The wavelength or λ (lambda) is the distance traveled by a waveform during one period. Since electricity travels at the speed of light (186,000 miles/second or 300,000,000 meters/second), the wavelength of electrical waveforms equals the product of the period and the speed of light (c):

$$\lambda = tc$$

or:

$$\lambda = \frac{c}{f}$$

Where:

λ = wavelength (meters)
t = period (seconds)
c = speed of light (meters/second)
f = frequency (Hz or cps)

3.3.0 Peak Value

The peak value is the maximum value of voltage (V_M) or current (I_M). For example, specifying that a sine wave has a **peak voltage** of 170V applies to either the positive or the negative peak. To include both peak amplitudes, the peak-to-peak (p–p) value may be specified. In the above example, the peak-to-peak value is 340V, double the peak value of 170V, because the positive and negative peaks are symmetrical. However, the two opposite peak values cannot occur at the same time. Furthermore, in some waveforms the two peaks are not equal. The positive peak value and peak-to-peak value of a sine wave are shown in *Figure 5*.

3.4.0 Average Value

The average value is calculated from all the values in a sine wave for one alternation or half cycle. The half cycle is used for the average because over a full cycle the average value is zero, which is useless for comparison purposes. If the sine values for all angles up to 180° in one alternation are added and then divided by the number of values, this average equals 0.637.

Since the peak value of the sine is 1 and the average equals 0.637, the average value can be calculated as follows:

Average value = 0.637 × peak value

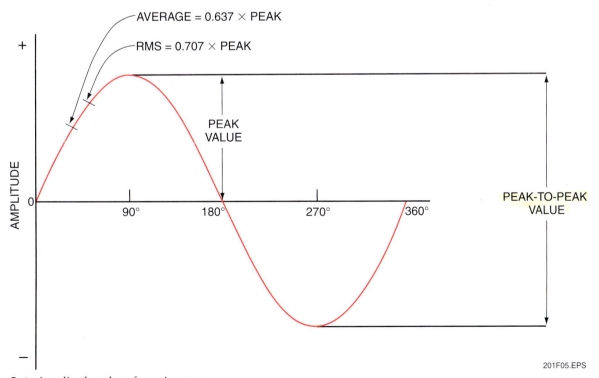

Figure 5 ♦ Amplitude values for a sine wave.

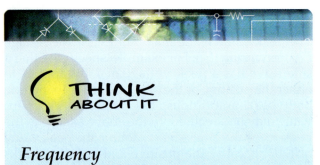

Frequency

The frequency of the utility power generated in the United States is normally 60Hz. In some European countries and elsewhere, utility power is often generated at a frequency of 50Hz. Which of these frequencies (60Hz or 50Hz) has the shortest period?

For example, with a peak of 170V, the average value is 0.637 × 170V, which equals approximately 108V. *Figure 5* shows where the average value would fall on a sine wave.

3.5.0 Root-Mean-Square or Effective Value

Meters used in AC circuits indicate a value called the effective value. The effective value is the value of the AC current or voltage wave that indicates the same energy transfer as an equivalent direct current (DC) or voltage.

The direct comparison between DC and AC is in the heating effect of the two currents. Heat produced by current is a function of current amplitude only and is independent of current direction. Thus, heat is produced by both alternations of the AC wave, although the current changes direction during each alternation.

In a DC circuit, the current maintains a steady amplitude. Therefore, the heat produced is steady and is equal to I^2R. In an AC circuit, the current is continuously changing; periodically high, periodically low, and periodically zero. To produce the same amount of heat from AC as from an equivalent amount of DC, the instantaneous value of the AC must at times exceed the DC value.

By averaging the heating effects of all the instantaneous values during one cycle of alternating current, it is possible to find the average heat produced by the AC current during the cycle. The amount of DC required to produce that heat will be equal to the effective value of the AC.

The most common method of specifying the amount of a sine wave of voltage or current is by stating its value at 45°, which is 70.7% of the peak. This is its **root-mean-square (rms)** value. Therefore,

Value of rms = 0.707 × peak value

For example, with a peak of 170V, the rms value is 0.707 × 170, or approximately 120V. This is the voltage of the commercial AC power line, which is always given in rms value.

4.0.0 ◆ AC PHASE RELATIONSHIPS

In AC systems, phase is involved in two ways: the location of a point on a voltage or current wave with respect to the starting point of the wave or with respect to some corresponding point on the same wave. In the case of two waves of the same frequency, it is the time at which an event of one takes place with respect to a similar event of the other.

Often, the event is the starting of the waves at zero or the points at which the waves reach their maximum values. When two waves are compared in this manner, there is a phase lead or lag of one with respect to the other unless they are alternating in unison, in which case they are said to be in phase.

4.1.0 Phase Angle

Suppose that a generator started its cycle at 90° where maximum voltage output is produced instead of starting at the point of zero output. The two output voltage waves are shown in *Figure 6*. Each is the same waveform of alternating voltage, but wave B starts at the maximum value while wave A starts at zero. The complete cycle of wave B through 360° takes it back to the maximum value from which it started.

Wave A starts and finishes its cycle at zero. With respect to time, wave B is ahead of wave A in its values of generated voltage. The amount it leads in time equals one quarter revolution, which is 90°. This angular difference is the phase angle between waves B and A. Wave B leads wave A by the phase angle of 90°.

The 90° phase angle between waves B and A is maintained throughout the complete cycle and in all successive cycles as long as they both have the same frequency. At any instant in time, wave B has the value that A will have 90° later. For instance, at 180°, wave A is at zero, but B is already at its negative maximum value, the point where wave A will be later at 270°.

To compare the phase angle between two waves, both waves must have the same frequency. Otherwise, the relative phase keeps changing. Both waves must also have sine wave variations, because this is the only kind of waveform that is measured in angular units of time. The amplitudes can be different for the two waves. The phases of two voltages, two currents, or a current with a voltage can be compared.

Left-Hand Rule for Generators

Hand rules for generators and motors give direction to the basic principles of induction. For a generator, if you move a conductor through a magnetic field made up of flux lines, you will induce an EMF, which drives current through a conductor. The left-hand rule for generators will help you determine which direction the current will flow in the conductor. It states that if you hold the thumb, first, and middle fingers of the left hand at right angles to one another with the first finger pointing in the flux direction (from the north pole to the south pole), and the thumb pointing in the direction of motion of the conductor, the middle finger will point in the direction of the induced voltage (EMF). The polarity of the EMF determines the direction in which current will flow as a result of this induced EMF. The left-hand rule for generators is also called Fleming's first rule.

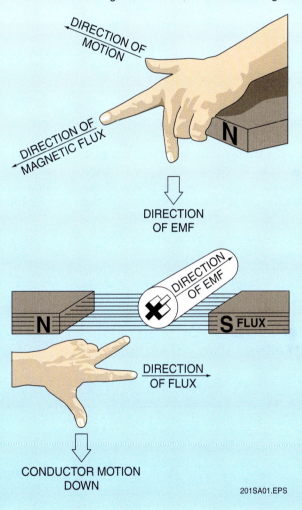

4.2.0 Phase Angle Diagrams

To compare AC phases, it is much more convenient to use vector diagrams corresponding to the voltage and current waveforms, as shown in *Figure 6*. V_A and V_B represent the vector quantities corresponding to the generator voltage.

A vector is a quantity that has magnitude and direction. The length of the arrow indicates the magnitude of the alternating voltage in rms, peak, or any AC value as long as the same measure is used for all the vectors. The angle of the arrow with respect to the horizontal axis indicates the phase angle.

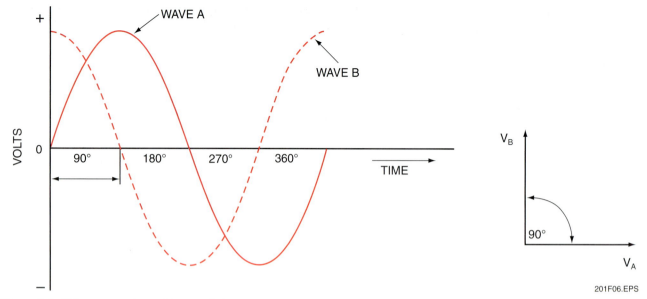

Figure 6 ◆ Voltage waveforms 90° out of phase.

In *Figure 6*, the vector V_A represents the voltage wave A, with a phase angle of 0°. This angle can be considered as the plane of the loop in the rotary generator where it starts with zero output voltage. The vector V_B is vertical to show the phase angle of 90° for this voltage wave, corresponding to the vertical generator loop at the start of its cycle. The angle between the two vectors is the phase angle.

The symbol for a phase angle is θ (theta). In *Figure 7*, $\theta = 0°$. *Figure 7* shows the waveforms and phasor diagram of two waves that are in phase but have different amplitudes.

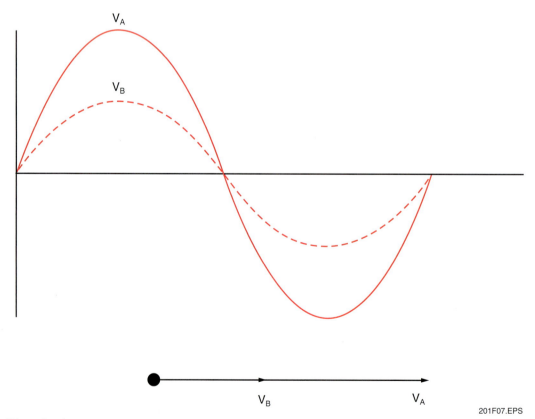

Figure 7 ◆ Waves in phase.

1.8 ELECTRICAL LEVEL TWO ◆ TRAINEE GUIDE

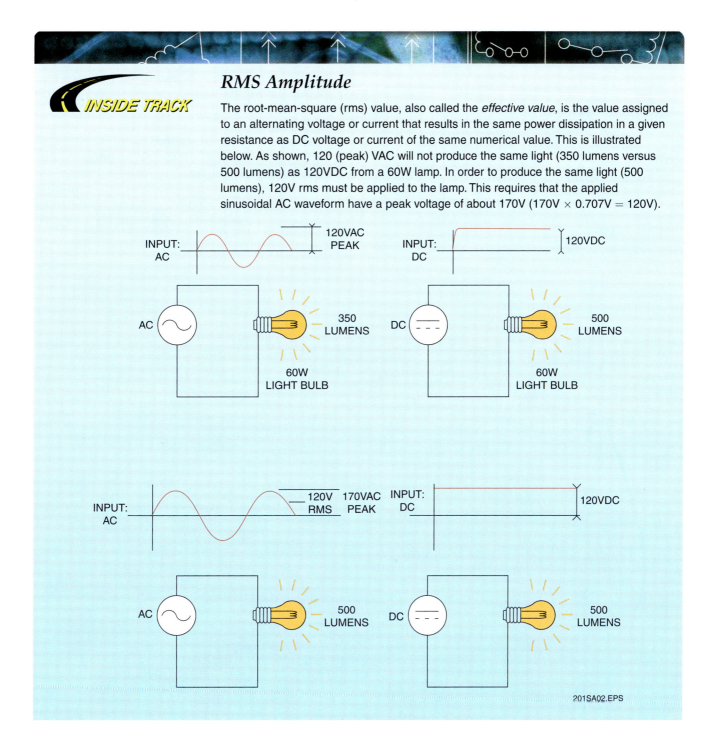

5.0.0 ♦ NONSINUSOIDAL WAVEFORMS

The sine wave is the basic waveform for AC variations for several reasons. This waveform is produced by a rotary generator, as the output is proportional to the angle of rotation. Because of its derivation from circular motion, any sine wave can be analyzed in angular measure, either in degrees from 0° to 360° or in radians from 0 to 2π radians.

In many electronic applications, however, other waveshapes are important. Any waveform that is not a sine (or cosine) wave is a nonsinusoidal waveform. Common examples are the square wave and sawtooth wave in *Figure 8*.

With nonsinusoidal waveforms for either voltage or current, there are important differences and

MODULE 26201-08 ♦ ALTERNATING CURRENT 1.9

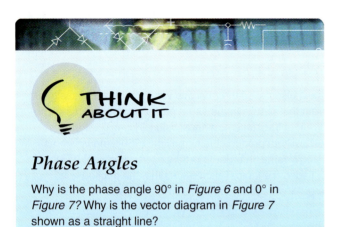

Phase Angles

Why is the phase angle 90° in *Figure 6* and 0° in *Figure 7*? Why is the vector diagram in *Figure 7* shown as a straight line?

similarities to consider. Note the following comparisons with sine waves:

- In all cases, the cycle is measured between two points having the same amplitude and varying in the same direction. The period is the time for one cycle.
- Peak amplitude is measured from the zero axis to the maximum positive or negative value. However, peak-to-peak amplitude is better for measuring nonsinusoidal waveshapes because they can have asymmetrical peaks, as with the rectangular wave in *Figure 8*.

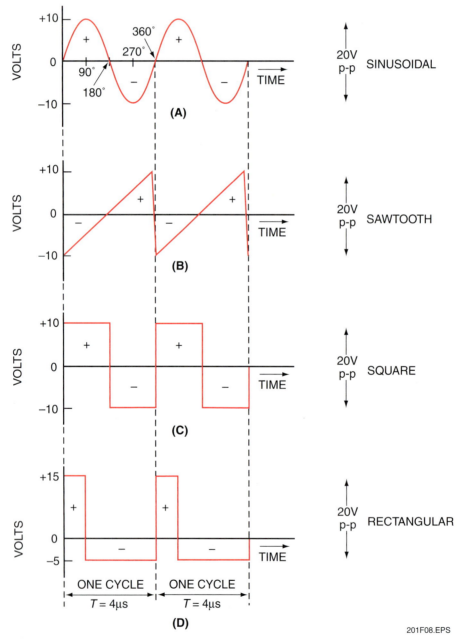

Figure 8 ◆ AC waveforms.

- The rms value 0.707 of peak applies only to sine waves, as this factor is derived from the sine values in the angular measure used only for the sine waveform.
- Phase angles apply only to sine waves, as angular measure is used only for sine waves. Note that the phase angle is indicated only on the sine wave of *Figure 8*.

6.0.0 ◆ RESISTANCE IN AC CIRCUITS

An AC circuit has an AC voltage source. Note the circular symbol with the sine wave inside it shown in *Figure 9*. It is used for any source of sine wave alternating voltage. This voltage connected across an external load resistance produces alternating current of the same waveform, frequency, and phase as the applied voltage.

According to Ohm's law, current (I) equals voltage (E) divided by resistance (R). When E is an rms value, I is also an rms value. For any instantaneous value of E during the cycle, the value of I is for the corresponding instant of time.

In an AC circuit with only resistance, the current variations are in phase with the applied voltage, as shown in *Figure 9*. This in-phase relationship between E and I means that such an AC circuit can be analyzed by the same methods used for DC circuits since there is no phase angle to consider. Components that have only resistance include resistors, the filaments for incandescent light bulbs, and vacuum tube heaters.

In purely resistive AC circuits, the voltage, current, and resistance are related by Ohm's law because the voltage and current are in phase.

$$I = \frac{E}{R}$$

Unless otherwise noted, the calculations in AC circuits are generally in rms values. For example, in *Figure 9*, the 120V applied across the 10Ω

What's wrong with this picture?

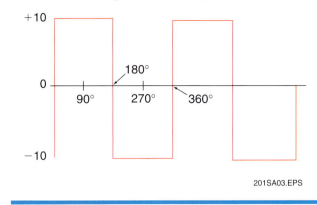

resistance R_L produces an rms current of 12A. This is determined as follows:

$$I = \frac{E}{R_L} = \frac{120V}{10\Omega} = 12A$$

Furthermore, the rms power (true power) dissipation is I^2R or:

$$P = (12A)^2 \times 10\Omega = 1{,}440W$$

Figure 10 shows the relationship between voltage and current in purely resistive AC circuits. The voltage and current are in phase, their cycles begin and end at the same time, and their peaks occur at the same time.

The value of the voltage shown in *Figure 10* depends on the applied voltage to the circuit. The value of the current depends on the applied voltage and the amount of resistance. If resistance is changed, it will affect only the magnitude of the current.

The total resistance in any AC circuit, whether it is a series, parallel, or series-parallel circuit, is calculated using the same rules that were learned and applied to DC circuits with resistance. Power computations are discussed later in this module.

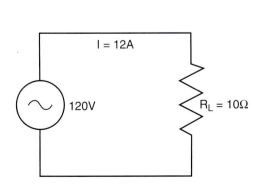

Figure 9 ◆ Resistive AC circuit.

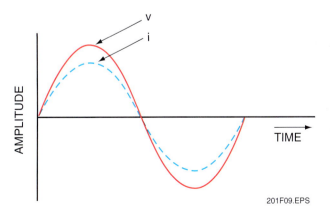

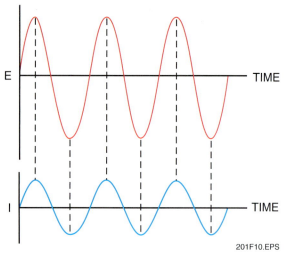

Figure 10 ♦ Voltage and current in a resistive AC circuit.

7.0.0 ♦ INDUCTANCE IN AC CIRCUITS

Inductance is the characteristic of an electrical circuit that opposes the change of current flow. It is the result of the expanding and collapsing field caused by the changing current. This moving flux cuts across the conductor that is providing the current, producing induced voltage in the wire itself. Furthermore, any other conductor in the field, whether carrying current or not, is also cut by the varying flux and has induced voltage. This induced current opposes the current flow that generated it.

In DC circuits, a change must be initiated in the circuit to cause inductance. The current must change to provide motion of the flux. A steady DC of 10A cannot produce any induced voltage as long as the current value is constant. A current of 1A changing to 2A does induce voltage. Also, the faster the current changes, the higher the induced voltage becomes, because when the flux moves at a higher speed it can induce more voltage.

However, in an AC circuit the current is continuously changing and producing induced voltage. Lower frequencies of AC require more inductance to produce the same amount of induced voltage as a higher frequency current. The current can have any waveform as long as the amplitude is changing.

The ability of a conductor to induce voltage in itself when the current changes is its **self-inductance,** or simply inductance. The symbol for inductance is L and its unit is the henry (H). One henry is the amount of inductance that allows one volt to be induced when the current changes at the rate of one ampere per second.

7.1.0 Factors Affecting Inductance

An inductor is a coil of wire that may be wound on a core of metal or paper, or it may be self-supporting. It may consist of turns of wire placed side by side to form a layer of wire over the core or coil form. The inductance of a coil or inductor depends on its physical construction. Some of the factors affecting inductance are:

- *Number of turns* – The greater the number of turns, the greater the inductance. In addition, the spacing of the turns on a coil also affects inductance. A coil that has widely-spaced turns has a lower inductance than one that has the same number of more closely-spaced turns. The reason for this higher inductance is that the closely-wound turns produce a more concentrated magnetic field, causing the coil to exhibit a greater inductance.
- *Coil diameter* – The inductance increases directly as the cross-sectional area of the coil increases.
- *Length of the core* – When the length of the core is decreased, the turn spacing is decreased, increasing the inductance of the coil.
- *Core material* – The core of the coil can be either a magnetic material (such as iron) or a non-magnetic material (such as paper or air). Coils wound on a magnetic core produce a stronger magnetic field than those with non-magnetic cores, giving them higher values of inductance. Air-core coils are used where small values of inductance are required.
- *Winding the coil in layers* – The more layers used to form a coil, the greater the effect the magnetic field has on the conductor. Layering a coil can increase the inductance.

Factors affecting the inductance of a coil can be seen in *Figure 11*.

7.2.0 Voltage and Current in an Inductive AC Circuit

The self-induced voltage across an inductance L is produced by a change in current with respect to time ($\Delta i / \Delta t$) and can be stated as:

$$V_L = L \frac{\Delta i}{\Delta t}$$

Where:

Δ = change

V_L = volts

L = henrys

$\Delta i / \Delta t$ = amperes per second

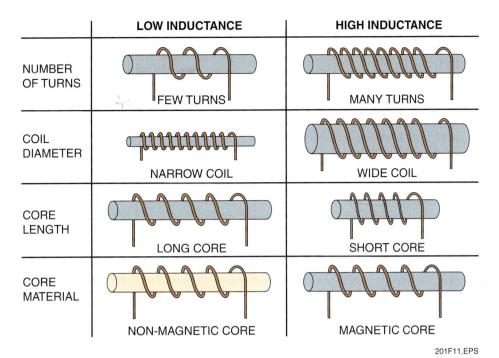

Figure 11 ♦ Factors affecting the inductance of a coil.

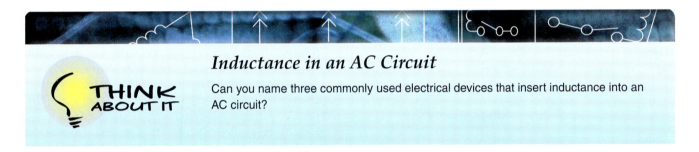

Inductance in an AC Circuit

Can you name three commonly used electrical devices that insert inductance into an AC circuit?

This gives the voltage in terms of how much magnetic flux is cut per second. When the magnetic flux associated with the current varies the same as I, this formula gives the same results for calculating induced voltage. Remember that the induced voltage across the coil is actually the result of inducing electrons to move in the conductor, so there is also an induced current.

For example, what is the self-induced voltage V_L across a 4h inductance produced by a current change of 12A per second?

$$V_L = L\frac{\Delta i}{\Delta t}$$
$$V_L = 4h \times \frac{12A}{1}$$
$$V_L = 4 \times 12$$
$$V_L = 48V$$

The current through a 200 microhenry (µh) inductor changes from 0 to 200 milliamps (mA) in 2 microseconds (µsec). (The prefix **micro** means one-millionth.) What is the V_L?

$$V_L = L\frac{\Delta i}{\Delta t}$$
$$V_L = (200 \times 10^{-6})\frac{200 \times 10^{-3}}{2 \times 10^{-6}}$$
$$V_L = 20V$$

The induced voltage is an actual voltage that can be measured, although V_L is produced only while the current is changing. When $\Delta i/\Delta t$ is present for only a short time, V_L is in the form of a voltage pulse. With a sine wave current that is always changing, V_L is a sinusoidal voltage that is 90° out of phase with I_L.

The current that flows in an inductor is induced by the changing magnetic field that surrounds the inductor. This changing magnetic field is produced by an AC voltage source that is applied to the inductor. The magnitude and polarity of the induced current depend on the field strength, direction, and rate at which the field cuts the inductor windings. The overall effect is that the current is out of phase and lags the applied voltage by 90°.

At 270° in *Figure 12*, the applied electromotive force (EMF) is zero, but it is increasing in the positive direction at its greatest rate of change. Likewise, electron flow due to the applied EMF is also increasing at its greatest rate. As the electron flow increases, it produces a magnetic field that is building with it. The lines of flux cut the conductor as they move outward from it with the expanding field.

As the lines of flux cut the conductor, they induce a current into it. The induced current is at its maximum value because the lines of flux are expanding outward through the conductor at their greatest rate. The direction of the induced current is in opposition to the force that generated it. Therefore, at 270° the applied voltage is zero and is increasing to a positive value, while the current is at its maximum negative value.

At 0° in *Figure 12*, the applied voltage is at its maximum positive value, but its rate of change is zero. Therefore, the field it produces is no longer expanding and is not cutting the conductor. Because there is no relative motion between the field and conductor, no current is induced. Therefore, at 0° voltage is at its maximum positive value, while current is zero.

At 90° in *Figure 12*, voltage is once again zero, but this time it is decreasing toward negative at its greatest rate of change. Because the applied voltage is decreasing, the magnetic field is collapsing inward on the conductor. This has the effect of reversing the direction of motion between the field and conductor that existed at 0°.

Therefore, the current will flow in a direction opposite of what it was at 0°. Also, because the applied voltage is decreasing at its greatest rate, the field is collapsing at its greatest rate. This causes the flux to cut the conductor at the greatest rate, causing the induced current magnitude to be maximum. At 90°, the applied voltage is zero decreasing toward negative, while the current is maximum positive.

At 180° in *Figure 12*, the applied voltage is at its maximum negative value, but just as at 0°, its rate of change is zero. At 180°, therefore, current will be zero. This explanation shows that the voltage peaks positive first, then 90° later the current peaks positive. Current thus lags the applied voltage in an inductor by 90°. This can easily be remembered using the phrase "ELI the ICE man." ELI represents voltage (E), inductance (L), and current (I). In an inductor, the voltage leads the current just as the letter E leads or comes before the letter I. The word ICE will be explained in the section on capacitance.

7.3.0 Inductive Reactance

The opposing force that an inductor presents to the flow of alternating current cannot be called resistance since it is not the result of friction within a conductor. The name given to this force is

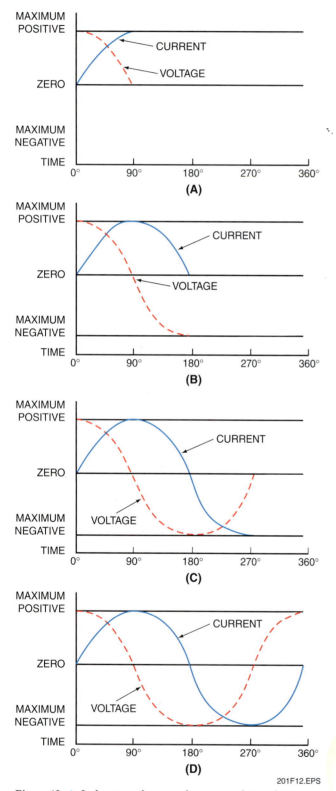

Figure 12 ♦ Inductor voltage and current relationship.

INSIDE TRACK

ELI in ELI the ICE Man

Remembering the phrase "ELI" as in "ELI the ICE man" is an easy way to remember the phase relationships that always exist between voltage and current in an inductive circuit. An inductive circuit is a circuit that has more inductive reactance than capacitive reactance. The L in ELI indicates inductance. The E (voltage) is stated before the I (current) in ELI, meaning that the voltage leads the current in an inductive circuit.

---

inductive reactance because it is the reaction of the inductor to alternating current. Inductive reactance is measured in ohms and its symbol is X_L.

Remember that the induced voltage in a conductor is proportional to the rate at which magnetic lines of force cut the conductor. The greater the rate or higher the frequency, the greater the counter-electromotive force (CEMF). Also, the induced voltage increases with an increase in inductance; the more turns, the greater the CEMF. Reactance then increases with an increase of frequency and with an increase in inductance. The formula for inductive reactance is as follows:

$$X_L = 2\pi f L$$

Where:

X_L = inductive reactance in ohms

2π = a constant in which the Greek letter pi (π) represents 3.14 and 2 × pi = 6.28

f = frequency of the alternating current in hertz

L = inductance in henrys

For example, if f is equal to 60Hz and L is equal to 20h, find X_L:

$$X_L = 2\pi f L$$
$$X_L = 6.28 \times 60Hz \times 20h$$
$$X_L = 7{,}536\,\Omega$$

Once calculated, the value of X_L is used like resistance in a form of Ohm's law:

$$I = \frac{E}{X_L}$$

Where:

I = effective current (amps)

E = effective voltage (volts)

X_L = inductive reactance (ohms)

Unlike a resistor, there is no power dissipation in an ideal inductor. An inductor limits current, but it uses no net energy since the energy required to build up the field in the inductor is given back to the circuit when the field collapses.

8.0.0 ◆ CAPACITANCE

A capacitor is a device that stores an electric charge in a dielectric material. Capacitance is the ability to store a charge. In storing a charge, a capacitor opposes a change in voltage. *Figure 13* shows a simple capacitor in a circuit, schematic representations of two types of capacitors, and a photo of common capacitors.

Figure 14(A) shows a capacitor in a DC circuit. When voltage is applied, the capacitor begins to charge, as shown in *Figure 14(B)*. The charging continues until the potential difference across the capacitor is equal to the applied voltage. This charging current is transient or temporary since it flows only until the capacitor is charged to the applied voltage. Then there is no current in the circuit. *Figure 14(C)* shows this with the voltage across the capacitor equal to the battery voltage or 10V.

The capacitor can be discharged by connecting a conducting path across the dielectric. The stored charge across the dielectric provides the potential difference to produce a discharge current, as shown in *Figure 14(D)*. Once the capacitor is completely discharged, the voltage across it equals zero, and there is no discharge current.

In a capacitive circuit, the charge and discharge current must always be in opposite directions. Current flows in one direction to charge the capacitor and in the opposite direction when the capacitor is allowed to discharge.

Current will flow in a capacitive circuit with AC voltage applied because of the capacitor charge and discharge current. There is no current through the dielectric, which is an insulator. While the capacitor is being charged by increasing applied voltage, the charging current flows in one direction to the plates. While the capacitor is discharging as the applied voltage decreases, the discharge current flows in the reverse direction. With alternating voltage applied, the capacitor alternately charges and discharges.

First, the capacitor is charged in one polarity, and then it discharges; next, the capacitor is charged in the opposite polarity, and then it discharges again.

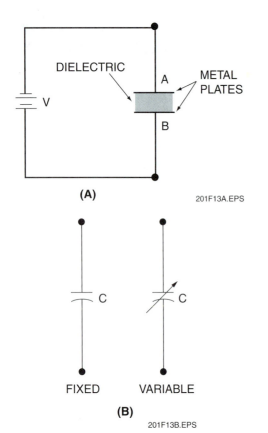

Figure 13 ◆ Capacitors.

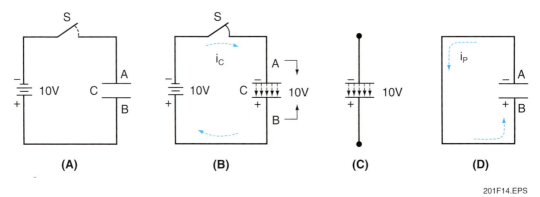

Figure 14 ◆ Charging and discharging a capacitor.

The cycles of charge and discharge current provide alternating current in the circuit at the same frequency as the applied voltage. The amount of capacitance in the circuit will determine how much current is allowed to flow.

Capacitance is measured in farads (F), where one farad is the capacitance when one coulomb is stored in the dielectric with a potential difference of one volt. Smaller values are measured in microfarads (µF). A small capacitance will allow less charge and discharge current to flow than a larger capacitance. The smaller capacitor has more opposition to alternating current, because less current flows with the same applied voltage.

In summary, capacitance exhibits the following characteristics:

- DC is blocked by a capacitor. Once charged, no current will flow in the circuit.
- AC flows in a capacitive circuit with AC voltage applied.
- A smaller capacitance allows less current.

8.1.0 Factors Affecting Capacitance

A capacitor consists of two conductors separated by an insulating material called a dielectric. There are many types and sizes of capacitors with

Capacitance

INSIDE TRACK

The concept of capacitance, like many electrical quantities, is often hard to visualize or understand. A comparison with a balloon may help to make this concept clearer. Electrical capacitance has a charging effect similar to blowing up a balloon and holding it closed. The expansion capacity of the balloon can be changed by changing the thickness of the balloon walls. A balloon with thick walls will expand less (have less capacity) than one with thin walls. This is like a small 10μF capacitor that has less capacity and will charge less than a larger 100μF capacitor.

---

different dielectric materials. The capacitance of a capacitor is determined by three factors:

- *Area of the plates* – The initial charge displacement on a set of capacitor plates is related to the number of free electrons in each plate. Larger plates will produce a greater capacitance than smaller ones. Therefore, the capacitance of a capacitor varies directly with the area of the plates. For example, if the area of the plates is doubled, the capacitance is doubled. If the size of the plates is reduced by 50%, the capacitance would also be reduced by 50%.
- *Distance between plates* – As two capacitor plates are brought closer together, more electrons will move away from the positively charged plate and move into the negatively charged plate. This is because the mutual attraction between the opposite charges on the plates increases as we move the plates closer together. This added movement of charge is an increase in the capacitance of the capacitor. In a capacitor composed of two plates of equal area, the capacitance varies inversely with the distance between the plates. For example, if the distance between the plates is decreased by one-half, the capacitance will be doubled. If the distance between the plates is doubled, the capacitance would be one-half as great.
- *Dielectric permittivity* – Another factor that determines the value of capacitance is the permittivity of the dielectric. The dielectric is the material between the capacitor plates in which the electric field appears. Relative permittivity expresses the ratio of the electric field strength in a dielectric to that in a vacuum. Permittivity has nothing to do with the dielectric strength of the medium or the breakdown voltage. An insulating material that will withstand a higher applied voltage than some other substance does not always have a higher dielectric permittivity. Many insulating materials have a greater dielectric permittivity than air. For a given applied voltage, a greater attraction exists between the opposite charges on the capacitor plates, and an electric field can be set up more easily than when the dielectric is air. The capacitance of the capacitor is increased when the permittivity of the dielectric is increased if all the other parameters remain unchanged.

8.2.0 Calculating Equivalent Capacitance

Connecting capacitors in parallel is equivalent to adding the plate areas. Therefore, the total capacitance is the sum of the individual capacitances, as illustrated in *Figure 15*.

A 10μF capacitor in parallel with a 5μF capacitor, for example, provides a 15μF capacitance for the parallel combination. The voltage is the same across the parallel capacitors. Note that adding parallel capacitance is opposite to the case of inductances in parallel and resistances in parallel.

Connecting capacitances in series is equivalent to increasing the thickness of the dielectric. Therefore, the combined capacitance is less than the smallest individual value. The combined equivalent capacitance is calculated by the reciprocal formula, as shown in *Figure 16*.

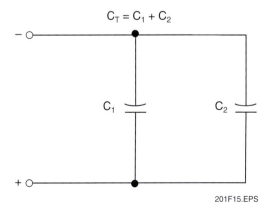

Figure 15 ♦ Capacitors in parallel.

$$C_T = \cfrac{1}{\cfrac{1}{C_1} + \cfrac{1}{C_2}}$$

Figure 16 ♦ Capacitors in series.

Capacitors connected in series are combined like resistors in parallel. Any of the shortcut calculations for the reciprocal formula apply. For example, the combined capacitance of two equal capacitances of 10µF in series is 5µF.

Capacitors are used in series to provide a higher voltage breakdown rating for the combination. For instance, each of three equal capacitances in series has one-third the applied voltage.

In series, the voltage across each capacitor is inversely proportional to its capacitance. The smaller capacitance has the larger proportion of the applied voltage. The reason is that the series capacitances all have the same charge because they are in one current path. With equal charge, a smaller capacitance has a greater potential difference.

8.3.0 Capacitor Specifications

This specifies the maximum potential difference that can be applied across the plates without puncturing the dielectric.

8.3.1 Voltage Rating

Usually, the voltage rating is for temperatures up to about 60°C. High temperatures result in a lower voltage rating. Voltage ratings for general-purpose paper, mica, and ceramic capacitors are typically 200V to 500V. Ceramic capacitors with ratings of 1 to 5kV are also available.

Electrolytic capacitors are commonly used in 25V, 150V, and 450V ratings. In addition, 6V and 10V electrolytic capacitors are often used in transistor circuits. For applications where a lower voltage rating is permissible, more capacitance can be obtained in a smaller physical size.

The potential difference across the capacitor depends on the applied voltage and is not necessarily equal to the voltage rating. A voltage rating higher than the potential difference applied across the capacitor provides a safety factor for long life in service. With electrolytic capacitors, however, the actual capacitor voltage should be close to the rated voltage to produce the oxide film that provides the specified capacitance.

The voltage ratings are for applied DC voltage. The breakdown rating is lower for AC voltage because of the internal heat produced by continuous charge and discharge.

8.3.2 Leak Resistance

Consider a capacitor charged by a DC voltage source. After the charging voltage is removed, a perfect capacitor would keep its charge indefinitely. After a long period of time, however, the charge will be neutralized by a small leakage current through the dielectric and across the insulated case between terminals, because there is no perfect insulator. For paper, ceramic, and mica capacitors, the leakage current is very slight, or inversely, the leakage resistance is very high. For paper, ceramic, or mica capacitors, R_1 is 100MΩ or more. However, electrolytic capacitors may have a leakage resistance of 0.5MΩ or less.

8.4.0 Voltage and Current in a Capacitive AC Circuit

In a capacitive circuit driven by an AC voltage source, the voltage is continuously changing. Thus, the charge on the capacitor is also continuously changing. The four parts of *Figure 17* show the variation of the alternating voltage and current in a capacitive circuit for each quarter of one cycle.

The solid line represents the voltage across the capacitor, and the dotted line represents the current. The line running through the center is the zero or reference point for both the voltage and

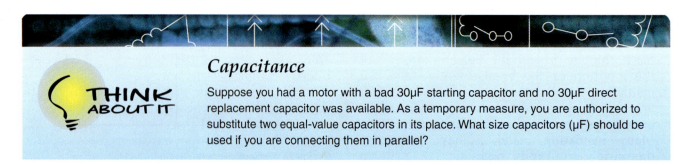

Capacitance

Suppose you had a motor with a bad 30µF starting capacitor and no 30µF direct replacement capacitor was available. As a temporary measure, you are authorized to substitute two equal-value capacitors in its place. What size capacitors (µF) should be used if you are connecting them in parallel?

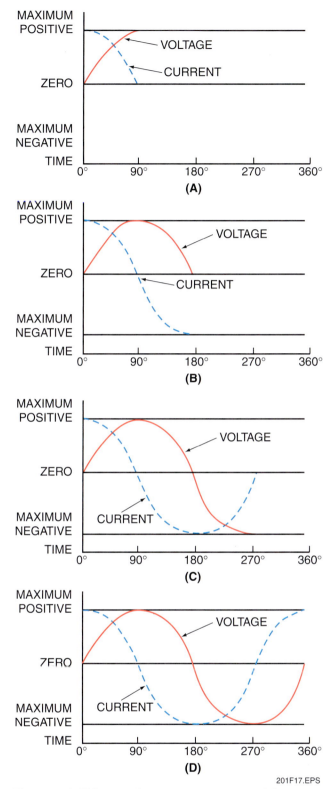

Figure 17 ♦ Voltage and current in a capacitive AC circuit.

At the beginning of the first quarter-cycle (0° to 90°), the voltage has just passed through zero and is increasing in the positive direction. Since the zero point is the steepest part of the sine wave, the voltage is changing at its greatest rate.

The charge on a capacitor varies directly with the voltage; therefore, the charge on the capacitor is also changing at its greatest rate at the beginning of the first quarter-cycle. In other words, the greatest number of electrons are moving off one plate and onto the other plate. Thus, the capacitor current is at its maximum value.

As the voltage proceeds toward maximum at 90°, its rate of change becomes lower and lower, making the current decrease toward zero. At 90°, the voltage across the capacitor is maximum, the capacitor is fully charged, and there is no further movement of electrons from plate to plate. That is why the current at 90° is zero.

At the end of the first quarter-cycle, the alternating voltage stops increasing in the positive direction and starts to decrease. It is still a positive voltage; but to the capacitor, the decrease in voltage means that the plate that has just accumulated an excess of electrons must lose some electrons. The current flow must reverse its direction. The second part of the figure shows the current curve to be below the zero line (negative current direction) during the second quarter-cycle (90° to 180°).

At 180°, the voltage has dropped to zero. This means that, for a brief instant, the electrons are equally distributed between the two plates; the current is maximum because the rate of change of voltage is maximum.

Just after 180°, the voltage has reversed polarity and starts building to its maximum negative peak, which is reached at the end of the third quarter-cycle (180° to 270°). During the third quarter-cycle, the rate of voltage change gradually decreases as the charge builds to a maximum at 270°. At this point, the capacitor is fully charged and carries the full impressed voltage. Because the capacitor is fully charged, there is no further exchange of electrons and the current flow is zero at this point. The conditions are exactly the same as at the end of the first quarter-cycle (90°), but the polarity is reversed.

Just after 270°, the impressed voltage once again starts to decrease, and the capacitor must lose electrons from the negative plate. It must discharge, starting at a minimum rate of flow and rising to a maximum. This discharging action continues through the last quarter-cycle (270° to 360°) until the impressed voltage has reached zero. The beginning of the entire cycle is 360°, and everything starts over again.

In *Figure 17*, note that the current always arrives at a certain point in the cycle 90° ahead of

the current. The bottom line marks off the time of the cycle in terms of electric degrees. Assume that the AC voltage has been acting on the capacitor for some time before the time represented by the starting point of the sine wave.

the voltage because of the charging and discharging action. This voltage-current phase relationship in a capacitive circuit is exactly opposite to that in an inductive circuit. The current through a capacitor leads the voltage across the capacitor by 90°. A convenient way to remember this is the phrase "ELI the ICE man" (ELI refers to inductors, as previously explained). ICE pertains to capacitors as follows:

$$I = \text{current}$$
$$C = \text{capacitor}$$
$$E = \text{voltage}$$

In capacitors (C), current (I) leads voltage (E) by 90°.

It is important to realize that the current and voltage are both going through their individual cycles at the same time during the period the AC voltage is impressed. The current does not go through part of its cycle (charging or discharging) and then stop and wait for the voltage to catch up. The amplitude and polarity of the voltage and the amplitude and direction of the current are continually changing.

Their positions, with respect to each other and to the zero line at any electrical instant or any degree between 0° and 360°, can be seen by reading upward from the time-degree line. The current swing from the positive peak at 0° to the negative peak at 180° is not a measure of the number of electrons or the charge on the plates. It is a picture of the direction and strength of the current in relation to the polarity and strength of the voltage appearing across the plates.

8.5.0 Capacitive Reactance

Capacitors offer a very real opposition to current flow. This opposition arises from the fact that, at a given voltage and frequency, the number of electrons that go back and forth from plate to plate is limited by the storage ability or the capacitance of the capacitor. As the capacitance is increased, a greater number of electrons changes plates every cycle. Since current is a measure of the number of electrons passing a given point in a given time, the current is increased.

Increasing the frequency will also decrease the opposition offered by a capacitor. This occurs because the number of electrons that the capacitor is capable of handling at a given voltage will change plates more often. As a result, more electrons will pass a given point in a given time (greater current flow). The opposition that a capacitor offers to AC is therefore inversely proportional to frequency and capacitance. This opposition is called capacitive reactance. Capacitive reactance decreases with increasing frequency or, for a given frequency, the capacitive reactance decreases with increasing capacitance. The symbol for capacitive reactance is X_C. The formula is:

$$X_C = \frac{1}{2\pi fC}$$

Where:

X_C = capacitive reactance in ohms
f = frequency in hertz
C = capacitance in farads
$2\pi = 6.28\ (2 \times 3.14)$

For example, what is the capacitive reactance of a 0.05μF capacitor in a circuit whose frequency is 1 megahertz?

$$X_C = \frac{1}{2\pi fC} = \frac{1}{(6.28)(10^6 \text{ hertz})(5 \times 10^{-8} \text{ farads})}$$

$$X_C = \frac{1}{3.14 \times 10^{-1}} = \frac{1}{0.314} = 3.18 \text{ ohms}$$

The capacitive reactance of a 0.05μF capacitor operated at a frequency of 1 megahertz is 3.18 ohms. Suppose this same capacitor is operated at a lower frequency of 1,500 hertz instead of 1 megahertz. What is the capacitive reactance now? Substituting where $1{,}500 = 1.5 \times 10^3$ hertz:

$$X_C = \frac{1}{2\pi fC} = \frac{1}{(6.28)(1.5 \times 10^3 \text{ hertz})(5 \times 10^{-8} \text{ farads})}$$

$$X_C = \frac{1}{4.71 \times 10^{-4}} = 2{,}123 \text{ ohms}$$

ICE in ELI the ICE Man

Remembering the phrase "ICE" as in "ELI the ICE man" is an easy way to remember the phrase relationships that always exist between voltage and current in a capacitive circuit. A capacitive circuit is a circuit in which there is more capacitive reactance than inductive reactance. This is indicated by the C in ICE. The I (current) is stated before the E (voltage) in ICE, meaning that the current leads the voltage in a capacitive circuit.

Frequency and Capacitive Reactance

A variable capacitor is used in the tuner of an AM radio to tune the radio to the desired station. Will its capacitive reactance value be higher or lower when it is tuned to the low end of the frequency band (550kHz) than it would be when tuned to the high end of the band (1,440kHz)?

Note a very interesting point from these two examples. As frequency is decreased from 1 megahertz to 1,500 hertz, the capacitive reactance increases from 3.18 ohms to 2,123 ohms. Capacitive reactance increases as the frequency decreases.

9.0.0 ♦ LC AND RLC CIRCUITS

AC circuits often contain inductors, capacitors, and/or resistors connected in series or parallel combinations. When this is done, it is important to determine the resulting phase relationship between the applied voltage and the current in the circuit. The simplest method of combining factors that have different phase relationships is vector addition with the trigonometric functions. Each quantity is represented as a vector, and the resultant vector and phase angle are then calculated.

In purely resistive circuits, the voltage and current are in phase. In inductive circuits, the voltage leads the current by 90°. In capacitive circuits, the current leads the voltage by 90°. *Figure 18* shows the phase relationships of these components used in AC circuits. Recall that these characteristics are summarized by the phrase "ELI the ICE man."

$$ELI = E \text{ Leads } I \text{ (inductive)}$$
$$ICE = I \text{ Capacitive (leads) } E$$

The **impedance** Z of a circuit is defined as the total opposition to current flow. The magnitude of the impedance Z is given by the following equation in a series circuit:

$$Z = \sqrt{R^2 + X^2}$$

Where:

Z = impedance (ohms)
R = resistance (ohms)
X = net reactance (ohms)

The current through a resistance is always in phase with the voltage applied to it; thus resistance is shown along the 0° axis. The voltage across an inductor leads the current by 90°; thus inductive reactance is shown along the 90° axis. The voltage across a capacitor lags the current by 90°; thus capacitive reactance is shown along the −90° axis. The net reactance is the difference between the inductive reactance and the capacitive reactance:

X = net reactance (ohms)
X_L = inductive reactance (ohms)
X_C = capacitive reactance (ohms)

The impedance Z is the vector sum of the resistance R and the net reactance X. The angle, called the phase angle, gives the phase relationship between the applied voltage and current.

9.1.0 RL Circuits

RL circuits combine resistors and inductors in a series, parallel, or series-parallel configuration. In a pure inductive circuit, the current lags the voltage by an angle of 90°. In a circuit containing both resistance and inductance, the current will lag the voltage by some angle between zero and 90°.

9.1.1 Series RL Circuit

Figure 19 shows a series RL circuit. Since it is a series circuit, the current is the same in all portions of the loop. Using the values shown, the circuit will be analyzed for unknown values such as X_L, Z, I, E_L, and E_R.

The solution would be worked as follows:

Step 1 Compute the value of X_L.

$$X_L = 2\pi f L$$
$$X_L = 6.28 \times 100 \times 4 = 2,512 \text{ ohms}$$

Step 2 Draw vectors R and X_L as shown in *Figure 19*. R is drawn horizontally because the circuit current and voltage across R are in phase. It therefore becomes the reference line from which other angles are measured. X_L is drawn upward at 90° from R because voltage across X_L leads circuit current through R.

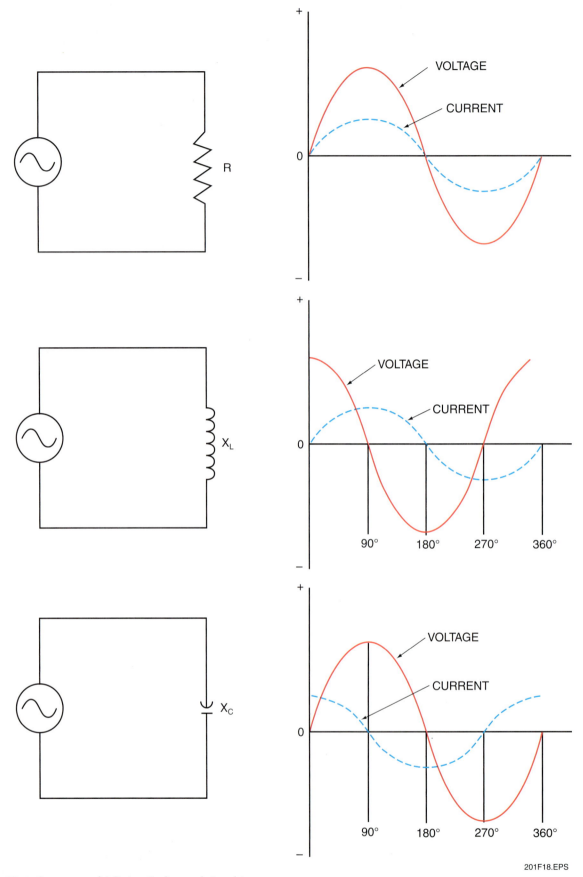

Figure 18 ◆ Summary of AC circuit phase relationships.

1.22 ELECTRICAL LEVEL TWO ◆ TRAINEE GUIDE

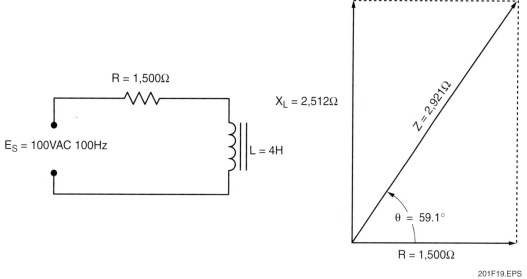

Figure 19 ◆ Series RL circuit and vector diagram.

Step 3 Compute the value of circuit impedance Z, which is equal to the vector sum of X_L and R.

$$\tan = \frac{X_L}{R} = \frac{2{,}512}{1{,}500} = 1.67$$
$$\arctan 1.67 = 59.1°$$
$$\cos 59.1° = 0.5135$$

Find Z using the cosine function:

$$\cos = \frac{R}{Z}$$
$$Z = \frac{R}{\cos}$$
$$Z = \frac{1{,}500}{0.5135} = 2{,}921 \text{ ohms}$$

Step 4 Compute the circuit current using Ohm's law for AC circuits.

$$I = \frac{E}{Z} = \frac{100\,V}{2.921\,\Omega} = 0.034\,A$$

Step 5 Compute voltage drops in the circuit.

$$E_L = IX_L = 0.034 \times 2{,}512 = 85 \text{ volts}$$
$$E_R = IR = 0.034 \times 1{,}500 = 51 \text{ volts}$$

Note that the voltage drops across the resistor and inductor do not equal the supply voltage because they must be added vectorially. This would be done as follows (because of rounding, numbers are not exact):

$$\tan = \frac{E_L}{E_R} = \frac{85}{51} = 1.67$$
$$\arctan 1.67 = 59.1°$$
$$\cos = \frac{E_R}{E_Z} \quad E_Z = \frac{E_R}{\cos}$$
$$E_Z = \frac{51}{0.5135} = \text{approx. } 100V = E_S$$

In this inductive circuit, the current lags the applied voltage by an angle equal to 59.1°.

NOTE
Since we are dealing with right triangles, we could also use the Pythagorean theorem (discussed later) to find this answer.

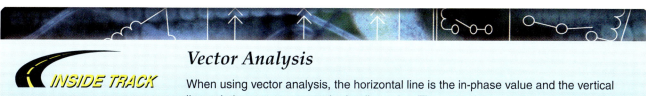

Vector Analysis
When using vector analysis, the horizontal line is the in-phase value and the vertical line pointing up represents the leading value. The vertical line pointing down represents the lagging value.

Figure 20 shows another series RL circuit, its associated waveforms, and vector diagrams. This circuit is used to summarize the characteristics of a series RL circuit:

- The current I flows through all the series components.
- The voltage across X_L, labeled V_L, can be considered an IX_L voltage drop, just as V_R is used for an IR voltage drop.
- The current I through X_L must lag V_L by 90°, as this is the angle between current through an inductance and its self-induced voltage.
- The current I through R and its IR voltage drop have the same phase. There is no reactance to sine wave current in any resistance. Therefore, I and IR have the same phase, or this phase angle is 0°.
- V_T is the vector sum of the two out-of-phase voltages V_R and V_L.
- Circuit current I lags V_T by the phase angle.
- Circuit impedance is the vector sum of R and X_L.

In a series circuit, the higher the value of X_L compared with R, the more inductive the circuit is. This means there is more voltage drop across the inductive reactance, and the phase angle increases toward 90°. The series current lags the applied generator voltage.

Several combinations of X_L and R in series are listed in *Table 1* with their resultant impedance and phase angle. Note that a ratio of 10:1 or more for X_L/R means that the circuit is practically all inductive. The phase angle of 84.3° is only slightly less than 90° for the ratio of 10:1, and the total impedance Z is approximately equal to X_L. The voltage drop across X_L in the series circuit will be equal to the applied voltage, with almost none across R.

At the opposite extreme, when R is 10 times as large as X_L, the series circuit is mainly resistive. The phase angle of 5.7° means the current has almost the same phase as the applied voltage, the total impedance Z is approximately equal to R, and the voltage drop across R is practically equal to the applied voltage, with almost none across X_L.

9.1.2 Parallel RL Circuit

In a parallel RL circuit, the resistance and inductance are connected in parallel across a voltage source. Such a circuit thus has a resistive branch and an inductive branch.

The 90° phase angle must be considered for each of the branch currents, instead of voltage drops in a series circuit. Remember that any series circuit has different voltage drops, but one common

Table 1 Series R and X_L Combinations

R (Ω)	X_L (Ω)	Z (Ω) (Approx.)	Phase Angle (θ) (°)
1	10	$\sqrt{101} = 10$	84.3°
10	10	$\sqrt{200} = 14$	45°
10	1	$\sqrt{101} = 10$	5.7°

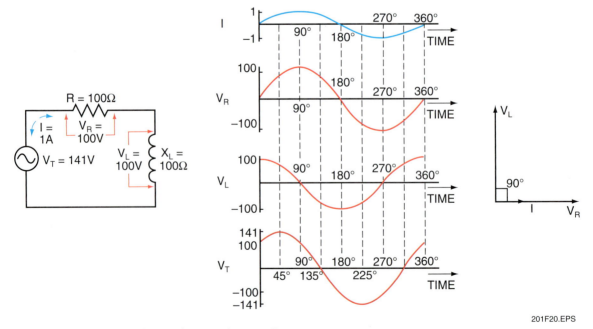

Figure 20 ◆ Series RL circuit with waveforms and vector diagram.

current. A parallel circuit has different branch currents, but one common voltage.

In the parallel circuit in *Figure 21*, the applied voltage V_A is the same across X_L, R, and the generator, since they are all in parallel. There cannot be any phase difference between these voltages. Each branch, however, has its individual current. For the resistive branch $I_R = V_A/R$; in the inductive branch $I_L = V_A/X_L$.

The resistive branch current I_R has the same phase as the generator voltage V_A. The inductive branch current I_L lags V_A, however, because the current in an inductance lags the voltage across it by 90°.

The total line current, therefore, consists of I_R and I_L, which are 90° out of phase with each other. The phasor sum of I_R and I_L equals the total line current I_T. These phase relations are shown by the waveforms and vectors in *Figure 21*. I_T will lag V_A by some phase angle that results from the vector addition of I_R and I_L.

The impedance of a parallel RL circuit is the total opposition to current flow by the R of the resistive branch and the X_L of the inductive branch. Since X_L and R are vector quantities, they must be added vectorially.

If the line current and the applied voltage are known, Z can also be calculated by the equation:

$$Z = \frac{V_A}{I_{Line}}$$

The Z of a parallel RL circuit is always less than the R or X_L of any one branch. The branch of a parallel RL circuit that offers the most opposition to current flow has the lesser effect on the phase angle of the current.

Several combinations of X_L and R in parallel are listed in *Table 2*. When X_L is 10 times R, the parallel circuit is practically resistive because there is little inductive current in the line. The small value of I_L results from the high X_L. The total impedance of the parallel circuit is approximately equal to the resistance then, since the high value of X_L in a parallel branch has little effect. The phase angle of −5.7° is practically 0° because almost all the line current is resistive.

As X_L becomes smaller, it provides more inductive current in the main line. When X_L is $\frac{1}{10}$R, practically all the line current is the I_L component. Then, the parallel circuit is practically all inductive, with a total impedance practically equal to X_L. The phase angle of −84.3° is almost −90° because the line current is mostly inductive. Note that these conditions are opposite from the case of X_L and R in series.

9.2.0 RC Circuits

In a circuit containing resistance only, the current and voltage are in phase. In a circuit of pure capacitance, the current leads the voltage by an angle of 90°. In a circuit that has both resistance

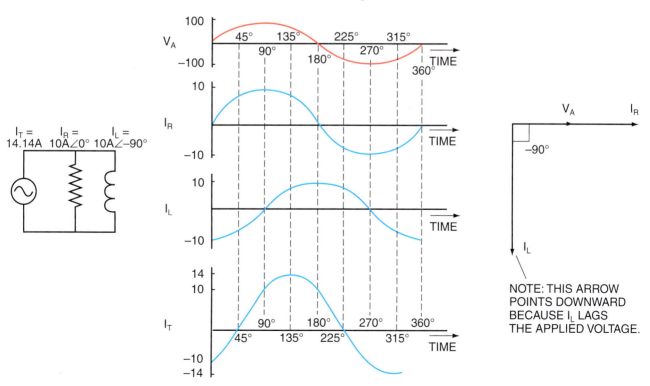

Figure 21 ♦ Parallel RL circuit with waveforms and vector diagram.

Table 2 Parallel R and X_L Combinations

R (Ω)	X_L (Ω)	I_R (A)	I_L (A)	I_T (A) (Approx.)	$Z_T = V_A/I_T$ (Ω)	Phase Angle (θ)$_I$ (°)
1	10	10	1	$\sqrt{101} = 10$	1	−5.7°
10	10	1	1	$\sqrt{2} = 1.4$	7.07	−45°
10	1	1	10	$\sqrt{101} = 10$	1	−84.3°

and capacitance, the current will lead the voltage by some angle between 0° and 90°.

9.2.1 Series RC Circuit

Figure 22A shows a series RC circuit with resistance R in series with capacitive reactance X_C. Current I is the same in X_C and R since they are in series. Each has its own series voltage drop, equal to IR for the resistance and IX_C for the reactance.

In *Figure 22B*, the phasor is shown horizontal as the reference phase, because I is the same throughout the series circuit. The resistive voltage drop IR has the same phase as I. The capacitor voltage IX_C must be 90° clockwise from I and IR, as the capacitive voltage lags. Note that the IX_C phasor is downward, exactly opposite from an IX_L phasor, because of the opposite phase angle.

If the capacitive reactance alone is considered, its voltage drop lags the series current I by 90°. The IR voltage has the same phase as I, however, because resistance provides no phase shift. Therefore, R and X_C combined in series must be added by vectors because they are 90° out of phase with each other, as shown in *Figure 22C*.

As with inductive reactance, θ (theta) is the phase angle between the generator voltage and its series current. As shown in *Figure 22B* and *Figure 22C*, θ can be calculated from the voltage or impedance triangle.

With series X_C the phase angle is negative, clockwise from the zero reference angle of I because the X_C voltage lags its current. To indicate the negative phase angle, this 90° phasor points downward from the horizontal reference, instead of upward as with the series inductive reactance.

In series, the higher the X_C compared with R, the more capacitive the circuit. There is more voltage drop across the capacitive reactance, and the phase angle increases toward −90°. The series X_C always makes the current lead the applied voltage. With all X_C and no R, the entire applied voltage is across X_C and equals −90°. Several combinations of X_C and R in series are listed in *Table 3*.

9.2.2 Parallel RC Circuit

In a parallel RC circuit, as shown in *Figure 23A*, a capacitive branch as well as a resistive branch are connected across a voltage source. The current

Table 3 Series R and X_C Combinations

R (Ω)	X_C (Ω)	Z (Ω) (Approx.)	Phase Angle (θ)$_Z$ (°)
1	10	$\sqrt{101} = 10$	84.3°
10	10	$\sqrt{200} = 14$	45°
10	1	$\sqrt{101} = 10$	5.7°

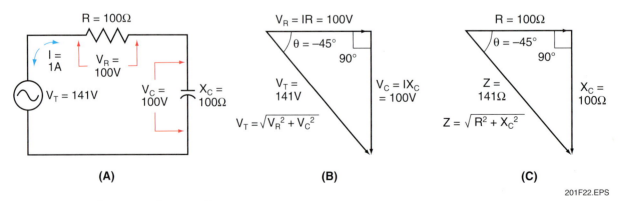

Figure 22 ◆ Series RC circuit with vector diagrams.

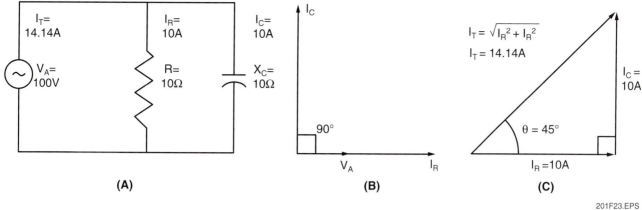

Figure 23 ◆ Parallel RC circuit with vector diagrams.

that leaves the voltage source divides among the branches, so there are different currents in each branch. The current is therefore not a common quantity, as it is in the series RC circuit.

In a parallel RC circuit, the applied voltage is directly across each branch. Therefore, the branch voltages are equal in value to the applied voltage and all voltages are in phase. Since the voltage is common throughout the parallel RC circuit, it serves as the common quantity in any vector representation of parallel RC circuits. This means the reference vector will have the same phase relationship or direction as the circuit voltage. Note in *Figure 23B* that V_A and I_R are both shown as the 0° reference.

Current within an individual branch of an RC parallel circuit is dependent on the voltage across the branch and on the R or X_C contained in the branch. The current in the resistive branch is in phase with the branch voltage, which is the applied voltage. The current in the capacitive branch leads V_A by 90°. Since the branch voltages are the same, I_C leads I_R by 90°, as shown in *Figure 23B*. Since the branch currents are out of phase, they have to be added vectorially to find the line current.

The phase angle, θ, is 45° because R and X_C are equal, resulting in equal branch currents. The phase angle is between the total current I_T and the generator voltage V_A. However, the phase of V_A is the same as the phase of I_R. Therefore, θ is also between I_T and I_R.

The impedance of a parallel RC circuit represents the total opposition to current flow offered by the resistance and capacitive reactance of the circuit. The equation for calculating the impedance of a parallel RC circuit is:

$$Z = \frac{RX_C}{\sqrt{I_R^2 + I_C^2}} \text{ or } Z = \frac{V_A}{I_T}$$

For the example shown in *Figure 23*, Z is:

$$Z = \frac{V_A}{I_T} = \frac{100}{14.14A} = 7.07\Omega$$

This is the opposition in ohms across the generator. This Z of 7.07Ω is equal to the resistance of 10Ω in parallel with the reactance of 10Ω. Notice that the impedance of equal values of R and X_C is not one-half, but equals 70.7% of either one.

When X_C is high relative to R, the parallel circuit is practically resistive because there is little leading capacitive current in the main line. The small value of I_C results from the high reactance of shunt X_C. The total impedance of the parallel circuit is approximately equal to the resistance, since the high value of X_C in a parallel branch has little effect.

As X_C becomes smaller, it provides more leading capacitive current in the main line. When X_C is very small relative to R, practically all the line current is the I_C component. The parallel circuit is practically all capacitive, with a total impedance practically equal to X_C.

The characteristics of different circuit arrangements are shown in *Table 4*.

9.3.0 LC Circuits

An LC circuit consists of an inductance and a capacitance connected in series or in parallel with a voltage source. There is no resistor physically in an LC circuit, but every circuit contains some resistance. Since the circuit resistance of the wiring and voltage source is usually so small, it has little or no effect on circuit operation.

In a circuit with both X_L and X_C, the opposite phase angles enable one to cancel the effect of the other. For X_L and X_C in series, the net reactance is the difference between the two series reactances, resulting in less reactance than either one. In

Table 4 Parallel R and X_C Combinations

R (Ω)	X_C (Ω)	I_R (A)	I_C (A)	I_T (A) (Approx.)	Z_T (Ω) (Approx.)	Phase Angle (θ) (°)
1	10	10	1	$\sqrt{101} = 10$	1	5.7°
10	10	1	1	$\sqrt{2} = 1.4$	7.07	45°
10	1	1	10	$\sqrt{101} = 10$	1	84.3°

parallel circuits, the I_L and I_C branch currents cancel. The net line current is then the difference between the two branch currents, resulting in less total line current than either branch current.

9.3.1 Series LC Circuit

As in all series circuits, the current in a series LC circuit is the same at all points. Therefore, the current in the inductor is the same as, and in phase with, the current in the capacitor. Because of this, on the vector diagram for a series LC circuit, the direction of the current vector is the reference or in the 0° direction, as shown in *Figure 24*.

When there is current flow in a series LC circuit, the voltage drops across the inductor and capacitor depend on the circuit current and the values of X_L and X_C. The voltage drop across the inductor leads the circuit current by 90°, and the voltage drop across the capacitor lags the circuit current by 90°. Using Kirchhoff's voltage law, the source voltage equals the sum of the voltage drops across the inductor and capacitor, with respect to the polarity of each.

Since the current through both is the same, the voltage across the inductor leads that across the capacitor by 180°. The method used to add the two voltage vectors is to subtract the smaller vector from the larger, and assign the resultant the direction of the larger. When applied to a series LC circuit, this means the applied voltage is equal to the difference between the voltage drops (E_L and E_C), with the phase angle between the applied voltage (E_T) and the circuit current determined by the larger voltage drop.

In a series LC circuit, one or both of the voltage drops are always greater than the applied voltage. Remember that although one or both of the voltage drops are greater than the applied voltage, they are 180° out of phase. One of them effectively cancels a portion of the other so that the total voltage drop is always equal to the applied voltage.

Recall that X_L is 180° out of phase with X_C. The impedance is then the vector sum of the two reactances. The reactances are 180° apart, so their vector sum is found by subtracting the smaller one from the larger.

Unlike RL and RC circuits, the impedance in an LC circuit is either purely inductive or purely capacitive.

9.3.2 Parallel LC Circuit

In a parallel LC circuit there is an inductance and a capacitance connected in parallel across a voltage source. *Figure 25* shows a parallel LC circuit with its vector diagram.

As in any parallel circuit, the voltage across the branches is the same as the applied voltage. Since they are actually the same voltage, the branch voltages and applied voltage are in phase. Because of this, the voltage is used as the 0° phase reference and the phases of the other circuit quantities are expressed in relation to the voltage.

The currents in the branches of a parallel LC circuit are both out of phase with the circuit voltage. The current in the inductive branch (I_L) lags the voltage by 90°, while the current in the capacitive

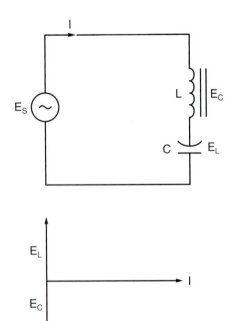

Figure 24 ◆ Series LC circuit with vector diagram.

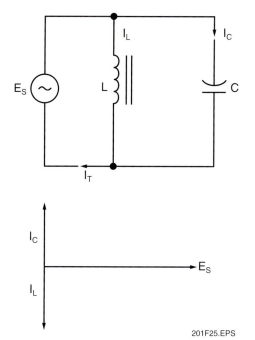

Figure 25 ◆ Parallel LC circuit with vector diagram.

branch (I_C) leads the voltage by 90°. Since the voltage is the same for both branches, currents I_L and I_C are therefore 180° out of phase. The amplitudes of the branch currents depend on the value of the reactance in the respective branches.

With the branch currents being 180° out of phase, the line current is equal to their vector sum. This vector addition is done by subtracting the smaller branch current from the larger.

The line current for a parallel LC circuit, therefore, has the phase characteristics of the larger branch current. Thus, if the inductive branch current is the larger, the line current is inductive and lags the applied voltage by 90°; if the capacitive branch current is the larger, the line current is capacitive, and leads the applied voltage by 90°.

The line current in a parallel LC circuit is always less than one of the branch currents and sometimes less than both. The reason that the line current is less than the branch currents is because the two branch currents are 180° out of phase. As a result of the phase difference, some cancellation takes place between the two currents when they combine to produce the line current. The impedance of a parallel LC circuit can be found using the following equations:

$$Z = \frac{X_L \times X_C}{X_L - X_C} \text{ (for } X_L \text{ larger than } X_C\text{)}$$

or:

$$Z = \frac{X_L \times X_C}{X_C - X_L} \text{ (for } X_C \text{ larger than } X_L\text{)}$$

When using these equations, the impedance will have the phase characteristics of the smaller reactance.

9.4.0 RLC Circuits

RLC circuits can be divided into two main categories: series RLC circuits and parallel RLC circuits. These circuits are described in the following sections.

9.4.1 Series RLC Circuit

Circuits in which the inductance, capacitance, and resistance are all connected in series are called series RLC circuits. The fundamental properties of series RLC circuits are similar to those for series LC circuits. The differences are caused by the effects of the resistance. Any practical series LC circuit contains some resistance. When the resistance is very small compared to the circuit reactance, it has almost no effect on the circuit and can be considered as zero. When the resistance is appreciable, though, it has a significant effect on the circuit operation and therefore must be considered in any circuit analysis. In a series RLC circuit, the same current flows through each component. The phase relationships between the voltage drops are the same as they were in series RC, RL, and LC circuits. The voltage drops across the inductance and capacitance are 180° out of phase. With current the same throughout the circuit as a reference, the inductive voltage drop (E_L) leads the resistive voltage drop (E_R) by 90°, and the capacitive voltage drop (E_C) lags the resistive voltage drop by 90°.

Figure 26 shows a series RLC circuit and the vector diagram used to determine the applied voltage. The vector sum of the three voltage drops is equal to the applied voltage. However, to calculate this vector sum, a combination of the methods learned for LC, RL, and RC circuits must be used. First, calculate the combined voltage drop of the two reactances. This value is designated E_X and is found as in pure LC circuits by subtracting the smaller reactive voltage drop from the larger. This is shown in *Figure 26* as E_X. The result of this calculation is the net reactive voltage drop and is either inductive or capacitive, depending on which of the individual voltage drops is larger. In *Figure 26*, the net reactive voltage drop is inductive since $E_L > E_C$. Once the net reactive voltage drop is known, it is added vectorially to the voltage drop across the resistance.

The angle between the applied voltage E_A and the voltage across the resistance E_R is the same as the phase angle between E_A and the circuit

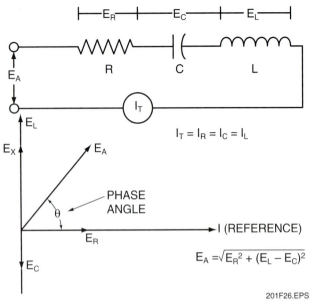

Figure 26 ♦ Series RLC circuit and vector diagram.

current. The reason for this is that E_R and I are in phase.

The impedance of a series RLC circuit is the vector sum of the inductive reactance, the capacitive reactance, and the resistance. This is done using the same method as for voltage drop calculations.

When X_L is greater than X_C, the net reactance is inductive, and the circuit acts essentially as an RL circuit. Similarly, when X_C is greater than X_L, the net reactance is capacitive, and the circuit acts as an RC circuit.

The same current flows in every part of a series RLC circuit. The current always leads the voltage across the capacitance by 90° and is in phase with the voltage across the resistance. The phase relationship between the current and the applied voltage, however, depends on the circuit impedance. If the impedance is inductive (X_L greater than X_C), the current is inductive and lags the applied voltage by some phase angle less than 90°. If the impedance is capacitive (X_C greater than X_L), the current is capacitive, and leads the applied voltage by some phase angle also less than 90°. The angle of the lead or lag is determined by the relative values of the net reactance and the resistance.

The greater the value of X or the smaller the value of R, the larger the phase angle, and the more reactive (or less resistive) the current. Similarly, the smaller the value of X or the larger the value of R, the more resistive (or less reactive) the current. If either R or X is 10 or more times greater than the other, the circuit will essentially act as though it is purely resistive or reactive, as the case may be.

9.4.2 Parallel RLC Circuit

A parallel RLC circuit is basically a parallel LC circuit with an added parallel branch of resistance. The solution of a parallel circuit involves the solution of a parallel LC circuit, and then the solution of either a parallel RL circuit or a parallel RC circuit. The reason for this is that a parallel combination of L and C appears to the source as a pure L or a pure C. So by solving the LC portion of a parallel RLC circuit first, the circuit is reduced to an equivalent RL or RC circuit.

The distribution of the voltage in a parallel RLC circuit is no different from what it is in a parallel LC circuit, or in any parallel circuit. The branch voltages are all equal and in phase, since they are the same as the applied voltage. The resistance is simply another branch across which the applied voltage appears. Because the voltages throughout the circuit are the same, the applied voltage is again used as the θ phase reference.

Figure 27 shows the current relationship in a parallel RLC circuit.

The three branch currents in a parallel RLC circuit are an inductive current I_L, a capacitive current I_C, and a resistive current I_R. Each is independent of the others, and depends only on the applied voltage and the branch resistance or reactance.

The three branch currents all have different phases with respect to the branch voltages. I_L lags the voltage by 90°, I_C leads the voltage by 90°, and I_R is in phase with the voltage. Since the voltages are the same, I_L and I_C are 180° out of phase with each other, and both are 90° out of phase with I_R. Because I_R is in phase with the voltage, it has the same zero-reference direction as the voltage. So I_C leads I_R by 90°, and I_L lags I_R by 90°.

The line current (I_T), or total current, is the vector sum of the three branch currents, and can be calculated by adding I_L, I_C, and I_R vectorially. Whether the line current leads or lags the applied voltage depends on which of the reactive branch currents (I_L or I_C) is the larger. If I_L is larger, I_T lags the applied voltage. If I_C is larger, I_T leads the applied voltage.

To determine the impedance of a parallel RLC circuit, first determine the net reactance X of the inductive and capacitive branches. Then use X to determine the impedance Z, the same as in a parallel RL or RC circuit.

Whenever Z is inductive, the line current will lag the applied voltage. Similarly, when Z is capacitive, the line current will lead the applied voltage.

AC Circuits

The photo below shows a simple series circuit comprised of an ON/OFF switch, small lamp, motor, and capacitor. How would you classify this circuit? When energized, which components insert resistance, inductive reactance, and capacitive reactance into the circuit?

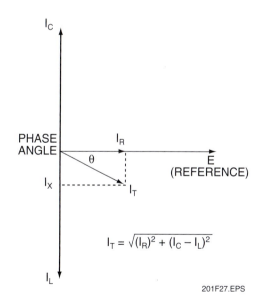

$E_A = E_R = E_C = E_L$

$I_T = \sqrt{(I_R)^2 + (I_C - I_L)^2}$

Figure 27 ♦ Parallel RLC circuit and vector diagram.

MODULE 26201-08 ♦ ALTERNATING CURRENT 1.31

10.0.0 ♦ POWER IN AC CIRCUITS

In DC circuits, the power consumed is the sum of all the I^2R heating in the resistors. It is also equal to the power produced by the source, which is the product of the source voltage and current. In AC circuits containing only resistors, the above relationship also holds true.

10.1.0 True Power

The power consumed by resistance is called true power and is measured in units of watts. True power is the product of the resistor current squared and the resistance:

$$P_T = I^2R$$

This formula applies because current and voltage have the same phase across a resistance. To find the corresponding value of power as a product of voltage and current, this product must be multiplied by the cosine of the phase angle θ:

$$P_T = I^2R \text{ or } P_T = EI \cos \theta$$

Where E and I are in rms values to calculate the true power in watts, multiplying I by the cosine of the phase angle provides the resistive component for true power equal to I^2R.

For example, a series RL circuit has 2A through a 100Ω resistor in series with the X_L of 173Ω. Therefore:

$$P_T = I^2R$$
$$P_T = 4 \times 100$$
$$P_T = 400W$$

Furthermore, in this circuit the phase angle is 60° with a cosine of 0.5. The applied voltage is 400V. Therefore:

$$P_T = EI \cos \theta$$
$$P_T = 400 \times 2 \times 0.5$$
$$P_T = 400W$$

In both cases, the true power is the same (400W) because this is the amount of power supplied by the generator and dissipated in the resistance. Either formula can be used for calculating the true power.

10.2.0 Apparent Power

In ideal AC circuits containing resistors, capacitors, and inductors, the only mechanism for power consumption is $I^2_{eff}R$ heating in the resistors. Inductors and capacitors consume no power. The only function of inductors and capacitors is to store and release energy. However, because of the phase shifts that are introduced by these elements, the power consumed by the resistors is not equal to the product of the source voltage and current. The product of the source voltage and current is called apparent power and has units of volt-amperes (VA).

The apparent power is the product of the source voltage and the total current. Therefore, apparent power is actual power delivered by the source. The formula for apparent power is:

$$P_A = (E_A)(I)$$

Figure 28 shows a series RL circuit and its associated vector diagram.

This circuit is used to calculate the apparent power and compare it to the circuit's true power:

$$P_A = (E_A)(I) \qquad P_T = EI \cos \theta$$
$$P_A = (400V)(2A) \qquad \theta = \frac{R}{X_L} = \frac{100}{173} = 60°$$
$$P_A = 800VA \qquad P_T = (400V)(2A)(\cos 60°)$$
$$\qquad P_T = (400V)(2A)(0.5)$$
$$\qquad P_T = 400W$$

Note that the apparent power formula is the product of EI alone without considering the cosine of the phase angle.

10.3.0 Reactive Power

Reactive power is that portion of the apparent power that is caused by inductors and capacitors in the circuit. Inductance and capacitance are always present in real AC circuits. No work is performed

INSIDE TRACK

Third-Order Harmonics

Harmonics are frequencies that are multiples of the basic frequency. For example, 180Hz is three times the frequency of 60Hz and is therefore known as a third-order harmonic. Harmonics are caused by a variety of devices, such as personal computers, uninterruptible power supplies, adjustable speed drives, and electronic ballasts. Third-order harmonics, also known as triplen harmonics or triplens, may cause transformer overheating and should be a consideration when sizing transformers.

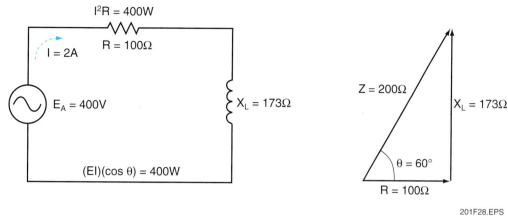

Figure 28 ♦ Power calculations in an AC circuit.

by reactive power; the power is stored in the inductors and capacitors, then returned to the circuit. Therefore, reactive power is always 90° out of phase with true power. The units for reactive power are volt-amperes-reactive (VARs).

In general, for any phase angle θ between E and I, multiplying EI by sine θ gives the vertical component at 90° for the value of the VARs. In *Figure 28*, the value of sine 60° is 800 × 0.866 = 692.8 VARs.

Note that the factor sine θ for the VARs gives the vertical or reactive component of the apparent power EI. However, multiplying EI by cosine θ as the power factor gives the horizontal or resistive component for the real power.

10.4.0 Power Factor

Because it indicates the resistive component, cosine θ is the power factor (pf) of the circuit, converting the EI product to real power. For series circuits, use the formula:

$$pf = \cos \theta = \frac{R}{Z}$$

For parallel circuits, use the formula:

$$pf = \cos \theta = \frac{I_R}{I_T}$$

In *Figure 28* as an example of a series circuit, R and Z are used for the calculations:

$$pf = \cos \theta = \frac{R}{Z} = \frac{100\Omega}{200\Omega} = 0.5$$

The power factor is not an angular measure but a numerical ratio with a value between 0 and 1, equal to the cosine of the phase angle. With all resistance and zero reactance, R and Z are the same for a series circuit of I_R and I_T and are the same for a parallel circuit. The ratio is 1. Therefore, unity power factor means a resistive circuit. At the opposite extreme, all reactance with zero resistance makes the power factor zero, meaning that the circuit is all reactive.

The power factor gives the relationship between apparent power and true power. The power factor can thus be defined as the ratio of true power to apparent power:

$$pf = \frac{P_T}{P_A}$$

For example, calculate the power factor of the circuit shown in *Figure 29*.

The true power is the product of the resistor current squared and the resistance:

$$P_T = I^2 R$$
$$P_T = 10A^2 \times 10\Omega$$
$$P_T = 1,000W$$

The apparent power is the product of the source voltage and total current:

$$P_A = (I_T)(E)$$
$$P_A = 10.2A \times 100V$$
$$P_A = 1,020 VA$$

Calculating total current:

$$I_T = \sqrt{I_R^2 + (I_C - I_L)^2} = \sqrt{10A^2 + (4A - 2A)^2}$$
$$I_T = 10.2A$$

The power factor is the ratio of true power to apparent power:

$$pf = \frac{P_T}{P_A}$$
$$pf = \frac{1,000}{1,020}$$
$$pf = 0.98$$

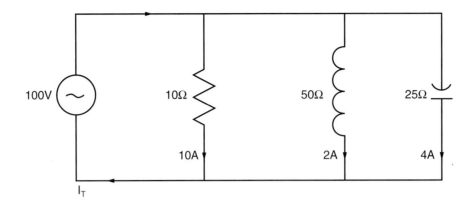

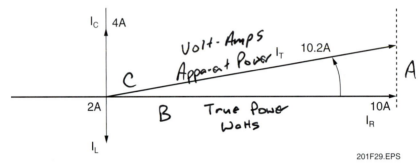

Figure 29 ◆ RLC circuit calculation.

As illustrated in the previous example, the power factor is determined by the system load. If the load contained only resistance, the apparent power would equal the true power and the power factor would be at its maximum value of one. Purely resistive circuits have a power factor of unity or one. If the load is more inductive than capacitive, the apparent power will lag the true power and the power factor will be lagging. If the load is more capacitive than inductive, the apparent power will lead the true power and the power factor will be leading. If there is any reactive load on the system, the apparent power will be greater than the true power and the power factor will be less than one.

10.5.0 Power Triangle

The phase relationships among the three types of AC power are easily visualized on the power triangle shown in *Figure 30*. The true power (W) is the horizontal leg, the apparent power (VA) is the hypotenuse, and the cosine of the phase angle between them is the power factor. The vertical leg of the triangle is the reactive power and has units of volt-amperes-reactive (VARs).

As illustrated on the power triangle (*Figure 30*), the apparent power will always be greater than the true power or reactive power. Also, the apparent power is the result of the vector addition of true and reactive power. The power magnitude relationships shown in *Figure 30* can be derived from the Pythagorean theorem for right triangles:

$$c^2 = a^2 + b^2$$

Therefore, c also equals the square root of $a^2 + b^2$, as shown below:

$$c = \sqrt{a^2 + b^2}$$

- Apparent power is always greater than true power.

- True Power is the power actually being used.

- True power is watts.

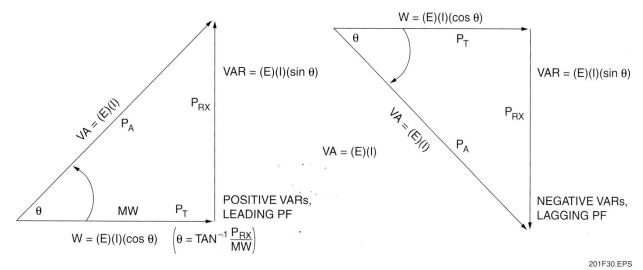

Figure 30 ◆ Power triangle.

11.0.0 ◆ TRANSFORMERS

A transformer is a device that transfers electrical energy from one circuit to another by electromagnetic induction (transformer action). The electrical energy is always transferred without a change in frequency, but may involve changes in the magnitudes of voltage and current. Because a transformer works on the principle of electromagnetic induction, it must be used with an input source voltage that varies in amplitude.

11.1.0 Transformer Construction

Figure 31 shows the basic components of a transformer. In its most basic form, a transformer consists of:

- A primary coil or winding
- A secondary coil or winding
- A core that supports the coils or windings

A simple transformer action is shown in *Figure 32*. The primary winding is connected to a 60Hz

- Transformers are a good example of inductance.

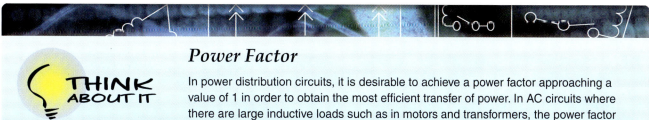

Power Factor

In power distribution circuits, it is desirable to achieve a power factor approaching a value of 1 in order to obtain the most efficient transfer of power. In AC circuits where there are large inductive loads such as in motors and transformers, the power factor can be considerably less than 1. For example, in a highly inductive motor circuit, if the voltage is 120V, the current is 12A, and the current lags the voltage by 60°, the power factor is 0.5 or 50% (cosine of 60° = 0.5). The apparent power is 1,440VA (120V × 12A), but the true power is only 720W [120V × (0.5 × 12A) = 720W]. This is a very inefficient circuit. What would you do to this circuit in order to achieve a circuit having a power factor as close to 1 as possible?

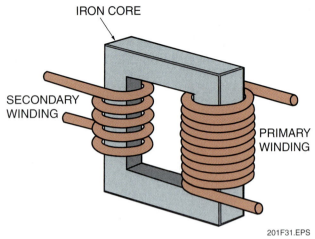

Figure 31 ◆ Basic components of a transformer.

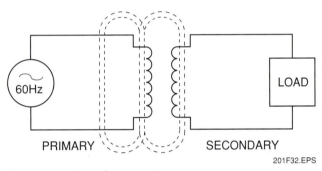

Figure 32 ◆ Transformer action.

AC voltage source. The magnetic field or flux builds up (expands) and collapses (contracts) around the primary winding. The expanding and contracting magnetic field around the primary winding cuts the secondary winding and induces an alternating voltage into the winding. This voltage causes AC to flow through the load. The voltage may be stepped up or down depending on the design of the primary and secondary windings.

11.1.1 Core Characteristics

Commonly used core materials are air, soft iron, and steel. Each of these materials is suitable for particular applications and unsuitable for others. Generally, air-core transformers are used when the voltage source has a high frequency (above 20kHz). Iron-core transformers are usually used when the source frequency is low (below 20kHz). A soft-iron transformer is very useful where the transformer must be physically small yet efficient. The iron-core transformer provides better power transfer than the air-core transformer. Laminated sheets of steel are often used in a transformer to reduce one type of power loss known as eddy currents. These are undesirable currents, induced into the core, which circulate around the core. Laminating the core reduces these currents to smaller levels. These steel laminations are insulated with a nonconducting material, such as varnish, and then formed into a core as shown in *Figure 33*. It takes about 50 such laminations to make a core one inch thick. The most efficient transformer core is one that offers the best path for the most lines of flux, with the least loss in magnetic and electrical energy.

11.1.2 Transformer Windings

A transformer consists of two coils called windings, which are wrapped around a core. The transformer operates when a source of AC voltage is connected to one of the windings and a load device is connected to the other. The winding that is connected to the source is called the primary winding. The winding that is connected to the load is called the secondary winding. *Figure 34* shows a cutaway view of a typical transformer.

The wire is coated with varnish so that each turn of the winding is insulated from every other turn. In a transformer designed for high-voltage

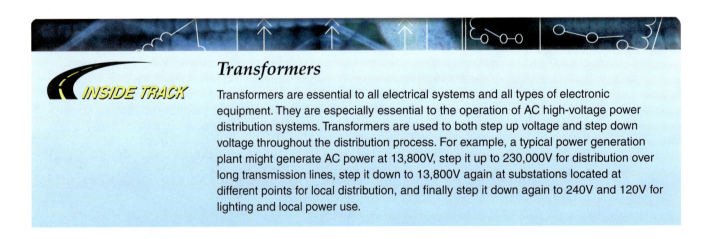

Transformers

Transformers are essential to all electrical systems and all types of electronic equipment. They are especially essential to the operation of AC high-voltage power distribution systems. Transformers are used to both step up voltage and step down voltage throughout the distribution process. For example, a typical power generation plant might generate AC power at 13,800V, step it up to 230,000V for distribution over long transmission lines, step it down to 13,800V again at substations located at different points for local distribution, and finally step it down again to 240V and 120V for lighting and local power use.

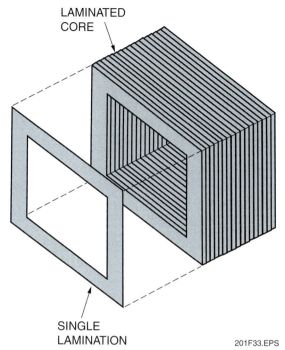

Figure 33 ◆ Steel laminated core.

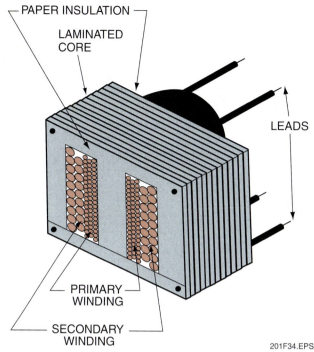

Figure 34 ◆ Cutaway view of a transformer core.

applications, sheets of insulating material such as paper are placed between the layers of windings to provide additional insulation.

When the primary winding is completely wound, it is wrapped in insulating paper or cloth. The secondary winding is then wound on top of the primary winding. After the secondary winding is complete, it too is covered with insulating paper. Next, the iron core is inserted into and around the windings as shown.

Sometimes, terminals may be provided on the enclosure for connections to the windings. The figure shows four leads, two from the primary and two from the secondary. These leads are to be connected to the source and load, respectively.

11.2.0 Operating Characteristics

The operating characteristics of transformers are determined by the applied voltage and the winding design. These affect both the exciting current (no-load condition) and the phase relationship between the windings during transformer operation.

11.2.1 Energized with No Load

A no-load condition is said to exist when a voltage is applied to the primary, but no load is connected to the secondary. Assume the output of the secondary is connected to a load by a switch that is open. Because of the open switch, there is no current flowing in the secondary winding. With the switch open and an AC voltage applied to the primary, there is, however, a very small amount of current, called exciting current, flowing in the primary. Essentially, what this current does is excite the coil of the primary to create a magnetic field. The amount of exciting current is determined by three factors: the amount of voltage applied (E_A); the resistance (R) of the primary coil's wire and core losses; and the X_L, which is dependent on the frequency of the exciting current. These factors are all controlled by transformer design.

This very small amount of exciting current serves two functions:

- Most of the exciting energy is used to support the magnetic field of the primary.
- A small amount of energy is used to overcome the resistance of the wire and core. This is dissipated in the form of heat (power loss).

Exciting current will flow in the primary winding at all times to maintain this magnetic field, but no transfer of energy will take place as long as the secondary circuit is open.

11.2.2 Phase Relationship

The secondary voltage of a simple transformer may be either in phase or out of phase with the primary voltage. This depends on the direction in which the windings are wound and the arrangement of the connection to the external circuit (load). Simply, this means that the two voltages may rise and fall together, or one may rise while the other is falling. Transformers in which the secondary voltage is in phase with the primary are referred to as like-wound transformers, while those in which the voltages are 180° out of phase are called unlike-wound transformers.

Dots are used to indicate points on a transformer schematic symbol that have the same instantaneous polarity (points that are in phase). The use of phase-indicating dots is illustrated in *Figure 35*. In the first part of the figure, both the primary and secondary windings are wound from top to bottom in a clockwise direction, as viewed from above the windings. When constructed in this manner, the top lead of the primary and the top lead of the secondary have the same polarity. This is indicated by the dots on the transformer symbol.

The second part of the figure illustrates a transformer in which the primary and secondary are wound in opposite directions. As viewed from above the windings, the primary is wound in a clockwise direction from top to bottom, while the secondary is wound in a counterclockwise direction. Notice that the top leads of the primary and secondary have opposite polarities. This is indicated by the dots being placed on opposite ends of the transformer symbol. Thus, the polarity of voltage at the terminals of the transformer secondary depends on the direction in which the secondary is wound with respect to the primary.

11.3.0 Turns and Voltage Ratios

To understand how a transformer can be used to step up or step down voltage, the term turns ratio must be understood. The total voltage induced into the secondary winding of a transformer is determined mainly by the ratio of the number of turns in the primary to the number of turns in the secondary, and by the amount of voltage applied to the primary. Therefore, to set up a formula:

$$\text{Turns ratio} = \frac{\text{Number of turns in the primary}}{\text{Number of turns in the secondary}}$$

The first transformer in *Figure 36* shows a transformer whose primary consists of 10 turns of wire, and whose secondary consists of a single turn of wire. As lines of flux generated by the primary expand and collapse, they cut both the 10 turns of the primary and the single turn of the secondary. Since the length of the wire in the secondary is approximately the same as the length of the wire in each turn of the primary, the EMF induced into the secondary will be the same as the EMF induced into each turn of the primary.

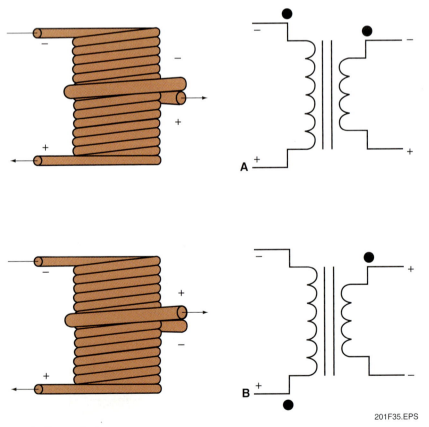

Figure 35 ◆ Transformer winding polarity.

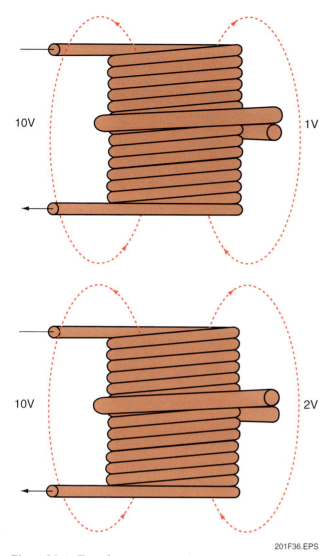

Figure 36 ♦ Transformer turns ratio.

This means that if the voltage applied to the primary winding is 10 volts, the CEMF in the primary is almost 10 volts. Thus, each turn in the primary will have an induced CEMF of approximately 1/10 of the total applied voltage, or one volt. Since the same flux lines cut the turns in both the secondary and the primary, each turn will have an EMF of one volt induced into it. The first transformer in *Figure 36* has only one turn in the secondary, thus, the EMF across the secondary is one volt.

The second transformer represented in *Figure 36* has a 10-turn primary and a two-turn secondary. Since the flux induces one volt per turn, the total voltage across the secondary is two volts. Notice that the volts per turn are the same for both primary and secondary windings. Since the CEMF in the primary is equal (or almost) to the applied voltage, a proportion may be set up to express the value of the voltage induced in terms of the voltage applied to the primary and the number of turns in each winding. This proportion also shows the relationship between the number of turns in each winding and the voltage across each winding, and is expressed by the equation:

$$\frac{E_S}{E_P} = \frac{N_S}{N_P}$$

Where:

N_P = number of turns in the primary
E_P = voltage applied to the primary
E_S = voltage induced in the secondary
N_S = number of turns in the secondary

The equation shows that the ratio of secondary voltage to primary voltage is equal to the ratio of secondary turns to primary turns. The equation can be written as:

$$E_P N_S = E_S N_P$$

For example, a transformer has 100 turns in the primary, 50 turns in the secondary, and 120VAC applied to the primary (E_P). What is the voltage across the secondary (E_S)?

$$N_P = 100 \text{ turns} \quad E_P = 120\text{VAC}$$
$$N_S = 50 \text{ turns}$$
$$\frac{E_S}{E_P} = \frac{N_S}{N_P} \text{ or } E_S = \frac{E_P N_S}{N_P}$$
$$E_S = \frac{120\text{V} \times 50 \text{ turns}}{100 \text{ turns}} = 60\text{VAC}$$

The transformers in *Figure 36* have fewer turns in the secondary than in the primary. As a result, there is less voltage across the secondary than across the primary. A transformer in which the voltage across the secondary is less than the voltage across the primary is called a step-down transformer. The ratio of a 10-to-1 step-down transformer is written as 10:1.

A transformer that has fewer turns in the primary than in the secondary will produce a greater voltage across the secondary than the voltage applied to the primary. A transformer in which the voltage across the secondary is greater than the voltage applied to the primary is called a step-up transformer. The ratio of a 1-to-4 step-up transformer should be written 1:4. Notice in the two ratios that the value of the primary winding is always stated first.

11.4.0 Types of Transformers

Transformers are widely used to permit the use of trip coils and instruments of moderate current and voltage capacities and to measure the characteristics of high-voltage and high-current circuits.

Turns and Voltage Ratios

What is the magnitude of the voltage and current supplied by the secondary of the transformer in the circuit shown below?

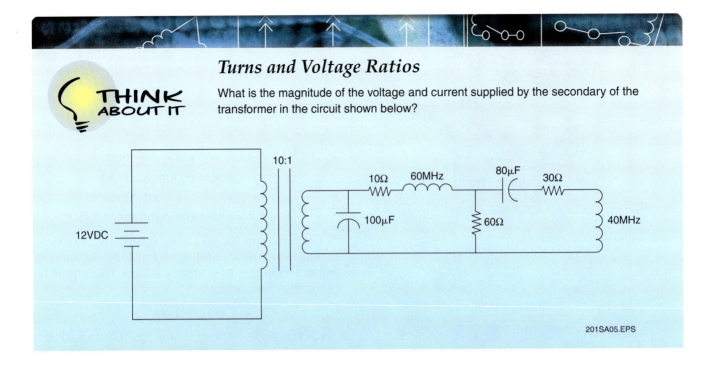

Since secondary voltage and current are directly related to primary voltage and current, measurements can be made under the low-voltage or low-current conditions of the secondary circuit and still determine primary characteristics. Tripping transformers and instrument transformers are examples of this use of transformers.

The primary or secondary coils of a transformer can be tapped to permit multiple input and output voltages. *Figure 37* shows several tapped transformers. The center-tapped transformer is particularly important because it can be used in conjunction with other components to convert an AC input to a DC output.

11.4.1 Isolation Transformer

Isolation transformers are wound so that their primary and secondary voltages are equal. Their purpose is to electrically isolate a piece of electrical equipment from the power distribution system.

Many pieces of electronic equipment use the metal chassis on which the components are mounted as part of the circuit (*Figure 38*). Personnel

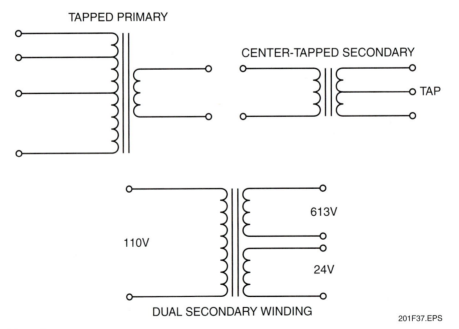

Figure 37 ◆ Tapped transformers.

1.40 ELECTRICAL LEVEL TWO ◆ TRAINEE GUIDE

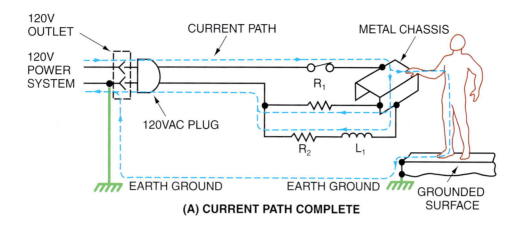

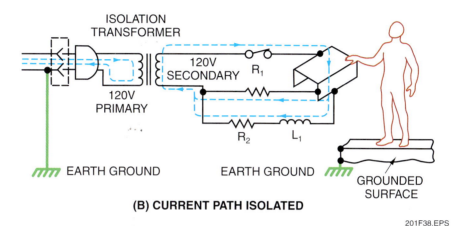

Figure 38 ◆ Importance of an isolation transformer.

working with this equipment may accidentally come in contact with the chassis, completing the circuit to ground, and receive a shock as shown in *Figure 38(A)*. If the resistances of their body and the ground path are low, the shock can be fatal. Placing an isolation transformer in the circuit as shown in *Figure 38(B)* breaks the ground current path that includes the worker. Current can no longer flow from the power supply through the chassis and worker to ground; however, the equipment is still supplied with the normal operating voltage and current.

11.4.2 Autotransformer

In a transformer, it is not necessary for the primary and secondary to be separate and distinct windings. *Figure 39* is a schematic diagram of what is known as an autotransformer. Note that a single coil of wire is tapped to produce what is electrically both a primary and a secondary winding.

The voltage across the secondary winding has the same relationship to the voltage across the primary that it would have if they were two distinct windings. The movable tap in the secondary is used to select a value of output voltage either higher or lower than E_P, within the range of the transformer. When the tap is at Point A, E_S is less than E_P; when the tap is at Point B, E_S is greater than E_P.

Autotransformers rely on self-induction to induce their secondary voltage. The term autotransformer can be broken down into two words: auto, meaning self; and transformer, meaning to change potential. The autotransformer is made of

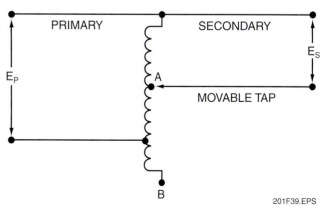

Figure 39 ◆ Autotransformer schematic diagram.

MODULE 26201-08 ◆ ALTERNATING CURRENT 1.41

one winding that acts as both a primary and a secondary winding. It may be used as either a step-up or step-down transformer. Some common uses of autotransformers are as variable AC voltage supplies and fluorescent light ballast transformers, and to reduce the line voltage for various types of low-voltage motor starters.

11.4.3 Current Transformer

A current transformer differs from other transformers in that the primary is a conductor to the load and the secondary is a coil wrapped around the wire to the load. Just as any ammeter is connected in line with a circuit, the current transformer is connected in series with the current to be measured. *Figure 40* is a diagram of a current transformer.

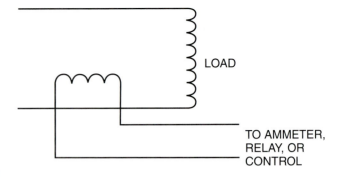

Figure 40 ◆ Current transformer schematic diagram.

 WARNING!
Do not open a current transformer under load.

Since current transformers are series transformers, the usual voltage and current relationships do not apply. Current transformers vary considerably in rated primary current, but are usually designed with ampere-turn ratios such that the secondary delivers five amperes at full primary load.

Current transformers are generally constructed with only a few turns or no turns in the primary. The voltage in the secondary is induced by the changing magnetic field that exists around a single conductor. The secondary is wound on a circular core, and the large conductor that makes up the primary passes through the hole in its center. Because the primary has few or no turns, the secondary must have many turns (providing a high turns ratio) in order to produce a usable voltage. The advantage of this is that you get an output off the secondary proportional to the current flowing through the primary, without an appreciable voltage drop across the primary. This is because the primary voltage equals the current times the impedance. The impedance is kept near zero by using no or very few primary turns. The disadvantage is that you cannot open the secondary circuit with the primary energized. To do so would cause the secondary current to drop rapidly to zero. This would cause the magnetic field generated by the secondary current to collapse rapidly. The rapid collapse of the secondary field through the many turns of the secondary winding would induce a dangerously high voltage in the secondary, creating an equipment and personnel hazard.

Because the output of current transformers is proportional to the current in the primary, they are most often used to power current-sensing meters and relays. This allows the instruments to respond to primary current without having to handle extreme magnitudes of current.

11.4.4 Potential Transformer

The primary of a potential transformer is connected across or in parallel with the voltage to be measured, just as a voltmeter is connected across a circuit. *Figure 41* shows the schematic diagram for a potential transformer.

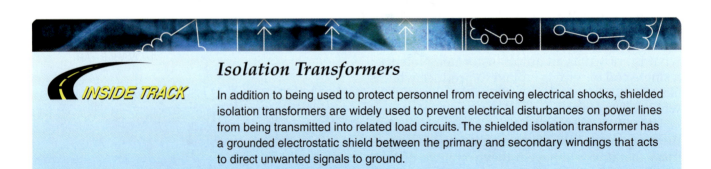

Isolation Transformers

INSIDE TRACK In addition to being used to protect personnel from receiving electrical shocks, shielded isolation transformers are widely used to prevent electrical disturbances on power lines from being transmitted into related load circuits. The shielded isolation transformer has a grounded electrostatic shield between the primary and secondary windings that acts to direct unwanted signals to ground.

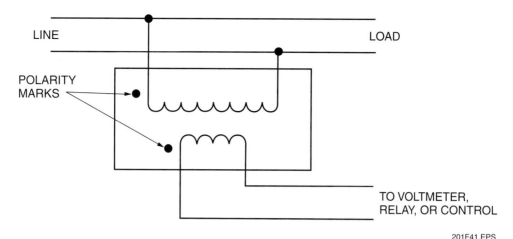

Figure 41 ◆ Potential transformer.

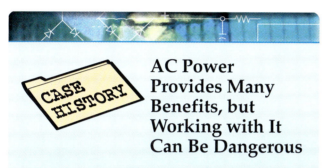

AC Power Provides Many Benefits, but Working with It Can Be Dangerous

Working with AC power can be dangerous unless proper safety methods and procedures are followed. The National Institute for Occupational Safety and Health (NIOSH) investigated 224 incidents of electrocutions that resulted in occupational fatalities. One hundred twenty-one of the victims were employed in the construction industry. Two hundred twenty-one of the incidents (99%) involved AC. Of the 221 AC electrocutions, 74 (33%) involved AC voltages less than 600V and 147 (66%) involved 600V or more. Forty of the lower-voltage electrocutions involved 120/240V.

Factors relating to the causes of these electrocutions included the lack of enforcement of existing employer policies including the use of personal protective equipment and the lack of supervisory intervention when existing policies were being violated. Of the 224 victims, 194 (80%) had some type of electrical safety training. Thirty-nine victims had no training at all. It is notable that 100 of the victims had been on the job less than one year.

The Bottom Line: Never assume that you are safe when working at lower voltages. All voltage levels must be considered potentially lethal. Also, safety training does no good if you don't put it into practice every day. Always put safety first.

Potential transformers are basically the same as any other single-phase transformer. Although primary voltage ratings vary widely according to the specific application, secondary voltage ratings are usually 120V, a convenient voltage for meters and relays.

Because the output of potential transformers is proportional to the phase-to-phase voltage of the primary, they are often used to power voltage-sensing meters and relays. This allows the instruments to respond to primary voltage while having to handle only 120V. Also, potential transformers are essentially single-phase step-down transformers. Therefore, power to operate low-voltage auxiliary equipment associated with high-voltage switchgear can be supplied off the high-voltage lines that the equipment serves via potential transformers.

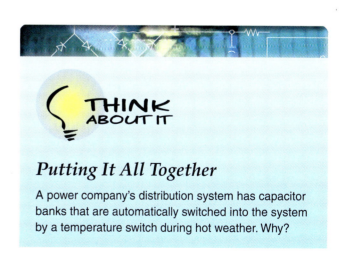

Putting It All Together

A power company's distribution system has capacitor banks that are automatically switched into the system by a temperature switch during hot weather. Why?

Current Transformers

Although the use of a current transformer completely isolates the secondary and the related ammeter from the high-voltage lines, the secondary of a current transformer should never be left open circuited. To do so may result in dangerously high voltage being induced in the secondary.

Potential Transformers

In addition to being used to step down high voltages for the purpose of safe metering, potential transformers are widely used in all kinds of control devices where the condition of high voltages must be monitored. One such example involves the use of a potential transformer-operated contactor in emergency lighting standby generator circuits. Under normal conditions with utility power applied, the contactor is energized and its normally closed contacts are open. If the power fails, the contactor de-energizes, causing its contacts to close and activating the standby generator circuit.

Review Questions

1. An electric current always produces _____.
   a. mutual inductance
   b. a magnetic field
   c. capacitive reactance
   d. high voltage

2. The number of cycles an alternating electric current undergoes per second is known as _____.
   a. amperage
   b. frequency
   c. voltage
   d. resistance

3. What is the peak voltage in a circuit with an rms voltage of 120VAC?
   a. 117 volts
   b. 120 volts
   c. 150 volts
   d. 170 volts

4. Which of the following conditions exist in a circuit of pure resistance?
   a. The voltage and current are in phase.
   b. The voltage and current are 90° out of phase.
   c. The voltage and current are 120° out of phase.
   d. The voltage and current are 180° out of phase.

5. In a purely resistive AC circuit where 240V is applied across a 10-ohm resistor, the amperage is _____.
   a. 10A
   b. 24A
   c. 60A
   d. 120A

6. When the current increases in an AC circuit, what role does inductance play?
   a. It increases the current.
   b. It plays no role at all.
   c. It causes the overcurrent protection to open.
   d. It reduces the current.

7. All of the following factors increase the inductance of a coil, *except* a _____.
   a. greater number of turns
   b. longer core
   c. wider coil
   d. magnetic core

8. Which of the following conditions exist in a circuit of pure inductance?
   a. The voltage and current are in phase.
   b. The voltage and current are 90° out of phase.
   c. The voltage and current are 120° out of phase.
   d. The voltage and current are 180° out of phase.

9. True or False? Reactance increases with an increase in inductance.
   a. True
   b. False

10. Capacitance is measured in _____.
    a. farads
    b. joules
    c. henrys
    d. amps

11. The total capacitance of two 15µF capacitors connected in parallel is _____.
    a. 5µF
    b. 10µF
    c. 15µF
    d. 30µF

12. The opposition to current flow offered by the capacitance of a circuit is known as _____.
    a. mutual inductance
    b. pure resistance
    c. inductive reactance
    d. capacitive reactance

Review Questions

13. The total opposition to current flow in an AC circuit is known as ____.
    a. resistance
    b. capacitive reactance
    c. inductive reactance
    d. impedance

14. A power factor is not an angular measure, but a numerical ratio with a value between 0 and 1, equal to the ____ of the phase angle.
    a. sine
    b. tangent
    c. cosine
    d. cotangent

15. The two windings of a conventional transformer are known as the ____ windings.
    a. mutual and inductive
    b. high and low voltage
    c. primary and secondary
    d. step-up and step-down

Summary

The process by which current is produced electromagnetically is called induction. As the conductor moves across the magnetic field, it cuts the lines of force, and electrons within the conductor flow, creating an electromotive force (EMF). EMF is also known as voltage. There are three conditions that must exist before a current can be produced in this way:

- There must be a magnetic field through which the conductor can pass.
- There must be a conductor in which the voltage will be produced, and the conductor should be perpendicular to the field.
- There must be motion. Either the magnetic field or the conductor must move.

Several factors control the magnitude of the induced current. Voltage will be increased if:

- The speed with which the conductor cuts through the magnetic field is increased (the faster the conductor cuts through the field, the greater the current pulse)
- The strength of the field is increased (the stronger the field, the greater the current pulse)
- The conductor is wound to form a coil (the voltage increases directly with the number of turns of the coil)

A decrease in voltage occurs as the conductor intersects the magnetic field at an angle less than 90°. The greatest current is produced when the conductor intersects the magnetic field at right angles (perpendicular) to the flux lines.

It should be emphasized that a current may be induced by using the magnetic field of a permanent magnet or the magnetic field of another current-carrying conductor (electromagnet).

The magnetic field is among the reasons why phases of current-carrying conductors should not be separated in a raceway; all phase conductors (including the neutral) should be contained in the same raceway. For example, if phase A is separated from phase B and phase C by a metal enclosure, conduit wall, etc., the magnetic field around the conductors will cut across the conduit, causing the conduit to heat up.

Notes

Trade Terms Introduced in This Module

Capacitance: The storage of electricity in a capacitor; capacitance produces an opposition to voltage change. The unit of measurement for capacitance is the farad (F) or microfarad (µF).

Frequency: The number of cycles an alternating electric current, sound wave, or vibrating object undergoes per second.

Hertz (Hz): A unit of frequency; one hertz equals one cycle per second.

Impedance: The opposition to current flow in an AC circuit; impedance includes resistance (R), capacitive reactance (X_C), and inductive reactance (X_L). Impedance is measured in ohms.

Inductance: The creation of a voltage due to a time-varying current; also, the opposition to current change, causing current changes to lag behind voltage changes. The unit of measure for inductance is the henry (H).

Micro: Prefix designating one-millionth of a unit. For example, one microfarad is one-millionth of a farad.

Peak voltage: The peak value of a sinusoidally varying (cyclical) voltage or current is equal to the root-mean-square (rms) value multiplied by the square root of two (1.414). AC voltages are usually expressed as rms values; that is, 120 volts, 208 volts, 240 volts, 277 volts, 480 volts, etc., are all rms values. The peak voltage, however, differs. For example, the peak value of 120 volts (rms) is actually $120 \times 1.414 = 169.71$ volts.

Radian: An angle at the center of a circle, subtending (opposite to) an arc of the circle that is equal in length to the radius.

Reactance: The imaginary part of impedance. Also, the opposition to alternating current (AC) due to capacitance (X_C) and/or inductance (X_L).

Root-mean-square (rms): The square root of the average of the square of the function taken throughout the period. The rms value of a sinusoidally varying voltage or current is the effective value of the voltage or current.

Self-inductance: A magnetic field induced in the conductor carrying the current.

Additional Resources

This module is intended to present thorough resources for task training. The following reference works are suggested for further study. These are optional materials for continued education rather than for task training.

Introduction to Electric Circuits, Latest Edition. New York: Prentice Hall.

Principles of Electric Circuits, Latest Edition. New York: Prentice Hall.

CONTREN® LEARNING SERIES – USER UPDATE

NCCER makes every effort to keep these textbooks up-to-date and free of technical errors. We appreciate your help in this process. If you have an idea for improving this textbook, or if you find an error, a typographical mistake, or an inaccuracy in NCCER's Contren® textbooks, please write us, using this form or a photocopy. Be sure to include the exact module number, page number, a detailed description, and the correction, if applicable. Your input will be brought to the attention of the Technical Review Committee. Thank you for your assistance.

Instructors – If you found that additional materials were necessary in order to teach this module effectively, please let us know so that we may include them in the Equipment/Materials list in the Annotated Instructor's Guide.

Write: Product Development and Revision
National Center for Construction Education and Research
3600 NW 43rd St., Bldg. G, Gainesville, FL 32606

Fax: 352-334-0932

E-mail: curriculum@nccer.org

Craft _____ Module Name _____

Copyright Date _____ Module Number _____ Page Number(s) _____

Description
_____
_____
_____
_____

(Optional) Correction
_____
_____
_____

(Optional) Your Name and Address
_____
_____
_____

Motors: Theory and Application

Procter & Gamble Olean Facility

This refinery produces the P&G patented no-fat Olean cooking oil in Cincinnati, Ohio. It won an ABC National Award of Excellence and involved the installation of over 300,000 feet of conduit, 10,000 feet of cable tray, 2,000,000 feet of power cable and wire, 100,000 feet of heat trace, and 2,700 light fixtures.

26202-08

26202-08
Motors: Theory and Application

Topics to be presented in this module include:

1.0.0	Introduction	2.2
2.0.0	DC Motor Principles	2.2
3.0.0	Types of DC Motors	2.12
4.0.0	Alternating Current Motors	2.16
5.0.0	Multiple-Speed Induction Motors	2.37
6.0.0	Variable-Speed Drives	2.40
7.0.0	Motor Enclosures	2.45
8.0.0	NEMA Frame Designations	2.46
9.0.0	Motor Ratings and Nameplate Data	2.49
10.0.0	Connections and Terminal Markings for AC Motors	2.57
11.0.0	*NEC*® Requirements	2.60
12.0.0	Braking	2.62
13.0.0	Motor Installation	2.63

Overview

Motor windings are electromagnets in which like polarities repel one another and unlike polarities attract each other. The rotor in any motor is the part that turns and is used to drive the load. The stationary windings are typically mounted in a doughnut-like arrangement into which the rotor is installed.

Windings are built into the rotor, and current flows through these windings. Likewise, a current flow is created through the stationary windings. Any time current flows through a conductor, a magnetic field is created around that conductor. Since both windings have current flowing through them at the same time, magnetic fields are generated around both windings. As these magnetic fields interact, they either repel or attract each other. Manipulating the current flow through the rotor and the secondary windings enables the rotor to turn at controlled speeds or in different directions. This is the basic operation of any motor. Motors can be designed to supply high speed, high torque, or both, depending on the demands of the load.

Note: *NFPF 70*®, *National Electrical Code*®, and *NEC*® are registered trademarks of the National Fire Protection Association, Inc., Quincy, MA 02269. All *National Electrical Code*® and *NEC*® references in this module refer to the 2008 edition of the *National Electrical Code*®.

Objectives

Upon completion of this module, you will be able to do the following:

1. Define the following terms:
   - Controller
   - Duty cycle
   - Full-load amps
   - Interrupting rating
   - Thermal protection
   - NEMA design letter
   - Overcurrent
   - Overload
   - Power factor
   - Rated full-load speed
   - Rated horsepower
   - Service factor
2. Describe the various types of motor enclosures.
3. Explain the relationships among speed, frequency, and the number of poles in a three-phase induction motor.
4. Define percent slip and speed regulation.
5. Explain how the direction of a three-phase motor is changed.
6. Describe the component parts and operating characteristics of a three-phase wound-rotor induction motor.
7. Describe the component parts and operating characteristics of a three-phase synchronous motor.
8. Describe the design and operating characteristics of various DC motors.
9. Describe the methods for determining various motor connections.
10. Describe general motor protection requirements as delineated in the *National Electrical Code® (NEC®)*.
11. Define the braking requirements for AC and DC motors.

ELECTRICAL LEVEL TWO

- 26211-08 Control Systems and Fundamental Concepts
- 26210-08 Circuit Breakers and Fuses
- 26209-08 Grounding and Bonding
- 26208-08 Conductor Terminations and Splices
- 26207-08 Cable Tray
- 26206-08 Conductor Installations
- 26205-08 Pull and Junction Boxes
- 26204-08 Conduit Bending
- 26203-08 Electric Lighting
- 26202-08 Motors: Theory and Application
- 26201-08 Alternating Current

ELECTRICAL LEVEL ONE

CORE CURRICULUM: Introductory Craft Skills

202CMAP.EPS

Trade Terms

Armature
Branch circuit
Brush
Circuit breaker
Commutator
Continuous duty
Controller
Duty
Equipment
Field poles
Horsepower
Hours
Intermittent duty
Overcurrent
Overload
Periodic duty
Revolutions per minute (rpm)
Rotation
Synchronous speed
Thermal protector
Varying duty

Required Trainee Materials

1. Pencil and paper
2. Appropriate personal protective equipment
3. Copy of the latest edition of the *National Electrical Code®*

MODULE 26202-08 ◆ MOTORS: THEORY AND APPLICATION 2.1

Prerequisites

Before you begin this module, it is recommended that you successfully complete *Core Curriculum; Electrical Level One;* and *Electrical Level Two*, Module 26201-08.

This course map shows all of the modules in *Electrical Level Two*. The suggested training order begins at the bottom and proceeds up. Skill levels increase as you advance on the course map. The local Training Program Sponsor may adjust the training order.

1.0.0 ♦ INTRODUCTION

The electric motor is the workhorse of modern industry. Its functions are almost unlimited. To control the motors that drive machinery and equipment, we must have electrical supply circuits that perform certain functions. They must provide electrical current to cause the motor to operate in the manner needed to make it perform its intended function. They must also provide protection for the motor from adverse mechanical and electrical conditions. These functions are frequently combined within electrical equipment that we classify as motor control centers.

A thorough understanding of the functions of the various components of a motor control center is desirable from both a maintenance and a troubleshooting standpoint. Properly-maintained motor control centers ensure a minimum of downtime for unscheduled repairs, increase productivity, and contribute to a safer working environment.

2.0.0 ♦ DC MOTOR PRINCIPLES

When a bar of magnetic material is given an induced magnetic charge, a field of magnetic force is developed around the bar. Picture this field as consisting of magnetic lines of force (flux) that exist in the space surrounding the bar. These lines of force appear to leave one end of the bar and extend outside the bar to the opposite end.

The end of the bar magnet where magnetic lines appear to start is called the north pole of the magnet, while the end where these lines reenter the magnet is called the south pole. Actually, the lines extend inside the magnet from the south pole to the north pole, completing a closed loop. This principle is shown in *Figure 1*.

There are several characteristics of these magnetic lines of force that must be remembered when dealing with electric motors. They are:

- Magnetic lines of force are continuous and always form closed loops.
- Magnetic lines of force do not cross.
- Magnetic lines of force with polarities in the same direction repel each other. In other words, a north pole will repel a north pole and a south pole will repel a south pole.
- Magnetic lines of force having polarities in opposite directions tend to attract each other and combine. In other words, a south pole will be attracted by a north pole and vice versa.
- Magnetic lines of force tend to shorten themselves. Therefore, the magnetic lines of force existing between two unlike poles cause the poles to tend to pull together.
- Magnetic lines of force pass through all known materials, magnetic or nonmagnetic. Some materials provide a much easier path for these lines than others. These materials have high permeability and low reluctance. (Reluctance is discussed later in this module.)

2.1.0 DC Motor Components

A DC motor consists of a few major components, each with a specific purpose in the motor's operation.

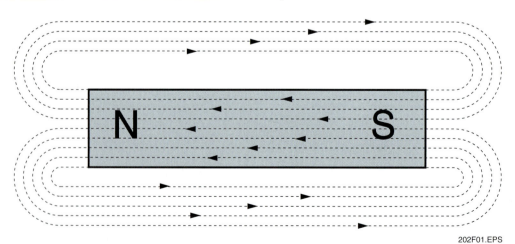

Figure 1 ♦ Magnetic field.

Early Electric Motors

The first U.S. patent for a motor was issued to Thomas Davenport in 1837. He reported that he used silk from his wife's wedding gown as insulation for the conductors, but despite this sacrifice, his motor was not commercially successful. Practical electric motors, like the practical light bulb, did not appear until the late 19th century.

The armature is a movable electromagnet located between the poles of another fixed permanent (field) magnet, as shown in *Figure 2*.

The magnetic field from the armature conductors interacts with the magnetic field from the field magnet. The result of the field interaction is motor action.

Current in a conductor also has its associated magnetic field. When a conductor is placed in another magnetic field from a separate source, the two fields can react to produce motor action. The conductor must be perpendicular to the magnetic field, as illustrated in *Figure 3*. This way, the perpendicular magnetic field of the current is in the same plane as the external magnetic field.

Unless the two fields are in the same plane, they cannot affect each other. In the same plane, however, lines of force in the same direction reinforce to make a stronger field, while lines in the opposite direction cancel and result in a weaker field. The stronger field tends to move the conductor toward the weaker field, as illustrated in *Figure 3*.

These directions are summarized as follows:

- With the conductor at 90°, or perpendicular to the external field, the reaction between the two magnetic fields is at its maximum.
- With the conductor at 0°, or parallel to the external field, there is no effect between them.
- When the conductor rise is at an angle between 0° and 90°, only the perpendicular component is effective.

In motor action, the wire only moves in a straight line, and it stops moving once out of the field, even though current still exists. A practical motor must develop continuous rotary motion. To produce this, a twisting force called torque must be developed.

Torque is produced by mounting a loop in a fixed magnetic field. Current is applied and the flux lines along both sides of the loop interact, causing the loop to act like a lever with a force pushing on its two sides in opposite directions. This is shown in *Figure 4*.

The combined forces result in a turning force or torque, because the rotor or armature is arranged to pivot on its axis. The overall turning force on the armature depends on several factors, including field strength, armature current strength, and the physical construction of the armature, especially the distance from the loop sides to the axis lines.

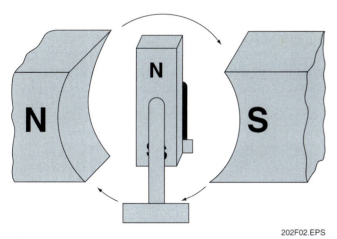

Figure 2 ◆ Basic motor action.

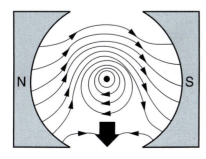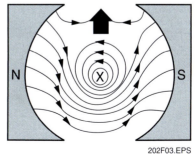

Figure 3 ◆ Motor action.

Brushless DC Motors

Brushless DC motors do not produce carbon dust and are ideal in sensitive electronic applications. They are commonly used in personal computers and CD/DVD players.

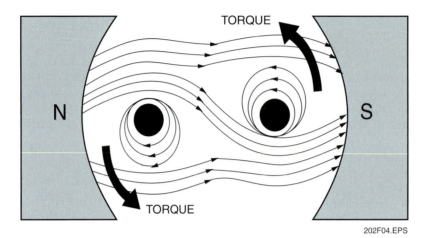

Figure 4 ◆ Torque.

Because of the lever action, the forces on the sides of the armature loop will increase as the loop sides are farther from the axis; therefore, larger armatures will produce greater torques.

In the practical motor, the torque determines the energy available for doing useful work. The greater the torque, the greater the energy. If a motor does not develop enough torque to turn its load, it stalls.

To get continuous rotation, the armature must be kept moving in the same direction. This requires reversing the direction of current through the armature for every 180° of revolution. A commutator is used to provide this switching action. This is shown in *Figure 5*.

The commutator on a DC motor is a conducting ring that is split into two segments, with each segment connected to an end of the armature loop.

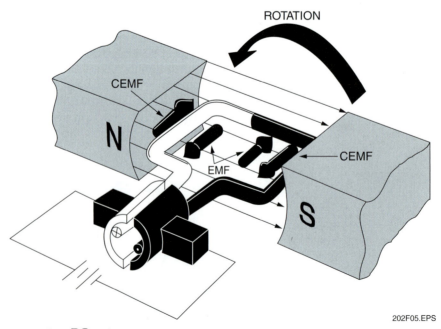

Figure 5 ◆ Single-loop armature DC motor.

Current enters the side of the armature closest to the south pole of the field and leaves the side closest to the north pole of the field. The interaction of the two fields produces a torque, and the armature rotates in that direction.

A brush makes contact with each segment of the commutator, providing a connection between the movable commutator and the stationary DC power source. *Figure 6* shows various brushes and commutator connections used in DC motors.

The problem of switching commutator segments in a simple single-loop motor is that when the motor stops, there is no way of predicting the position of the armature at rest. If the armature stops in a position where the commutator is in the middle of switching, the motor will not start unless you physically turn the armature.

This problem can be overcome by winding more coils on the armature and by using more commutator segments. This will produce a self-starting motor. The motor shown in *Figure 7* uses three armature coils and three commutator segments. Regardless of where the armature comes to

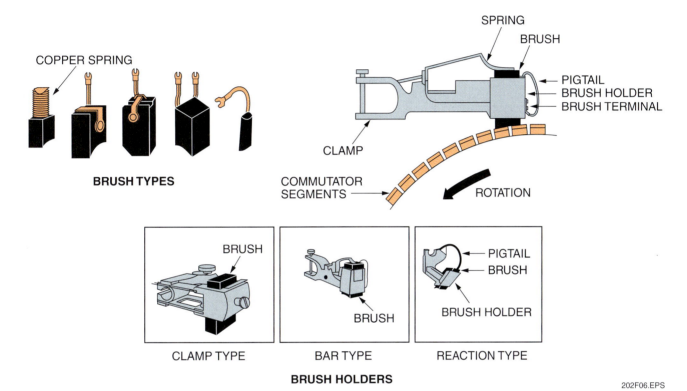

Figure 6 ◆ Brushes, brush rigging, and commutator connections.

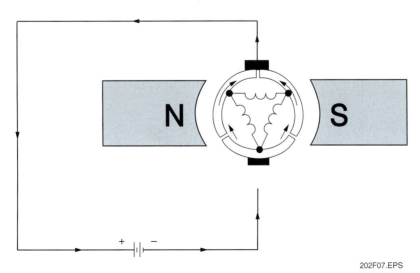

Figure 7 ◆ Self-starting motor.

rest, there is always a path for current that will produce torque to rotate the armature.

2.2.0 The Neutral Plane

The armature turns when torque is produced, and torque is produced as long as the fields of the magnet and armature interact. When the loop reaches a position perpendicular to the field, the interaction of the magnetic fields stops. This position is the neutral plane, shown in *Figure 8*.

In the neutral plane, no torque is produced and the rotation of the armature should stop. However, inertia tends to keep the armature in motion even after the prime moving force is removed; thus the armature tends to rotate past the neutral plane. At the neutral position, the commutator disconnects from the brushes, and once the armature goes past neutral, the sides of the loop reverse positions. The switching action of the commutator maintains the direction of current through the armature. Current still enters the armature side that is closest to the south pole.

Since the magnet's field direction remains the same throughout, the interaction of fields after commutation keeps the torque going in the original direction; thus, continuous rotation is maintained. See *Figure 9*.

Although such an elementary DC motor can be built and operated, it has two serious shortcomings that prevent it from being useful: first, such a motor cannot always start by itself; and second, once started, it operates very irregularly.

When the elementary DC motor runs, its operation is erratic because it produces torque irregularly. Maximum torque is produced only when the

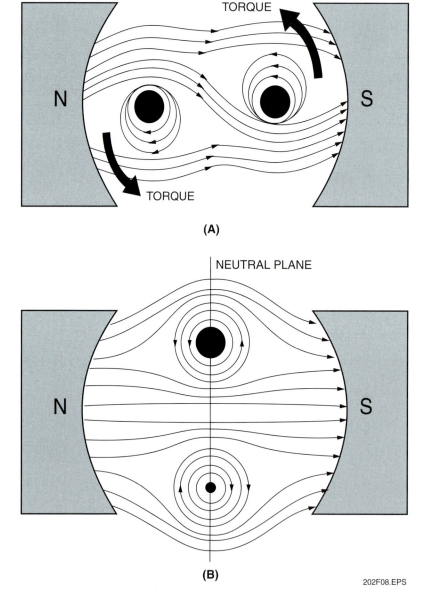

Figure 8 ◆ Neutral plane.

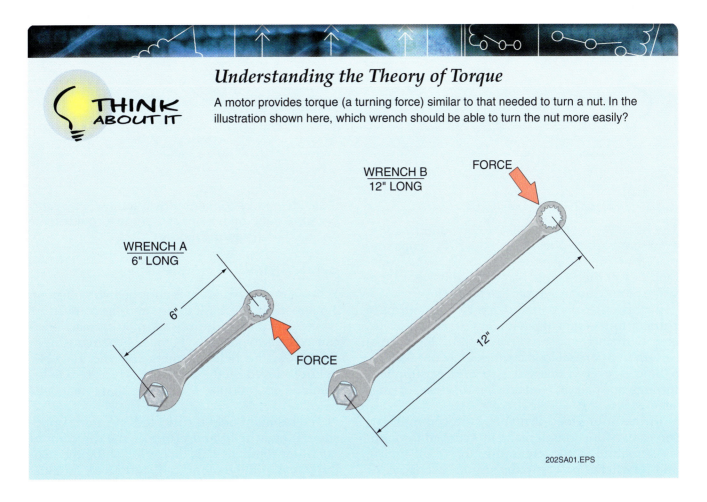

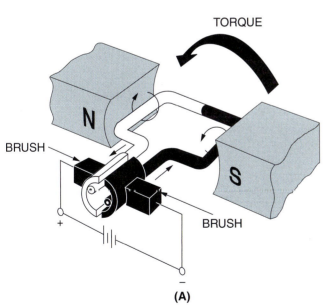
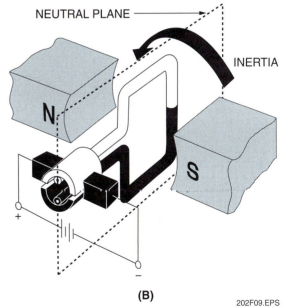

Figure 9 ◆ Neutral plane in a DC motor.

plane of the single-loop armature is parallel with the plane of the field. This is the position at right angles to the neutral plane. Once the armature passes this plane of maximum torque, less and less torque is developed until it arrives at the neutral plane again. Inertia carries the armature past the neutral plane and so the motor continues to turn. Its irregularity in producing torque, however, prevents the single-loop elementary DC motor from being used for practical jobs.

Two-Loop DC Motors

Using the basic principles of motor action, explain why the two-loop DC motor is self-starting and why its torque is erratic.

2.3.0 Two-Loop DC Motors

The basic DC motor is improved by building the armature with two or more loops. The loops are placed at right angles to each other; when one loop lies in the neutral plane, the other is in the plane of maximum torque. See *Figure 10*.

In this case, the commutator is split into two pairs or four segments, with one segment associated with each end of each armature loop. This sets up two parallel loop circuits. Only one loop at a time is ever connected if power is supplied through one pair of fixed brushes to one set of ring segments.

In this multi-loop armature, the commutator serves two functions: it maintains current through the armature in the same direction at all times, and it switches power to the armature loop nearing the maximum torque position.

This motor is self-starting because at least one winding will have interaction with the main field. With this two-loop system, the torque developed is steadier and stronger but still somewhat erratic, because only one loop at a time provides the torque that drives the motor.

2.4.0 Armature Reaction

When a motor armature is supplied with current, a magnetic flux is built up around the conductors of the armature windings. Armature reaction is caused by two magnetic fields: the main magnetic field from the field magnets and the magnetic field produced by the armature. These two fields combine to produce a new resultant magnetic field.

The resultant field is distorted and shifts opposite the main field and opposite the direction of armature rotation. This distortion shifts the neutral plane of the motor. See *Figure 11* for an illustration of armature reaction.

The amount of armature reaction determines how far the neutral plane is shifted. The amount of armature reaction depends on the amount and

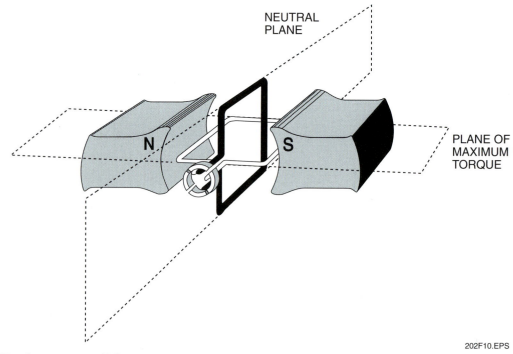

Figure 10 ♦ Two-loop armature DC motor.

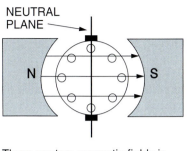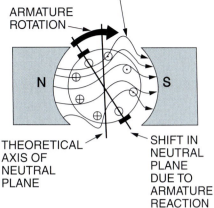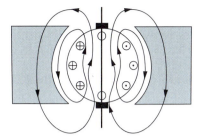

Figure 11 ♦ Armature reaction.

direction of the armature current. The concern over the neutral plane shift occurs because of the need for commutation.

Commutation, or the switching of the armature polarity, must take place at the neutral plane in order to allow the output current from the machine to remain in the same direction without arcing. When commutation takes place anywhere other than the neutral plane, it is like a switch that is opened during high current—it will draw an arc.

This armature reaction can be overcome by installing interpole windings. Interpoles are special electromagnetic pole pieces that are connected in series with the armature winding. The armature current causes a magnetic field to form around the windings. Their action is self-regulating, and the interpole field will apply the proper amount of cancellation field for any set of conditions. For a high armature reaction, the canceling field is strong. For a low armature reaction, the canceling field is weaker. See *Figure 12*.

2.5.0 Counter-Electromotive Force (CEMF)

When a DC motor is in operation, it acts much like a DC generator. A magnetic field is produced by the field poles, and a loop of wire in the armature turns and cuts this magnetic field. To understand counter-electromotive force (CEMF), first disregard the fact that external current is being applied to the rotor via the carbon brushes on the commutator segments. As the armature wires rotate and cut the magnetic field of the field poles, a voltage is induced in them similar to that which was discussed in induction motors. This induced voltage (EMF) causes a current to flow in them and a resulting magnetic field is created.

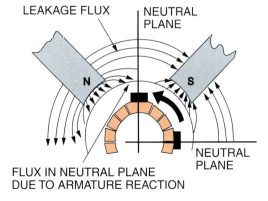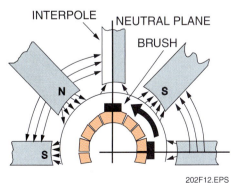

Figure 12 ♦ Interpoles.

MODULE 26202-08 ♦ MOTORS: THEORY AND APPLICATION 2.9

Before analyzing the relative direction between the current induced in the armature windings and the current that caused it in the field poles, first remember the left-hand rule. Using your left hand, hold it such that your index finger points in the direction of the magnetic field (north to south) and your thumb points in the direction of rotational force on a given conductor. Your middle finger will now point in the direction of current flow for that conductor. This current would be in opposition to the current that is flowing from the battery. Since this induced voltage and induced current are opposite to those of the battery, they are called CEMF. The two currents are flowing in opposite directions. This would mean that the battery voltage and the CEMF are opposite in polarity. See *Figure 13*.

When first discussing CEMF, we disregarded the fact that external DC was being applied to the armature via the brushes. The induced voltage and resulting current flow was then shown to flow opposite to the externally applied current. This was an oversimplification, since only one current flows. Since the CEMF can never become as large as the external applied voltage, and since they are opposite in polarity, the CEMF works to cancel only a part of the applied voltage. The single current that flows is smaller due to the CEMF.

Since the CEMF of a motor is generated by the action of the armature windings cutting the lines of force set up by the field poles, the value of it will depend on the field strength and the armature speed. The effective voltage acting in the armature is the terminal voltage minus the CEMF. Ohm's law gives the value of armature current by:

$$\text{Armature } (I_A) = \frac{\text{terminal voltage} - \text{CEMF}}{\text{armature resistance } (R_A)}$$

Where CEMF = terminal voltage − $(I_A \times R_A)$

Example:

Find the value of CEMF of a DC motor when the terminal voltage is 240V and the armature current is 60 amps. The armature resistance has been measured at 0.08 ohm.

CEMF = terminal voltage − $(I_A \times R_A)$
CEMF = 240 − (60 × 0.08) = 240 − 4.8 = 235.2V

CEMF acts as an automatic current limiter that reduces armature current to a level adequate to drive the motor but not great enough to heat the armature to where it is in danger of burning out. CEMF acts as a load for the DC power supply feeding the motor, so that the low-resistance motor windings do not draw excessive amounts of current.

If we stalled the armature so that no CEMF was produced, we would find that the motor draws so much current that it heats up. This reaction is shown in *Figure 14*. CEMF is present in all motors and is necessary for a motor's operation.

We have now covered the basic principles and major components of the DC motor. However, there are many types of DC motors, and several of them will be covered later in this module.

2.6.0 Starting Resistance

Large DC motors require that a starting resistance be inserted in series with the motor armature. The

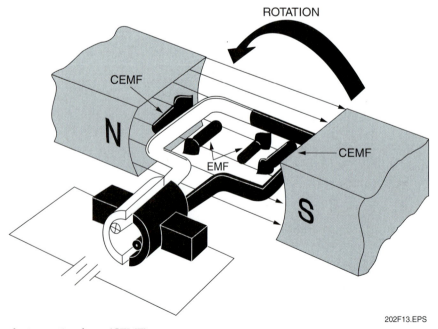

Figure 13 ◆ Counter-electromotive force (CEMF).

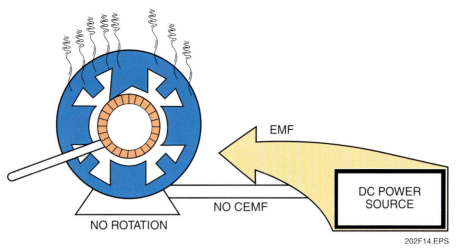

Figure 14 ◆ No CEMF.

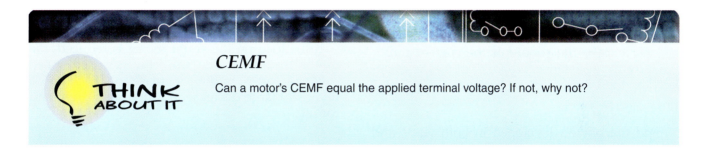

CEMF

Can a motor's CEMF equal the applied terminal voltage? If not, why not?

current drawn by the armature is governed by CEMF and the armature resistance. When starting, CEMF will be zero because the rotor is at a standstill. There is also no inductive reactance, as in AC induction motors. This means that the starting current will be abnormally high unless limited by external starting resistance.

Figure 15 shows a shunt motor that is connected directly across a 250V line. The armature resistance is known to be 0.5 ohm. The full-load current of the motor is known to be 25 amps and the shunt field current is one amp. The resulting armature current under full-load conditions would therefore be 24 amps.

If starting resistance is not used, the value of the armature current (I_A) can be found using the following equation:

$$I_A = \frac{\text{terminal voltage} - \text{CEMF}}{R_A}$$

$$I_A = \frac{250V - 0V}{0.5 \text{ ohm}}$$

$$I_A = 500 \text{ amps}$$

This amount of starting current is too high and may result in excessive torque and heat that may cause damage to the motor. When starting resistance is added in series with the armature, the

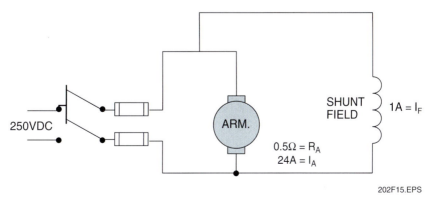

Figure 15 ◆ Shunt motor.

starting current can be limited to 1.5 times the full-load current value. After starting, this external resistance can be removed from service.

If we want to limit the starting armature current to 1.5 times the full-load value, we can solve for the size of resistance that would be required using the previous equations.

Where:

Starting I_A = 1.5 × steady state
= 1.5 × 24 amps
= 36 amps

At the moment of motor start, when the rotor is at a standstill and the CEMF is zero, the series resistance will be:

$$R_{starting} = \frac{(250V - 0V) - (36A)(0.5\Omega)}{36A} = 6.44\Omega$$

To find the wattage required in the starting resistance, take the square of the current multiplied by the resistance, where watt loss is calculated by the I^2R method.

Example:

Find the power developed in both watts and horsepower in a DC motor that has a terminal voltage of 240V and an armature current of 60A. The armature resistance is known to be 0.08Ω.

CEMF = $V_T - (I_A \times R_A)$
CEMF = 240 − (60 × 0.08) = 235.2V
Power = EI
Power = 235.2V × 60A = 14,112 watts

We now convert to horsepower. This means we are converting an electrical term into a mechanical term.

$$\text{Horsepower} = \frac{\text{watts}}{746}$$
$$= \frac{14{,}112 \text{ watts}}{746}$$
$$= 18.92 \text{hp}$$

$$R_{starting} = \frac{(\text{terminal voltage} - \text{CEMF}) - (I_A \times R_A)}{I_A}$$

3.0.0 ♦ TYPES OF DC MOTORS

There are two basic types of motor connections that are in common use. They are the series motor and the shunt motor. The series motor is so called because the field is connected in series with the armature winding. The shunt motor has the field coils connected in parallel with the armature (rotor) winding. One additional type of motor is a compound motor. This motor has both a series- and a shunt-connected field. *Figure 16* shows a typical DC motor.

3.1.0 Shunt Motors

The field circuit of a shunt motor is connected across the supply line and is in parallel with the armature. A shunt motor connection is shown in *Figure 17*.

When an external load is applied to the shunt motor, it tends to slow down. The slight decrease in speed causes a corresponding decrease in CEMF. Since the armature resistance is low, the resulting increases in armature current and torque are relatively large. Therefore, the torque is increased until it matches the opposing torque of the load. The speed of the motor then remains constant at the new value as long as the load is constant.

If the load on the shunt motor is reduced, the motor tends to speed up. The increased speed causes a corresponding increase in CEMF and a relatively large decrease in armature current and torque.

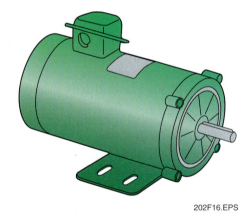

Figure 16 ♦ Typical DC motor.

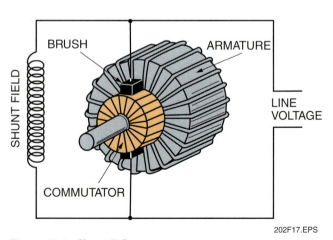

Figure 17 ♦ Shunt DC motor.

Thus, it may be seen that the amount of current through the armature of a shunt motor depends largely upon the load on the motor. The larger the load, the larger the armature current; the smaller the load, the smaller the armature current. The change in speed causes a change in CEMF and armature current in each case.

The main advantage of a shunt-wound motor is that its speed is fairly constant, changing only a few revolutions per minute (rpm) when the amount of load changes. The main disadvantage of this connection is that the motor does not develop much torque when it is first started. If a motor is to be started with a large load, it is generally series connected.

It is important to note that the shunt field circuit of a DC motor should never be opened when the motor is operating, especially when unloaded. This is because an open field may cause the motor to rotate at dangerously high speeds. Large DC shunt motors have a field rheostat with a no-field release feature that disconnects the motor from the power source if the field circuit opens.

3.1.1 Torque

A DC shunt motor has high torque at any rated speed. At startup, a DC shunt motor can develop up to 150% of its normal running torque as long as the resistors in the starting circuit can withstand the heating effect of the current.

3.1.2 Speed Control

DC motors have excellent speed control. To operate the motor above rated speed, a field rheostat is used to reduce the field current and field flux. To operate below rated speed, resistors are used to reduce the armature voltage.

3.1.3 Speed Regulation

The speed regulation of a shunt motor drops from 5% to 10% from no-load to full-load. As a result, a shunt motor is superior to the series DC motor but is inferior to a differential compound-wound DC motor.

3.2.0 Series Motors

The field coils of a series motor are connected in series with the armature (*Figure 18*). The value of current through the armature and the field is the same. Hence, if the armature current changes, the field current must also change.

As the motor speeds up, the armature current and field current decrease. With a weaker field, the armature speed will increase still more. The limiting factor on the speed is the load.

If there is no load on the motor, the armature will speed up to such an extent that the windings might be thrown from the slots and the commutator destroyed by the excessive centrifugal forces. For this reason, series motors are not belt-connected to their loads. The belt might break, allowing the motor to overspeed and destroy itself. Series motors are usually connected to their loads directly or through gears.

The series motor is used where there is a wide variation in both torque and speed requirements, such as traction equipment, blowers, hoists, cranes, and so forth.

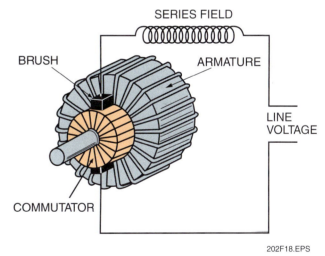

Figure 18 ♦ Series DC motor.

DC Motor Applications

DC motors were developed before AC motors and one of their first uses was in electric trolleys. The DC motor is still widely used in applications that require accurate speed control or high starting torque. For example, in an elevator, the motor must start under a heavy load and accelerate smoothly. It must also stop precisely and reverse direction easily. A DC motor is a good choice for this application.

3.2.1 Torque

The DC series motor develops 500% of its full-load torque at starting. Therefore, this type of motor is used in applications where large amounts of starting torque are needed, such as cranes, railway applications, and other equipment with high starting torque demands. With a series motor, any increase in load causes an increase in both the armature current and the field current. Since torque depends on the interaction of these two flux fields, the torque increases as the square of the value of the current increases. Therefore, series motors produce greater torque than shunt motors for the same increase in current. The series motor shows a greater reduction in speed for an equal change in load.

3.2.2 Speed Control and Speed Regulation

The speed control of a series motor is poorer than that of a shunt motor because if the load is reduced, a simultaneous reduction of current occurs in both the armature and field windings, and therefore, there is a greater increase in speed than there would be in a shunt-wound motor.

If the mechanical load were to be disconnected completely from a series motor, the motor would continue to accelerate until the motor armature self-destructed. For this reason, series-wound motors are always permanently connected to their loads.

The speed of a series DC motor is controlled by varying the applied voltage. A series motor controller is usually designed to start, stop, reverse, and regulate speed. The direction of rotation of a series motor is changed by reversing either the armature or field winding current flow.

3.3.0 Compound Motors

Compound DC motors are used whenever it is necessary to obtain speed regulation characteristics not obtainable with either the shunt- or series-wound motor. Because many applications require high starting torque and constant speed under load, the compound motor is used. Some industrial applications include drives for elevators, stamping presses, rolling mills, and metal shears. The compound motor has a normal shunt winding and a series winding on each field pole. They may be connected as a long shunt, as shown in *Figure 19(A)*, or a short shunt, as shown in *Figure 19(B)*. When the series winding is connected to aid the shunt winding, the machine is known as a cumulative compound motor. When the series field opposes the shunt field, the machine is known as a differential compound motor.

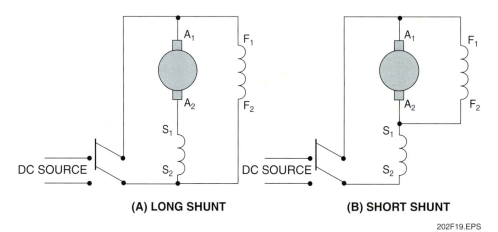

Figure 19 ♦ Long and short shunts.

Inside Track: Compound DC Motors

The compound motor avoids some of the limitations of the series-wound and shunt-wound motors. The shunt field has a constant current, so the motor will not self-destruct like a series motor. The series winding, on the other hand, provides strong torque.

Permanent Magnet DC Motors

Many ¼hp to 3hp variable-speed DC motors available for constant or diminishing torque applications use permanent magnets for the field poles instead of shunt or series windings. They employ variable DC armature voltages up to 90V or 180V for speed control. However, they are inefficient if they use only rheostat control of the armature voltage.

3.3.1 Torque

The operating characteristics of a cumulative compound-wound motor are a combination of the series motor and the shunt motor. A cumulative compound-wound motor develops high torque for sudden increases in load.

3.3.2 Speed

Unlike the series motor, the cumulative compound-wound motor has definite no-load speeds and will not build up self-destructive speeds if the load is removed. Speed control of a cumulative compound-wound motor can be controlled by inserting resistors in the armature circuit to reduce the applied voltage. When the motor is to be used for installations where the rotation must be reversed frequently, such as in elevators, hoists, and railways, the controller should have voltage dropping resistors and switching arrangements to accomplish reversal.

3.3.3 Speed Regulation

The speed regulation of a cumulative compound-wound motor is inferior to that of a shunt motor and superior to that of a series motor.

3.4.0 Operating Characteristics

Different types of motors have different operating characteristics. Therefore, the proper type of DC motor should be selected when the load to be driven is known. *Figure 20* shows the operating characteristics of a typical DC shunt motor.

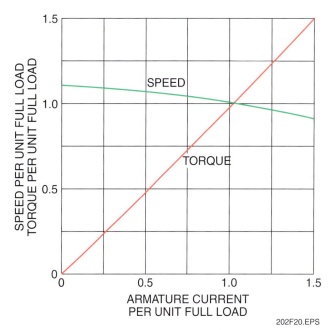

Figure 20 ◆ Operating characteristics of a typical DC shunt motor.

Notice that the motor speed is relatively independent of the torque (load applied) from 0 to 150% of the rated capacity of the motor. Such motors find application where relatively constant speed over a wide load range is required. *Figure 21* shows the operating characteristics of a typical DC series motor.

Notice that the motor speed varies greatly with respect to the torque (load applied). With less than half of its rated load applied, the motor operates at more than 150% of its rated speed. When 150%

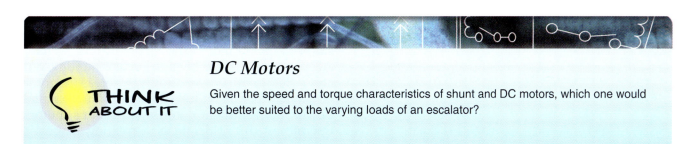

DC Motors

Given the speed and torque characteristics of shunt and DC motors, which one would be better suited to the varying loads of an escalator?

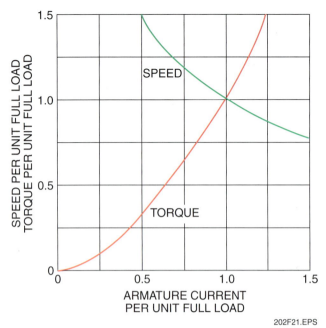

Figure 21 ♦ Operating characteristics of a typical series motor.

of the rated load is applied to the motor, it drops to 75% of its rated speed. Such motors find application where a constant heavy load exists or where great speed variations are tolerable.

Figure 22 shows the operating characteristics of a DC compound motor. Notice that the motor speed is relatively constant over the operating range. Its speed does vary with the torque somewhat more than the shunt motor, but will not run away or markedly decrease, as with the series motor. Such motors find application where the load is not known exactly or where some speed variation is tolerable with load variation.

3.5.0 Brushless DC Motors

The brushless DC motor was developed to eliminate commutator problems in missiles and spacecraft operating above the Earth's atmosphere. Two general types of brushless motors are in use: the inverter-induction motor and a DC motor with an electronic commutator.

4.0.0 ♦ ALTERNATING CURRENT MOTORS

Alternating current motors can be divided into two major types: single-phase motors and polyphase motors. The single-phase motor is normally limited to fractional horsepower ratings up to about five horsepower. They are commonly used to power such things as fans, small pumps, appliances, and other devices not requiring a great amount of power. Single-phase motors are not likely to be connected to complicated motor control circuitry.

Polyphase motors make up the majority of motors needed to drive large machinery such as pumps, large fans, and compressors. These motors have several advantages over single-phase motors in that they do not require a separate winding or other device to start the motor. They have relatively high starting torque and good speed regulation for most applications.

There are two classes of polyphase motors: induction and synchronous. The rotor of a synchronous motor revolves at synchronous speed, or the speed of the revolving magnetic field in the stator. The rotor of an induction motor revolves at a speed somewhat less than synchronous speed. The differences in rotor speed are due to differences in construction and operation. Both will be discussed in depth after a review of motor theory.

4.1.0 Polyphase Motor Theory

AC motors consist of two parts: the stator, or stationary part; and the rotor, or revolving part. The stator is connected to the incoming three-phase AC power. The rotor in an induction motor is not connected to the power supply, whereas the rotor of a synchronous motor is connected to external power. Both induction and synchronous motors operate on the principle of a rotating magnetic field.

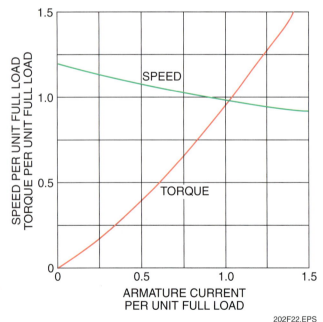

Figure 22 ♦ Operating characteristics of a typical DC compound motor.

4.1.1 Rotating Fields

This section shows how the stator windings can be connected to a three-phase AC input to create a magnetic field that rotates. Another magnetic field in the rotor can be made to chase it by being attracted and repelled by the stator field. Because the rotor is free to turn, it follows the rotating magnetic field in the stator.

Polyphase AC is brought into the stator and connected to windings that are physically displaced 120° apart. These windings are connected to form north and south magnetic poles, as shown in *Figure 23*. An analysis of the electromagnetic polarity of the poles at points 1 through 7 in *Figure 23* shows how the three-phase AC creates magnetic fields that rotate.

At point 1, the magnetic field in coil (pole) 1–1A is at its maximum. Negative voltages are shown in 1–2A and 3–3A. The negative voltages in these windings create smaller magnetic fields that will tend to aid the field set up in 1–1A.

At point 2, phase 3 creates a maximum negative flux in 3–3A windings. This strong negative field is aided by the weaker magnetic fields in 1–1A and 1–2A.

The three-phase AC input rises and falls with each cycle. Analyzing each point on the voltage graph shows that the resultant magnetic field rotates clockwise. When the three-phase input completes a full cycle at point 7, the magnetic field has completed an entire revolution of 360°.

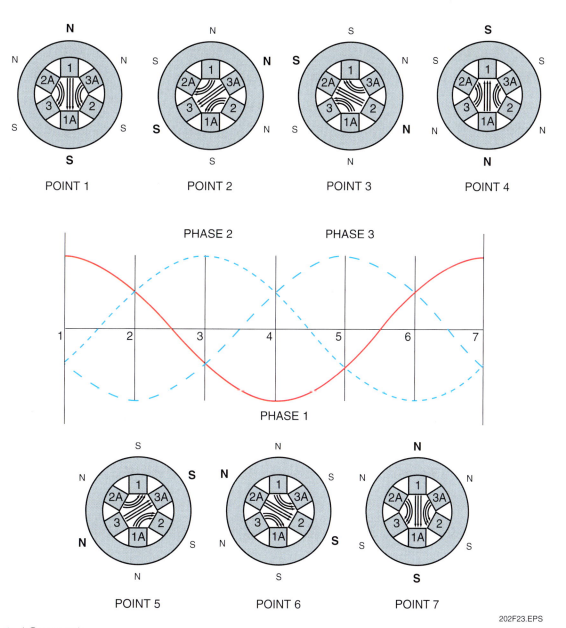

Figure 23 ♦ AC generation.

4.1.2 Rotor Behavior in a Rotating Field

An oversimplification of rotor behavior shows how the magnetic field of the stator influences the rotor. Assume that a simple bar magnet is placed in the center of the stator diagrams shown in *Figure 23*. Also assume that the bar magnet is free to rotate. It has been aligned such that at point 1, its south pole is opposite the large north of the stator field.

Unlike poles attract and like poles repel. As the AC completes a cycle, going from point 1 to point 7, the stator field rotates and pulls the bar magnet with it because of the attraction of unlike poles and the repulsion of like poles. The bar magnet is rotating at the same speed as the revolving flux of the stator. The speed of the revolving flux is known as synchronous speed. The synchronous speed of a motor is given by the equation:

$$N = \frac{120f}{P}$$

Where:

- N = speed in rpm
- f = frequency in cycles per second
- P = number of magnetic poles
- 120 = a constant derived by multiplying 60 seconds/minute (to convert speed in rpm to seconds) by two for one pair of poles

4.1.3 Induction

Current flowing through a conductor sets up a magnetic field around the length of the conductor. Conversely, a conductor in a magnetic field will produce a current when the magnetic lines of flux cut across the conductor. This action is called induction because there is no physical connection between the magnetic field and the conductor. Current is induced in the conductor.

4.2.0 Three-Phase Induction Motors

In a three-phase induction motor, the driving torque is caused by the reaction of a current-carrying conductor in a magnetic field. In induction motors, the rotor currents are supplied by electromagnetic induction. The stator windings are supplied with three-phase power and produce a rotating magnetic field.

The rotor is not electrically connected to the power supply. The induction motor derives its name from the mutual inductance taking place between the stator and the rotor under operating conditions. The rotating field produced by the stator cuts the rotor conductors, inducing a voltage into the conductors. The induced voltage causes rotor current. This develops motor torque due to the reaction of a current-carrying conductor in a magnetic field. This torque causes the rotor to rotate. This principle is shown in *Figure 24*.

The three-phase (3φ) induction motor has a frame, or stationary part, which is the stator. The stator is made of laminated steel rings with slots on the inside circumference. The motor stator windings are the phase windings. They are symmetrically placed on the stator and may be either wye- or delta-connected. Depending on how the stator is wound, it may have two, four, or any even number of poles.

There are two varieties of three-phase induction motors: the squirrel cage rotor motor and the wound rotor motor.

4.2.1 Squirrel Cage Induction Motor

The squirrel cage is the most popular rotor in use. Three-phase squirrel cage induction motors consist of a stator, a rotor, and two end shields that house the bearings that support the rotor shaft. The frame is usually made of cast steel. The stator core is pressed into the frame. In this rotor, the bars are connected together at the ends by short-

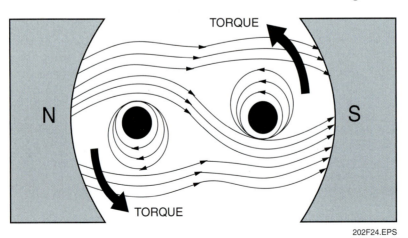

Figure 24 ◆ Producing torque.

Squirrel Cage Motor Applications

Squirrel cage induction motors offer many advantages. For example, maintenance costs are low because these motors have no brushes or slip rings, but work entirely through induction. They also have a high starting torque, so they are useful in common applications such as overhead doors, large compressors, fans, and printing presses.

ing rings made of similar material. The conductor bars carry large currents at low voltages. The bearings can be either sleeve or ball bearings. *Figure 25* shows the main components of an induction motor.

It is not necessary to insulate the bars from the core because the current will follow the path of least resistance and is confined to the cage windings. *Figure 26* shows how a squirrel cage rotor is constructed.

The squirrel cage rotor induction motor has a fixed rotor circuit. The resistance and reactance of the windings are determined when the motor is designed. The standard cage rotor motor is a general-purpose motor. It is used to drive loads that require variable torque at relatively constant speed with high full-load efficiency. Some examples are blowers, centrifugal pumps, and fans. Due to the absence of any moving electrical contacts, they are suitable for use where they are exposed to flammable dust or gas.

If the load requires special operating characteristics, such as high starting torque, the squirrel cage rotor can be designed to have high resistance bars for a starting circuit and low resistance bars for running operation. A rotor of this type is called a double squirrel cage rotor.

4.2.2 Wound Rotor Induction Motor

A wound rotor (*Figure 27*) has a winding that is similar to the three-phase stator windings. The rotor windings are usually wye-connected with the free ends of the windings connected to three slip rings mounted on the rotor shaft. The slip rings are shown physically mounted on the end of the rotor shaft in *Figure 27*. They are used with brushes to form an electromechanical connection to the rotor.

Slip rings are contact surfaces mounted on the shaft of a motor or generator to which the rotor windings are connected and against which the brushes ride. The brushes are sliding contacts, usually made of carbon, that make continuous electrical connection to the rotating part of a motor or generator.

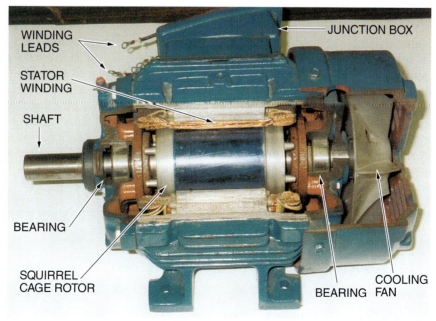

Figure 25 ◆ Main components of an induction motor.

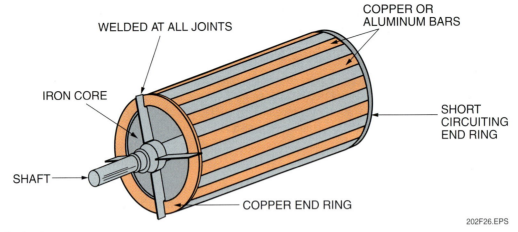

Figure 26 ◆ Squirrel cage rotor.

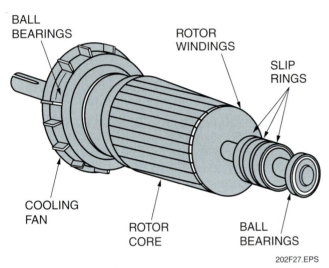

Figure 27 ◆ Wound rotor.

The wound rotor motor often uses an external wye-connected resistor connected to the rotor through slip rings. The resistor provides a means of varying the rotor resistance. This can be used when the motor is started to produce a high starting torque. As the motor accelerates, the resistance is reduced. When the motor has reached full speed, the slip rings are short circuited, and the operation is similar to that of a squirrel cage rotor induction motor. A schematic representation of this is shown in *Figure 28*.

The wound rotor induction motor is used when it is necessary to vary the rotor resistance, to limit starting current, or to vary the motor speed. Speed can be varied by as much as 50% to 75%; the greater the resistance inserted in the rotor circuit, the lower the speed will be below synchronous speed. When the motor is operating below full speed, the percent slip is increased and the motor is operating at reduced efficiency and horsepower. When all resistance is cut completely out, the speed is somewhat less than that obtained with squirrel cage rotors. Because the rotor circuit heat generation is largely external to the rotor windings, the wound rotor motor is used for applications that require frequent starts without overheating the motor.

4.2.3 Wound Rotor Speed Control

The insertion of resistance in the rotor circuit not only limits the starting surge of current, but also produces a high starting torque and provides a means of adjusting the speed. If the full resistance of the speed controller is cut into the rotor circuit when the motor is running, the rotor current decreases and the motor slows down. As the rotor speed decreases, more voltage is induced in the rotor windings and more rotor current is developed to create the necessary torque at the reduced rotor speed.

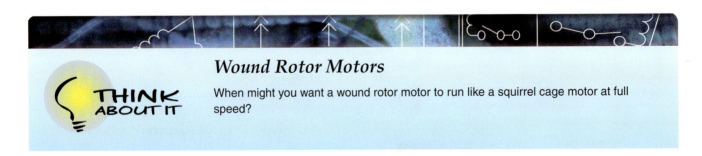

Wound Rotor Motors
When might you want a wound rotor motor to run like a squirrel cage motor at full speed?

2.20 ELECTRICAL LEVEL TWO ◆ TRAINEE GUIDE

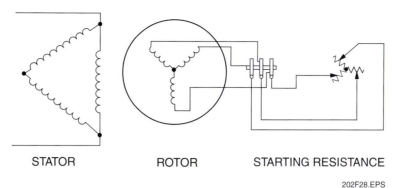

Figure 28 ◆ Wound rotor motor circuit.

If all the resistance is removed from the rotor circuit, both the current and motor speed will increase. However, the rotor speed will always be less than the synchronous speed of the field developed by the stator windings. Recall that this is also true of the squirrel cage induction motor. The speed of a wound rotor motor can be controlled manually or automatically with timing relays, contactors, and pushbutton speed selection.

The advantages of the wound rotor motor are high starting torque with moderate starting current, smooth acceleration under heavy load, no excessive heating during starting, good running characteristics, and adjustable speed control. The chief disadvantage is that both initial and maintenance costs are greater than those of the squirrel cage rotor motor.

4.2.4 Torque

The torque on the rotor of an induction motor tends to turn the rotor in the same direction as the rotating field. If the motor is not driving a load, it will accelerate to nearly the same speed as the rotating field. As the rotor accelerates, the magnitude of the induced voltage in the rotor decreases. This is because the relative motion between the rotating field and the rotor conductors is reduced. It is impossible for an induction motor to operate at synchronous speed because there would be no relative motion between the rotating field and the rotor. Thus, there would be no induced voltage, no rotor current, no rotor magnetic field, and no torque.

4.2.5 Slip

In an induction motor, the rotor always rotates at a speed less than the synchronous speed. The rotor speed is such that sufficient torque is produced to balance the restraining torque caused by motor friction and mechanical load. The difference between the synchronous speed and the rotor speed is known as slip. Slip is expressed mathematically as follows:

$$S = \frac{N - N_R}{N} \times 100\%$$

Where:

S = slip
N = synchronous speed
N_R = rotor speed

To express the quantity as a percent, multiply by 100.

Example:

A four-pole, 208V, 2hp, 60Hz, three-phase induction motor has a no-load speed of 1,790 rpm and a full-load speed of 1,650 rpm.

Find the percent slip for each case below:

- No-load condition
- Full-load condition
- Locked-rotor condition (standstill)

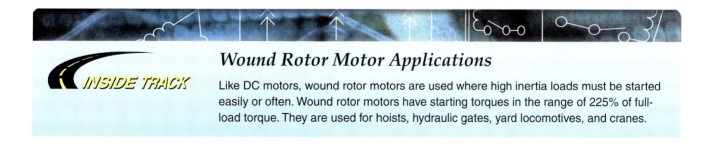

Wound Rotor Motor Applications

Like DC motors, wound rotor motors are used where high inertia loads must be started easily or often. Wound rotor motors have starting torques in the range of 225% of full-load torque. They are used for hoists, hydraulic gates, yard locomotives, and cranes.

Before any calculations can be made, we must first calculate synchronous speed.

$$N = \frac{120f}{P}$$

$$N = \frac{120 \times 60}{4}$$

$$N = 1{,}800 \text{ rpm}$$

- At no-load condition:

$$S = \frac{N - N_R}{N} \times 100\%$$

$$S = \frac{1{,}800 - 1{,}790}{1{,}800} \times 100\%$$

$$S = 0.556\%$$

- At full-load condition:

$$S = \frac{N - N_R}{N} \times 100\%$$

$$S = \frac{1{,}800 - 1{,}650}{1{,}800} \times 100\%$$

$$S = 8.33\%$$

- At locked-rotor condition:

$$S = \frac{N - N_R}{N} \times 100\%$$

$$S = \frac{1{,}800 - 0}{1{,}800} \times 100\%$$

$$S = 100\%$$

Figure 29 shows how torque relates to speed over the operating range of a motor. Note that speed is proportional to torque on the left side up to pullout torque. Beyond this point, however, torque decreases as speed increases.

Slip is the difference between the synchronous speed and the actual speed of the rotor in an induction motor. Slip is necessary to permit motor action to occur. Under increasing load, the rotor torque increases. Since percent slip is proportional to torque, the amount of slip will increase. This increase means a higher current draw by the motor due to the greater difference between the rotor and the magnetic field. Motor supply voltages, current, torque, speed, and rotor impedance are closely related. By changing the resistance and reactance of the rotor, the characteristics of the motor can be changed; however, for any one rotor design these characteristics are fixed.

4.2.6 Starting Current

At the moment a three-phase induction motor is started, the current supplied to the motor stator terminals may be as high as six times the motor full-load current. This is because at starting, the rotor is at rest; therefore, the rotating magnetic field of the stator cuts the squirrel cage rotor at the maximum rate, inducing large amounts of EMF in the rotor.

This results in proportionally high currents at the input terminals of the motor as was previously discussed. Because of this high inrush, current starting protection as high as 300% of full-load current must be provided to allow the motor to start and come up to speed.

Because 100% slip exists at the instant the motor is energized (see *Figure 29*), the rotor current lags the rotor EMF by a large angle. This means that the maximum current flow occurs in a rotor conductor at a time after the maximum amount of stator flux has passed by. This results in a high starting current at a low power factor, which results in a low value of starting torque.

As the rotor speeds up, the rotor frequency and reactance decrease, causing the torque to increase up to its maximum value, then decrease to the value needed to carry the load.

4.2.7 Loaded Torque

If a load is now placed on the shaft, the rotor will tend to slow down. As it slows down, more flux lines are cut until enough torque is developed to overcome the load placed on the shaft.

The motor now runs under load at a slower speed than before the load was placed on the shaft. This normal range of operation is shown in the lower right corner of *Figure 29* as the rated or full-load torque.

In this range, the slip will vary from 2% to 10%, depending on the load applied and the motor. Rated slip will occur at the point where 100% rated load is applied. Increased load means increased slip, which means the rotor is now rotating slower. An induction motor is considered to be a constant speed motor. We will now examine how much speed fluctuates from no-load speed to full-load speed.

Example:

A two-pole induction motor has a no-load slip of 2% and a full-load slip of 8%. What are the no-load speed, full-load speed, and percent speed change?

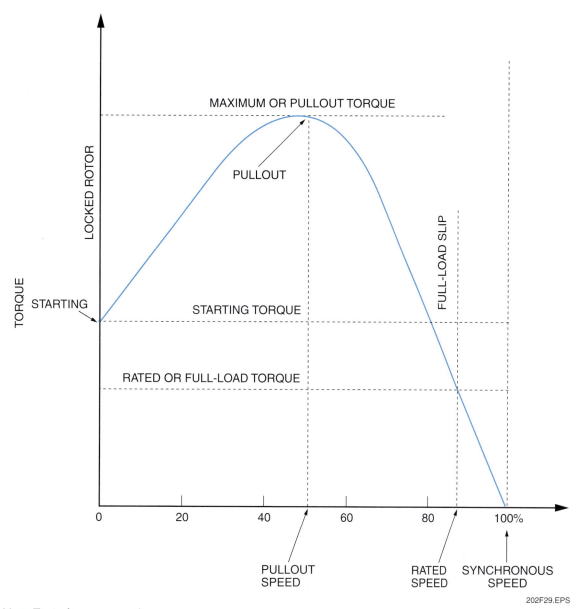

Figure 29 ♦ Typical torque-speed curve.

Nominal $= \dfrac{120 \times 60}{2} = 3{,}600$ rpm

No-load speed $= \dfrac{100\% - 2\%}{100\%} \times 3{,}600 \text{ rpm} = 3{,}528 \text{ rpm}$

Full-load speed $= \dfrac{100\% - 8\%}{100\%} \times 3{,}600 \text{ rpm} = 3{,}312 \text{ rpm}$

Percent speed change
$= \dfrac{\text{no-load speed} - \text{full-load speed}}{\text{no-load speed}} \times 100\%$

$= \dfrac{3{,}528 - 3{,}312}{3{,}528} \times 100\%$

$= 6.12\%$

4.2.8 Overload Condition

If the load is increased above full-rated load, everything happens as stated before to increase torque up to a certain point. *Figure 30* shows typical torque and current curves.

In *Figure 30*, note how the torque climbs as the load is increased.

This will continue as load is increased until the pullout torque point is reached. Beyond this point, the torque decreases and the motor will quickly stall. A typical situation is when a bench circular saw or a lathe stalls on a heavy cut. The machine will slow down as its cutting load is increased until it suddenly stalls and hums or growls loudly.

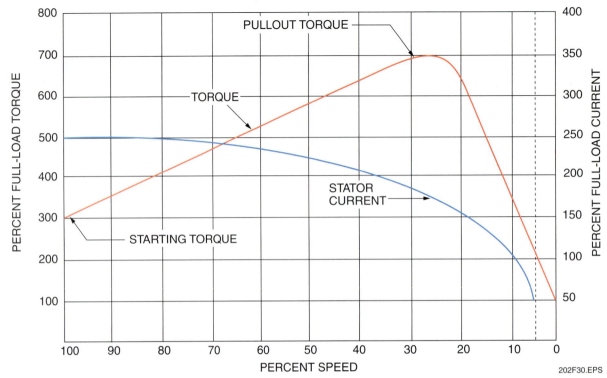

Figure 30 ♦ Torque and current curves.

The condition will persist until the load is relieved or a fuse blows or a breaker trips. The motor has simply reached a point where it cannot continue to increase its torque. Any further increase in load will cause a stall.

4.2.9 Power Factor

The power factor of a squirrel cage induction motor is poor at no-load and low-load conditions. At no-load conditions, the power factor can be as low as 15% lagging. However, as load is increased, the power factor increases. At high-rated load, the power factor may be as high as 85% to 90% lagging.

The power factor at no-load speed is low because the magnetizing component of input current is a large part of the total input current of the motor. When the load on the motor is increased, the in-phase current supplied to the motor increases, but the magnetizing component of current remains practically the same. This means that the resultant line current is more nearly in phase with the voltage, and the power factor is improved when the motor is loaded compared with an unloaded motor, which chiefly draws its magnetizing current.

Figure 31 shows the increase in power factor from no-load conditions to full-load conditions. In the no-load diagram, the in-phase current (I_{ENERGY}) is small when compared to the magnetizing current (I_M); thus, the power factor is poor at no-load conditions. In the full-load diagram, the in-phase current has increased, while the magnetizing current remains the same. As a result, the angle of lag of the line current decreases, and the power factor increases.

4.2.10 Speed Control

The speed of a three-phase squirrel cage induction motor depends on the frequency of the applied voltage and the number of poles. As a result, these motors are used in applications where speed remains constant or where it can be controlled by other means such as variable frequency drives.

4.2.11 Reversing Rotation

The direction of rotation of a three-phase induction motor can be readily reversed. The motor will rotate in the opposite direction if any two of the three incoming leads are reversed, as shown in *Figure 32*.

4.3.0 Synchronous Motors

The synchronous motor is a three-phase motor that operates at synchronous speed from no-load conditions to full-load conditions.

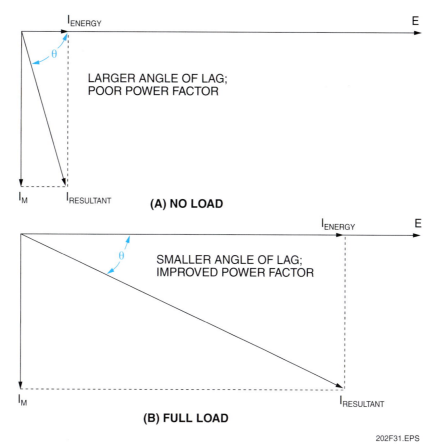

Figure 31 ♦ Power factor versus load for an induction motor.

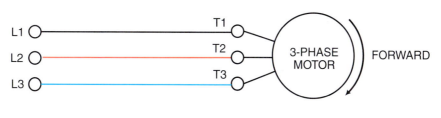

(A) ROTATION BEFORE CONNECTIONS ARE CHANGED

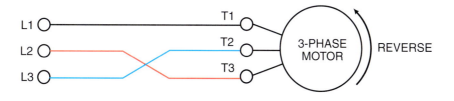

(B) ROTATION AFTER CONNECTIONS ARE CHANGED

Figure 32 ♦ Three-phase induction motor rotational direction change.

Variable-Speed Drives

Variable-speed drives, known as VSDs or ASDs (adjustable-speed drives), are powerful electronic devices that are available for virtually any size motor in all types of applications. Of the various types of VSDs available, the most efficient versions for AC motors are VFDs (variable-frequency drives), which control both the frequency and the voltage applied to the motor. By changing the frequency of the rotating stator field, you change the speed of the rotor, thus changing the speed of the motor. VSDs are often used in energy management systems to conserve energy by supporting variable loads such as those that occur in heating, ventilating, and air conditioning systems. In addition to controlling the speed of a motor, VSDs are available to control both motor starting and stopping functions, as well as to provide controlled acceleration and deceleration.

4.3.1 Characteristics

This type of motor has a revolving field that is energized from a source separate from the stator winding. The rotor is excited by a DC source. The magnetic field set up by the direct current on the rotor then locks in with the rotating magnetic field of the stator and causes the rotor to revolve at synchronous speed. By changing the magnitude of DC excitation, the power factor of the motor can be changed over a wide variety of power factors from leading to lagging. Because of the unique ability of synchronous motors to change power factors, they are often used as power-factor correctors. They are most often used in applications that require constant speed from no-load conditions to full-load conditions.

4.3.2 Construction

The construction of synchronous motors is essentially the same as the construction of three-phase generators. They have three stator windings that are 120° apart and a wound rotor that is connected to slip rings where the rotor excitation current is applied.

When three-phase AC is applied to the stator, a revolving magnetic field is created just as it is in induction motors. The rotor is energized with DC, which creates a magnetic field around the rotor. The strong rotating magnetic field of the stator attracts the rotor field. This results in a strong turning force on the rotor shaft.

This is how the synchronous motor works once it is started. However, one of the disadvantages of this type of motor is that it cannot be started just by applying AC to the stator. When AC is applied to the stator, the high-speed rotating magnetic field rushes past the rotor poles so quickly that the rotor does not have a chance to get started. The rotor is locked; it is repelled in one direction and then in another direction. In its purest form, the synchronous motor has no starting torque.

This is more easily understood using *Figure 33*. When the stator and rotor fields are energized, the poles of the rotating field approach the rotor poles of opposite polarity. The attracting force

Synchronous Motors

Three-phase synchronous motors can be used in industrial applications to correct the low power factor of a number of induction motors or other inductive devices that are operating at less than their rated load levels. Synchronous motors can accomplish power factor correction while driving their own mechanical loads. Correcting a low power factor created by inductive loads through the use of synchronous motors reduces energy costs by making efficient use of the power supplied to the industrial facility. The use of synchronous motors can eliminate the need for dedicated capacitor banks or switched capacitor banks and the surges caused by them.

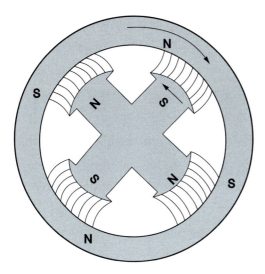

TENDENCY OF ROTOR TO
TURN COUNTERCLOCKWISE

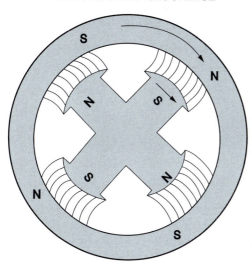

TENDENCY OF ROTOR TO
TURN CLOCKWISE

Figure 33 ◆ Synchronous motor operation at start.

The number of field poles must equal the number of stator poles. The rotor field windings are brought out to slip rings that are mounted on the rotor shaft. The field current is supplied through carbon brushes to the field windings. *Figure 34* shows a simplification of a synchronous motor. *Figure 35* shows the construction of the rotor pole assembly.

4.3.3 Principles of Operation

When a synchronous motor is started, current is first applied to the stator windings. Current is induced in the amortisseur winding and the motor starts as an induction motor. The motor then comes up to near-synchronous speed (about 5% to 10% slip). At that point, the field is excited, and the motor, turning at high speed, pulls into synchronism. When this occurs, the rotor is turning at synchronous speed, and the squirrel cage winding will not be generating any current, and therefore will not affect the synchronous motor's operation. The amortisseur windings serve an additional purpose. When the load changes frequently, the motor speed is not steady because the torque angle (discussed later) oscillates (or hunts) back and forth, trying to settle at its required value. This momentary change in speed creates a current due to induction, and there will be torque in the amortisseur winding. This momentary torque serves to dampen or stabilize the oscillating torque angle. That is why amortisseur windings are sometimes referred to as damper windings.

will tend to turn the rotor in a direction opposite the rotating field. As the rotor starts to move in that direction, the rotating field moves past the rotor poles and tends to pull the rotor in the same direction as the rotating field. The result is no starting torque.

To allow this type of motor to start, a squirrel cage winding is added to the rotor to cause it to start like an induction motor. This winding is called an amortisseur winding. The rotor windings are constructed so that definite north and south poles are created and these poles, when excited by DC, will lock in with the revolving field. The rotor windings are wound about the salient field poles, which are connected in series for opposite polarity.

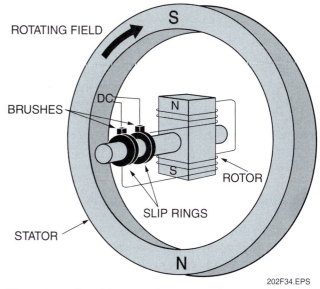

Figure 34 ◆ Simplification of a synchronous motor.

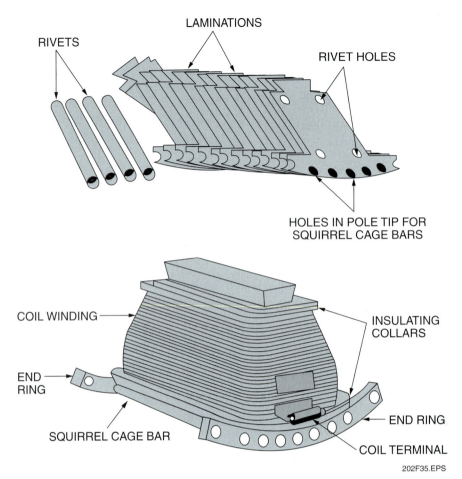

Figure 35 ◆ Pole assembly.

4.3.4 Rotor Field Excitation

The rotor must be excited from an external DC source. *Figure 36* shows a simplified synchronous motor excitation circuit. Notice that the DC field current can be varied by the rheostat; however, this does not change the speed of the motor. It only changes the power factor of the motor stator circuit. If full resistance is applied to the rotor field circuit, then the field strength of the rotor is at its minimum and the power factor is extremely lagging. As the DC field strength is increased, the power factor improves. If current is increased sufficiently, the power factor can be increased to near unity or 100%. This value of field current is referred to as normal excitation. By increasing the rotor field strength further, the power factor decreases but in a leading direction; that is, the stator circuit becomes capacitive and the motor is said to be overexcited. The synchronous motor can be used to counteract the lagging power factor in circuits by adding capacitive reactance to the circuit, thereby bringing the overall power factor closer to unity.

If the rotor DC field windings of a synchronous motor are open when the stator is energized, a high AC voltage will be induced in it because the rotating field sweeps through the large number of turns at synchronous speed.

It is therefore necessary to connect a resistor of low resistance across the rotor DC field winding during the starting period. During the starting period, the DC field winding is disconnected from the source, and the resistor is connected across the field terminals. This permits alternating current to flow in the DC field winding. Because the impedance of the winding is high compared with the inserted external resistance, the internal voltage drop limits the terminal voltage to a safe value.

4.3.5 Synchronous Motor Pullout

When a synchronous motor loses synchronism with the system to which it is connected, it is said to be out of step. This occurs when the following take place singly or in combination:

- Excessive load applied to the shaft
- Supply voltage reduced excessively
- Motor excitation lost or too low

Torque pulsations applied to the shaft of a synchronous motor are also a possible cause of loss of

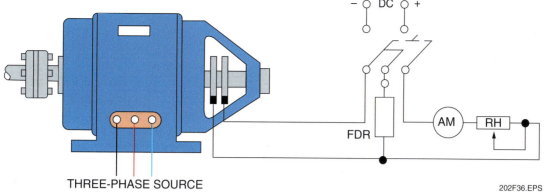

Figure 36 ♦ Simplified synchronous motor excitation circuit.

synchronism if the pulsations occur at an unfavorable period relative to the natural frequency of the rotor with respect to the power system.

A prevalent cause of loss of synchronism is a fault occurring on the supply system. Underexcitation of the rotor is also a distinct possibility.

Synchronous motor pullout is significant in that the squirrel cage or amortisseur winding is designed for starting only. They are not as hardy as those found in induction motors. The amortisseur winding will not overheat if the motor starts, accelerates, and reaches synchronous speed within a time interval determined to be normal for the motor. However, the motor must continue to operate at synchronous speed. If the motor operates at a speed less than synchronous, the amortisseur winding may overheat and suffer damage.

Protection against a synchronous motor losing synchronism can be provided by polarized field frequency relays and out-of-step relays as well as various digital methods.

4.3.6 Synchronous Motor Torque Angle

Once the rotor is brought up to high speed (close to synchronous speed) it will lock on to the rotating magnetic field. Under these conditions, a running torque will be developed. The rotor will rotate at synchronous speed in a direction and at a speed determined by synchronous speed.

While the motor is running, the two rotating fields will line up perfectly. The rotor pole will always lag behind the stator pole by some angle. This angle is called the torque angle and is shown in *Figure 37*.

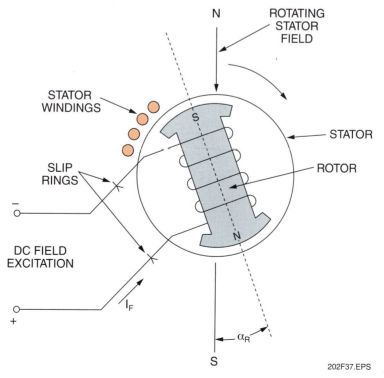

Figure 37 ♦ Torque angle.

As the load on the shaft increases, the torque angle increases even though the rotor continues to turn at synchronous speed. This behavior continues until the torque angle is approximately 90°. At that point, the motor is developing a maximum torque. Any further increase in load will cause either of the following to occur:

- If the increase in load is momentary or very small, the rotor will slip a pole. In other words, the stator field will lose hold of the rotor and grab onto it again the next time around.
- If the increase in load is large enough and is not momentary, the motor will lose synchronism and will either stall or cause the rotor to suffer thermal damage.

In both cases, a noticeable straining sound will be heard.

The synchronous motor should not be used where fluctuations in torque are violent. As a rule, it is also not used in small sizes (under 50hp), because it requires DC excitation. It is more difficult to start than induction motors and falls out of step quite readily when system disturbances occur. Its common applications are in motor generator sets, air compressors, and compressors in refrigerating plants.

4.4.0 Single-Phase AC Motors

Single-phase motors operate on a single-phase power supply. This is important because in the typical home or office and many areas of industrial plants, the only power source available is single-phase AC. Not only do single-phase AC motors eliminate the need for three-phase AC lines, but they are also easier to manufacture in small sizes and are, therefore, less expensive.

Examples of the many applications of single-phase AC motors today are the following: refrigerators, freezers, washers, dryers, power tools, typewriters, copying machines, heating systems, water pumps, computer peripherals, and various small appliances.

There are two basic types of single-phase motors. First, there is the single-phase induction motor. Its theory of operation is similar to that of the three-phase induction motor; hence, it runs at a speed slightly lower than synchronous speed. Second, there is the single-phase synchronous motor.

4.4.1 Single-Phase Induction Motors

Single-phase AC induction motors are extremely popular. Unlike polyphase induction motors, the stator field in the single-phase motor does not rotate. Instead, it simply alternates polarity between poles as the AC voltage changes polarity.

Voltage is induced in the rotor, and a magnetic field is produced around the rotor. This field will always be in opposition to the stator field. However, the interaction between the rotor and stator fields will not produce rotation (see *Figure 38*). Because this force is across the rotor and through the pole pieces, there is no rotary motion, just a push and/or pull along this line.

INSIDE TRACK

Single-Phase Induction Motors

Outside of large industrial and commercial facilities, single-phase induction motors are the most common type of motor used. While they are initially less expensive than polyphase motors, they are also less efficient and more costly to maintain. They are typically available in small sizes from ⅛hp to 1hp (previously referred to as fractional horsepower sizes) and in sizes up to 10hp.

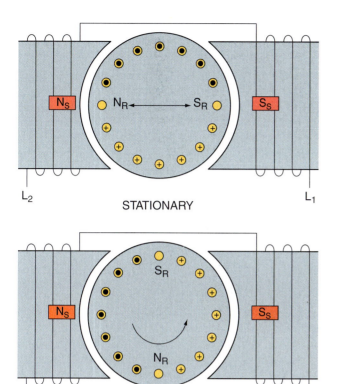

Figure 38 ♦ AC induction motor.

If the rotor is rotated by some outside force (a twist of your hand, for example), the push-pull along the line is disturbed. Look at the fields shown in *Figure 38* as the motor begins to rotate. At this instant, the south pole on the rotor is being attracted to the left-hand pole. The north rotor pole is being attracted to the right-hand pole. All of this is a result of the rotor being rotated 90° by the outside force.

The pull that now exists between the two fields becomes a rotary force, turning the rotor toward magnetic correspondence with the stator. Because the two fields continuously alternate, they will never actually line up and the rotor will continue to turn once started.

Since a single-phase rotor will rotate if it has a rotating magnetic field present, all that remains is to find a means of generating a rotating field at the start. There are a number of practical means for generating a rotating field. All the methods used for single-phase induction motors involve the simulation of a second phase for a starting circuit. In this module, we will discuss the following types of motors: split-phase, capacitor-start, capacitor-run, shaded-pole, and repulsion-start.

4.4.2 Split-Phase Induction Motor

The split-phase motor, shown schematically in *Figure 39*, has a stator composed of slotted laminations that contain an auxiliary (starting) winding and a running (main) winding. The axes of these two windings are displaced by an angle of 90 electrical degrees. The starting winding has fewer turns and smaller wire than the running winding and, therefore, has different electrical characteristics. The main winding occupies the lower half of the slots and the starting winding occupies the upper half. The two windings are connected in parallel across the single-phase line supplying the motor. The motor derives its name from the action of the stator during the starting period.

When energized with single-phase AC, the two windings are physically different enough in position and construction to produce a magnetic revolving field that rotates around the stator air gap at synchronous speed. As the rotating field moves around the air gap, it cuts across the rotor conductors and induces a voltage in them. The interaction between the rotor and stator causes the rotor to accelerate in the direction in which the stator field is rotating.

When the rotor has come up to about 75% of synchronous speed, a centrifugally operated switch disconnects the starting winding from the line supply, and the motor continues to run on the main winding alone. As the motor ages, the centrifugal switch contacts pit and corrode. When this happens, they may get stuck in the closed position. To safeguard against the winding burning up, a thermal relay is also used. If the motor draws the high starting current for more than 5 or 10 seconds, the relay will de-energize.

In a split-phase motor, the starting torque is 150% to 200% of the full-load torque, and the starting current is six to eight times the full-load current. Fractional-horsepower split-phase motors are used in a variety of devices such as washers,

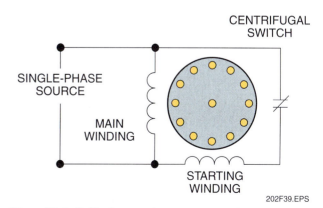

Figure 39 ♦ Split-phase motor.

Modern Split-Phase Induction Motor

Photo (A) shows a centrifugally-actuated start winding switch that is closed when the motor is at rest. The start winding switch is opened and closed by the movement of an actuator disk against a contact lever. The actuator disk is moved back and forth on the rotor shaft by a centrifugal weight assembly. When the motor is at rest, springs retract the weights and cause the actuator disk to move toward the bearing at the end of the shaft. This pushes on the contact lever, closing the switch contacts. After the motor starts and reaches about 75% of its rated speed, the weights swing out against the spring tension and retract the contact disk from the contact lever. This opens the switch contacts, removing the starting winding from the circuit. If the switch does not open after starting, the motor will operate at a reduced speed until it overheats the starting winding and activates the thermal relay. Photo (B) shows the starting winding (green) and the running winding (copper).

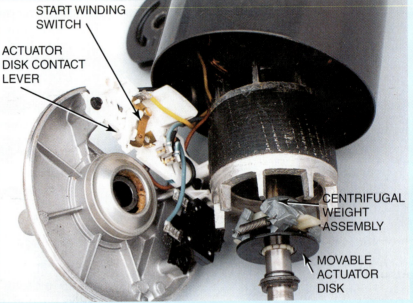

(A)

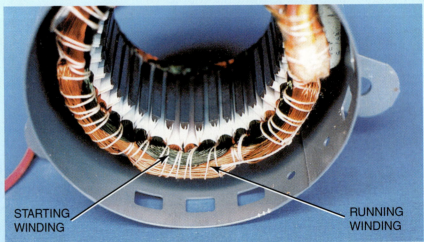

(B)

2.32   ELECTRICAL LEVEL TWO ◆ TRAINEE GUIDE

oil burners, and ventilating fans. The direction of rotation of the split-phase motor can be reversed by interchanging the starting winding leads.

4.4.3 Capacitor-Type Induction Motor

The capacitor-type motor is a modified form of split-phase motor. A typical capacitor-type motor is shown in *Figure 40*. The capacitor is located on top of the motor.

To develop a larger starting torque than that available with a standard split-phase motor, a capacitor is placed in series with the auxiliary winding of a split-phase motor, as shown in *Figure 41*. This is called a capacitor-start motor. The capacitor tends to create a greater electrical phase separation of the two windings. Also, because the reactance of a capacitor is 180° out of phase with the inductive reactance of the motor windings when they are combined, they yield a lower total impedance. This allows a larger current to produce a greater magnetic field.

The net effect of the capacitor is to give its motor a starting torque of about four times its rated torque. The split-phase motor, on the other hand, produces a starting torque of about one to two times its rated torque. Once the capacitor motor has come up to speed and the starting winding has been disconnected, it will have the same running characteristics as the split-phase motor.

Figure 40 ◆ Capacitor motor.

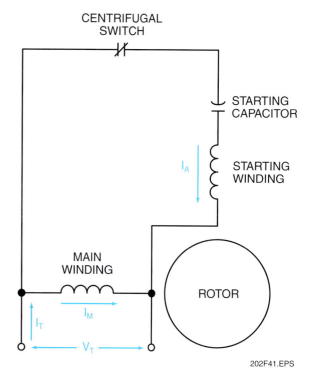

Figure 41 ◆ Capacitor-start motor schematic.

To reverse the direction of rotation of the capacitor-start motor and split-phase motor, the connection of either winding would have to be reversed. Since the starting winding is disconnected at a high speed, this reversal can be accomplished only at standstill or at low speeds when the centrifugal switch is still closed.

The capacitor-start motor is made in sizes from ¼ to 10hp (150W to 7.5kW). The starting capacitor is the dry-type electrolytic capacitor made for AC use. Typical values are from 200 to 600 microfarads (µF). *Figure 42* shows a comparison of torque slip curves for a split-phase and capacitor-start motor. It also shows the typical effect of the starting capacitor.

A variation of the capacitor-start motor is one in which the capacitor and auxiliary winding are not disconnected. The centrifugal switch in *Figure 41* is eliminated, and the auxiliary winding is left in all the time. This motor is called a capacitor-run motor.

The capacity used for running under load is not the same as that needed for starting. Furthermore, the capacitors used for starting cannot be used for continuous operation. Since the capacitor used in this motor is in all the time, it must be of a different type; that is, one capable of operating continuously. The net result is that the motor has improved running characteristics; however, it does not provide a starting torque as large as that of the capacitor-start motor.

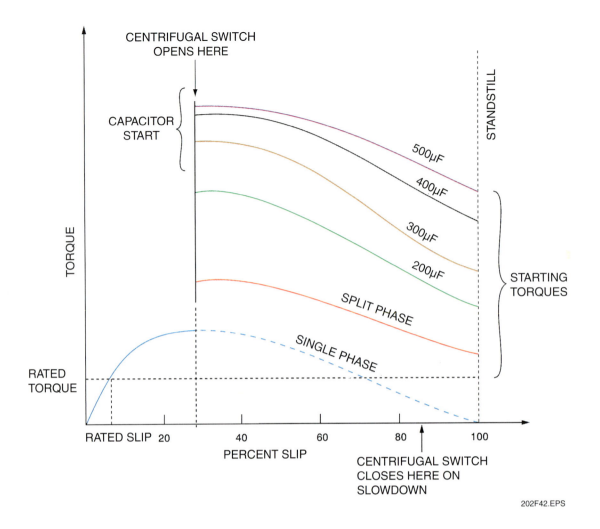

Figure 42 ◆ Torque-slip curves.

Among the improvements are higher efficiency and power factor at rated load, lower line current, and very quiet operation. It should also be pointed out that the start winding must be designed for continuous operation. This makes the motor somewhat more costly.

Another variation is the capacitor-start, capacitor-run motor. This motor combines the useful features of the capacitor-start and the capacitor-run motors by using two different capacitors, as shown in *Figure 43*.

4.4.4 Shaded-Pole Induction Motor

The shaded-pole motor employs a salient-pole stator and a cage rotor. The projecting poles on the stator resemble those of DC machines, except that the entire magnetic circuit is laminated and a portion of each pole is split to accommodate a short circuited copper strap called a shading coil. This motor is generally manufactured in very small sizes and runs up to 1/20hp. A four-pole motor of this type is illustrated in *Figure 44*.

The shading coils are placed around the leading pole tip, and the main pole winding is concentrated and wound around the entire pole. The four coils that make up the main winding are connected in series across the motor terminals. An inexpensive type of two-pole motor that uses shading coils is illustrated in *Figure 45*.

Referring to *Figure 45*, we see that during part of the cycle when the main pole flux (ϕ_1) is increasing, the shading coil is cut by the flux, and the resulting induced EMF and current in the shading coil tend to prevent the flux from rising readily through it. Thus, the greater portion of the flux rises in the portion of the pole that is not in the vicinity of the shading coil ($\phi_1 > \phi_2$). When the flux reaches its maximum value, the rate of change of flux is zero, and the voltage and current in the shading coil are also at zero. At this time, the flux is distributed more uniformly over the entire pole face ($\phi_1 = \phi_2$).

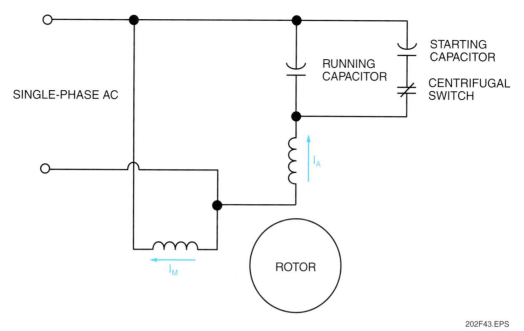

Figure 43 ◆ Capacitor-start, capacitor-run motor schematic.

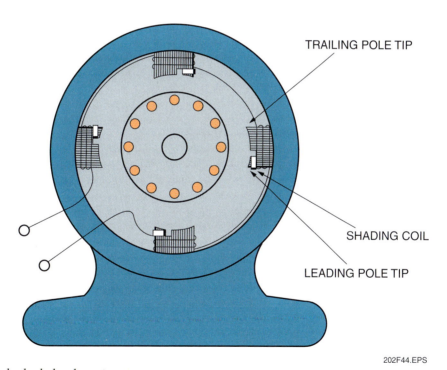

Figure 44 ◆ Four-pole shaded-pole motor.

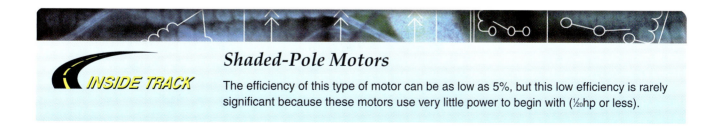

Shaded-Pole Motors

The efficiency of this type of motor can be as low as 5%, but this low efficiency is rarely significant because these motors use very little power to begin with (1/20hp or less).

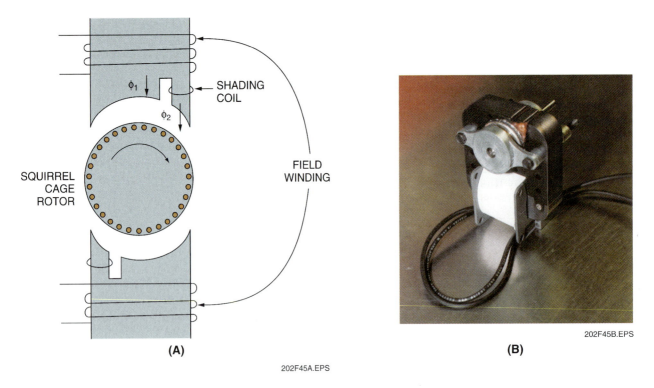

Figure 45 ◆ Two-pole shaded-pole motor.

Capacitor-Start, Capacitor-Run Motors

INSIDE TRACK

These motors run quietly and smoothly and have a high starting torque. Their construction is similar to a split-phase motor in that they use a centrifugal switch to remove only the start capacitor from the starting winding while leaving the run capacitor connected to the winding. These motors have a higher power factor than ordinary split-phase motors.

2.36 ELECTRICAL LEVEL TWO ◆ TRAINEE GUIDE

As the main flux decreases toward zero, the induced voltage and current in the shading coil reverse their polarity, and the resulting force tends to prevent the flux from collapsing through the iron in the region of the shading coil ($\phi_2 > \phi_1$). The result is that the main flux rises first in the unshaded portion of the pole and later in the shaded portion. This action is equivalent to a sweeping movement of the field across the pole face in the direction of the shaded pole. The cage rotor conductors are cut by this moving field, and the force exerted on them causes the rotor to turn in the direction of the sweeping field.

Most shaded-pole motors have only one edge of the pole split, and therefore, the direction of rotation is not reversible. However, some shaded-pole motors have both leading and trailing pole tips split to accommodate shading coils. The leading pole tip shading coils form one series group, and the trailing pole tip shading coils form another series group. Only the shading coils in one group are simultaneously active, while those in the other group are on an open circuit.

The shaded-pole motor is similar in operating characteristics to the split-phase motor. It has the advantages of simple construction and low cost. It has no sliding electrical contacts and is reliable in operation. However, it has low starting torque, low efficiency, and a high noise level. It is normally used to operate small fans. The shading coil and split pole are also used in timers to make them self-starting.

4.4.5 Single-Phase Synchronous Motor

The single-phase synchronous motor, as its name implies, runs at synchronous speed. It finds use where a constant speed is needed, such as in turntables and clocks. It is started in the same way as any of the single-phase induction motors and therefore has a rotating field. By having a modified rotor, the motor pulls into synchronism and runs at synchronous speed.

5.0.0 ♦ MULTIPLE-SPEED INDUCTION MOTORS

The speed of an induction motor depends on the power supply frequency and the number of pairs of poles used in the motor. Obviously, to alter motor speed it is merely necessary to change one of these two factors. By far the most common method used involves changing the number of poles, generally at some type of external controller.

There are two types of multiple-speed squirrel cage induction motors in common use: the multiple-winding motor and the consequent-pole motor. Both feature poles that may be changed, as required, by shifting key external connections, and in this way they provide for operating the motor at a limited number of different speeds.

5.1.0 Multiple-Winding Motor

In the multiple-winding motor, two or more separate windings are placed in the stator core slots, one over the other, as shown in *Figure 46*. For example, a four-pole winding can be positioned in the core slots and have a two-pole winding placed on top of it. The windings are insulated from each other and arranged so that only one winding at a time can be energized. Switching speeds is normally accomplished by switching contacts that are in the motor controller external to the motor itself.

5.2.0 Consequent-Pole Motor

In the consequent-pole motor, there are two speeds. The motor is constructed to have a certain number of poles for high-speed operation and

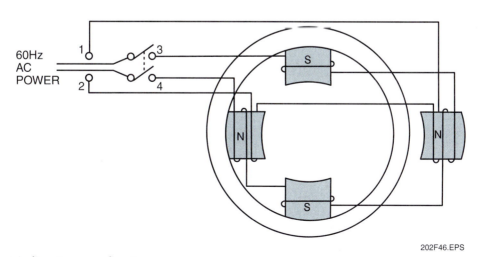

Figure 46 ♦ Two-winding, two-speed motor.

Inside Track

Hysteresis Synchronous Motor

This unusual synchronous motor has an external rotor. These motors have low noise levels, high efficiency, and constant speed, and are used for applications such as phonograph turntables.

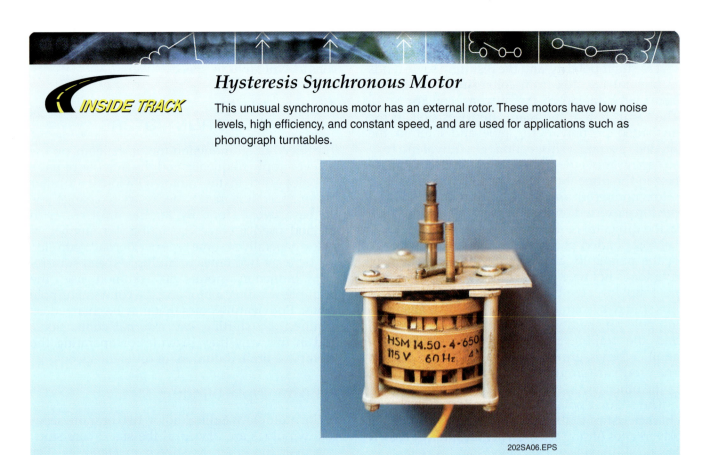

then, by a switching action, double this number of poles to give low-speed operation. The switching action is illustrated by the use of the two-phase motor in *Figure 47*. If you trace the wiring in *Figure 47*, you can see how the system is phased so that both magnetic north and south poles are produced at the winding projections. With two-phase power applied to the two-pole motor (two poles per phase), a rotating magnetic field of 3,600 rpm is produced.

In *Figure 48*, the connections are changed so that the system is phased to produce four magnetic

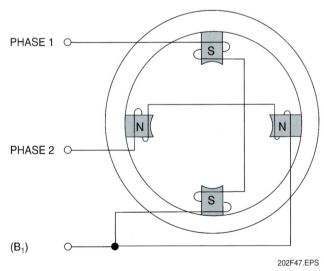

Figure 47 ◆ High-speed consequent-pole motor.

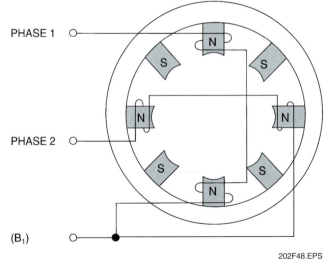

Figure 48 ◆ Low-speed consequent-pole motor.

2.38 ELECTRICAL LEVEL TWO ◆ TRAINEE GUIDE

north poles at the winding projections. Since every north pole must have a south pole, consequent south poles are produced between the projecting north poles as a consequence of having formed north poles. Accordingly, in *Figure 48*, there are twice as many pole groups as in *Figure 47*. Therefore, a four-pole rotating magnetic field of 1,800 rpm is produced.

Figure 49 shows the short-jumpering arrangement of the consequent-pole motor. In this, all windings are in series, and alternate north and south poles are produced. To produce consequent poles, the series connection is replaced with a parallel connection accomplished by the long-jumpering arrangement. By connecting the motor in this manner, four salient monopoles are produced and, as a result, create four opposite consequent poles. In the practical consequent-pole motor, all necessary internal connection rearrangements are accomplished at an external control panel.

Consequent-pole motor characteristics depend on the intended application. In a constant-horsepower motor, torque varies inversely with speed. It is used for driving machine tools.

In a constant-torque motor, horsepower varies directly with speed. It is used to drive pumps and air compressors, as well as in constant-pressure blowers. In a variable-torque, variable-horsepower motor, both torque and horsepower change with changes in speed. This is the type of motor found in household fans and air conditioners.

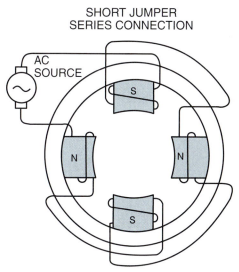
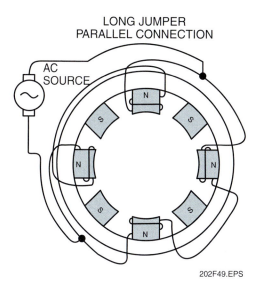

Figure 49 ◆ Single-phase consequent-pole motor.

6.0.0 ♦ VARIABLE-SPEED DRIVES

The use of adjustable speed in industrial equipment is increasing due to the need for better equipment control and for energy savings where partial power is required. AC drives compete with DC drives, eddy current drives, and mechanical and hydraulic systems as methods to control speed. Reliability, cost, and control capabilities are the major factors in system selection.

A drive system includes both the drive controller and the motor being driven. This module focuses specifically on the electronic drive components and covers various types of control for both DC and AC drives. This section provides a basic review of some fundamental principles that are important to understand when starting up, operating, or troubleshooting a variable-speed drive system.

6.1.0 Types of Adjustable Speed Loads

Most drive controllers can be adjusted or modified to optimize performance and provide the most efficient and cost-effective drive, depending on the load characteristics of the application.

It is important to understand the speed and torque characteristics as well as the maximum horsepower requirements for the type of load to be considered. Based on this, either a constant-torque controller or a variable-torque controller is selected. The most common types of loads are shown in *Figure 50*. Note that a load requires the same amount of torque at low speed as at high speed.

For a constant-torque load, the torque remains constant throughout the speed range, and the horsepower increases and decreases in direct proportion to the speed. This applies to applications such as conveyors, as well as applications in which shock loads, overloads, or high inertia loads are encountered.

A variable-torque load requires much lower torque at low speeds than at high speeds. Horsepower varies approximately as the cube of the speed, and the torque varies approximately as the square of the speed. This applies to applications such as centrifugal fans, pumps, and blowers.

A constant-horsepower load requires high torque at low speeds, low torque at high speeds, and thus constant horsepower at any speed. It applies to applications such as lathes requiring slow speeds for deep cuts and high speeds for finishing. Usually, very high starting torques are required.

6.2.0 Motor Considerations

For industrial applications, motors are required to function at varying torques and speeds, and in forward and reverse directions. Besides operating as a motor, the machine may also function as a brake or a generator for short periods.

The various operating modes for industrial drives are shown in *Figure 51*. Positive and negative speed (rotation) are plotted on the horizontal axis and the torques are plotted on the vertical axis. The four quadrants of operation are labeled 1, 2, 3, and 4.

A machine operating in quadrant 1 has positive torque and speed, which means that they both act in the same direction (in this case, clockwise). A machine in this quadrant is functioning as a motor. It delivers mechanical power to a load. The machine will also act as a motor in quadrant 3, but torque and speed are reversed from quadrant 1 (counterclockwise).

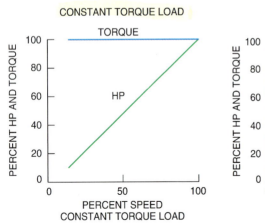
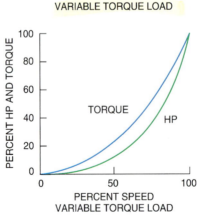
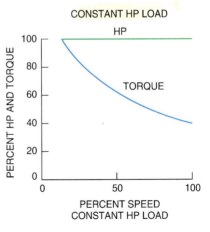

Figure 50 ♦ Types of adjustable speed loads.

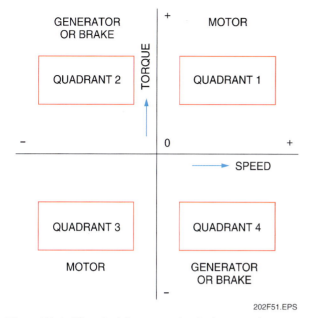

Figure 51 ◆ Electric drive operation in four quadrants.

While operating in quadrant 2, a machine will develop a positive torque and a negative speed. The torque is acting clockwise, and the speed is counterclockwise. In this quadrant, the machine is absorbing mechanical power from the load and functions as a generator. This mechanical power is converted into electric power and is generally transmitted back into the line. The electric power may also be dissipated in an external resistor, which is known as dynamic braking.

Depending on its connections, a machine may also be used as a brake while operating in quadrant 2. Absorbed mechanical power is converted to electric power, then converted into heat. If the machine absorbs electric line power as it is converting mechanical power into electric power, it functions as a brake. Both power inputs are dissipated as heat. Large power drives seldom use the brake mode of operation as it is very inefficient. The circuitry is generally chosen so that the machine will function as a generator when it is operating in quadrant 2. Quadrant 4 operation is identical to quadrant 2 except that speed and torque are reversed.

6.2.1 Typical Torque-Speed Curves

A three-phase motor has a torque-speed curve that is a good example of an electrical machine's behavior as a generator brake. The solid curve in *Figure 52* is the torque-speed curve for a machine acting as a motor in quadrant 1, a brake in quadrant 2, and a generator in quadrant 4.

If the stator leads are reversed, the torque-speed curve is shown by the dotted curve. Now the motor operates as a motor in quadrant 3, a generator in quadrant 2, and a brake in quadrant

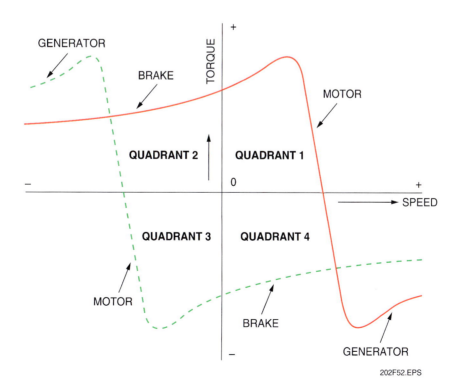

Figure 52 ◆ Four-quadrant operation for a squirrel cage motor.

4. The machine functions as a brake or a generator in quadrants 2 and 4, but it always runs as a motor in quadrants 1 and 3.

Figure 53 shows the torque-speed curve of a DC shunt motor. Motor, generator, and brake modes are apparent. The dotted curve represents reversed armature leads.

Variable-speed electric drives are designed to vary speed and torque in a smooth and continuous manner so as to satisfy load requirements. Typically, this is accomplished by shifting the torque-speed characteristic back and forth along the horizontal axis. The torque-speed characteristic of the motor is shifted by varying the armature voltage. Also, the curve of an induction motor can be shifted by varying the voltage and frequency applied to the stator.

In describing the various methods of motor control, only the behavior of power circuits will be discussed. The many ways of shaping and controlling triggering pulses will not be covered. They constitute a complex subject that involves sophisticated electronics, logic circuits, integrated circuits, and microprocessors.

6.2.2 Motor Heating

Since a variable-speed drive system includes both the drive controller and the motor, the design engineer should always consider the capabilities of the motor to perform acceptably under the desired operating conditions. One of the factors that you should be aware of is motor heating. When operating a motor at reduced speeds, the ability to dissipate heat is also reduced due to the slower cooling fan speed. This factor should be considered when maintaining the motor, modifying its enclosure or surrounding area, or troubleshooting the drive system.

6.3.0 Motor Speed Control

It is important to understand how DC or AC motor speed can be varied in order to understand how a drive controller accomplishes that task. This section reviews the fundamentals of DC and AC motor speed control.

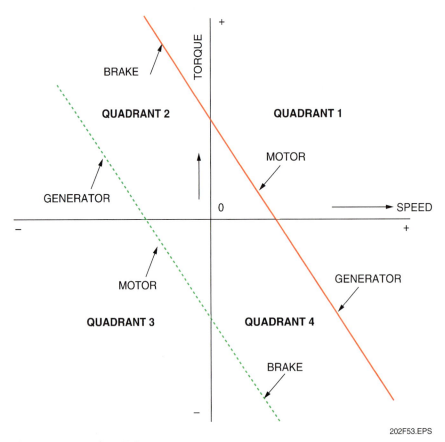

Figure 53 ◆ Four-quadrant operation for a DC motor.

Using a Motor as Both a Generator and a Brake

DC motors that power subway cars are also used for regenerative braking. When driven by the train's momentum, such as when slowing down upon approaching a station, the motor acts as a generator and puts current back into the system. Thus, while the train's motors are creating a significant amount of power from the train's movement, the train makes use of the resistive torque created by the power generation to slow down the train. The same theory applies to many hybrid cars.

6.3.1 Varying the Speed of a DC Shunt Motor

A DC shunt motor is shown in *Figure 54(A)*. Basically, there are two ways of varying the running speed of a DC shunt motor:

- Adjusting the voltage (and current) applied to the field winding. As the field voltage is increased, the motor slows down. This method is suggested by *Figure 54(B)*.
- Adjusting the voltage (and current) applied to the armature. As the armature voltage is increased, the motor speeds up. This method is suggested by *Figure 54(C)*.

6.3.2 Field Control

Here is how method 1, adjusting the field voltage, works. As the field voltage is increased, by reducing R_V in *Figure 54(B)*, for example, the field current is increased. This results in a stronger magnetic field, which induces a greater CEMF in the armature winding. The greater CEMF tends to oppose the applied DC voltage and thus reduces the armature current, I_A. Therefore, an increased field current causes the motor to slow down until the induced CEMF has returned to near its normal value.

Going in the other direction, if the field current is reduced, the magnetic field gets weaker. This causes a reduction in CEMF created by the rotating armature winding. The armature current increases, forcing the motor to spin faster, until the CEMF is once again approximately equal to what it was before. The reduction in magnetic field strength is compensated for by an increase in armature speed.

This method of speed control has certain positive features. It can be accomplished by a small, inexpensive rheostat, since the current in the field winding is fairly low. Also, because of the low value of the field current, I_F, the rheostat R_V does not dissipate very much energy. Therefore, this method is energy efficient.

However, there is one major drawback to speed control from the field winding: to increase the speed, you must reduce I_F and weaken the magnetic field, thereby lessening the motor's torque-producing ability. The ability of a motor to create torque depends on two things: the current in the armature conductors and the strength of the magnetic field. As I_F is reduced, the magnetic field is weakened, and the motor's torque-producing ability declines. Unfortunately, it is at this point that the motor needs all the torque-producing ability it can get, since it probably requires greater torque to drive the load at a faster speed.

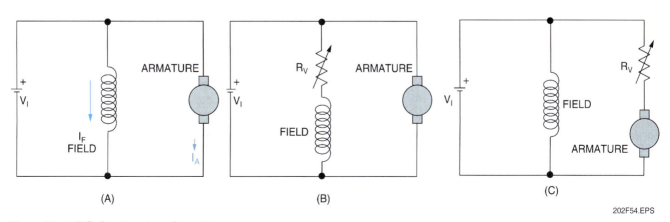

Figure 54 ♦ DC shunt motor schematic.

6.3.3 Armature Control

From the torque-producing point of view, method 2, armature control, is much better. As the armature voltage and current are increased by reducing R_V, the motor starts running faster, which normally requires more torque. The reason for the rise in speed is that the increased armature voltage demands an increased CEMF to limit the increase in armature current to a reasonable amount. The only way the CEMF can increase is for the armature winding to spin faster, since the magnetic field strength is fixed. In this instance, the ingredients are all present for increased torque production, since the magnetic field strength is kept constant and I_A is increased.

The problem with the armature control method of *Figure 54(C)* is that R_V, the rheostat, must handle the armature current, which is relatively large. Therefore, the rheostat must be physically large and expensive, and it will waste a considerable amount of energy.

6.3.4 Varying the Speed of an AC Motor

The principle of speed control for adjustable-frequency drives is based on the following fundamental formula for a standard AC motor:

$$N_s = \frac{120f}{P}$$

Where:

N_s = synchronous speed (rpm)
f = frequency
P = number of poles
120 = constant

The number of poles of a particular motor is set in its design and manufacture.

The adjustable-frequency system controls the frequency (f) applied to the motor. The speed (N_s) of the motor is then proportional to this applied frequency. Control frequency is adjusted by means of a potentiometer or external signal, depending on the application.

The frequency output of the controller is adjustable over its design speed range. Therefore, the speed of the motor is adjustable over this same range. Because an electronic means of generating variable frequencies is being used, the speed range often exceeds the 60 hertz (Hz) rated speed of the motor.

When variable-frequency speed control is employed, the motor supply voltage cannot be allowed to remain at a steady value. The magnitude of the motor voltage must be increased or decreased in proportion to the frequency. That is, the voltage-to-frequency ratio, V/f, must remain approximately constant.

For instance, if the motor has a nameplate rating of 240V at 60Hz, the voltage-to-frequency ratio is 4 (240 ÷ 60 = 4). If the motor is speeded up by adjusting its variable-frequency inverter to 90Hz, the voltage magnitude must be increased to 360V, since $4 \times 90 = 360$. If the motor is slowed down by adjusting the inverter frequency to 45Hz, the voltage magnitude must be decreased to 180V, since $4 \times 45 = 180$.

The stator's magnetic field strength must remain constant under all operating conditions. If the stator field strength should happen to rise much above the design value, the motor's core material would go into magnetic saturation. This would effectively lower the core's permeability, thereby inhibiting proper induction of voltage and current in the rotor loops (or bars), thus detracting from the torque-producing capability of the motor. On the other hand, if the stator field strength should happen to fall much below the design value, the weakened magnetic field would simply induce lower values of voltage and current in the rotor loops. This would also detract from the torque-producing ability of the motor.

Therefore, the magnetic field produced by the stator windings must hold a constant rms value, regardless of frequency. The magnetizing current of an induction motor is the current that flows through the stator winding when the rotor is spinning at steady-state speed with no torque load. The magnetizing current for an induction motor is given by Ohm's law:

$$I_{mag} = \frac{V}{X_L}$$

Where:

V = rms value of the applied stator voltage
X_L = inductive reactance of the stator winding

In the equation, X_L does not remain constant as the supply frequency is adjusted; it varies in proportion to the frequency ($X_L = 2\pi fL$). Therefore V must also be varied in proportion to the frequency, so that the Ohm's law division operation yields an unvarying value of magnetizing current.

Alternatively, using $X_L = 2\pi fL$, we can rewrite the equation:

$$I_{mag} = \frac{V}{X_L}$$

$$I_{mag} = \frac{V}{2\pi fL}$$

$$I_{mag} = \frac{1}{2\pi L} \times \frac{V}{f}$$

Since $1 \div 2\pi L$ is a constant determined by the motor's construction, the magnetizing current is kept constant by maintaining the V/f ratio.

The controller can automatically maintain the required volts/cycle (V/Hz) ratio to the motor at any speed. This provides maximum motor capability throughout the speed range.

The V/Hz setting is typically preset at the factory. However, on many controllers it can be adjusted or changed to fine-tune controller operation.

7.0.0 ◆ MOTOR ENCLOSURES

Motors are usually designed with covers over the moving parts. These covers, called enclosures, are classified by NEMA (National Electrical Manufacturers Association) according to the degree of environmental protection provided and the method of cooling. If the cover has openings, the motor is classified as an open motor; if the enclosure is complete, the motor is classified as an enclosed motor. Each of these types of motors has many modifications. *Table 1* lists the various types for both open and totally enclosed motors.

The different types of open motors as defined by NEMA are as follows:

- *General purpose* – Has ventilating openings which permit the passage of external cooling air over and around the windings of the machine.
- *Drip-proof* – Ventilating openings are so constructed that successful operation is not interfered with when drops of liquid or solid particles strike or enter the enclosure at any angle from 0° to 15° downward from the vertical.
- *Splash-proof* – Ventilating openings are so constructed that successful operation is not interfered with when drops of liquid or solid particles strike or enter the enclosure at any angle not greater than 100° downward from the vertical.
- *Guarded* – Openings giving direct access to live metal or rotating parts (except smooth surfaces) are limited in size by the structural parts or by screens, baffles, grills, expanded metal, or other means to prevent accidental contact with hazardous parts.

Table 1 Motor Enclosure Types

Open	Totally Enclosed
General purpose	Nonventilated
Drip-proof	Fan-cooled
Splash-proof	Fan-cooled guarded
Guarded	Explosion-proof
Semi-guarded	Dust- and ignition-proof
Drip-proof guarded	Pipe-ventilated
Externally ventilated	Water-cooled
Pipe-ventilated	Water-to-air-cooled
Weather-protected (Type I & Type II)	
Encapsulated windings	
Sealed windings	

- *Semi-guarded* – Some of the ventilating openings, usually in the top half, are guarded as in the case of a guarded machine, but the others are left open.
- *Drip-proof guarded* – This type of machine has ventilating openings as in a guarded machine.
- *Externally ventilated* – Designating a machine that is ventilated by a separate motor-driven blower mounted on the machine enclosure. Mechanical protection may be as defined above. This machine is sometimes known as a blower-ventilated or force-ventilated machine.
- *Pipe-ventilated* – Openings for the admission of ventilating air are so arranged that inlet ducts or pipes can be connected to them.
- *Weather-protected* – Type I: Ventilation passages are so designed as to minimize the entrance of rain, snow, and airborne particles to the electrical parts. Type II: In addition to the enclosure described for a Type I machine, ventilating passages at both intake and discharge are so arranged that high-velocity air and airborne particles blown into the machine by storms or high winds can be discharged without entering the internal ventilating passages leading directly to the electric parts.
- *Encapsulated windings* – An AC squirrel cage machine having random windings filled with an insulating resin, which also forms a protective coating.

Self-Cooling Motors

Conventional squirrel cage fan-cooled motors may overheat when operated at reduced speeds. Many manufacturers now offer inverter duty-rated motors with increased self-cooling capability.

- *Sealed windings* – An AC squirrel cage machine making use of form-wound coils and an insulation system that, through the use of materials, processes, or a combination of materials and processes, results in a sealing of the windings and connections against contaminants.

The different types of totally enclosed motors as defined by NEMA are as follows:

- *Nonventilated* – Not equipped for cooling by means external to the enclosing parts.
- *Fan-cooled* – Equipped for exterior cooling by means of a fan or fans that are integral with the machine but external to the enclosing parts.
- *Fan-cooled guarded* – All openings giving direct access to the fan are limited in size by design of the structural parts or by screens, grills, expanded metal, etc. to prevent accidental contact with the fan.
- *Explosion-proof* – Designed and constructed to withstand an explosion of a specified gas or vapor that may occur within it and to prevent the ignition of the specified gas or vapor surrounding the machine by sparks, flashes, or explosions of the specified gas or vapor that may occur within the machine casing.
- *Dust- and ignition-proof* – Designed and constructed in a manner that will exclude ignitable amounts of dust or amounts which might affect performance or rating, and that will not permit arcs, sparks, or heat otherwise generated or liberated inside the enclosure to cause ignition of exterior accumulations or atmospheric suspensions of a specific dust on or in the vicinity of the enclosure.
- *Pipe-ventilated* – Openings are so arranged that when inlet and outlet ducts or pipes are connected to them there is no free exchange of the internal air and the air outside the case.
- *Water-cooled* – Cooled by circulating water, with the water or water conductors coming in direct contact with the machine parts.
- *Water-to-air-cooled* – Cooled by circulating air, which in turn is cooled by circulating water.

7.1.0 Open Motor

The most common type of motor is the open motor. It has ventilating openings that permit the passage of external cooling air over and around its windings. If these are limited in size and shape, the motor is called a protected motor, since it is protected from any large pieces of material that may somehow enter the motor, thus damaging its internal parts. A protected motor also prevents a person from touching the rotating or electrically-energized parts of the motor. Drip-proof and splash-proof motors are constructed such that drops of liquid cannot enter the motor.

7.2.0 Enclosed Motor

The totally enclosed motor is designed to prevent the free exchange of air between the inside and outside of the actual motor housing. It is used where hostile environmental conditions and the motor application require maximum protection of the internal parts of the motor.

8.0.0 ◆ NEMA FRAME DESIGNATIONS

Frame sizes were developed by NEMA to ensure interchangeability of motors among manufacturers. They appear on motor nameplates to give information about the machine's physical dimensions. Key dimensions are shown in *Figures 55* and *56*. A few of these are:

- Distance from motor feet to shaft centerline, known as the D dimension
- Bolt-hole center-to-center distance between front and back feet, known as the 2F dimension
- Exposed shaft distance from shaft end to shaft shoulder, known as the N-W dimension

Manufacturer tables are available to correlate frame size to dimensions. An example is shown in *Table 2*. The system for designating the frames of motors and generators consists of a series of numbers in combination with letters.

More compact design, better ventilation, and insulation systems with higher temperature ratings have enabled manufacturers to house motors in increasingly smaller frame sizes. NEMA re-rates occurred in 1952 and 1964. Motors manufactured before 1952 are generally referred to as pre-U-frame motors. Those manufactured between 1952 and 1964 are called U-frame motors; those manufactured since 1964 are called T-frame motors.

8.1.0 Small Machines

The frame number for small machines is the D dimension in inches multiplied by 16. The following letters shall immediately follow the frame number to denote variations:

B – Carbonator pump motors

C – Type C face-mounting motors

G – Gasoline pump motors

H – A frame having an F dimension larger than that of the same frame without the suffix H

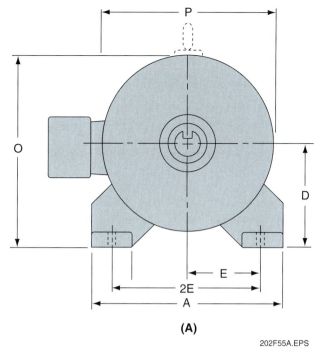

Figure 55 ♦ End view of a foot-mounted motor.

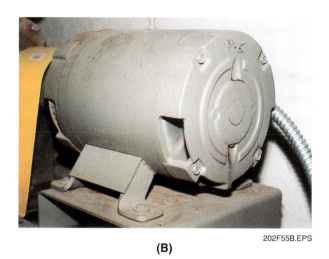

8.2.0 Medium Machines

The system for numbering frames of medium machines is as follows:

- The first two digits of the frame number are equal to four times the D dimension in inches. (If this product is not a whole number, the first two digits of the frame number shall be the next higher whole number.)
- The third and, when required, the fourth digit of the frame number are obtained from the value of 2F in inches.

Figure 55 shows a typical end view of a foot-mounted machine. The many different dimensions can be found on a dimension sheet for that machine. The NEMA frame designation will provide information relating to both the D and 2F dimensions.

Table 2 may be used to determine the D dimension and 2F dimension for medium-size motors. The D dimension is the distance from the centerline of the shaft to the bottom of the feet. The 2F dimension is the distance between the centerlines of the mounting holes in the feet or base of the machine.

Medium machines also use letters that denote variations. These letters follow the frame number. Since there are many more varieties of medium-size machines, the letter relates to the different aspects of mounting and shaft orientation.

Figure 56 ♦ Lettering of dimension sheets for foot-mounted machines (side view).

J – Jet pump motors
K – Sump pump motors
M – Oil burner motors
N – Oil burner motors
Y – Special mounting dimensions (must obtain dimensional diagram from manufacturer)
Z – All mounting dimensions are standard except the shaft extension

Table 2 Frame Dimension Chart

Frame Number Series	Third/Fourth Digit in Frame Number							
	D	1	2	3	4	5	6	7
	2F Dimensions							
140	3.50	3.00	3.50	4.00	4.50	5.00	5.50	6.25
160	4.00	3.50	4.00	4.50	5.00	5.50	6.25	7.00
180	4.50	4.00	4.50	5.00	5.50	6.25	7.00	8.00
200	4.50	4.50	5.00	5.50	6.50	7.00	8.00	9.00
210	5.00	4.50	5.00	5.50	6.50	7.00	8.00	9.00
220	5.50	5.00	5.50	6.25	6.75	7.50	9.00	10.00
250	6.25	5.50	6.25	7.00	8.25	9.00	10.00	11.00
280	7.00	6.25	7.00	8.00	9.50	10.00	11.00	12.50
320	8.00	7.00	8.00	9.00	10.50	11.00	12.00	14.00
360	9.00	8.00	9.00	10.00	11.25	12.25	14.00	16.00
400	10.00	9.00	10.00	11.00	12.25	13.75	16.00	18.00
440	11.00	10.00	11.00	12.50	14.50	16.50	18.00	20.00
500	12.50	11.00	12.50	14.00	16.00	18.00	20.00	22.00
580	14.50	12.50	14.00	16.00	18.00	20.00	22.00	25.00
680	17.00	16.00	18.00	20.00	22.00	25.00	28.00	32.00

Frame Number Series	Third/Fourth Digit in Frame Number							
	D	8	9	10	11	12	13	14	15
	2F Dimensions							
140	3.50	7.00	8.00	9.00	10.00	11.00	12.50	14.00	16.00
160	4.00	8.00	9.00	10.00	11.00	12.50	14.00	16.00	18.00
180	4.50	9.00	10.00	11.00	12.50	14.00	16.00	18.00	20.00
200	5.00	10.00	11.00	...	...	...	...	...	...
210	5.25	10.00	11.00	12.50	14.00	16.00	18.00	20.00	22.00
220	5.50	11.00	12.50	...	...	...	...	...	...
250	6.25	12.50	14.00	16.00	18.00	20.00	22.00	25.00	28.00
280	7.00	14.00	16.00	18.00	20.00	22.00	25.00	28.00	32.00
320	8.00	16.00	18.00	20.00	22.00	25.00	28.00	32.00	36.00
360	9.00	18.00	20.00	22.00	25.00	28.00	32.00	36.00	40.00
400	10.00	20.00	22.00	25.00	28.00	32.00	36.00	40.00	45.00
440	11.00	22.00	25.00	28.00	32.00	36.00	40.00	45.00	50.00
500	12.50	25.00	28.00	32.00	36.00	40.00	45.00	50.00	56.00
580	14.50	28.00	32.00	36.00	40.00	45.00	50.00	56.00	63.00
680	17.00	36.00	40.00	45.00	50.00	56.00	63.00	71.00	80.00

For example, to understand the NEMA frame designation, we will take a typical motor frame designation and determine the D and 2F dimensions. Then we will use a different set of dimensions to determine the frame designation number.

Example 1:

A typical medium-size frame number is a 256T. Since we know this is a medium frame, we divide the first two digits by 4:

$$25 \div 4 = 6.25$$

Therefore, the D dimension is 6.25 inches.

To determine the 2F dimension, we use *Table 2* and the third digit in the frame number. The third digit is 6 and the frame is a 250 series. Using the table, the 2F dimension is 10 inches. The T in the frame number is included as part of a frame designation for which standard dimensions have been established.

Medium-size frames can have multiple letters that denote a variety of different applications and arrangements. A 256AT has the same dimensions, with the A added to denote an industrial DC machine.

Example 2:

A frame has a D dimension of 3.5 inches and a 2F dimension of 4 inches, and all standard dimensions have been established. Multiplying the D dimension by 4 will give the first two digits of the frame designation:

$$3.5 \times 4 = 14$$

Using the table, a 140 frame series and a 2F dimension of 4 inches provides a third digit of 3. Since it is a standard dimension frame, the letter will be the suffix. This frame has a designation of 143T.

Full-load torque (rather than horsepower) determines the frame size required to house the motor. Thus, a motor developing a large amount of horsepower at high speed will have the same frame size as a machine developing less horsepower at a slower speed.

9.0.0 ♦ MOTOR RATINGS AND NAMEPLATE DATA

The ratings of an electric motor include:

- Voltage
- Full-load current
- Speed
- Number of phases and frequency
- Full-load horsepower
- Service classification

Except for full-load horsepower and service classification, these are self-explanatory units. The horsepower rating that is stamped on the motor nameplate by the manufacturer is the horsepower load the motor will carry without damaging any part of the motor.

Electric motor service classification depends on the type of service for which the motor is designed. A motor will usually fall into one of two classifications. General-purpose motors are those motors designed for use without restriction to a particular application. They meet certain specifications as standardized by NEMA. A definite-purpose motor is one that is designed in standard ratings and with standard operating characteristics for use under service conditions other than usual or for use on a particular type of application. A special-purpose motor is one with special operating characteristics or special mechanical construction, or both, that is designed for a particular application and that does not meet the definition of a general-purpose or a definite-purpose motor.

The most common machine rating is the continuous duty rating defining the output (in kilowatts for DC generators, kilovolt-amperes at a specified power factor for AC generators, and horsepower for motors) that can be carried indefinitely without exceeding established limitations. For intermittent duty, periodic duty, or varying duty, a machine may be given a short-time rating defining the load that can be carried for a specific time. Standard periods for short-time ratings are 5, 15, 30, and 60 minutes. Speeds, voltages, and frequencies are also specified in ratings, and provision is made for possible variations in voltage and frequency.

For example, motors must operate successfully at voltages 10% above and below rated voltage and, for AC motors, at frequencies 5% above and below rated frequency; the combined variation of voltage and frequency may not exceed 10%. Other performance conditions are so established that reasonable short-time overloads can be carried. Thus, the user of a motor can expect to be able to apply an overload of 25% for a short time at 90% of normal voltage with an ample margin of safety.

9.1.0 Nameplate Data

NEC Section 430.7 has specified information that must be listed on a motor nameplate based on its type. Requirements can also be found in NEMA Standards MG-1 and MG-2. Required information plus additional information is shown on the nameplate in *Figure 57*.

9.1.1 Rated Voltage

Power plant induction motors are designed to operate with a balanced three-phase voltage source applied at the terminals. The rated voltage on the nameplate is usually lower than the voltage of the electrical system. For example, a 460V motor is designed to operate in a 480V system. Here, an assumption is made by motor manufacturers that there will be a voltage drop of 20V from the transformer down to the motor terminals (see *Table 3*). The rated or nameplate voltage is the voltage at which the motor will operate most effectively. When other than rated voltage is applied, performance will change and motor life may be reduced.

Many three-phase motors have two voltages listed on the nameplate. For example, 230/460V means the motor can be connected for either 230V or 460V operation. In these cases, a connection diagram is usually found on the nameplate, as shown in *Figure 58*. These diagrams refer to low-voltage and high-voltage connections.

9.1.2 Full-Load Amps (FLA)

The FLA rating appearing on the nameplate indicates the current the motor will draw at nameplate

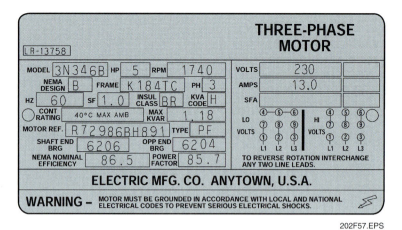

Figure 57 ♦ Nameplate data.

Table 3	Induction Motor Voltages
System Voltage	**Rated Voltage**
216	208
240	230
480	460
600	575
2,400	2,300
4,160	4,000
4,800	4,600
6,900 and 7,200	6,600
13,200 and 13,800	13,200

The FLA ratings are guaranteed if the induction motor is operating at full-load conditions and the applied voltage and frequency are the same as stated on the nameplate. When voltage or frequency are not the same, however, the current drawn by the motor at full-load conditions will be different from the nameplate indication (see *Table 4*). It is possible to damage a motor operated below its rated voltage or frequency, since the current the motor draws at full-load conditions increases in both cases. If the overload protective device is not sized properly, motor life may be shortened by this overcurrent condition.

horsepower, frequency, and voltage. Most manufacturers test to determine this value on a periodic basis during production, ensuring reasonable accuracy. The *NEC*® requires that the rated full-load current be the basis for determining the proper sizing of cable, overload protective devices, and other overcurrent protection in the motor circuit. Since many motors can be connected for one of two voltage ratings, they have two FLA ratings.

9.1.3 Rated Full-Load Speed

The rated full-load speed is the value indicated in rpm on the nameplate. It is the speed at which the shaft will turn at the nameplate horsepower when supplied with power at the nameplate voltage and frequency. If the driven load is less than the nameplate horsepower, the shaft will turn faster than full-load speed.

If the motor is operating unloaded, the shaft will turn very close to synchronous speed. For

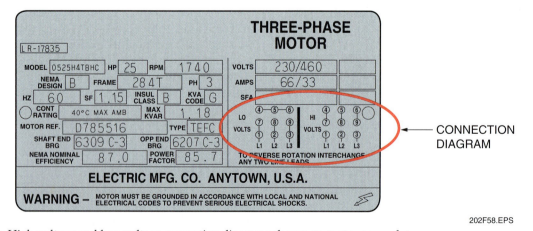

Figure 58 ♦ High-voltage and low-voltage connection diagrams shown on motor nameplate.

Table 4	Motor Operation
Mode	Full-Load Current
110% of rated volts	7% decrease
90% of rated volts	11% increase
105% of rated frequency	5%–6% decrease
95% of rated frequency	5%–6% increase

example, with a full-load speed of 1,750 rpm, it can be inferred that the motor's synchronous speed is 1,800 rpm. The machine will operate from close to 1,800 rpm down to 1,750 rpm, from no-load to full-load conditions.

Common synchronous speeds are 3,600, 1,800, 1,200, 900, and 600 rpm. Synchronous speed is rarely found on the motor nameplate unless the machine has been retrofitted and has not yet been tested for new full-load speed.

9.1.4 Rated Horsepower

An induction motor is really a torque generator. It delivers a needed torque to a driven machine at a certain speed. Thus:

$$\text{Horsepower} = \frac{\text{load torque in ft-lbs} \times \text{rpm}}{5,250}$$

For induction motors that are built to NEMA standards, the ratings will range from ½hp to 400hp, with 24 categories in all. If horsepower requirements fall between any two ratings, the larger motor size should be selected.

Remember, an induction motor will try to deliver any amount of horsepower the load requires. If properly sized, most motors operate at something less than the motor nameplate horsepower. Standard motors are designed to operate at nameplate values from sea level up to an altitude of 3,300 feet if the ambient temperature does not exceed 104°F (40°C). Above this altitude, the nameplate horsepower no longer applies.

NEMA standards provide a method for determining the proper temperature rise, or the new maximum ambient temperature, at higher elevations. However, the standards do not provide a direct method for deriving the horsepower. Several methods are available to estimate true motor horsepower output.

9.1.5 Duty or Time Rating

All polyphase induction motors have either a duty or a time rating, which is the elapsed time the motor can operate at nameplate horsepower without shortening its life. The time rating of a

Motor Horsepower and Speed

Can you replace a fan motor rated at a specific horsepower and speed with a motor rated at the same horsepower but a higher speed to increase airflow? If not, why not?

motor is determined by operating the machine at full-load conditions and measuring the time it takes for the windings to heat up to the temperature rating of the insulation.

Standard time ratings are 5, 15, 30, and 60 minutes, and continuous or 24 hours. A motor with a time (or duty) rating other than continuous is a smaller motor that is given a higher horsepower rating for a shorter period of time, thus reducing size and cost. In power plant uses, most motor ratings are continuous at 104°F (40°C).

9.1.6 NEMA Design Letters

The NEMA design letter defines the starting torque characteristics of an induction motor. It is one of the most important pieces of information on the nameplate; unfortunately, when a motor is replaced, the NEMA design letter is usually ignored, often leading to misapplication of the new machine. For fans or centrifugal pumps, starting torque requirements increase with the square of the change in speed. For mixers or loaded conveyor belts, however, starting torque requirements change very little with speed.

To account for these differences, NEMA has formulated design letters A, B, C, D, and F. The difference between motors with these letters is mainly in the design of the rotor, although there are also a few external differences. Design A and B motors are intended to drive conventional loads such as fans, blowers, and centrifugal pumps. About 80% of industrial motors are NEMA Design B.

Generally, the starting current is about five to seven times the rated full-load current. From *Table 5*, it can be seen that for larger motors, the starting current can be very significant, and across-the-line starting of larger motors could result in objectionable line-voltage dips. These voltage dips could result in other control equipment dropping out on low voltage and could even cause lights to dim.

Table 5 Typical Currents for 220V, 60-Cycle Squirrel Cage Motors

HP	Rated Full-Load Current	Starting (Maximum) Current	
		Classes B, C, D	Class F
½	2.0	12	—
1	3.5	24	—
1½	5.0	35	—
2	6.5	45	—
3	9	60	—
5	15	90	—
7½	22	120	—
10	27	150	—
15	40	220	—
20	52	290	—
25	64	365	—
30	78	435	270
40	104	580	360
50	125	725	450
60	150	870	540
75	185	1,085	675
100	246	1,450	900
125	310	1,815	1,125
150	360	2,170	1,350
200	480	2,900	1,800

The insulation class is a NEMA designation that identifies the class of material used to insulate the windings. Four letters designate the four classifications. They are A, B, F, and H. The insulation class defines the temperature that the insulation can be subjected to without suffering damage. The insulation class is shown on the nameplate in *Figure 59*.

Class A is now obsolete insofar as industrial motors are concerned. Class A was once the most common classification for motor insulation, especially for small motors. Class A comprises materials or combinations of materials such as cotton or paper, when suitably impregnated or coated, or other materials capable of operation at the temperature rise assigned for Class A insulation for the particular machine.

Class B is the predominant class of insulation used in motor manufacturing and rewinding today. This class is the basic standard of the industry. It includes materials such as mica, glass fiber, polyester, and aramid laminates, etc., with suitable bonding substances, or other materials, not necessarily inorganic, capable of operation at the temperature rise assigned for Class B insulation for the particular machine. (The insulation class may be designated more specifically by the use of additional letters, such as the BR shown in *Figure 59*.)

Class F incorporates materials that are similar to those in Class B but are capable of operation at the temperature rise assigned for Class F for the particular machine.

Class H insulation systems comprise materials or combinations of materials such as silicone elastomer, mica, glass fiber, polyester, and aramid laminates, etc., with suitable bonding substances such as silicone resins, or other materials capable of operation at the temperature rise assigned for Class H insulation for the particular machine.

NOTE
Per *NEC Table 430.251(B)*, Design A motors are not limited to a maximum starting current.

9.1.7 Insulation Class

The electrical insulation system in a motor determines the machine's ultimate life span more than any other component. By some estimates, over 60% of all motors brought to repair shops are there because of premature failure of the insulation system.

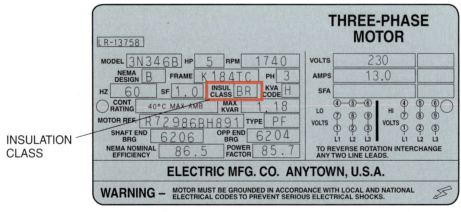

Figure 59 ♦ Nameplate showing insulation class.

When replacing motors, ensure that the insulation class is equal to or better than that of the motor removed from service.

9.1.8 Nominal Rated Voltage

Nominal rated voltage is defined as the voltage rating at which the motor is designed to operate.

9.1.9 Minimum Starting Voltage

Minimum starting voltage may be defined as the lowest voltage at which a motor will start without drawing an excessive/trip current.

9.1.10 Frequency

Frequency is given for AC motors in hertz or cycles per second. Standard frequencies for AC motors are 50Hz and 60Hz. Alternating current in the U.S. is 60Hz.

9.1.11 Service Factor

The service factor is a multiplier for the nameplate horsepower rating that determines the amount of overload the motor can withstand. This extra horsepower is available if the motor is already operating at rated voltage and frequency and is in an environment that does not exceed the ambient temperature rating. The most common service factor appearing on a motor nameplate is 1.15.

9.1.12 NEMA Code Letters

The high current draw of the motor during the first moments of startup is called the in-rush or locked-rotor current. It can be derived from the kVA code letter on the motor nameplate. The letter corresponds to the kilovolt-amperes per rated horsepower (kVA/hp) required during the first moments of motor startup. *Table 6* provides the kVA/hp value for each kVA code—a letter from A to V, excluding I, O, and Q. The locked-rotor current is required when sizing fuses or determining a circuit breaker setting in an induction motor circuit.

9.1.13 Bearings

Polyphase induction motors require either anti-friction or sleeve bearings. Anti-friction bearings are standard in medium (integral) horsepower motor sizes through 125hp/1,800 rpm. They are optional in 150 to 600hp/1,800 rpm sizes. Sleeve bearings are standard in 500hp/3,600 rpm and larger sizes.

Since radial loads are higher at the drive end of the motor, the drive-end bearing has a higher load rating than the bearing at the opposite end. A typ-

Table 6 Locked-Rotor Code Letters
[Data from *NEC Table 430.7(B)*]

Code Letter	kVA Per Horsepower with Locked Rotor
A	0–3.14
B	3.15–3.54
C	3.55–3.99
D	4.0–4.49
E	4.5–4.99
F	5.0–5.59
G	5.6–6.29
H	6.3–7.09
J	7.1–7.99
K	8.0–8.99
L	9.0–9.99
M	10.0–11.19
N	11.2–12.49
P	12.5–13.99
R	14.0–15.99
S	16.0–17.99
T	18.0–19.99
U	20.0–22.39
V	22.4–AND UP

Reprinted with permission from *NFPA 70®*, the *National Electrical Code®*. Copyright © 2007, National Fire Protection Association, Quincy, MA 02269. This reprinted material is not the complete and official position of the National Fire Protection Association on the referenced subject, which is represented only by the standard in its entirety.

ical nameplate (*Figure 60*) might depict both bearing duties as:

- Shaft end brg: 6,206
- Opp end brg: 6,204

Bearing internal clearances are: C1 and C2 (smaller-than-normal clearance); standard clearance (normal); and C3, C4, and C5 (larger-than-normal clearance). Electric motors usually require a C3 internal clearance. Some bearing manufacturers have a different designation for motor bearings that have a larger-than-normal internal clearance.

9.1.14 Rated Amperage

Rated amperage may be defined as the full-load current required to produce full-rated horsepower at the motor's rated voltage and frequency. *Figure 60* shows the amperage for a typical motor.

9.1.15 Rated Horsepower

Horsepower is a rating used to specify the capacity of an electric motor to produce mechanical power to drive a specific piece of equipment (see *Figure 60*).

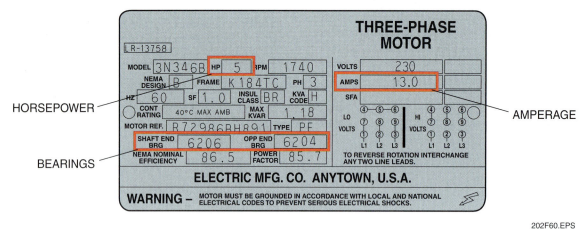

Figure 60 ♦ Nameplate showing bearings, horsepower, and amperage.

9.1.16 Locked-Rotor Current

The locked-rotor current is the steady-state current of a motor with the rotor locked and with rated voltage applied at rated frequency. NEMA has designated a set of code letters (*NEC Section 430.7*) to define locked-rotor kilovolt-amperes (kVA) per horsepower (*Table 6*). This code letter appears on the nameplate of all AC squirrel cage induction motors. The kVA rating is an indication of the current draw and, indirectly, the impedance of the locked rotor.

The current drawn by the motor under stall conditions can be calculated using the values given in *Table 6*. The current drawn by the motor under stalled conditions must be considered when selecting the motor protection and starting package and in coordination with the power system protective devices.

9.1.17 Starting Current

The total instantaneous starting current comprises the locked-rotor current plus the transient in-rush that flows until the motor magnetic circuit stabilizes.

9.1.18 Temperature Rise

The temperature rise may be defined as the measure of the heat produced by the operation of the motor. Several conditions contribute to temperature rise. Examples include running current, hysteresis losses, and friction of rotating parts.

9.1.19 Power Factor

The power factor (pf) is the ratio of active power of an alternating or pulsating current (when measured with a wattmeter) to the apparent power indicated by an ammeter and voltmeter. It is also referred to as the phase factor. The power factor is the measure of the system or equipment efficiency.

9.2.0 Motor Protection

Fuses are normally used for motor overload protection. When used, fuses must be provided in each ungrounded conductor and also the grounded conductor of a three-wire, three-phase AC system with one conductor grounded. When non-fuse overload protective devices are used (*NEC Section 430.37*), follow the guidelines as stated in *Table 7*. Note that each motor winding must be individually protected against short circuits and ground faults.

In general, when providing overcurrent protection for motor circuits against overcurrents due to grounds and short circuits (*NEC Section 430.52*), follow the guidance listed in *Table 8*. *Table 8* provides time delay and instantaneous trip values for various types of motors as a percentage of full-load current.

When sizing the overload protection for continuous duty motors larger than 1hp, refer to *NEC Section 430.32(A)*. Motors that have a service factor of 1.15 and/or a maximum temperature rise of 40°C shall be provided with overload protection limited to 125% of the full-load current of the motor (this also applies to the secondary circuit of a wound rotor motor). All other motors shall be limited to 115% of the full-load current. Motors that are rated at less than 1hp have exceptions to this rule. Refer to *NEC Article 430* for these exceptions.

If the desired overload ratings are not available when sizing overload protection for the motor, then use the next highest available overload rating.

Table 7 Minimum Number of Overload Units (Data from NEC Table 430.37)

Type of Motor	Supply System	Number and Location of Overload Units (such as trip coils or relays)
Single-phase AC or DC	Two-wire, single-phase AC or DC, ungrounded	One in either conductor
Single-phase AC or DC	Two-wire, single-phase AC or DC, one grounded conductor	One in ungrounded conductor
Single-phase AC or DC	Three-wire, single-phase AC or DC, grounded neutral	One in either ungrounded conductor
Single-phase AC	Any three-phase supply	One in ungrounded conductor
Two-phase AC	Three-wire, two-phase AC, ungrounded	Two, one in each phase
Two-phase AC	Three-wire, two-phase AC, one grounded conductor	Two, one in each ungrounded conductor
Two-phase AC	Four-wire, two-phase AC, grounded or ungrounded	Two, one per phase in ungrounded conductors
Two-phase AC	Five-wire, two-phase AC, grounded neutral or ungrounded	Two, one per phase in any ungrounded phase wire
Three-phase AC	Any three-phase supply	Three, one in each phase*

*Exception: An overload unit in each phase shall not be required where overload protection is provided by other approved means.

Reprinted with permission from *NFPA 70*®, the *National Electrical Code*®. Copyright © 2007, National Fire Protection Association, Quincy, MA 02269. This reprinted material is not the complete and official position of the National Fire Protection Association on the referenced subject, which is represented only by the standard in its entirety.

Table 8 Motor Protection (Data from NEC Table 430.52)

	Percent of Full-Load Current			
Type of Motor	Nontime Delay Fuse	Dual Element (Time Delay) Fuse**	Instantaneous Trip Breaker	Inverse Time Breaker*
Single-phase motors	300	175	800	250
AC polyphase motors other than wound rotor:				
Squirrel cage:				
Other than Design D, energy efficient	300	175	800	250
Design B, energy efficient	300	175	1,100	250
Synchronous†	300	175	800	250
Wound rotor	150	150	800	150
Direct current (constant voltage)	150	150	250	150

For certain exceptions to the values specified, see NEC Section 430.54.

*The values given in the last column also cover the ratings of nonadjustable inverse time types of circuit breakers that may be modified per NEC Section 430.52.

**The values in the Nontime Delay Fuse column apply to time delay Class CC fuses.

†Synchronous motors of the low-torque, low-speed type (usually 450 rpm or lower), such as are used to drive reciprocating compressors, pumps, etc., that start unloaded, do not require a fuse rating or circuit breaker setting in excess of 200% of the full-load current.

Reprinted with permission from *NFPA 70*®, the *National Electrical Code*®. Copyright © 2007, National Fire Protection Association, Quincy, MA 02269. This reprinted material is not the complete and official position of the National Fire Protection Association on the referenced subject, which is represented only by the standard in its entirety.

This is allowed provided that 140% of full-load current is not exceeded by motors that have a service factor of 1.15 and/or a maximum temperature rise of 40°C. For all other motors, this maximum value would be 130% rather than 115% as stated earlier.

Additionally, in certain situations, motors with installed overloads rated as discussed previously may not start or be able to carry system load. In these instances, it is permissible to increase the overload settings to the respective 140%/130% values. When motor starting is still a problem, the overload protective device may be shorted out during the equipment startup sequence provided that the circuit breaker or fuse protecting the branch is not set at greater than 400% of the full-load current value and the motor does not have an automatic starter.

Overload protection for adjustable speed drives is based on the rated input to the power conversion equipment. If overload protection is supplied with the equipment, then no further overload protection is required. The rating of this disconnecting means shall be no less than 115% of the power conversion equipment rated input current and it shall be physically located in the incoming line. Overload protection, if not shunted, should allow a sufficient time delay for the motor to start and accelerate.

9.2.1 Thermal Protectors

Thermal protectors that are integral with the motor are often used to protect the motor from overloads and starting failures. All motors with a voltage rating greater than 600V must have a thermal protector and its overload must not have an automatic reset feature. They shall trip no higher than the following percentage of full-load current:

- Motor full-load current not exceeding 9 amps—170%
- Motor full-load current between 9.1 and 20 amps—156%
- Motor full-load current greater than 20.1 amps—140%

This requirement is based on the maximum full-load motor current as listed in the tables provided in NEC Article 430.

Motors over 1,500hp include a device that is set to de-energize the motor once the actual temperature rise of the motor equals the rated temperature rise of the motor insulation. Thermal protectors are usually sized and installed by the motor manufacturer.

9.2.2 Branch Considerations

According to NEC Section 430.22, when a single motor used in a continuous duty application is supplied from a branch circuit, the ampacity of the branch circuit must be not less than 125% of the motor full-load current as determined by NEC Section 430.6(A)(1). If a multiple-speed motor is used, then the ampacity shall be based on the highest of the full-load current ratings on the motor nameplate. Where motors have unusual duty cycle requirements, use the requirements listed in Table 9, as referenced in NEC Section 430.22(E).

Table 9 Duty Cycle Service [Data from NEC Table 430.22(E)]

	Percentages of Nameplate Current Rating			
Classification of Service	5-Minute Rated Motor	15-Minute Rated Motor	30- and 60-Minute Rated Motor	Continuous Rated Motor
Short-Time Duty Operating valves, raising or lowering rolls, etc.	110	120	150	—
Intermittent Duty Freight and passenger elevators, tool heads, pumps, drawbridges, turntables, etc. For arc welders, see NEC Section 630.11	85	85	90	140
Periodic Duty Rolls, ore- and coal-handling machines, etc.	85	90	95	140
Varying Duty	110	120	150	200

Any motor application shall be considered as continuous duty unless the nature of the apparatus it drives is such that the motor will not operate continuously with load under any condition of use.

Reprinted with permission from NFPA 70®, the National Electrical Code®. Copyright © 2007, National Fire Protection Association, Quincy, MA 02269. This reprinted material is not the complete and official position of the National Fire Protection Association on the referenced subject, which is represented only by the standard in its entirety.

Per *NEC Section 430.24*, when sizing conductors supplying several motors, the capacity shall not be less than 125% of the largest motor plus the sum of the full-load current ratings of all other motors in the group. Values for the full-load amps are taken from *NEC Tables 430.247 through 430.250*. Several motors or loads are permitted to be provided for on one branch circuit if:

- The system voltage is less than 600 volts.
- The branch protective device protects the smallest installed motor.
- All motors on the circuit must not be over 1hp and less than 20A (15A) on 120V (600V) circuits where each motor draws less than 6A, overloads must be installed on the motors, and short circuit current and ground fault current must not exceed the branch circuit rating.
- It is part of a factory-listed assembly.

In instances where taps are used, short circuit current and ground fault current protection may not be required for the taps used. This is true provided that the tap used has the same ampacity as the branch circuit it is connected to. Additionally, the tap cannot be longer than 25' and it must also be physically protected from damage.

10.0.0 ◆ CONNECTIONS AND TERMINAL MARKINGS FOR AC MOTORS

The markings on the external leads of an induction motor are sometimes missing or illegible, and proper identification must be made before the motor can be connected to the line. This section describes the procedures for identifying leads in either a wye-connected or delta-connected, three-phase, nine-lead motor.

The required materials for this procedure are:

- Appropriate personal protective equipment
- 12V battery such as an automotive battery
- Analog meter with a large scale and low range (digital meters may not clearly capture the voltage kick)
- Test leads and jumpers
- Momentary contact, normally open (N.O.) pushbutton switch
- Labels to mark leads as they are identified
- Three-phase, nine-lead induction motor

Before starting, identify whether the motor to be tagged is wye-connected or delta-connected. Compare the two connection diagrams in *Figure 61* and *Figure 62*. You will see that both types of motors have nine leads and six coils. In a wye-connected motor, three coils are connected together and three

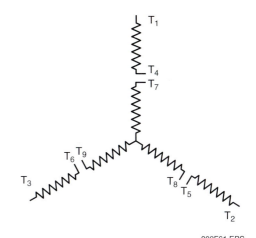

Figure 61 ◆ Dual-voltage, three-phase wye connection.

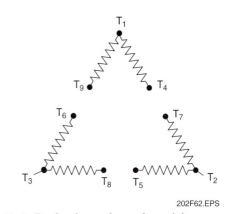

Figure 62 ◆ Dual-voltage, three-phase delta connection.

coils are isolated. A delta-connected motor has three sets of two coils connected together. Using an ohmmeter, insulate all motor leads from one another and check for continuity between each lead. Start by placing one probe on one lead and check through the remaining leads. As you identify which leads show continuity, group them together. Continue this procedure with each lead until all leads are grouped. When you have completed this procedure, you should have either three wires in one group and three sets of two wires grouped together, which would indicate a wye-connected motor, or three sets of three wires grouped together, indicating a delta-connected motor.

NOTE

If the leads are partially tied together, mark or identify them in a way that will allow you to reconnect them after the testing procedure is completed.

10.1.0 Identifying the Terminals of Wye-Connected Motors

Figure 63 shows the coil arrangement for a three-phase, wye-connected motor. To identify the terminals of a wye-connected motor, proceed as follows:

Step 1 Taking the group of three common leads, arbitrarily identify them as leads 7, 8, and 9.

Step 2 Using the diagram in *Figure 64* as a guide, connect the positive lead from the battery to lead 7. Connect the lead from the battery negative through the N.O. switch to leads 8 and 9 simultaneously.

Step 3 Connect one of the three remaining lead pairs to the voltmeter terminals.

Step 4 Close the N.O. switch while observing the DC voltmeter. Use the lowest scale practical without over-ranging the meter. If the meter deflection is upward, note the voltage reading. Observe that the reading occurs only on the initial energization of the windings and then decays. Note the peak reading only and ignore the deflection in the opposite direction that occurs when the switch is opened. If the meter initially deflects downward, reverse the test lead connections.

Step 5 Continue with the remaining two lead pairs. The pair with the highest voltage reading is the winding associated with lead 7. The lead with positive polarity is identified as lead 4 and the lead with negative polarity is lead 1.

Step 6 Repeat Step 3, but apply the positive lead of the battery to lead 8 and the negative lead of the battery to leads 7 and 9. The positive lead of the pair with the highest voltage is identified as lead 5 and the lead with negative polarity is lead 2.

Step 7 Repeat Step 3, but apply the positive lead of the battery to lead 9 and the negative lead of the battery to leads 7 and 8. The positive lead of the pair with the highest voltage is marked lead 6 and the negative lead is lead 3.

Step 8 To confirm that all leads are correctly identified, connect the motor to the circuit. Be sure to observe proper connection procedures for the applied voltage. Once connected to the circuit, start the motor and take current readings on all three lines. If the motor starts correctly and the current readings are approximately equal, the procedure was a success.

10.2.0 Identifying the Terminals of Delta-Connected Motors

A delta-connected motor has three sets of three leads. *Figure 65* shows how the coils are arranged in a delta-connected, three-phase motor. In this figure, the coils that are side by side are actually wound on the same poles on the motor. As discussed previously, this will allow some transformer interaction between adjacent coils. To identify the terminals of a delta-connected motor, proceed as follows:

Step 1 Using an ohmmeter on a low scale, measure the resistance between each of the

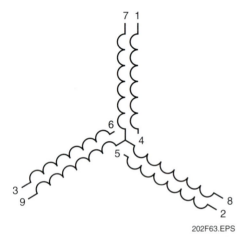

Figure 63 ◆ Coil arrangement in a wye-connected motor.

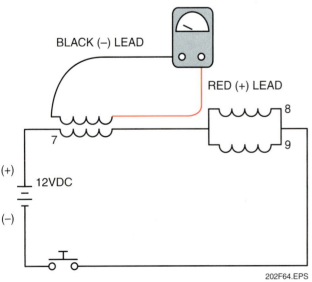

Figure 64 ◆ Battery hookup for wye-connected motor lead identification.

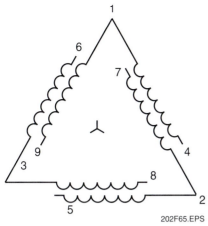

Figure 65 ♦ Coil arrangement in a delta-connected motor.

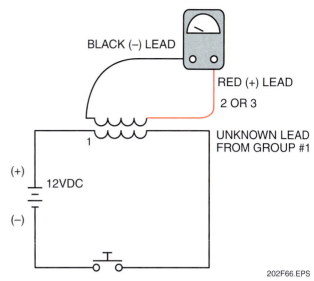

Figure 66 ♦ Battery hookup for delta-connected motor lead identification.

three leads in one group. When performing this measurement, you should see that the resistance between two of the leads is about twice that between either of those two and the third. The lead that shows the least resistance to the other two will be lead 1. Refer to all three wires in this set as set 1.

Step 2 Repeat Step 1 with the second set to identify the lead with the least resistance as lead 2. Refer to all three wires in this set as set 2.

Step 3 Repeat Step 1 with the final set of leads to identify the lead with the least resistance as lead 3. Refer to all three wires in this set as set 3.

Step 4 Using the diagram in *Figure 66* as a guide, connect lead 1 to the positive terminal of a DC voltage source and one of the remaining leads in that set (set 1) to the negative terminal. Attach the red lead of the voltmeter to lead 2 and the black lead to one of the two unknown leads in set 2. Press the pushbutton and observe the meter needle. If the correct leads have been selected, a voltage will be induced into this coil. If not, connect the second lead to the voltmeter and repeat the test. If there is still no induced voltage, disconnect the unknown lead in set 1 from the DC voltage and connect the remaining unknown lead to the negative source. Repeat the test until the leads with an induced voltage have been identified. Once these leads are located, identify the lead connected to the negative terminal of the DC voltage source as lead 4 and the lead connected to the negative voltmeter probe as lead 7. Identify the other lead in set 2 as lead 5.

Step 5 The remaining lead in set 1 will be lead 9. Leaving lead 1 on the positive DC terminal, connect the negative terminal to lead 9. Attach the red lead of the voltmeter to lead 3 and the black lead to one of the two unknown leads in set 3. Press the pushbutton and observe the meter needle. If the correct leads have been selected, a voltage will be induced into this coil. If not, connect the second lead to the voltmeter and repeat the test. Once these leads are located, identify the lead connected to the negative voltmeter probe as lead 6.

Step 6 The remaining lead in set 3 is lead 8. To verify this, connect lead 3 to the positive terminal of the DC voltage source and lead 8 to the negative terminal. Attach the red lead of the voltmeter to lead 2 and the black lead to lead 5. Press the pushbutton and observe the meter needle. If the results are correct, a voltage will be induced into this coil, resulting in meter needle deflection.

Step 7 To confirm that all leads are correctly identified, connect the motor to the circuit. Be sure to observe proper connection procedures for the applied voltage. Once connected to the circuit, start the motor and take current readings on all three lines. If the motor starts correctly and the current readings are approximately equal, the procedure was a success.

11.0.0 ◆ NEC® REQUIREMENTS

NEC Article 430 covers the application and installation of motors, motor circuits, and motor control connections, including conductors, short circuit and ground fault protection, starters, disconnects, and overload protection.

NEC Article 440 contains provisions for motor-driven equipment and for branch circuits and controllers for HVAC equipment.

All motors must be installed in a location that allows adequate ventilation to cool the motors. Furthermore, the motors should be located so that maintenance, troubleshooting, and repairs can be readily performed. Such work could consist of lubricating the motor bearings or perhaps replacing worn brushes. Testing the motor for open circuits and ground faults is also necessary from time to time.

When motors must be installed in locations where combustible material, dust, or similar material may be present, special precautions must be taken in selecting and installing the motors.

Any exposed live parts of motors operating at 50V or more between terminals must be guarded; that is, they must be installed in a room, enclosure, or location so as to allow access only by qualified persons (electrical maintenance personnel). If such a room, enclosure, or location is not feasible, an alternative is to elevate the motors not less than 8' above the floor. In all cases, adequate space must be provided around motors with exposed live parts, even when properly grounded, to allow for maintenance, troubleshooting, and repairs.

The chart in *Table 10* summarizes NEC® installation rules.

A summary of *NEC Article 430* is shown in *Figure 67*. Detailed information may be found in the NEC® under the articles or sections indicated.

Table 10 Summary of NEC® Requirements for Motor Installations

Application	Requirement	NEC® Reference
Location	Motors must be installed in areas with adequate ventilation. They must also be arranged so that sufficient work space is provided for replacement and maintenance.	NEC Section 430.14(A)
	Open motors must be located or protected so that sparks cannot reach combustible materials.	NEC Section 430.14(B)
	In locations where dust or flying material will collect on or in motors in such quantities as to seriously interfere with the ventilation or cooling of motors and thereby cause dangerous temperatures, suitable types of enclosed motors that will not overheat under the prevailing conditions must be used.	NEC Section 430.16
Disconnecting means	A motor disconnecting means must be within sight from the controller location (with exceptions) and disconnect both the motor and controller. The disconnect must be readily accessible and clearly indicate the OFF/ON positions (open/closed).	NEC Article 430, Part IX NEC Section 430.104
	Motor control circuits require a disconnecting means to disconnect them from all supply sources.	NEC Section 430.75
	The disconnecting means must be as specified in the code.	NEC Section 430.109
Wiring methods	Flexible connections such as Type AC cable, Greenfield, flexible metal tubing, etc., are standard for motor connections.	NEC Articles 300 and 430
Motor control circuits	All conductors of a remote motor control circuit outside of the control device must be installed in a raceway or otherwise protected. The circuit must be wired so that an accidental ground in the control device will not start the motor.	NEC Section 430.73
Guards	Exposed live parts of motors and controllers operating at 50 volts or more must be guarded by installation in a room, enclosure, or other location so as to allow access by only qualified persons, or elevated 8 feet or more above the floor.	NEC Section 430.232
Adjustable speed drive systems	Requirements for adjustable speed drives and their motors.	NEC Article 430, Part X
Motors operating over 600 volts	Special installation rules apply to motors operating at over 600 volts.	NEC Article 430, Part XI
Controller grounding	Motor controller enclosures must be grounded.	NEC Section 430.244

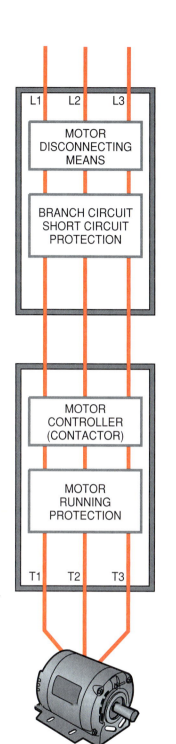

NEC Article 430, Part IX *Sections 430.101* *through 430.113*	**Disconnects motor and controllers from circuit.** 1. Continuous rating of 115% or more of motor FLC. Also see *NEC Article 430, Part II*. 2. All disconnecting means in the motor circuit must be per *NEC Sections 430.108, 430.109, and 430.110*. 3. Must be located in sight of motor location and driven machinery. The controller disconnecting means can serve as the disconnecting means if the controller disconnect is located in sight of the motor location and driven machinery.
NEC Article 430, Part IV *Sections 430.51* *through 430.58*	**Protects branch circuit from short circuits or grounds.** 1. Must carry starting current of motor. 2. Rating must not exceed values in *NEC Table 430.52* unless not sufficient to carry starting current of motor. 3. Values of branch circuit protective devices shall in no case exceed exceptions listed in *NEC Section 430.52*.
NEC Article 430, Part VII *Sections 430.81* *through 430.91*	**Used to start and stop motors.** 1. Must have current rating of 100% or more of motor FLC. 2. Must be able to interrupt LRC. 3. Must be rated as specified in *NEC Section 430.83*.

NEC Article 430, Part III
Sections 430.31
through 430.44
	Protects motor and controller against excessive heat due to motor overload.
	1. Must trip at following percent or less of motor FLC for continuous motors rated more than one horsepower.
	   a) 125% FLC for motors with a marked service factor of not less than 1.15 or a marked temperature rise of not over 40°C.
	   b) 115% FLC for all others. (See the *NEC®* for other types of protection.)
	2. Three thermal units required for any three-phase AC motor.
	3. Must allow motor to start.
	4. Select size from FLC on motor nameplate.

NEC Article 430, Part II *Sections 430.21* *through 430.29*	**Specifies the sizes of conductors capable of carrying the motor current without overheating.** 1. To determine the ampacity of conductors, switches, branch circuit overcurrent devices, etc., the full-load current values given in *NEC Tables 430.247 through 430.250* shall be used instead of the actual current rating marked on the motor nameplate. *(See NEC Section 430.6.)* 2. According to *NEC Section 430.22*, branch circuit conductors supplying a single motor used in a continuous duty application shall have an ampacity of not less than 125% of motor FLC, as determined by *NEC Section 430.6(A)(1)*.

Figure 67 ◆ Summary of requirements for motors, motor circuits, and controllers.

Motor Connections

On dual-voltage and/or multi-speed motors, always check the wiring connection diagrams given on the motor nameplate to wire the motor for the correct voltage and/or speed.

12.0.0 ◆ BRAKING

When we think of braking, we usually think in terms of mechanical friction braking, such as that used in cars. However, friction braking requires a physical connection between moving parts, which results in wear and maintenance. For many motors, it is often more effective to employ injection or dynamic braking.

12.1.0 DC Injection Braking

DC injection braking is a method of braking in which direct current (DC) is applied to the stationary windings of an AC motor after the AC voltage is removed. This is an efficient and effective method of stopping most AC motors. DC injection braking provides a quick and smooth braking action on all types of loads, including high-speed and high-inertia loads.

In an AC induction motor, when the AC voltage is removed, the motor will coast to a standstill over a period of time because there is no induced field to keep it rotating. Because the coasting time may be unacceptable, particularly in an emergency situation, electric braking can be used to provide a more immediate stop.

By applying a DC voltage to the stationary windings once the AC is removed, a magnetic field is created in the stator that will not change polarity. This constant magnetic field in the stator creates a magnetic field in the rotor. Because the magnetic field of the stator is not changing in polarity, it will attempt to stop the rotor when the magnetic fields are aligned. When the fields are aligned north and south, the two magnetic fields are attracted to each other. The resistive pull of the fields is used to stop the motor's rotational inertia. The same is true when the magnetic poles are the same, but instead of attraction, the magnetic forces repel one another, causing the motor to slow.

The stopping time is based on the amount of DC current applied. Normally, the applied current is three times the full load current of the motor. The higher the current level, the faster the motor will stop. For instance, in paper mills where sudden motor stopping time is an issue, the applied current is less than normal to avoid tearing the paper.

The only thing that can keep the rotor from stopping with the first alignment is the rotational inertia of the load connected to the motor shaft. However, because the braking action of the stator is present at all times, the motor is stopped quickly and smoothly to a standstill.

Because there are no parts that come in physical contact during braking, maintenance is kept to a minimum.

12.2.0 Dynamic Braking

Dynamic braking is another method for stopping a motor. It is achieved by reconnecting a running motor to act as a generator immediately after it is turned off, rapidly stopping the motor. The generator action converts the mechanical energy of rotation to electrical energy that can be dissipated as heat in a resistor or used as a load as in a generator. Using the motor as a generator is not very practical due to the fact that the motor stopping time is very limited.

Dynamic braking of a DC motor may be needed because DC motors are often used for lifting and moving heavy loads that may be difficult to stop. For example, forklifts often use dynamic braking.

There must be access to the rotor windings in order to reconnect the motor to act as a generator. On a DC motor, access is accomplished through the brushes on the commutator. The armature terminals of the DC motor are disconnected from the power supply and immediately connected across a resistor, which acts as a load. The smaller the resistance of the resistor, the greater the rate of energy dissipation and the faster the motor slows down.

The field windings of the DC motor are left connected to the power supply. The armature generates a voltage referred to as counter-electromotive force (CEMF). CEMF causes current to flow through the resistor and armature. The current causes heat to be dissipated in the resistor, removing energy from the system and slowing the motor rotation.

The generated CEMF decreases as the speed of the motor decreases. As the motor speed approaches zero, the generated voltage also approaches zero. This means that the braking action lessens as the speed of the motor decreases. As a result, a motor cannot be braked to a complete stop using dynamic braking. Dynamic braking also cannot hold a load once it is stopped because without the generated CEMF there is no more braking action.

For this reason, electromechanical friction brakes are sometimes used along with dynamic braking in applications that require the load to be held, or in applications where a large, heavy load is to be stopped. This is similar to using a parachute to slow a racecar before applying the brakes.

13.0.0 ◆ MOTOR INSTALLATION

The best motors on the market will operate improperly if they are installed incorrectly. Therefore, all personnel involved with the installation of electric motors should understand the procedures for installing the various types of motors that will be used.

WARNING!
When a motor is received at the job site, always refer to the manufacturer's instructions and follow them to the letter. Failure to do so could result in serious injury or death. Install and ground according to *NEC*® requirements and good practices. Consult qualified personnel with any questions or problems.

Keep the following in mind when installing new motors:

- *Uncrating* – Once the motor has been carefully uncrated, check to see if any damage has occurred during handling. Be sure that the motor shaft and armature turn freely. This is also a good time to check to determine if the motor has been exposed to dirt, grease, grit, or excessive moisture during shipment or storage. Motors in storage should have shafts turned over once each month to redistribute grease in the bearings. The measure of insulation resistance is a good dampness test. Clean the motor of any dirt or grit.

WARNING!
Never start a motor that has been wet until it has been completely dried and thoroughly tested.

- *Lifting* – Eyebolts or lifting lugs on motors are intended only for lifting the motor and factory motor-mounted standard accessories. These lifting provisions should never be used when lifting or handling the motor when the motor is attached to other equipment as a single unit. The eyebolt lifting capacity rating is based on a lifting alignment coincident with the eyebolt centerline. The eyebolt capacity reduces as deviation from this alignment increases.
- *Guards* – Rotating parts such as pulleys, couplings, external fans, and shaft extensions must be permanently guarded against accidental contact with clothing or body extremities.
- *Requirements* – All motors must be installed, protected, and fused in accordance with *NEC Article 430*. For general information on grounding, refer to *NEC Article 250* and *NEC Article 430, Part XIII*.
- *Thermal protector information* – The motor nameplate may or may not be stamped to indicate thermal protection.

Installing Motors

The shaft of this air conditioner motor must be aligned precisely with the shaft of the driven device. Note the micrometer attached to the motor and the load to achieve exact alignment.

Putting It All Together

Count the motors in your home. The typical home may easily have over 30 motors (including electronic equipment and tools). A century ago, a typical home might have had none. Determine what types of motors you have. Are they AC or DC? What identifying information can you determine by examining each motor nameplate?

Review Questions

1. The name of the motor part that rotates during operation is the ____.
   a. stator
   b. shunt/capacitor
   c. brushes/commutator
   d. armature/rotor

2. The electrical energy required to produce one horsepower of mechanical energy is ____.
   a. 476W
   b. 647W
   c. 746W
   d. 864W

3. When there is a wide variation in both torque and speed requirements, such as a blower or a hoist, use a ____ DC motor.
   a. brushless
   b. compound
   c. series
   d. shunt

4. The principal reason for developing the brushless DC motor was to ____.
   a. eliminate commutator problems
   b. improve efficiency
   c. increase horsepower
   d. improve airplanes

5. The name of the stationary motor part that produces the magnetic field during operation is the ____.
   a. stator
   b. shunt/capacitor
   c. brushes/commutator
   d. armature/rotor

6. The most popular rotor in use for polyphase AC motors is the ____ rotor.
   a. wound
   b. squirrel cage
   c. multiphase
   d. induction

7. The starting torque characteristics of an induction motor are defined by ____.
   a. rotor resistance
   b. starting current
   c. load
   d. stator resistance

8. A three-phase synchronous motor is started with a(n) ____.
   a. squirrel cage bar
   b. moveable actuator disc
   c. centrifugal weight assembly
   d. amortisseur winding

9. The speed of an induction motor depends on the power supply frequency and the ____.
   a. current
   b. voltage
   c. size
   d. number of poles

10. The speed on a multiple-winding motor is normally changed by ____.
    a. switching contacts in the motor controller
    b. reversing any two of the incoming leads
    c. an external wye-connected resistor
    d. timing relays

11. The two primary types of motor enclosures classified by NEMA are ____.
    a. general purpose and guarded
    b. open and totally enclosed
    c. guarded and semi-guarded
    d. nonventilated and externally ventilated

12. An open AC squirrel cage motor that makes use of form-wound coils and an insulation system is known as a(n) ____ type.
    a. semi-guarded
    b. sealed
    c. non-ventilated
    d. explosion-proof

13. The NEMA frame designation for gasoline pump motors is ____.
    a. B
    b. C
    c. G
    d. H

14. The letters FLA on a motor nameplate stand for ____.
    a. fused last application
    b. fast lower arm
    c. full-load amps
    d. full-load armature

Review Questions

15. To determine the horsepower rating of a motor that is already installed and in place, _____.
    a. multiply the rated voltage times the FLA
    b. divide the nominal efficiency by the power factor
    c. check the nameplate data
    d. use a torque wrench and horsepower data

16. The multiplier to the nameplate horsepower rating is called the _____.
    a. maximum kVAR
    b. service factor
    c. power factor
    d. NEMA nominal efficiency

17. A _____ is used to protect the motor from overloads and starting failures.
    a. fuse
    b. circuit breaker
    c. centrifugal switch
    d. thermal protector

18. The code requirements for motor disconnects are covered in _____.
    a. *NEC Article 430, Part VIII*
    b. *NEC Article 430, Part IX*
    c. *NEC Article 430, Part X*
    d. *NEC Article 430, Part XI*

19. The method of braking in which DC is applied to the motor windings after AC is removed is known as _____ braking.
    a. injection
    b. friction
    c. dynamic
    d. regenerative

20. Reconnecting a running motor to act as a generator is known as _____ braking.
    a. injection
    b. friction
    c. regenerative
    d. dynamic

Summary

This module discussed AC and DC motor theory, construction, and various motor types and applications. This discussion included torque, speed, and speed regulations as well as the fundamental concepts associated with variable speed drive systems. Motor enclosures were described, including open and totally enclosed motors and motor frame designations. Discussions of horsepower and calculation of load under various conditions were included. *NEC®* requirements and installation considerations were also discussed.

Notes

Trade Terms Introduced in This Module

Armature: The rotating windings of a DC motor.

Branch circuit: The circuit conductors between the final overcurrent device protecting the circuit and the outlet(s).

Brush: A conductor between the stationary and rotating parts of a machine. It is usually made of carbon.

Circuit breaker: A device designed to open and close a circuit by nonautomatic means and to open the circuit automatically on a predetermined overcurrent without injury to itself when properly applied within its rating.

Commutator: A device used on electric motors or generators to maintain a unidirectional current.

Continuous duty: Operation at a substantially constant load for an indefinitely long time.

Controller: A device that serves to govern, in some predetermined manner, the electric power delivered to the apparatus to which it is connected.

Duty: Describes the length of operation. There are four designations for circuit duty: continuous, periodic, intermittent, and varying.

Equipment: A general term including material, fittings, devices, appliances, fixtures, apparatus, and the like used as a part of, or in connection with, an electrical installation.

Field poles: The stationary portion of a DC motor that produces the magnetic field.

Horsepower: The rated output capacity of the motor. It is based on breakdown torque, which is the maximum torque a motor will develop without an abrupt drop in speed.

Hours: The duty cycle of a motor. Most fractional horsepower motors are marked continuous for around-the-clock operation at the nameplate rating in the rated ambient conditions. Motors marked one-half are for ½-hour ratings, and those marked one are for 1-hour ratings.

Intermittent duty: Operation for alternate intervals of (1) load and no load; or (2) load and rest; or (3) load, no load, and rest.

Overcurrent: Any current in excess of the rated current of equipment or the ampacity of a conductor. It may result from an overload, short circuit, or ground fault.

Overload: Operation of equipment in excess of the normal, full-load rating, or of a conductor in excess of rated ampacity, which, after a sufficient length of time, will cause damage or dangerous overheating. A fault, such as a short circuit or ground fault, is not an overload.

Periodic duty: Intermittent operation at a substantially constant load for a short and definitely specified time.

Revolutions per minute (rpm): The approximate full-load speed at the rated power line frequency. The speed of a motor is determined by the number of poles in the winding. A four-pole, 60Hz motor runs at an approximate speed of 1,725 rpm. A six-pole, 60Hz motor runs at an approximate speed of 1,140 rpm.

Rotation: For single-phase motors, the standard rotation, unless otherwise noted, is counterclockwise facing the lead or opposite shaft end. All motors can be reconnected at the terminal board for opposite rotation unless otherwise indicated.

Synchronous speed: When the speed of the rotor is equal to the speed of the stator. The speed is determined by multiplying 120 times the frequency divided by the number of poles.

Thermal protector: A protective device for assembly as an integral part of a motor or motor compressor that, when properly applied, protects the motor against dangerous overheating due to overload or failure to start.

Varying duty: Operation at varying loads and/or intervals of time.

Additional Resources

This module is intended to present thorough resources for task training. The following reference works are suggested for further study. These are optional materials for continued education rather than for task training.

American Electricians' Handbook, Latest Edition. New York: Croft and Summers, McGraw-Hill.

National Electrical Code® Handbook, Latest Edition. Quincy, MA: National Fire Protection Association.

CONTREN® LEARNING SERIES – USER UPDATE

NCCER makes every effort to keep these textbooks up-to-date and free of technical errors. We appreciate your help in this process. If you have an idea for improving this textbook, or if you find an error, a typographical mistake, or an inaccuracy in NCCER's Contren® textbooks, please write us, using this form or a photocopy. Be sure to include the exact module number, page number, a detailed description, and the correction, if applicable. Your input will be brought to the attention of the Technical Review Committee. Thank you for your assistance.

Instructors – If you found that additional materials were necessary in order to teach this module effectively, please let us know so that we may include them in the Equipment/Materials list in the Annotated Instructor's Guide.

Write: Product Development and Revision
National Center for Construction Education and Research
3600 NW 43rd St., Bldg. G, Gainesville, FL 32606

Fax: 352-334-0932

E-mail: curriculum@nccer.org

Craft

Module Name

Copyright Date   Module Number   Page Number(s)

Description

(Optional) Correction

(Optional) Your Name and Address

Electric Lighting

Walt Disney World's Wilderness Lodge

The Wilderness Lodge is modeled after the Old Faithful Inn that was built in Yellowstone National Park in 1902; it even includes a functional reproduction of Old Faithful Geyser. Installation of the electrical work involved highly detailed coordination to ensure that the various lighting and power systems would be concealed and not detract from the period feel of the building.

26203-08

26203-08
Electric Lighting

Topics to be presented in this module include:

1.0.0	Introduction	3.2
2.0.0	Human Vision	3.2
3.0.0	Light Characteristics	3.3
4.0.0	Lamps	3.6
5.0.0	Ballasts	3.17
6.0.0	Lighting Fixtures	3.24
7.0.0	Lighting Fixture Installation	3.28
8.0.0	Controls for Lighting	3.47
9.0.0	Energy Management Systems	3.49

Overview

Electric lighting is divided into four primary groups: incandescent, halogen, fluorescent, and high-intensity lighting. Incandescent lighting was the first type of lighting developed, but is the least energy efficient. A variation of standard incandescent lighting is the tungsten-halogen incandescent lamp, which provides up to 20% greater efficiency than incandescent lamps, as well as longer lamp life and improved light quality.

Fluorescent lighting provides a bright, even lighting and is much more energy efficient than incandescent lighting. In addition, fluorescent lamps may last up to 20 times longer than standard incandescent lamps. Fluorescent lamps are available in a variety of lamp shapes, wattages, and color-rendering qualities.

High-intensity discharge lamps (HID) demonstrate a high energy efficiency and long life, but the color of their light is not generally appealing to the eye. Three primary types of HIDs are mercury vapor, metal halide, and sodium, with each type displaying its own unique color of light.

Lighting fixtures, now referred to as luminaires in the *NEC®*, come in many designs and mounting configurations. They may be designed for recess, surface, pendant, lay-in, suspended, or track mounting. Lighting must be installed in accordance with the manufacturer's procedures and the *NEC®*.

Note: *NFPF 70®*, *National Electrical Code®*, and *NEC®* are registered trademarks of the National Fire Protection Association, Inc., Quincy, MA 02269. All *National Electrical Code®* and *NEC®* references in this module refer to the 2008 edition of the *National Electrical Code®*.

Objectives

When you have completed this module, you will be able to do the following:

1. Describe the characteristics of light.
2. Recognize the different kinds of lamps and explain the advantages and disadvantages of each type:
   - Incandescent
   - Halogen
   - Fluorescent
   - High-intensity discharge (HID)
3. Properly select and install various lamps into lighting fixtures.
4. Recognize and describe the installation requirements for various types of lighting fixtures:
   - Surface-mounted
   - Recessed
   - Suspended
   - Track-mounted
5. Recognize ballasts and describe their use in fluorescent and HID lighting fixtures.
6. Explain the relationship of Kelvin temperature to the color of light produced by a lamp.
7. Recognize basic occupancy sensors, photoelectric sensors, and timers used to control lighting circuits and describe how each device operates.

Trade Terms

Ballast
Color rendering index (CRI)
Dip tolerance
Efficacy
Incandescence
Incident light
Lumen (lm)
Lumen maintenance
Lumens per watt (LPW)
Luminaire (fixture)
Luminance
Reflected
Reflected light
Refracted/refraction
Starter
Troffer

Required Trainee Materials

1. Pencil and paper
2. Appropriate personal protective equipment
3. Copy of the latest edition of the *National Electrical Code®*

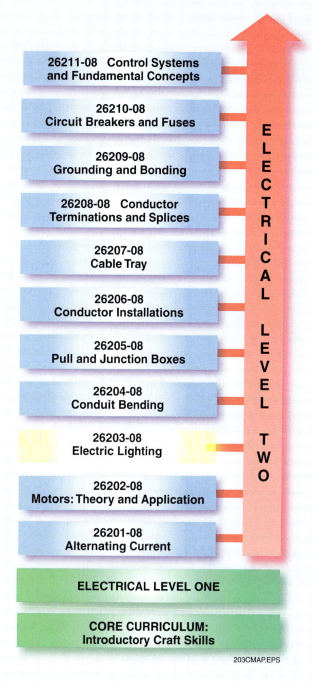

Prerequisites

Before you begin this module, it is recommended that you successfully complete *Core Curriculum; Electrical Level One;* and *Electrical Level Two*, Modules 26201-08 and 26202-08.

This course map shows all of the modules in *Electrical Level Two*. The suggested training order begins at the bottom and proceeds up. Skill levels increase as you advance on the course map. The local Training Program Sponsor may adjust the training order.

1.0.0 ♦ INTRODUCTION

Electric lighting is used extensively throughout residential structures, commercial businesses, industrial plants, and outdoor sites. Its use in residential structures can improve their appearance and that of the objects in them. In stores, hotels, and office buildings, electric lighting is used on a large scale to improve the efficiency of the employees and aid in selling merchandise. In industry, electric lighting helps to increase production, reduce errors, and increase safety. Regardless of the application, correct electric lighting serves to provide illumination for the performance of visual tasks with a maximum of comfort and a minimum of eyestrain and fatigue, allowing individuals to perform their daily living and work-related tasks more easily.

This module is the first of two modules that cover the subject of electric lighting. It introduces the basic principles of human vision and seeing and describes the characteristics of light. It also covers different kinds of lamps, lighting fixtures, and related components. The remainder of this module focuses on basic guidelines and procedural information for the receiving, storing, handling, and installation of electric lamps and lighting fixtures. It also provides an overview of common lighting circuit control devices and energy management systems. The second module, presented later in your training, will cover the practical applications of lighting fixtures.

2.0.0 ♦ HUMAN VISION

Human vision is a process that occurs partly in the eye and partly in the brain. Light **reflected** from objects stimulates the eye. This stimulation is conveyed to the brain, where it is registered as a conscious sensation.

The structure of the eye is similar in many ways to a camera, which consists of a lens system, a variable diaphragm, and film. The variable diaphragm is the iris of the eye (*Figure 1*), and the film is the retina. Light enters the eye through a transparent layer called the cornea. The amount of light that is allowed to strike the lens is controlled by the contraction and expansion of the iris. During low-light levels, the iris expands, and during high-light levels, it contracts.

Light passes through the pupil, which is the opening of the iris, and then through the lens, which is directly behind the iris. There, the light is focused to form an image that is passed through a transparent, jelly-like substance, called the vitreous humor, to the back wall or retina of the eye. The light on the retina stimulates nerve terminals

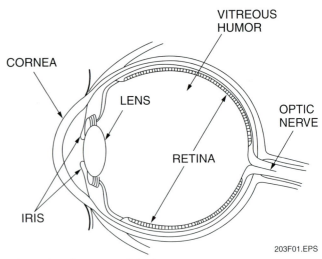

Figure 1 ♦ Structure of the human eye.

called rods and cones. These nerve terminals are connected to the brain by the optic nerve, which furnishes the path by which the light impulses are transferred from the eye to the brain.

The field of vision covers a wide angle of about 200° horizontally and 120° vertically. In the central region of the field of vision, the eye is responsive to color and detail, whereas in the outer region, it is chiefly sensitive to motion. The limits of vision are determined mainly by the following factors:

- *Intensity threshold* – The lowest brightness level that can stimulate the eye.
- *Contrast* – The difference in the degree of **luminance** (brightness).
- *Visual angle* – When an object is made smaller or is placed at a greater distance from the eye, the angle formed by the light rays from the extremities of the object to the eye becomes smaller (*Figure 2*), and vice versa.
- *Time threshold* – The minimum time during which a light stimulus must be present in order to be effective. If the interval is too short, the rods and cones of the eye do not have time to respond to an image on the retina. The time threshold is also dependent on the size, brightness, and color of the object being observed.

Other characteristics of vision are:

- High contrast, sharp edges, and motion increase the sensitivity to detail.
- Straight lines are easier to see than curved lines. Horizontal or vertical lines are more easily seen than diagonal lines.
- The background surrounding an object changes its appearance. For example, a gray object appears lighter when it is placed upon a black background, but it appears darker when placed on a white background.

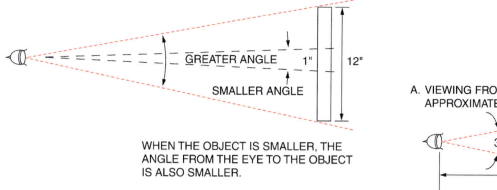
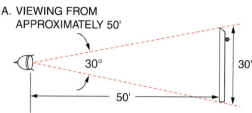
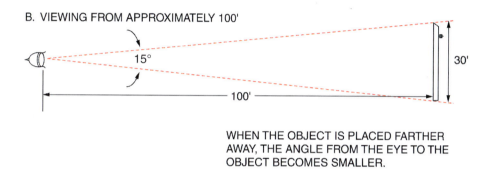

Figure 2 ♦ Visual angle.

3.0.0 ♦ LIGHT CHARACTERISTICS

When discussing light, we normally think of natural light from the sun or the light emitted from a light source, such as an electric light bulb. This type of light is called **incident light**. Another type of light is **reflected light**, which is light given off by an object when incident light strikes it. The difference between the two is that reflected light is dependent on the incident light. When light is not shining on an object, light will not be given off unless the object contains self-luminating properties.

3.1.0 Absorption, Reflection, and Refraction of Light

Three things can happen when light strikes objects or surfaces. Some of the light can be absorbed, some can be transmitted through the object media and **refracted** or diffused, and some can be reflected back from the surface (*Figure 3*). Absorption occurs when light rays pass through a transparent or translucent medium or meet a dense body, such as an opaque reflector surface. The amount of energy absorbed depends on the object's molecular construction, the wavelength

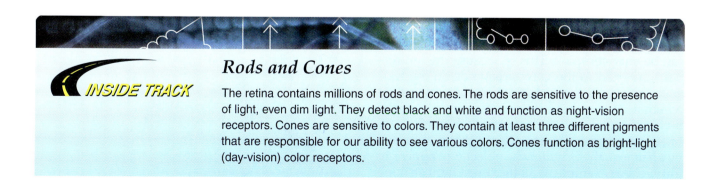

Rods and Cones

The retina contains millions of rods and cones. The rods are sensitive to the presence of light, even dim light. They detect black and white and function as night-vision receptors. Cones are sensitive to colors. They contain at least three different pigments that are responsible for our ability to see various colors. Cones function as bright-light (day-vision) color receptors.

Visible Light Spectrum

Can you think of one example that sometimes occurs in nature that confirms the fact that white light is actually made up of different colors?

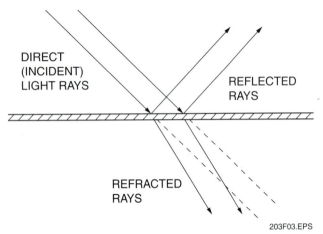

Figure 3 ♦ Simplified example of reflected and refracted light rays.

of the incident light striking it, and the angle at which the light strikes it. Light energy absorbed by an object is dissipated in the form of heat. This explains why walls, floors, metals, etc., feel warm or hot when exposed to direct sunlight or light from a strong artificial light source.

The energy not absorbed by an object when light strikes it can be transmitted through the object or reflected back from its surface, or both. If light rays strike the object perpendicular (at 90°) to its surface, the rays will be transmitted through it in a straight line and/or be reflected back from its surface in the same direction from which it came. If the light strikes the object's surface at an angle other than 90°, as shown in *Figure 3*, then the light transmitted through the object is bent in a different direction, called refraction, and/or the light is reflected back from the object in a different direction from that in which it came. It should be pointed out here that the resultant directions taken by reflected and/or refracted light waves can be very complex. This is because the redirection imparted to the waves depends on the type and density of the object's material, as well as whether its surface is smooth or rough, or if the material contains particles beneath its surface.

The larger the amount of light that is reflected by an object, the brighter the object will appear to the eye. For example, if all the direct light striking an object's surface is reflected from the surface, the object appears light; if it is entirely absorbed, the object appears dark. In addition, the more intense the light source, the brighter the object will become. For example, if you cast a shadow on a portion of an object and look at the difference in brightness between the two areas, the portion without the shadow will be brighter.

Some surfaces and materials reflect light better than others. Light-colored or highly-polished surfaces reflect more light than darker-colored or dull surfaces. This is because white and light-colored materials absorb less of the light energy than do dark or dull surfaces, thus leaving more energy available to be reflected. The color of the walls, ceilings, and floors and their reflecting ability are major considerations in interior lighting design. Typically, darker areas require the use of more artificial light. *Table 1* shows some typical examples of the percentages of light that will be reflected by common surface materials.

3.2.0 Light Colors

Any energy that travels by wave motion is considered radiant energy. Light is one of the many forms of radiant energy. Radio waves and X-rays are also forms of radiant energy. Visible light occupies only a small portion of the radiant energy spectrum. Light is made up of that portion of the

Table 1 Percentages of Light Reflected by Common Surface Materials

Surface Material	Percentage of Light Reflected
White plaster	90% to 92%
Mirrored glass	80% to 90%
White paint	75% to 90%
Metalized plastic	75% to 85%
Polished aluminum	75% to 80%
Stainless steel	55% to 65%
Limestone	35% to 65%
Marble (white)	45%
Concrete	40%
Dark red–glazed bricks	30%

spectrum encompassing wavelengths between 380 and 780 nanometers (*Figure 4*). A nanometer is one billionth of a meter.

When all the wavelengths of the light spectrum are presented to the eye in nearly equal proportions, white light is seen. This white light is made up of the wavelengths that create the different colors. This composition can be demonstrated by passing light through a prism. As shown in *Figure 5*, the light spectrum is broken up into its component wavelengths, with each representing a different color. The ability of the prism to separate the different wavelengths, and therefore colors, is because the white light is dispersed into its component wavelengths by refraction as it passes through the prism. The spectrum ranges from violet on the lower end to red on the upper end. In between fall blue, green, yellow, and orange. A total of six colors are visible when passing white light through a prism. Since the colors of the spectrum pass gradually from one to the other, the theoretical number of colors becomes infinite. It has been determined that about 125 colors can be identified over the visible spectrum.

As previously described, the amount of light energy absorbed by an object depends on the wavelengths of the incident light striking the object. All objects absorb light of different wavelengths in different proportions. This is called selective absorption. Practically all colored objects owe their color to selective absorption in some part of the visible light spectrum, with resulting reflection and transmission in other selected parts of the spectrum. It is this characteristic that gives us our color distinction when viewing different objects, surfaces, etc. An object's appearance results from the way it reflects the particular light that is falling on it. For example, suppose an apple is red. Under white light, the apple appears red because it tends to reflect light in the red portion of the spectrum and absorb light of other wavelengths (colors).

Now that we have covered the basic concepts of light and how it is perceived, we will discuss the

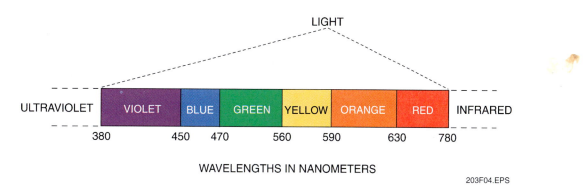

Figure 4 ◆ Visible light spectrum.

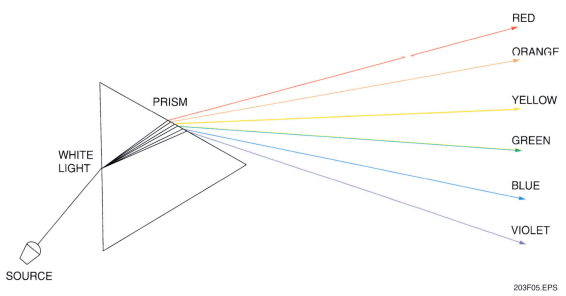

Figure 5 ◆ Light being separated into its component colors by refraction.

various electric lamps and luminaires (fixtures) that are used to provide light in residential, commercial, and industrial applications.

4.0.0 ◆ LAMPS

There are three main categories of lamps used for lighting: incandescent, fluorescent, and HID. Each category of lamp is made in a wide variety of shapes, sizes, finishes, and mounting bases.

4.1.0 Standard Incandescent Lamps

Incandescent lamps, also called filament lamps, are used for general lighting and typically provide a warm and natural light. Incandescent lamps were invented over a century ago. With some refinements, the basic construction of a standard incandescent lamp remains the same today. Incandescent lamps consist of a thin coiled or shaped tungsten-wire filament (*Figure 6*) supported inside an evacuated glass envelope (bulb) filled with an inert gas, typically a mix of argon and nitrogen. The inert gas helps to prevent the filament from combining with oxygen and burning out. The envelopes of most lamps are made of regular lead or soda lime (soft) glass. Envelopes of lamps that must withstand higher temperatures are typically made of borosilicate heat-resistant (hard) glass. The lamp's base supports the lamp envelope and filament and provides the electrical connection between the lamp and its power source.

In an incandescent lamp, light is generated by passing an electric current through the filament, and its resistance causes it to heat to incandescence. The hotter the filament gets, the more efficient it is in converting electricity to light output. Tungsten has a positive resistance characteristic that makes its resistance at operating temperature much greater than its cold resistance (typically 12 to 16 times greater). It should be pointed out that when a filament operates hotter, its life is shortened. This makes the design of each type of lamp a trade-off between efficiency and lamp life. This is why lamps of equal wattage may have different lumen (lm) and life ratings.

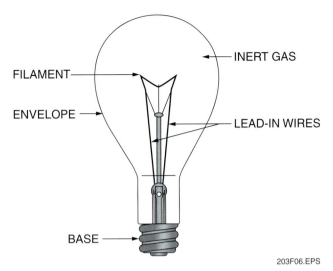

Figure 6 ◆ Components of an incandescent lamp.

Of all the lamp types, standard incandescent lamps are the most inefficient. Because they produce light by heating the filament until it glows, most of the energy they consume is given off as heat, resulting in a low efficiency (efficacy) of typically 5 to 22 lumens per watt (LPW). Incandescent lamps also have the shortest life expectancy of all lamp types, typically between 750 and 2,000 average hours, depending on the type. This is because tungsten from the filament evaporates over time and is deposited on the walls of the bulb, thus reducing the light output. Also, the filament gets thinner and thinner with use and eventually breaks, causing the lamp to fail.

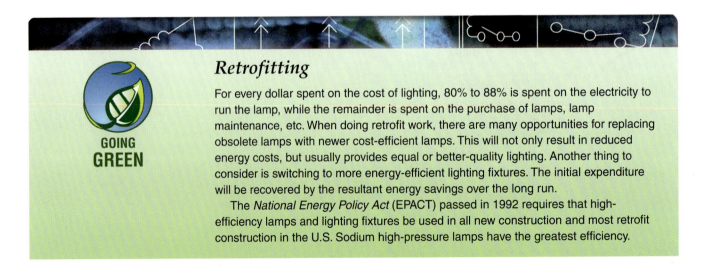

Retrofitting

For every dollar spent on the cost of lighting, 80% to 88% is spent on the electricity to run the lamp, while the remainder is spent on the purchase of lamps, lamp maintenance, etc. When doing retrofit work, there are many opportunities for replacing obsolete lamps with newer cost-efficient lamps. This will not only result in reduced energy costs, but usually provides equal or better-quality lighting. Another thing to consider is switching to more energy-efficient lighting fixtures. The initial expenditure will be recovered by the resultant energy savings over the long run.

The *National Energy Policy Act* (EPACT) passed in 1992 requires that high-efficiency lamps and lighting fixtures be used in all new construction and most retrofit construction in the U.S. Sodium high-pressure lamps have the greatest efficiency.

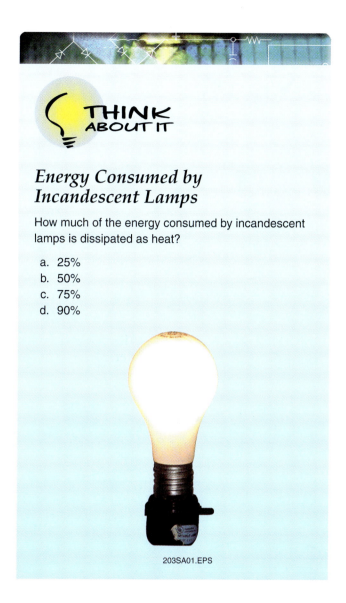

Think About It

Energy Consumed by Incandescent Lamps

How much of the energy consumed by incandescent lamps is dissipated as heat?

a. 25%
b. 50%
c. 75%
d. 90%

Incandescent lamps are made in numerous sizes and shapes, with different filament form arrangements and mounting bases. *Figure 7* shows some examples of typical lamp sizes and shapes. Lamps are identified by a letter referring to their shape and a number that indicates the maximum diameter stated in eighths of an inch. For example, A-40 identifies a lamp with an A-shape that is $^{40}/_{8}"$ (5') in diameter. Lamps with standard, tubular, and similar envelope shapes provide lighting in all directions (omnidirectional). Those with shapes designated as R, ER, and PAR are all reflector-type lamps. They direct their light out in front by reflecting it from their cone-shaped inside walls.

Figure 8 shows some examples of incandescent lamp filament forms used by one lamp manufacturer. The form of a filament is determined mainly by service requirements. Filament forms are identified by a letter or letters followed by an arbitrary number. Commonly used letters are C (coiled), indicating that the filament wire is wound into a helical coil; CC (coiled coil), indicating the coil itself is wound into a helical coil; and S (straight), indicating that the filament wire is uncoiled. The numbers shown in *Figure 8* indicate the arrangement of the filament on the supports.

The lamp base supports the lamp and provides the connection between the lamp and the power source. *Figure 9* shows some examples of common incandescent lamp bases.

Incandescent lamps and other types of lamps are available in many different voltage and wattage ratings. When installing incandescent lamps, it is important to select the correct voltage rating. This is because a small difference between the rating and the actual supply voltage has a great effect on the lamp life and light output. The wattage rating is an indication of the consumption of electrical energy used by a lamp to produce its rated light output. It is not a measure of the lamp's light output. For example, a standard 100W lamp may produce 1,600 lumens, while a typical 32W fluorescent lamp may produce about 2,600 lumens. *Figure 10* shows the relationships between watts, lumens, and lamp life for a typical lamp when operated at different percentages of its rated voltage. For example, reducing the supply voltage for a 120V lamp to approximately 94% (113V) will increase its life by 220%, but reduce the light output to 80% and the wattage to about 90%.

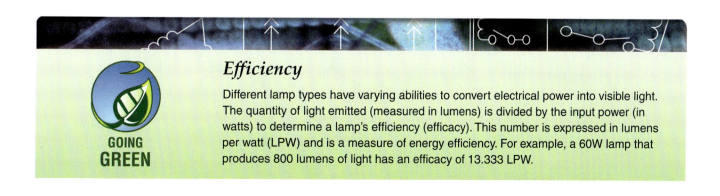

Efficiency

Different lamp types have varying abilities to convert electrical power into visible light. The quantity of light emitted (measured in lumens) is divided by the input power (in watts) to determine a lamp's efficiency (efficacy). This number is expressed in lumens per watt (LPW) and is a measure of energy efficiency. For example, a 60W lamp that produces 800 lumens of light has an efficacy of 13.333 LPW.

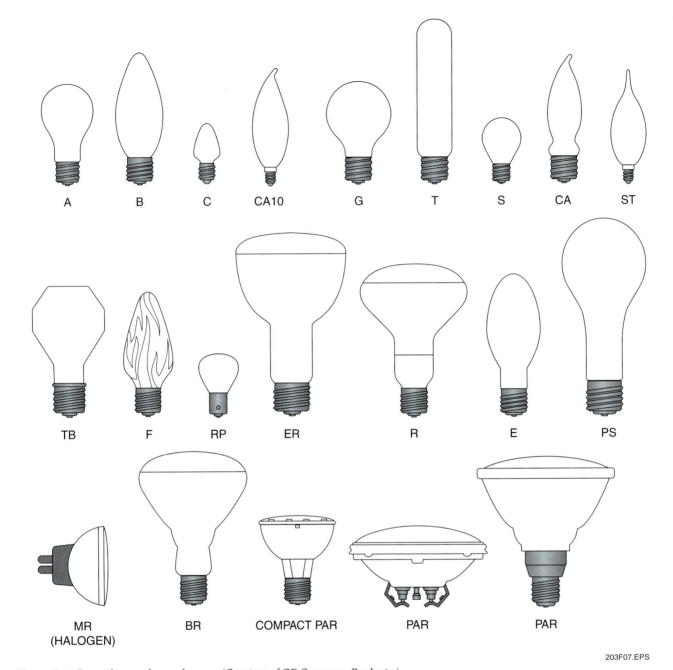

Figure 7 ♦ Incandescent lamp shapes. *(Courtesy of GE Consumer Products.)*

Figure 8 ♦ Examples of incandescent lamp filament forms. *(Courtesy of GE Consumer Products.)*

3.8 ELECTRICAL LEVEL TWO ♦ TRAINEE GUIDE

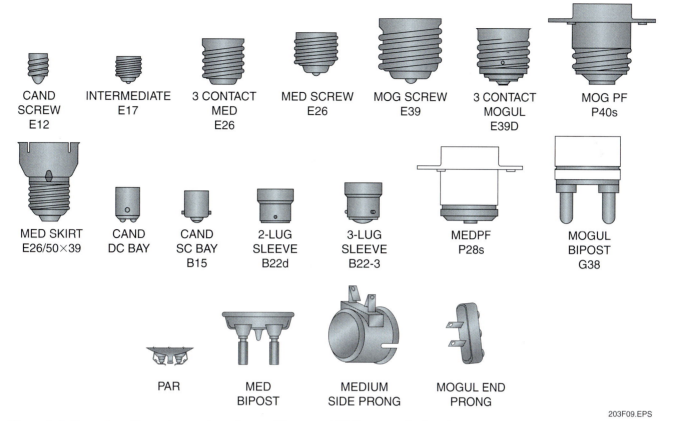

Figure 9 ♦ Examples of incandescent lamp bases. *(Courtesy of GE Consumer Products.)*

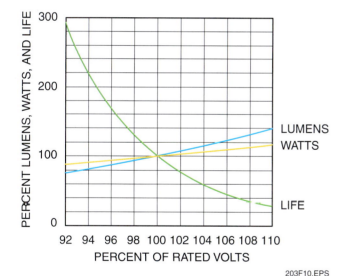

Figure 10 ♦ Relationship of rated lamp voltage to watts, lumens, and lamp life.

4.1.1 Tungsten Halogen Incandescent Lamps

Tungsten halogen incandescent lamps (halogen lamps) are a refinement over the standard incandescent lamp. Like standard incandescent lamps, halogen lamps are made in many sizes, shapes, and wattages. When compared to the standard incandescent lamp, they provide greater efficacy (12 to 36 LPW), longer service life (2,000 to 5,000 average hours), and improved light quality. Their light output contains more blue and less yellow than standard incandescent lamps, making their light appear whiter and brighter.

Halogen lamps (*Figure 11*) typically have a short, thick tungsten filament encased in a capsule filled with halogen gases, such as iodine or bromine, that allow the filaments to operate at higher temperatures than a standard incandescent lamp. This increases their efficacy (LPW) by more than 20%. The use of halogen gas in the lamp accounts for the longer life and excellent **lumen maintenance.** During operation, tungsten atoms evaporated from the filament combine with halogen atoms to form a gaseous compound that circulates inside the lamp, causing the tungsten atoms to be redeposited on the hot filament, rather than on the inside surface of the lamp envelope. The halogen atoms are then released, allowing them to combine with additional tungsten atoms, thus repeating the process. This action slows down any deterioration of the filament, thereby improving lumen maintenance and extending the lamp's life.

Because of the higher filament operating temperatures, there is more ultraviolet (UV) radiation general lighting applications, it is recommended that lighting fixtures for halogen lamps have a lens or cover glass that will, in addition to providing the required safety protection, filter out most of the UV radiation.

> **CAUTION**
> Operating halogen lamps at voltages above and below the manufacturer's recommendations can have adverse effects on the internal chemical process because the temperature will differ from the design value. Also, it is important to follow the manufacturer's instructions as to burning position, lamp handling, and lighting fixture temperatures.

4.2.0 Fluorescent Lamps

Fluorescent lamps are low-pressure mercury discharge lamps that are very energy efficient (75 to 100 LPW) and have a long service life (12,000 to more than 24,000 average hours). Each requires a **ballast** to effectively start the lamp and regulate its operation. Light is produced by passing an electric arc between two tungsten cathodes at opposite ends of a glass tube filled with a low-pressure mercury vapor and other gases (*Figure 12*). The arc excites the atoms of mercury. This generates UV radiation, which causes the phosphor coating on the inside of the tube to fluoresce and produce visible light. By using different phosphor coatings, the spectral light output of a fluorescent lamp can be made to produce warm, intermediate, or cool color temperatures. The color temperatures of lamps are covered later in this module.

Fluorescent bulbs are made in straight, U-bent, circular, and compact varieties, several of which are shown in *Figure 13a*. Not only do they come in

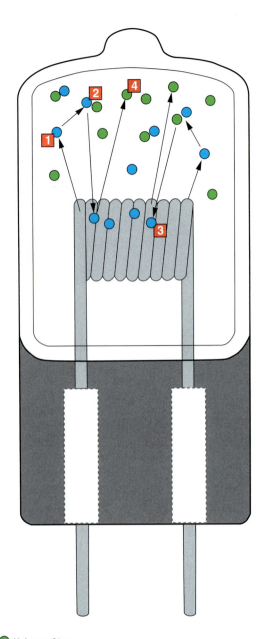

○ Halogen Atoms
○ Tungsten Atoms

1. Tungsten atoms evaporate from filament.
2. Tungsten atoms combine with halogen atoms.
3. Gaseous compound returns to hot filament, redepositing tungsten atoms.
4. Halogen atoms are released to combine with additional tungsten atoms.

Figure 11 ♦ Basic tungsten halogen lamp.

generated from a halogen lamp than from standard incandescent lamps. The amount of UV radiation emitted is determined by the lamp envelope material. Fused quartz and high-silica glass transmit most of the UV radiation emitted by the filament; special high-silica glass and aluminosilicate glasses absorb UV radiation. For

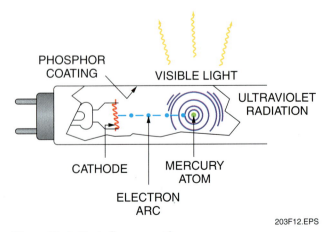

Figure 12 ♦ Basic fluorescent lamp.

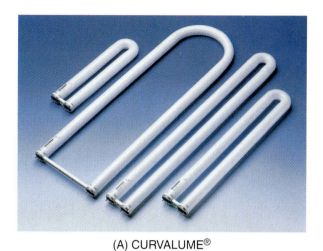
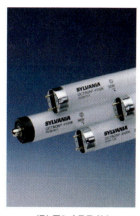

(A) CURVALUME®   (B) T8 ARRAY   (C) T12 ARRAY

Figure 13 ♦ Typical fluorescent lamps.

Bi-Pin Lamps

Most bi-pin lamps, like the one shown here, have indents or index marks to aid in the proper alignment when inserting into a fixture. When properly inserted, the indents or marks will be perpendicular to the lamp holders.

a wide variety of wattages, sizes, and bases, but they are also available in several color temperatures and color rendition capabilities. Fluorescent lamps are designated by the letter T followed by the diameter of the lamp tube expressed in eighths of an inch. They vary in diameter from T-5 (⅝") to PG (power groove)-17 (2⅛"). In overall length, straight fluorescent lamps range from 6" to 96". Higher wattages go with longer tubes. For example, a 20W straight T-12 tube is shorter than a 40W T-12 tube.

Fluorescent lamps have two electrical requirements. To start the lamp, a high-voltage surge is needed to establish an arc in the mercury vapor. Once the lamp is started, the gas offers a decreasing amount of resistance, which means that current must be regulated to match this drop. Otherwise, the lamp would draw more and more power and rapidly burn itself out. This is why fluorescent lamps are operated in lighting fixture circuits containing a ballast that provides the required voltage surge at startup and then controls the subsequent flow of current to the lamp. Ballasts are covered in detail later in this module.

There are three electrical classes of fluorescent lamps and lighting fixtures: preheat, rapid start, and instant start. The term preheat refers to a lighting fixture circuit used with fluorescent lamps wherein the lamp electrodes are heated or warmed to a glow stage by a replaceable **starter** separate from the ballast. When power is applied to the lighting fixture, the starter functions to preheat the lamp's cathodes before the lamp is started. Note that preheat lamps and lighting fixtures are nearly obsolete and are not used in new construction. The term rapid start refers to a lighting fixture circuit designed to start the lamp by continuously heating or preheating the lamp electrodes by means of heater windings built into the ballast. Unlike the

NOTE
Although the bases on HO and VHO lamps are the same, each must be matched to the correct ballast.

Fluorescent Lamps

The life of a fluorescent lamp is affected by the number of times the lamp is started. Frequent switching on and off of fluorescent lamps results in shorter lamp life, while continuous operation provides the longest lamp life. The typical fluorescent lamp ratings given by one major manufacturer of fluorescent lamps are based on three hours per start. For example, if a lamp is rated as having an average life of 12,000 hours, this is based on the lamp being turned on 4,000 times and being left on for a minimum of at least three hours.

As fluorescent lamps age, they tend to darken at the ends. This helps in identifying lamps for future replacement.

---

preheat circuit, the ballast does not require a separate starter, but the fixture must be properly grounded. During operation, standard rapid-start lamps draw about 430mA of current. The term instant start refers to a lighting fixture circuit used to start specially designed lamps without the aid of a starter. To strike the arc instantly, the circuit uses a higher open circuit voltage than is required for a preheat or rapid-start lamp of the same length, a voltage that is approximately three times the normal lamp operating voltage. Both preheat and rapid-start lamps have a bi-pin (2-pin) base at each end. Instant-start lamps have a single pin at each end of the lamp. Normally, lamps identified as preheat, rapid-start, or instant-start types should be used only with the corresponding type of ballast. Other terms commonly used to designate types of fluorescent lamps include:

- *Slimline lamps* – A group of instant-start lamps with single-pin bases.
- *High-output (HO) lamps* – A group of rapid-start lamps designed to operate at higher operating currents (800mA to 1,000mA) that produce higher levels of light output. Because of the higher operating currents involved, the lamps have a recessed double-contact base. HO lamps are typically used in industrial areas and retail stores with high ceilings.
- *Very high-output (VHO) lamps* – A group of rapid-start lamps designed to operate at high currents (1,500mA) in order to produce high light output levels. Because of the higher operating currents involved, the lamps have a recessed double-contact base. VHO lamps are typically used in factories, warehouses, gymnasiums, and open areas.
- *Compact lamps* – Lamps made of ½" to ⅝" single or multiple U-shaped tubes that terminate in a plastic base (*Figure 14a*). The base contains the cathodes and, in some versions, a magnetic or electronic ballast. Some have replaceable tubes.

A few of the newer versions are dimmable, and others look much like a standard lamp in size and shape. They are used both to replace incandescent lamps and in lighting fixtures designed for their use. They can provide up to 75% energy cost savings when compared to incandescent lamps of comparable light output. They also have a lifespan of up to 13 times longer than standard incandescent lamps.

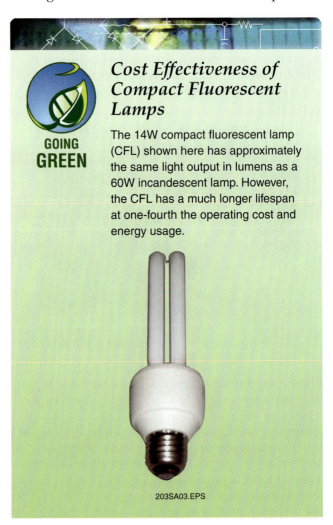

Cost Effectiveness of Compact Fluorescent Lamps

The 14W compact fluorescent lamp (CFL) shown here has approximately the same light output in lumens as a 60W incandescent lamp. However, the CFL has a much longer lifespan at one-fourth the operating cost and energy usage.

When fluorescent lamps are used in circuits providing an open circuit voltage in excess of 330V, or in circuits that may permit a lamp to ionize and conduct current with only one end inserted in the lamp holder, electrical codes require some automatic means for de-energizing the circuit when the lamp is removed. This is usually accomplished by the lamp holder so that upon removal, the ballast primary circuit is opened. Note that the use of recessed contact bases for HO and VHO lamps has eliminated the need for this disconnect feature in the lamp holders for these lamps.

4.3.0 High-Intensity Discharge (HID) Lamps

High-intensity discharge (HID) lamps provide long life and high efficiency (*Figure 16*). They are somewhat similar to fluorescent lamps in that they produce light when electricity excites specific gases in pressurized bulbs. An arc is established between two electrodes in a gas-filled tube, which causes mercury vapor to produce radiant energy. Unlike a fluorescent lamp, a combination of factors shifts the wavelength of much of the energy to within the visible range, so light is produced without phosphors. First, the electrodes are only a few inches apart at opposite ends of a sealed arc tube, and the gases in the tube are highly pressurized. This allows the arc to generate extremely high temperatures, causing metallic elements within the gas atmosphere to vaporize and release large amounts of visible radiant energy. Like fluorescent lamps, HID lamps must be used in matched lighting fixtures with a ballast specifically designed for the lamp type and wattage. In addition, HID lamps require a warmup period to achieve full light output.

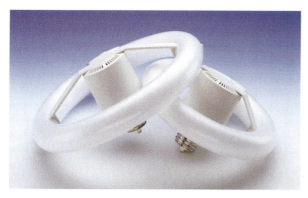

(A) CIRCLINE FLUORESCENT

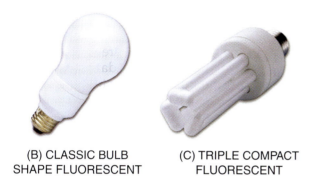

(B) CLASSIC BULB SHAPE FLUORESCENT

(C) TRIPLE COMPACT FLUORESCENT

203F14.EPS

Figure 14 ♦ Compact fluorescent lamps.

- *T-5 lamps* – Newer and more energy-efficient lamps than T-8 or T-12 lamps. They are the first type of linear lamp to use only electronic ballasts, and produce nearly twice the light output of a T-8 or T-12 system. In addition, they use an improved triphosphor coating that allows them to maintain higher light levels over the life of the lamp.

Lamp holders (*Figure 15a*) are made in several variations for each lamp base style to allow for various spacings and mounting methods in fixtures.

(A) BI-PIN SLIDE-ON AND SCREW MOUNT LAMP HOLDER

(B) HIGH-OUTPUT (HO) LAMP HOLDER

(C) SINGLE-PIN (SLIMLINE) LAMP HOLDER

203F15.EPS

Figure 15 ♦ Typical fluorescent lamp holders.

Figure 16 ♦ Typical high-intensity discharge lamps.

There are three types of HID lamps: mercury vapor, metal halide, and high-pressure sodium. The names refer to elements that are added to the gases in the lamp, which cause each type to have somewhat different color characteristics and efficiency. Mercury vapor lamps are the oldest HID technology. They are energy efficient (50 to 60 LPW) and have a long service life (12,000 to more than 24,000 average hours). These lamps produce light energy by radiation from excited mercury vapor in both the visible and ultraviolet range. They normally have specially formulated glass outer jackets to filter the UV energy. The phosphor coatings used in some mercury vapor lamp types add additional light and improve color rendering. Today, the use of mercury vapor lamps is limited mainly to the replacement of existing lamps and landscape lighting of evergreen trees. Other HID lamps that have better efficiency and color properties are being used for new construction.

Metal halide lamps are the most energy-efficient source of white light. They have high efficacy (80 to 115 LPW), excellent color rendition, long service life (10,000 to more than 20,000 average hours), and good lumen maintenance (longevity). The metal halide HID lamp combines mercury and metal halide atoms under high pressure (*Figure 17A*). In the arc stream, these atoms generate both UV radiation and visible light. A special glass bulb filters the UV radiation without affecting the visible light.

If the arc tube is ruptured in an open metal halide lamp, the outer bulb may explode, causing personal injury or a fire. Per *NEC Section 410.130(F)(5)*, all metal halide lamps except thick-glass PARs must either have a containment barrier (lens) or be Type O with a protected socket. Type O lamps have an internal shroud that protects the outer bulb from damage if the arc tube is ruptured (*Figure 17B*). The protected socket prevents the use of any lamp except a Type O lamp.

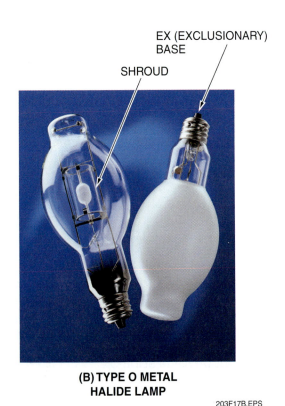

Figure 17 ♦ Basic metal halide lamp.

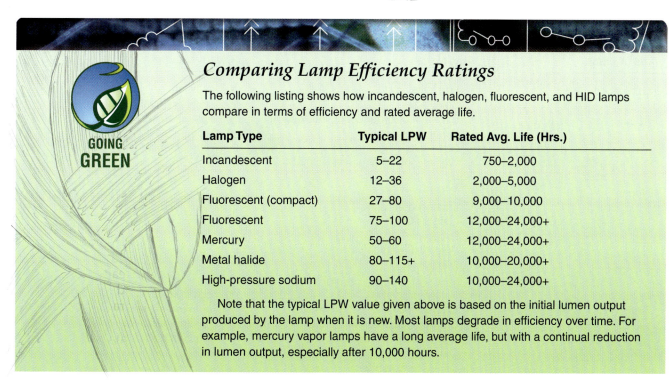

Comparing Lamp Efficiency Ratings

The following listing shows how incandescent, halogen, fluorescent, and HID lamps compare in terms of efficiency and rated average life.

Lamp Type	Typical LPW	Rated Avg. Life (Hrs.)
Incandescent	5–22	750–2,000
Halogen	12–36	2,000–5,000
Fluorescent (compact)	27–80	9,000–10,000
Fluorescent	75–100	12,000–24,000+
Mercury	50–60	12,000–24,000+
Metal halide	80–115+	10,000–20,000+
High-pressure sodium	90–140	10,000–24,000+

Note that the typical LPW value given above is based on the initial lumen output produced by the lamp when it is new. Most lamps degrade in efficiency over time. For example, mercury vapor lamps have a long average life, but with a continual reduction in lumen output, especially after 10,000 hours.

Low-pressure sodium lamps use sodium in a low-pressure arc stream and produce light that is limited to a single wavelength in the yellow portion of the spectrum. These lamps are the most efficient of any lamp type, but are used only where energy efficiency and long life are the only requirements. It should be pointed out here that technically speaking, low-pressure sodium lamps are not actually a type of HID lamp.

4.4.0 Lamp Color Rendering and Color Temperature Characteristics

Colors appear differently under various light sources. The color rendering index (CRI), a scale from 0 to 100, is used by lamp manufacturers to indicate how normal and natural a specific lamp makes objects appear. Generally, the higher the CRI, the better it makes people and objects appear. Note that the CRI of different lamps can be compared only if the sources have approximately the same color temperature. Also, CRI differences among lamps are not usually visible to the eye unless the difference is greater than three to five points.

Lamps can create atmospheres that are warm or cool in appearance. The color temperature, expressed in kelvins (K), is one way lamp manufacturers describe the color tone (warmth or coolness) produced by a lamp. For example:

- Color temperatures of 3,000K and lower are described as warm in tone and slightly enhance reds and yellows.
- A color temperature of 3,500K is considered moderate in tone, producing a balance between warmth and coolness.
- Color temperatures of 4,100K and higher are considered cool in tone, slightly biased toward blues and greens.

Some typical color temperatures are 2,200K for high-pressure sodium lamps, 2,800K for incandescent lamps, 3,000K for halogen lamps, 3,500K for metal halide lamps, 4,100K for cool white fluorescent lamps, and 5,000K for daylight-simulating fluorescent lamps.

Fluorescent lamps have more options in terms of light quality than any other lamp type. This is because of the variations (formulations) available in the composition of the phosphor coating on the inside of the lamp tube. Early fluorescent lamps used a single halophosphor coating and could offer improved color quality with only a decrease in efficacy (LPW). With the newer lamps, triphosphor coatings are used that allow precise control over the generation of red, green, and blue (the primary colors of light). This enables the manufacture of high-LPW lamps in a variety of color temperatures that provide excellent color rendition.

Table 2 provides a brief comparison of the major lamp types.

Table 2 Comparison of Major Lamp Types

Type of Lamp	Advantages	Disadvantages
Incandescent	• Low initial cost • Small size • Excellent color rendering index (CRI) • Variety of shapes • Dimmable • Wattage interchangeable	• Inefficient • Excessive heat output • High operating costs • Short service life • Glare potential
Halogen	• Small size • Increased efficiency* • Longer life* • Excellent CRI • Bright white light • Dimmable	• Excessive heat output • Glare potential
Fluorescent	• Highly efficient • Long service life • Choice of color temperatures and CRI • Low operating costs • Low heat output • Diffuse light source	• High initial cost • Temperature sensitive • Limited optical control • Requires matched fixture/ballast
Metal halide (HID)	• Highly efficient • Long service life • Low operating costs • Good color rendering	• High initial cost • Requires matched fixture/ballast • Long startup/restrike period • Glare potential
High-pressure sodium (HID)	• Long service life • Exceptionally efficient • Very low operating costs • High lumen maintenance	• High initial cost • Requires matched fixture/ballast • Long startup/restrike period • Poor color rendering • Glare potential

*Compared to standard incandescent lamps

4.5.0 Guidelines for Installing Lamps

Listed below are some general guidelines to follow when installing lamps in their related lighting fixtures.

- Make sure that the lamp and lighting fixture are compatible. Install incandescent lamps in incandescent lighting fixtures, fluorescent lamps in fluorescent fixtures, etc.
- Make sure that the lamp or lamps being installed in a lighting fixture have the correct voltage rating for use with the fixture. This is especially important when installing incandescent lamps because a small difference between the rating and the actual voltage has a great effect on a lamp's life and light (lumen) output.
- Make sure that you do not install a lamp, or combination of lamps, in a lighting fixture that exceeds the wattage rating for the lighting fixture; otherwise, the fixture can overheat and may cause a fire. In recessed lighting fixtures with built-in thermal protectors, it will cause the lamps to cycle on and off as the fixture overheats, then cools down.
- When installing fluorescent and HID lamps, make sure that the lamp type and wattage are matched to the lighting fixture and ballast being used. Always install the lamps specified by the lighting fixture manufacturer.
- When installing bi-pin fluorescent lamps, make sure to rotate them ¼ turn to seat and connect them.
- Do not install lamps that have scratched or otherwise damaged envelopes.
- Use a soft cloth or gloves to handle an unencapsulated halogen tubular lamp. This prevents oil from your hands from contacting the bulb. This is important because its quartz glass walls can withstand high operating temperatures, but may crack if etched with oil. Also make sure that no combustible materials can come in contact with the lamp.

- Unlike standard lamps, halogen and metal halide lamps may continue to light even if the exterior lamp glass is broken. If this occurs, replace the lamp immediately to prevent creating a hazard or personnel injury.
- In some applications, such as food processing plants, bakeries, dairies, schools, warehouses, etc., safety sleeves made to fit over fluorescent lamps should be used in order to retain broken glass and phosphors if the lamp ever breaks, falls, or bursts.

In general, energy legislation promotes replacing standard incandescent lamps with halogen and compact fluorescent lamps, replacing incandescent reflector lamps with halogen PARs, and using higher-efficiency fluorescent lamps, such as T-8s, instead of full wattage types. If it is necessary to relamp existing light fixtures designed for use with a discontinued or inefficient type of lamp, replace the lamp with an approved energy-efficient substitute. Lists of approved substitutes are available at all lighting and electrical distributors.

5.0.0 ◆ BALLASTS

Fluorescent and HID fixtures require the use of a ballast. The main function of a ballast is to provide the voltage needed to strike an arc between the electrodes.

5.1.0 Fluorescent Lighting Fixture Ballasts

In fluorescent lighting fixtures, the ballasts perform the following functions:

- Provide the proper voltage to establish an arc between two electrodes
- Regulate the electric current flowing through the lamp to stabilize the light output
- Supply the correct voltage required for proper lamp operation and compensate for voltage variations in the electrical current
- Provide continuous voltage to maintain heat in the lamp electrodes while the lamp operates (rapid-start circuits)

Fluorescent ballasts (*Figure 18*) are made to operate in the three basic fluorescent lighting fixture operating circuits: preheat, rapid start, and instant start.

5.1.1 Preheat Lamp and Ballast Operation

In preheat circuits, such as the one shown in *Figure 19(A)*, the lamp electrodes (cathodes) are heated before application of the high voltage across the lamp(s). The preheating requires a few seconds, and the necessary delay is provided by an automatic switch called a starter. When power is first applied to the lamp circuit, the starter places the lamp's electrodes in series across the ballast, causing current to flow through both electrodes, heating them. After the electrodes are sufficiently preheated, the switch opens and applies the voltage across the lamp. Because the switch opens under load, a transient voltage (inductive spike) is developed in the circuit, which aids in the ignition of the lamp. Note that the first fluorescent lamps developed were of the preheat type. This type of lamp is now obsolete and is seldom used except in smaller sizes, such as those used in desk lamps and similar luminaires.

5.1.2 Rapid-Start Lamp and Ballast Operation

This is probably the most common type of lamp and luminaire used today. Lamps designed for rapid-start operation, such as the one shown in *Figure 19(B)*, normally have low-resistance electrodes. These remain energized by low voltage applied from the ballast while the lamps are in operation. They usually start in one second, the time required to bring the electrodes up to proper temperature. The standard rapid-start circuit operates with a typical lamp current of about 430mA. Rapid-start circuits used with high-output (HO) and very high-output (VHO) lamps draw currents of about 800mA and 1500mA, respectively. In some energy-saving circuits, the electrode voltage is reduced or disconnected after the starting of the lamps. Heating is accomplished through low-voltage windings built into the ballast or through separate low-voltage transformers designed for this purpose. Fluorescent lamps

Figure 18 ◆ Fluorescent ballast.

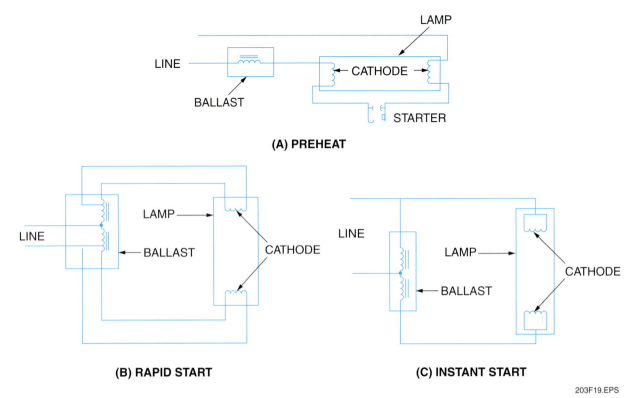

Figure 19 ◆ Basic fluorescent lighting fixture circuits.

used with rapid-start ballasts are bi-pin lamps. Rapid-start lamps can be dimmed using special dimming ballasts. These are covered later in this module.

Another version of a rapid-start ballast is the trigger-start ballast. Trigger-start ballasts are used with preheat fluorescent lamps up to 32W without the need for a starter.

5.1.3 Instant-Start Lamp and Ballast Operation

The lamp electrodes in instant-start lamps are not preheated. The ballasts provide a high voltage (100V to 1,000V) across the electrodes that causes electrons to be emitted from the electrodes, as shown in *Figure 19(C)*. These electrons flow through the tube, ionizing the gas and initiating an arc discharge. Thereafter, the arc current provides electrode heating. Because no preheating of electrodes is required, instant-start lamps need only a single contact at each end. Thus, a single pin is used on most instant-start lamps, commonly called slimline lamps. However, some bi-pin lamps can be operated with either rapid-start or instant-start electronic ballasts. When used with an instant-start ballast, the terminals in the lamp holders must be connected together. New fixtures come with the lamp holders wired in this way.

5.1.4 Types of Fluorescent Ballasts

Each fluorescent lamp must be operated by a ballast that is specifically designed to provide the proper starting and operating voltage required by the particular lamp. Lamp and lighting fixture manufacturers make a wide variety of ballasts designed for use in lighting fixtures that operate in all three of the lamp starting modes previously described. Ballasts are made for single-lamp, two-lamp, three-lamp, and four-lamp operation. The names used to identify the different types of ballasts can vary by manufacturer. Some common categories of ballasts are:

- *Standard ballast* – Lowest priced, least efficient, and highest wattage ballast. Their use is obsolete.
- *High-efficiency ballast* – Lower wattage, better efficiency, and longer life than a standard ballast.
- *Hybrid ballast* – Lower wattage, higher efficiency, and longer life than a standard high-efficiency ballast.
- *Electronic rapid-start ballast* – Low wattage and longest life available with various ballast factors and wattage packages for various lamp types.
- *Electronic instant-start ballast* – Lowest system wattage and highest system efficiency. The lamp life is slightly shorter than with rapid-start ballasts.

Rapid-Start Lamps
Why do some rapid-start lamps fail to ignite at low ambient temperatures?

High-efficiency ballasts can be of the magnetic or electronic type. The magnetic type usually contains coils, capacitors, transformers, and a thermal protector installed in a metal case. The coils and transformers are generally made with steel laminations and copper wire. Once assembled, the ballast components are encapsulated in the ballast case with a potting compound to improve the heat dissipation and reduce ballast noise.

If the ballasts are of the electronic type, they may be made with discrete electronic components and/or with integrated circuits. Some of those with integrated circuits are compatible for use with dimming systems, occupancy sensors, and daylight sensors. Electronic ballasts are quieter, more efficient, and weigh less, but are also more expensive. With electronic ballasts, the power input is 50Hz to 60Hz, but the ballast operates the lamps at 20kHz to 50kHz, with resulting improvements in ballast and lamp efficacy. The operating frequency is selected so that it is high enough to increase the lamp's efficacy and to shift the ballast noise to the inaudible range, but not so high as to cause electromagnetic interference (EMI) problems. Note that many electronic ballasts do not come equipped with wire leads. These ballasts are connected into the circuit using wire harnesses specifically designed for that purpose.

NEC Section 410.130(E)(1) requires that all fluorescent fixture ballasts used indoors, including any replacement ballasts, have a thermal protection device integral within the ballast. All such ballasts are marked Class P. The exceptions to this requirement include simple reactance ballasts

Electronic Fixture Ballasts
Modern electronic fixture ballasts offer superior application flexibility. These ballasts can be rated for two, three, or four lamps. In addition, electronic ballasts may also be used to operate less than the maximum number of lamps listed. For example, a ballast rated for four lamps may be used to operate a two- or three-lamp fixture. These ballasts also provide for the ability to use lamps of different lengths.

Replacing Ballasts
When replacing ballasts, be sure to use an exact equivalent. Nuisance tripping may occur if a non-Class P ballast is replaced with a Class P ballast.

Power Factor
The power factor of a ballast is not a measurement of the ballast's ability to supply light through the ballast but is an indication of power consumption. The power factor is based on a value of one. The higher the power factor, the lower the power consumption. Some ballasts offer power factors of greater than 90, which means that they are very efficient devices. Always look for a high power factor when selecting devices for a new installation.

used with fluorescent fixtures using straight tubular lamps, ballasts used with exit fixtures and so identified, and egress lighting energized only during an emergency. Class P ballasts with thermal protection operate to open the circuit at a predetermined temperature in order to prevent abnormal heat buildup caused by a fault in one or more of the ballast components, or by some lamp holder or wiring fault.

High-efficiency, energy-saving ballasts have a high power factor rating. The power factor is the ratio of watts to volt-amperes. To be classified as a high power factor ballast, a ballast must have a power factor of at least 90%. Anything less is considered a normal or low power factor. A ballast's power factor rating is marked on the ballast nameplate. Energy-saving high power factor ballasts cost more than low power factor ballasts, but over time, the savings in energy consumption far exceed the higher initial cost. In any application where there are to be a large number of ballasts, it is best to install ballasts with a high power factor. When compared to magnetic ballasts, electronic ballasts are more energy efficient.

Ballasts can emit a hum, especially the magnetic types. This is caused by magnetic vibration in the ballast core. Ballast manufacturers give their ballasts a sound rating ranging from A to F, with A being the quietest. The need for quiet ballast operation is determined mainly by the desired ambient noise level of the location where the ballast is to be installed. For example, a ballast with an A rating might be used in a doctor's office, while one with an F rating might be suitable for a factory application.

When installing a replacement ballast, make sure to dispose of the old ballast in a proper manner. Unless you see a label stamped *No PCBs* on the failed ballast you are disposing of, you must assume that it contains toxic PCBs and must be disposed of in accordance with the prevailing EPA and local requirements. Failure to do so can expose you and your employer to potential liability for cleanup in the event of PCB leakage.

5.1.5 Fluorescent Dimming Ballasts

Special dimming ballasts and dimmer switches are required to dim fluorescent lamps. The dimmer control allows the dimming ballast to maintain a voltage to the lamp's electrodes that will maintain the electrodes' proper operating temperature. It also allows the dimming ballast to vary the current flowing in the arc. This in turn varies the intensity of light coming from the lamp. Dimming fluorescent lamps differs from dimming incandescent lamps in two main ways: first, fluorescent dimmers do not provide dimming to zero light as do incandescent dimmers; and second, when dimming fluorescent lamps, the color temperature does not vary much over the dimming range. This is unlike incandescent lamps, which tend to turn yellower when dimmed.

Most fluorescent dimming ballasts are of the electronic type. However, older autotransformer magnetic types are also available. Electronic dimming ballasts are normally more efficient and less bulky than magnetic ballasts. *Figure 20* shows a wiring diagram for an electronic dimmer used with a dimming ballast for rapid-start lamps. It is important to point out that the performance of a dimming system may not be satisfactory if the lamp is not correctly matched with the dimming ballast and the controller. Also, when connecting such dimmer circuits, always check the dimmer

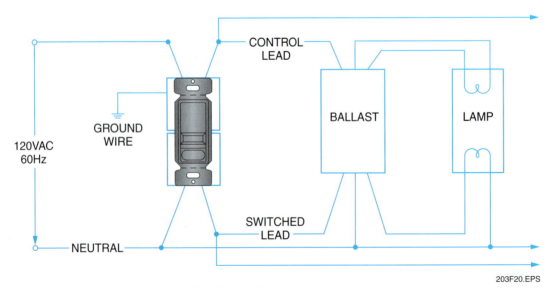

Figure 20 ◆ Dimmer circuit using dimming ballast for rapid-start lamp.

The Importance of Connecting Grounds

A painter in an industrial facility was standing on a 10' wooden ladder and painting steel I-beams. He received a fatal shock while leaning across one of many suspended fluorescent light fixtures, while touching a metal pipe with his other arm. Later investigation revealed that the ground wire in the fixture was disconnected. It is presumed that the ground wire had not been reconnected when the ballast was last replaced. Numerous burn marks were noted within the light fixture at the points where the conductors were connected to the ballast.

The Bottom Line: Don't forget to connect the green (ground) wire when replacing any electrical device.

INSIDE TRACK

Ballast Wiring

Manufacturers color-code ballast wiring for ease of installation. Always follow the manufacturer's color code when wiring ballasts.

and ballast manufacturer's wiring diagrams to determine the proper connections.

5.1.6 Emergency Lighting Ballasts

Special emergency ballasts with self-contained battery-operated power packs are made for use in some fluorescent lighting fixtures. Upon the loss of input power to the lighting fixture, these ballasts typically function to operate one 8' lamp at emergency lighting levels for about 90 minutes, or one 4' lamp for about 120 minutes.

5.2.0 HID Lighting Fixture Ballasts

HID lamps require the use of a ballast to provide enough voltage to strike the arc in the lamp. This function may be accomplished by the ballast itself or in conjunction with a separate electronic ignitor circuit. An ignitor (*Figure 21*) is an electronic device used in the circuitry for high-pressure sodium and some metal halide HID lamps. It provides a pulse of at least 2,500V peak root mean square (rms) to initiate the lamp arc. When the system is energized, the ignitor provides the required pulse until the lamp is completely lit and then automatically stops pulsing.

The HID lamp ballast also acts to control the arc wattage during warmup and normal operation. In addition, some ballasts may also provide a line voltage matching transformer function, enhance lamp wattage regulation with respect to changes in line voltage and/or lamp voltage, and dimming or other control interface functions.

Physically, there are numerous types and shapes of ballasts used with HID lamps (*Figure 22*). The same is true electrically. Some ballasts are made

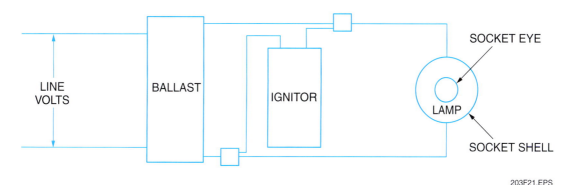

Figure 21 ◆ Simplified HID lamp ignitor circuit.

Figure 22 ♦ HID ballast and lamp.

with primary leads that allow the ballast to be connected to different supply voltages, such as 120V, 208V, 240V, or 277V. Such ballasts are called multi-tap ballasts. It is extremely important that only the proper voltage lead be connected to the supply voltage. The types of ballasts used by a major HID lighting fixture manufacturer (Hubbell) are described here. The ballasts used by other manufacturers are similar. HID ballasts can be grouped into three basic categories:

- Linear, nonregulating circuit ballasts
- Constant wattage autotransformer ballasts
- Three-coil ballasts

5.2.1 Linear, Nonregulating Circuit Ballasts

Linear, nonregulating circuits include reactor ballasts and auto-lag ballasts. These provide for the basic operation of some mercury and high-pressure sodium HID lamps. With the exception of the self-ballasted lamp, the reactor ballast is the most basic form of ballast. See *Figure 23(A)*. It consists of a single coil wound on an iron core placed in series with the lamp. The only function performed by the reactor is to limit the current delivered to the lamp. This type of ballast can only be used when the line voltage applied to the lamp is within the required starting voltage range of the lamp. The power factor for a reactor ballast is typically in the 40% to 50% range. However, a capacitor is normally added to the circuit to improve the power factor to better than 90%. Because this type of ballast provides for no line voltage regulation, outages due to line dips and brownouts are typical. The auto-lag ballast, as shown in *Figure 23(B)*, is a reactor ballast combined with a step-up or step-down autotransformer, which provides for some input voltage regulation.

5.2.2 Constant-Wattage Autotransformer Ballasts

The constant-wattage autotransformer (CWA) ballast, shown in *Figure 23(C)*, is a ballast circuit that uses magnetic saturation to maintain better lamp wattage regulation and improved **dip tolerance.** CWA ballasts are used mainly with mercury and high-pressure sodium lamps. A variation of the CWA, called the peaked lead auto-regulator (PLA), is used with metal halide lamps. Another variation, called the constant-wattage isolated (CWI) ballast, shown in *Figure 23(D)*, is an isolated winding version of the PLA. It is typically used with mercury lamps.

The term dip tolerance relates to the dips in line voltage experienced by all power systems as loads are switched in and out, or as other transitory conditions occur. A well-regulated distribution system will seldom experience voltage dips of more than 10%, but on some circuits, dips of 20% or more may occur. If a ballast is not capable of riding through the voltage dip and sustaining the lamp, it will extinguish and have to cool down before reignition. Lamp dropout due to line voltage dips generally increases with lamp age. The use of ballasts with improved dip tolerance may delay the onset of such lamp dropout problems.

5.2.3 Three-Coil Ballasts

Three-coil ballasts are isolated winding ballasts in which the input and lamp windings are separated (isolated) by a third winding, which helps to eliminate drastic changes in the demands of the lamp

INSIDE TRACK

Unused Wire Taps

The unused wire taps in multi-tap ballasts must be electrically insulated in order to avoid damage to the fixture.

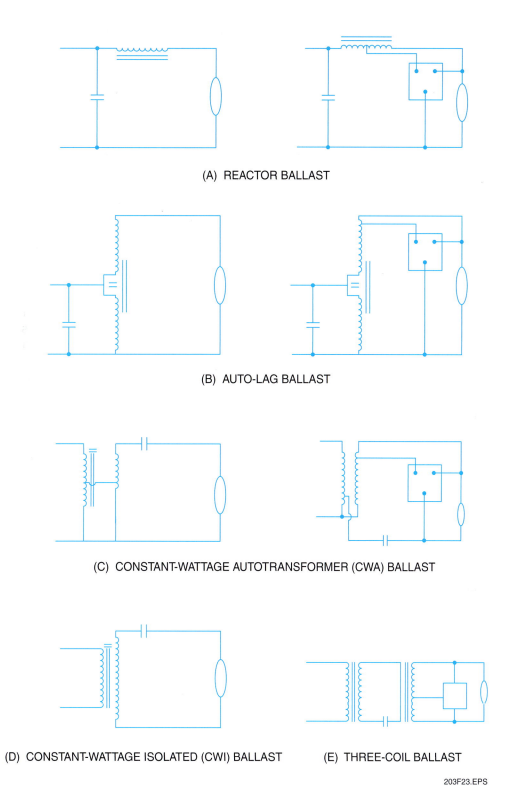

Figure 23 ◆ Simplified HID ballast circuits.

on the supply system and maintains lamp stability during supply system variations. See *Figure 23(E)*. This type of circuit provides the highest degree of lamp operating stability and waveform control, with the lowest harmonics and best performance consistency through the life of the lamp. A magnetic regulator version is used with high-pressure sodium lamps and an electro-regulator version is used with metal halide lamps.

5.2.4 Dimming HID Lamps

HID lamps can be controlled using equipment similar to that used for dimming fluorescent lamps. However, the long warmup and restrike times associated with HID lamps may limit their applications. Multi-level ballasts are made for HID lamps that allow their light to be reduced. This type of ballast is typically used in lighting fixtures for warehouses, parking garages, tunnels, and daytime lighting applications. Some equipment is made that allows HID lamps to be dimmed to less than 20% of their full light output; however, most lamp manufacturers will not guarantee full life expectancy if their lamps are operated below 50%. Also, color shifts in the light output of the lamps may limit their use in some applications.

6.0.0 ◆ LIGHTING FIXTURES

In the electrical trade, the terms lighting fixture and luminaire are used interchangeably (luminaire is an international term). A lighting fixture (luminaire) is defined as a complete lighting unit consisting of the lamp or lamps, together with the parts designed to distribute the light, position and protect the lamps, and connect the lamps to a power supply.

Lighting fixtures are classified in many ways, including: incandescent, fluorescent, and HID; indoor or outdoor; surface-mounted, recessed, or suspended; residential, commercial, or industrial. There are thousands of lighting fixtures manufactured in different styles for all of these applications. From an installation point of view, lighting fixtures can best be classified by the way in which they are mounted. These include:

- Surface-mounted
- Recessed
- Suspended
- Track-mounted

6.1.0 Surface-Mounted Lighting Fixtures

Surface-mounted lighting fixtures include those fixtures that are directly mounted on a ceiling or wall. They can be incandescent, fluorescent, or HID units. Some typical examples of surface-mounted indoor ceiling and wall lighting fixtures are shown in *Figures* 24 and 25, respectively. *Figure* 26 shows an example of a wall-mounted outdoor lighting fixture.

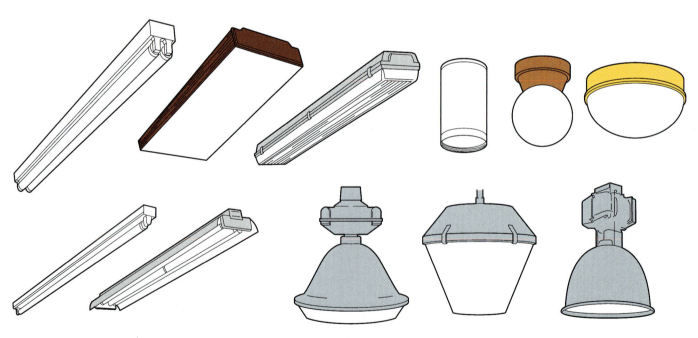

Figure 24 ◆ Surface-mounted ceiling light fixtures.

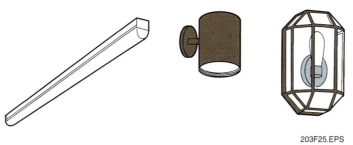

Figure 25 ◆ Surface-mounted indoor wall lighting fixtures.

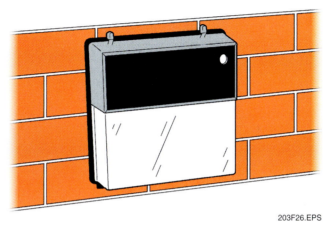

Figure 26 ◆ Surface-mounted outdoor wall fixture.

6.2.0 Recessed Lighting Fixtures

As the name implies, recessed lighting fixtures are mounted into a ceiling so that the main body of the fixture is not visible. *Figure 27* shows examples of recessed lighting fixtures typical of those used with incandescent, metal halide HID, or compact fluorescent lamps. As shown, a typical unit for new work consists of a housing, junction box, and a mounting frame used to fasten it to the ceiling structure. These components of the lighting fixture are all located above the ceiling line. Only the related lens/diffuser and trim are viewable at the ceiling surface.

Other recessed fluorescent fixtures, called **troffers,** are square or rectangular units installed above the ceiling (*Figure 28*). These are typically 2' × 4' or 2' × 2' units, containing two or three T-8, T-10, or T-12 lamps. Some troffers may incorporate a mounting flange or are used with a mounting flange kit, allowing them to be installed in ceiling recesses over plaster, drywall, or concealed-frame ceiling systems. Other troffers, called grid-type troffers, are designed for installation in grid T-bar suspended ceilings (*Figure 29*).

 CAUTION

Incandescent recessed lighting fixtures generate a considerable amount of heat within their enclosure and are a definite fire hazard if not installed properly. For this reason, several precautions and *NEC®* requirements must be observed when installing them. These precautions and requirements are covered later in the installation section of this module.

6.3.0 Suspended Lighting Fixtures

Suspended lighting fixtures are hung from a ceiling using a chain, aircraft wire, cord, or metal stem. Included in this group are chandeliers and pendant units (*Figure 30*).

6.4.0 Track-Mounted Lighting Fixtures

Track-mounted lighting systems (*Figure 31*) include track, an electrical feed box, and two or more fixtures that can be positioned along the

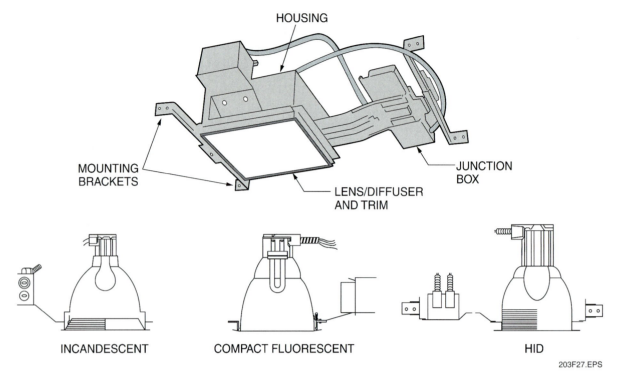

Figure 27 ◆ Typical recessed lighting fixtures.

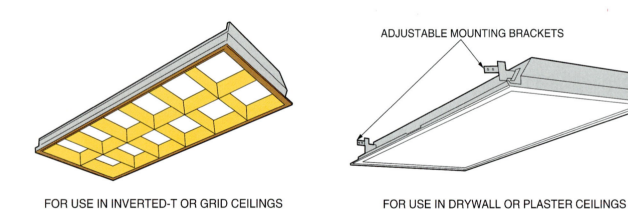

Figure 28 ◆ Typical recessed fluorescent troffers.

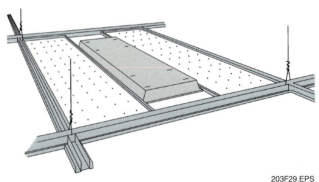

Figure 29 ◆ Fluorescent troffer mounted in a suspended ceiling.

track. Track lighting wiring is in the metal track channels. It can be single-circuit or multiple-circuit wiring to allow two or more sets of lights to be independently controlled.

The tracks are normally mounted on or recessed in the ceiling, but can also be mounted horizontally or vertically on walls.

6.5.0 Storage, Handling, and Security of Lamps and Lighting Fixtures

Prior to installation, it is extremely important that lamps and lighting fixtures be handled and stored properly, both while in transit and at the job site.

Figure 30 ◆ Suspended lighting fixtures.

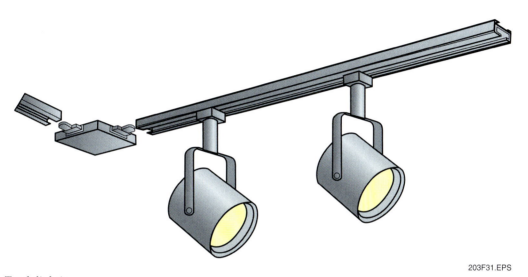

Figure 31 ◆ Track lighting.

Use of Ceiling Fixtures

For large residential rooms, consideration should be given to using more than one ceiling-mounted fixture to provide more efficient and even lighting throughout the space. A single fixture may be too bright in the center, leaving the walls and corners dark. A widely used alternative is to install perimeter recessed fixtures to light up the walls and corners of the room.

Failure to do so can result in broken and/or damaged lamps/lighting fixtures. This can result in expensive replacement costs and possible delays.

The proper care, handling, and storage of lamps/lighting fixtures begins by using some common sense. Manufacturers of lamps and lighting fixtures mark their shipping cartons, pallets, etc., with handling and storage instructions. The first thing you should always do is read and follow all such instructions. Obviously, lamps and lighting fixtures are fragile; therefore, pallets and/or cartons should not be dropped or otherwise handled roughly.

If you are responsible for receiving an order of lamps and/or lighting fixtures from a supplier or distributor, make sure that there is no obvious damage to the shipping cartons. Also make sure that the entire quantity that was ordered is delivered or that the delivery ticket shows what is missing.

If the lamps/lighting fixtures are not going to be installed immediately, store them in a dry place where they will be protected from loss or damage. If they must be stored in several locations, keep a record of what is stored at each of the locations.

7.0.0 ♦ LIGHTING FIXTURE INSTALLATION

Before starting the actual installation of lighting fixtures, it is extremely important to review the lighting floor plan and related lighting fixture schedules to thoroughly familiarize yourself with the different types of lighting fixtures to be installed and their locations. Given this information, you can distribute the different kinds of lighting fixtures to the areas within the building where they are to be installed. *Figure 32* shows an example of a typical lighting floor plan and lighting fixture schedule. *Appendix A* shows a summary of several symbols commonly used on floor and building plans to identify different kinds of lighting fixtures while *Appendix B* provides troubleshooting procedures for various fixtures. If available, shop drawings should be studied prior to installation, and if there are any questions, the fixture manufacturer or lighting designer should be consulted.

When opening lighting fixture shipping cartons, make sure to open the carton carefully so as not to damage the fixture. Do not rip or cut open the carton using knives with long blades, screwdrivers, or similar tools that can cut, scratch, gouge, or otherwise damage the fixture. When handling the fixture itself, wear gloves in order to prevent any grease, dirt, etc., on your hands from getting on its finished and/or decorative surfaces. Also, make sure not to lose or misplace the mount-

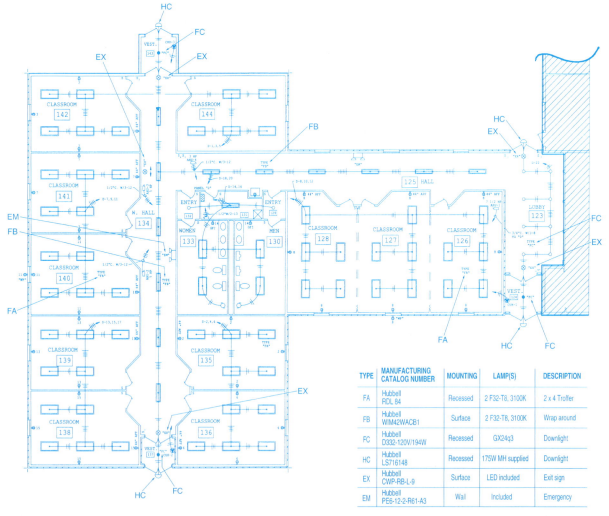

Figure 32 ♦ Typical lighting floor plan and lighting fixture schedule.

Tracking the Installation

When roughing-in lighting circuit wiring, how can you keep track of the work that has been accomplished?

---

ing hardware and insulation (if any) supplied with the fixture. Most lighting fixtures require a minimum amount of preparation before use. Depending on how they are purchased, some are shipped fully assembled, while others may require minor assembly, which should always be done according to the manufacturer's instructions.

Always read the labels on the lighting fixture and/or the installation instructions supplied with the fixture to familiarize yourself with its characteristics. Knowing this information can help prevent improper or inappropriate installation. *Figure 33* shows an example of one lighting manufacturer's installation instructions.

Typical information that can be found marked on lighting fixture labels and/or given in the installation instructions includes:

- Type(s) of lamps to use with the fixture and the maximum lamp wattage
- Approval for use in suspended ceilings, non-insulated ceilings (non-IC), or insulated ceilings (IC)
- Range of ambient temperatures for fixture operation
- Type of wire to use for fixture supply connections
- Thermal protection
- Suitability for wall or ceiling mount only
- Suitability for use in damp and/or wet locations
- Suitability for use as a raceway
- Suitability for installation in concrete
- Mechanical installation instructions
- Electrical connection instructions

NEC Article 410 sets forth most of the requirements pertaining to the installation of lighting fixtures. *NEC Article 411* covers lighting systems of 30V or less. You must meet all *NEC®* and/or local code requirements, including those for mounting, supporting, grounding, live-parts exposure, insulation clearances, supply conductor types, maximum lamp wattages, etc. Whenever you have questions or need clarification about the method to use, always refer to the appropriate codes.

It is important to understand how different types of lighting fixtures are safely and securely attached to a structure or outlet box. Code requirements for the installation of lighting fixtures, lighting outlet boxes, and related methods of support are covered in *NEC Sections 314.23(F), 314.27(A) and (B)*, and *NEC Article 410, Parts III and IV*. The specific methods and hardware used for mounting and supporting lighting fixtures depend on several factors:

- Type of lighting fixture
- Location of lighting fixture
- Construction of mounting surface
- Weight of lighting fixture

7.1.0 Mechanical Installation of Surface-Mounted Lighting Fixtures

Surface-mounted lighting fixtures include those that are directly mounted on a ceiling or wall. They can be incandescent, fluorescent, or HID units. The general methods for mechanically installing ceiling and wall surface-mounted lighting fixtures are basically the same; only the location of the outlet box is different. Normally, boxes for indoor and outdoor wall lighting fixtures are installed at heights of 6'-0" and 6'-6" above the floor, respectively. A lighting fixture should be mounted to its outlet box using the mounting hardware supplied with the fixture and installed according to the manufacturer's instructions. If you are not thoroughly familiar with the lighting fixture to be installed, first read and be sure that you understand the related manufacturer's product literature and installation instructions supplied either with the fixture or printed on the shipping carton.

In buildings of all types, branch circuit wiring for surface-mounted ceiling and wall lighting fixture outlets normally terminates in octagonal or round boxes (*Figure 34*). Some may terminate in square or rectangular boxes. Both metal and nonmetallic types of boxes may be used. These boxes are designed to be covered by the mounting base of a lighting fixture with a similar shape. For concealed wiring in wooden frame construction, the outlet boxes are fastened directly to

---

MODULE 26203-08 ◆ ELECTRIC LIGHTING 3.29

INSTALLATION INSTRUCTIONS FOR MSV

See warnings on carton.

WARNING
- Disconnect power before installation or servicing.
- Install, operate and maintain to meet all applicable codes.
- Labeled ballast voltage must match supply voltage.
- Protect all wiring connections with approved insulators (by others).
- Insure that **all** internal wiring, excluding lead wires from reflector, do not contact reflector assembly due to high lamp temperatures.
- Do not **overtighten** lens door hinge and retaining screws.

PRE-INSTALLATION
1. Remove lens door and set aside **(See Figure 1)**. To remove lens door, **loosen** lens door retaining screws (2) and one hinge screw (1).
2. Remove the two reflector retaining screws and lift out reflector assembly.
3. Remove shipping bolt located in arm mounting hole.
4. When a multi-tap ballast is supplied, the fixture is factory wired to the highest voltage. Verify that fixture wiring matches the supply voltage.

ARM INSTALLATION (See Figure 2)
NOTE: Guide electrical supply wires from fixture through the bearing plate, fixture, gasket, cover, positioning bracket, pole and bolt plate as each mounting step is completed.

1. Insert longer threaded end of hex rods (5) through lockwashers, positioning bracket (4) and pole.
2. Tighten securely to bolt plate (3).
3. Slide cover (6) over hex rods. Drain hole in cover must be on bottom.
4. Slide gasket (7) over threaded end of hex rods.
5. Place fixture onto hex rods.
6. Slide bearing plate (8) and lockwashers over end of hex rods on inside of fixture.
7. Install and tighten nuts securely.
8. Install supply wire (by others) thru the cord grip bracket in pole using appropriate supplied cord grip (9).
9. Make wiring connections.
10. Push wires and wirenuts into pole top and install cap (10).

FINAL ASSEMBLY
1. Re-install and secure reflector assembly using the reflector retaining screws. The reflector assembly is fully rotatable 360° by 90° increments.
2. Install proper lamp.
3. Re-install lens door and insure that all fasteners are securely tightened.

Hubbell Lighting
® A Division of Hubbell Lighting, Inc.
2000 Electric Way
Christiansburg, VA 24073-2500
(540) 382-6111
FAX (540) 382-1526

FIGURE 1
FIGURA 1

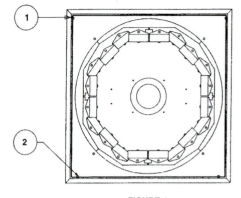

FIGURE 2
FIGURA 2

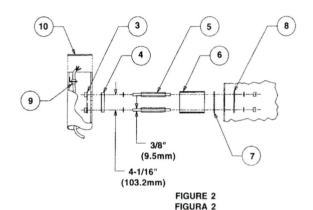

268-1072-9901

203F33.EPS

Figure 33 ◆ Example of lighting fixture installation instructions.

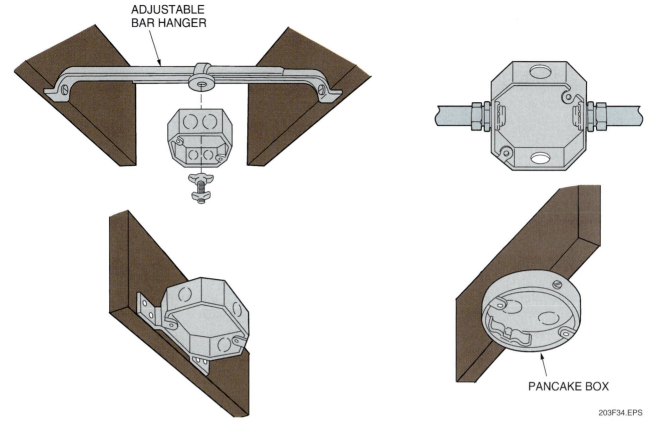

Figure 34 ◆ Typical outlet boxes used with surface-mounted lighting fixtures.

studs, joists, or other framing members using nails or screws. Some ceiling outlet boxes are mounted on an adjustable bar hanger so that the box can be installed in the space between two ceiling joists. The box is attached to the bar hanger by a fixture stud or two bolts. When properly installed, the front face of all concealed ceiling and wall boxes should be flush with the finished ceiling or wall. For exposed conduit installed on concrete or other masonry surfaces, the outlet boxes should be fastened to the ceiling or wall with screws and expansion shields, anchors, powder-actuated fasteners, or similar appropriate fasteners. On steel work, they should be bolted or clamped to the steel. Outlet boxes embedded in concrete are held securely in position by the hardened concrete.

The size and weight of the fixture determine what type of outlet box should be used with a fixture. When mounting lighting fixtures that weigh more than 50 pounds in a ceiling, *NEC Section 314.27(B)* requires that the fixture be supported independently of the box or mounted to a listed outlet box designed and marked to support a fixture weighing more than 50 pounds. Boxes used at luminaire or lampholder outlets in a wall must be designed for the purpose and marked to indicate the maximum weight of the fixture to be supported, if other than 50 pounds, per *NEC Section 314.27(A)*. In addition to complying strictly with the code, good judgment must also be exercised. Remember, codes only give the minimum requirements for electrical installations, not necessarily those that make a quality installation.

Installing Surface-Mounted Fixtures in Closets
What precautions should you take when installing surface-mounted ceiling or wall fixtures in a clothes closet?

NOTE

When a manufacturer supplies insulation with a surface-mounted lighting fixture, do not throw it away. It must be placed between the fixture and the mounting surface to insulate the mounting surface from the heat produced by the fixture. Make sure to install any such insulation as directed by the manufacturer.

7.1.1 Drum, Globe, Sconce, and Similar Lighting Fixtures

Figure 35 shows some common methods used to mount drum, globe, sconce, and similar types of indoor lighting fixtures to an outlet box. Usually, these types of lighting fixtures can be directly mounted to the outlet box with no additional support. As shown in *Figure 35(A)*, the common porcelain-type lamp holder fixture screws directly to the ears of the outlet box. Other ceiling fixtures and wall sconces are fastened to the outlet box using a mounting bar that is attached to a threaded stud in the box, as shown in *Figure 35(B)*. The mounting bar is positioned on the stud and fastened in place with a locknut. After the fixture wiring has been completed, the mounting holes in the base of the fixture are slipped over the machine screws installed in the mounting bar. Cap nuts are then installed on the machine screws and tightened until the fixture is drawn flush with the surface. Fixtures can be mounted to an outlet box without a box stud in a similar manner by screwing a mounting bar to the ears of the box, as shown in *Figure 35(C)*. Another common way some fixtures are installed is by screwing a reducing nut and nipple onto a box stud, as shown in *Figure 35(D)*. Following this, the hole in the center of the fixture base is placed over the nipple, then

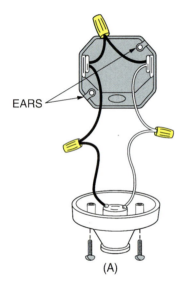

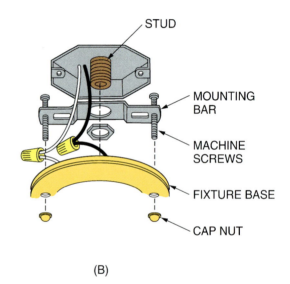

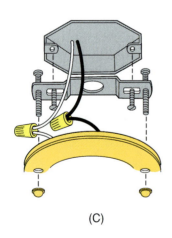

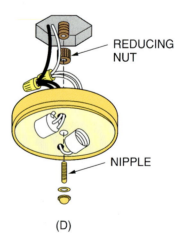

Figure 35 ◆ Common methods of installing surface-mounted lighting fixtures.

Installing Ceiling-Mounted Fluorescent Fixtures

While standing on a ladder to mount a fluorescent fixture on a ceiling, it is much easier to mark the mounting hole locations for the fixture on the ceiling using a cardboard template than it is to hold the actual fixture in place while marking the hole locations.

Use a portion of the cardboard carton that the fixture was shipped in to make a template of the fixture. Transfer the locations of all fixture mounting holes, and any other required holes, to the template. Cut out these holes, then use the template to locate and mark the mounting hole locations on the ceiling.

---

a cap nut is installed on the nipple and tightened until the fixture is drawn flush with the surface. A similar method is to screw a nipple into the threaded center hole of a mounting bar attached to the ears of the outlet box.

7.1.2 Fluorescent Lighting Fixtures

Some lightweight, surface-mounted fluorescent fixtures can be supported directly from an outlet box. Such types typically use a stud, reducing nut, and nipple in the box, with a mounting bar inside the housing, as shown in *Figure 36*. The assembly is held with a locknut. Heavier fluorescent fixtures are supported directly from the ceiling using toggle bolts, wood screws in joists, anchors in masonry, or other appropriate fasteners. These fixtures have an access opening that, when the fixture is in place, must be positioned directly under the outlet box. Many surface-mounted fluorescent fixtures have bumps or ridges on their mounting surfaces that are designed to hold the fixtures slightly off the ceiling in order to allow for the dissipation of heat.

7.1.3 Chandeliers and Pendants

Figure 37 shows some methods used for attaching chandeliers to an outlet box. Chandeliers are normally heavy, requiring that they be mounted on a box with a stud. As shown, a hickey (or a hickey together with a threaded adapter) is installed over the fixture chain and wires, then screwed onto the stud. The fixture base is then fastened to the hickey or adapter with a locknut or cap nut. Some are secured to a mounting bar by means of a nipple. Pendant lighting fixtures are typically supported by chains, metal stems, thin cables, or retractable cords. Most pendants are designed to be wired directly into a ceiling outlet box in a similar manner as that described for chandeliers.

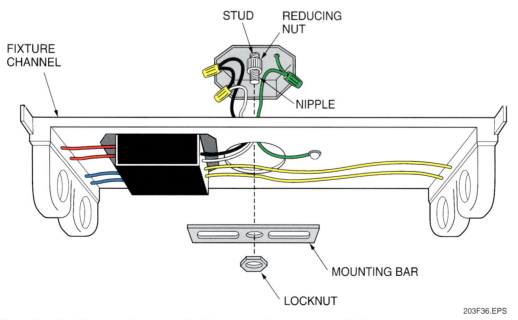

Figure 36 ◆ Typical method for mounting a smaller fluorescent fixture to an outlet box.

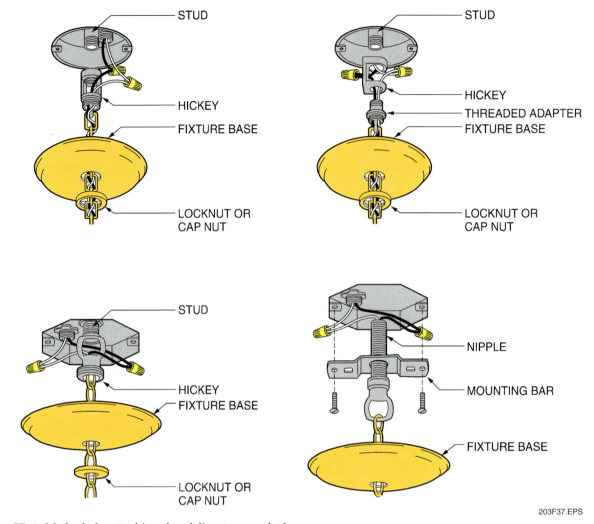

Figure 37 ◆ Methods for attaching chandeliers to an outlet box.

 WARNING!
NEC Section 410.10(D) prohibits pendants and other hanging-type fixtures from being installed within a zone measured 3' horizontally and 8' vertically from the top of a bathtub rim or shower stall threshold.

7.1.4 Ceiling Fans/Fixtures

NEC Section 422.18 covers the support of ceiling-suspended (paddle) fans. It refers to *NEC Section 314.27(D)*, which states that for fans weighing 35 pounds or less, the outlet box may be used as the only support of the fan if it is listed for such use. The code also requires that fans weighing over 35 pounds either be supported independently of the outlet box or the box must be listed for such use and marked with the maximum weight to be supported. Extra support can be provided by mounting the box on a fan joist hanger (*Figure 38*).

The specific method for mounting ceiling fans varies among ceiling fan manufacturers; therefore, the fan should be mounted to the outlet box per the manufacturer's directions. *Figure 38* shows an example of one mounting method. *NEC Section 410.10(D)* also prohibits ceiling-suspended (paddle) fans from being installed within a zone measured 3' horizontally and 8' vertically from the top of a bathtub rim or shower stall threshold.

7.1.5 Outdoor Lighting Fixtures

The methods for attaching outdoor surface-mounted lighting fixtures to outlet boxes are basically the same as those described for indoor surface-mounted fixtures. However, the installation must be done so that the final assembly is watertight.

Installing Chandeliers and Pendants

It is important to make sure that a chandelier or pendant fixture is installed at the proper height. For example, you should allow for door clearances when installing the fixture in a foyer or entrance hallway. Such a fixture should also be hung at a height of 7' or more so that it allows safe passage for tall people.

If the fixture is being hung over a table, its width should be at least 12" less than the width of the table (6" from each edge) so that people do not hit their heads on the fixture when sitting down at or getting up from the table. It should be hung at a height of about 30" to 36" above the table, unless the ceiling is higher than 8'. If the ceiling height is greater than 8', the fixture may look better raised even higher than 36". To provide light for demanding tasks, pendants should be hung about 15" above the work surface.

All fixtures installed in wet locations must be marked as suitable for wet locations. All fixtures installed in damp locations must be marked as either suitable for wet locations or suitable for damp locations. When mounting a lighting fixture to an outlet box that is fully exposed to the weather, the outlet box must be a metal weatherproof-type box.

All washers, gaskets, etc. required to make a watertight assembly must be properly installed as directed by the manufacturer. *Figure 39* shows an example of a common outdoor spotlight fixture. With this fixture, waterproofing fiber washers must be inserted into the fixture sockets and rubber gaskets installed over the end of each socket. In addition, a gasket must be installed between the fixture base and the face of the outlet box.

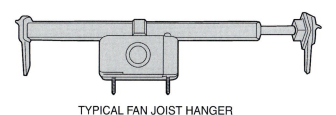

TYPICAL FAN JOIST HANGER

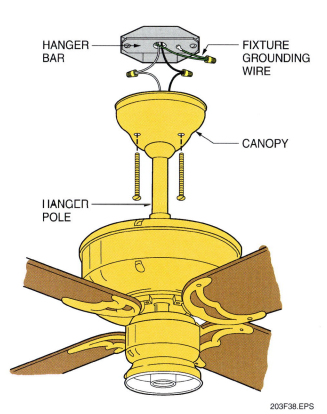

Figure 38 ◆ Typical ceiling fan/light mounting method.

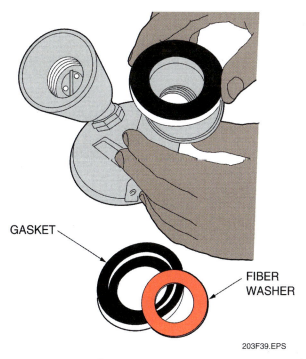

Figure 39 ◆ Installing washers and gaskets in outdoor lighting fixtures.

7.2.0 Mechanical Installation of Recessed Lighting Fixtures

Recessed lighting fixtures are mounted into ceiling recesses so that the main body of the fixture is not visible.

Figure 40 lists some of the code requirements for recessed fixtures.

> **WARNING!**
> Recessed lighting fixtures generate a considerable amount of heat within their enclosures and are a definite fire hazard if not installed properly. For this reason, the code requirements given in **NEC Article 410** must be strictly observed when installing these fixtures.

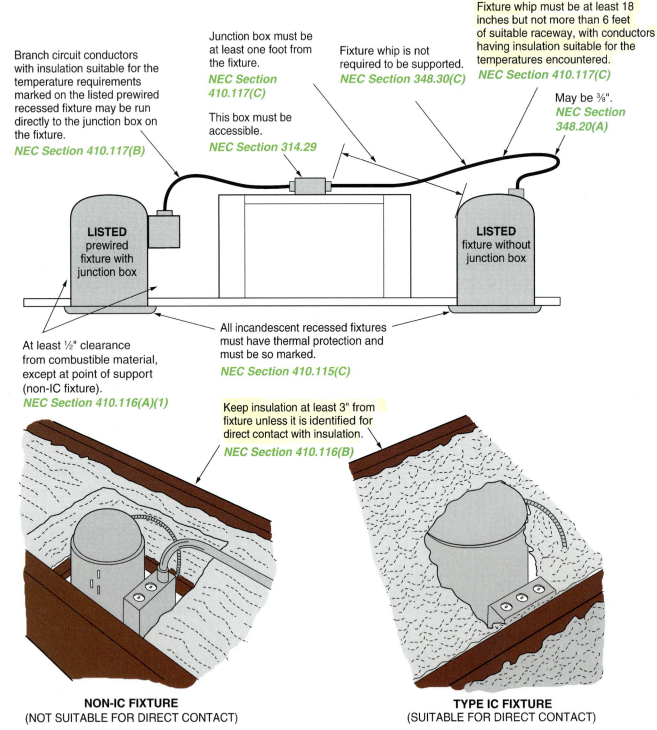

Figure 40 ◆ Summary of *NEC®* requirements for recessed fixtures.

What's wrong with this picture?

7.2.1 Incandescent, Compact Fluorescent, and HID Recessed Fixtures

Incandescent, compact fluorescent, and HID recessed lighting fixtures are all mounted in a ceiling in basically the same way. The rough-in for these types of fixtures must be completed before the ceiling material is installed. As shown in *Figure 41*, they typically include captive adjustable bar hangers for attachment to wooden joists or T-bar grid ceiling members. Recessed fixtures that are identified as acceptable for through wiring are equipped with feed-through junction boxes, which allow a series of them to be wired from one to another or branch circuits to be fed through the housing junction box. Recessed fixtures are typically wired with conduit, nonmetallic-sheathed cable, or armored cable, depending on the application.

A general installation procedure involves the following steps:

Step 1 Extend the fixture mounting bars to reach the framing members.

Step 2 Position the lighting fixture on the mounting bars to locate it properly.

Step 3 Align the bottom edges of the mounting bars with the bottom face of the framing members, then nail, screw, or otherwise attach the bars to the framing members.

Step 4 Remove the junction box cover and open one knockout for each cable entering the box.

Step 5 If required, install cable clamps for each open knockout and tighten the locknut.

Step 6 Run the branch circuit wiring through the cable clamps and into the fixture connection box.

Step 7 After the electrical connections are made and with the finished ceiling in place, install the lens/diffuser and/or trim on the fixture. The fixture trim combinations are marked on the label inside the fixture.

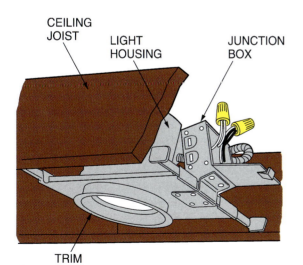

Figure 41 ◆ Typical mounting of a recessed lighting fixture.

CAUTION

The use of a trim type that is not listed on the fixture can result in overheating and possible on-off cycling of the fixture's thermal protective device. Mismatching fixtures and trim is a violation of *NEC Section 110.3(B)*, which states that "Listed or labeled equipment shall be installed and used in accordance with any instructions included in the listing or labeling."

7.2.2 Fluorescent Troffers

The code requirements for the installation of recessed fluorescent troffers are basically the same as those for other types of recessed fixtures. Recessed fluorescent troffers are inserted into the rough-in opening in the drywall, plaster, or other ceiling types and fastened in place using various devices. Some have mounting brackets used to fasten the fixture to the ceiling structure that are adjustable from within the installed fixture. Other types of fixtures are held in place by plaster frame kits or flange kits used to trim out an opening in the ceiling (*Figure 42*). Note that flange kits are used to convert grid troffers to flange troffer applications. Other fixtures with a flange on the housing are supported on the ceiling structure by swing-out supports adjusted in position by a

Installing Recessed Fixtures

When installing recessed fixtures, is it permissible to wire a series of lights from one to another or to run branch circuit wiring through a fixture junction box? If so, when?

screwdriver from below the ceiling. When in position, the adjacent ceiling structure is clamped between the supports and the housing flange holding the fixture in place. The screwdriver access holes are sealed after installation using the plugs supplied with the fixture.

If a suspended ceiling supports recessed fluorescent troffers or other lighting fixtures, *NEC Section 410.36(B)* requires that all the ceiling framing members be securely fastened together and to the building structure. This job is normally done by the installer of the ceiling grid. The proper support of suspended fixtures is a critical safety issue

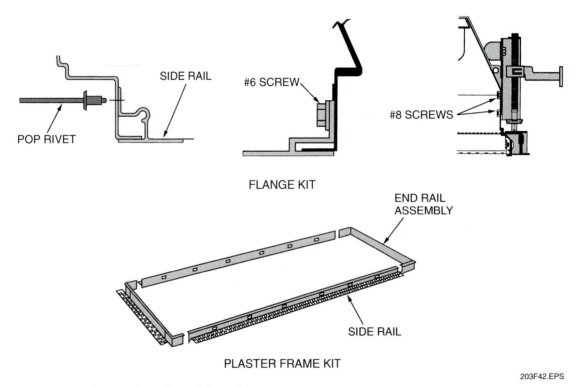

Figure 42 ◆ Fluorescent fixture plaster kit and flange kit.

for people that occupy the building and for any emergency personnel that may be called into the building in the event of an earthquake, fire, etc. For this reason all lighting fixtures, including any lay-in fixtures such as fluorescent troffers, must also be securely fastened to the main ceiling framing members by mechanical means such as bolts, screws, rivets, or listed clips identified for use with the type of ceiling framing members and fixtures involved. These are installed by the electrician. Most manufacturers now provide locking clips in the end plates of their troffers. These clips are simply bent perpendicular to the end plate so that as the troffer is lowered into the ceiling grid, the clip locks over the top of the T. *Figure 43* shows some types of listed clips commonly used to fasten lay-in troffers to suspended grid systems.

Wiring for lighting fixtures in a suspended ceiling is installed above the suspended ceiling using electrical metallic tubing (EMT), armored cable (AC), nonmetallic-sheathed cable (NM), or other acceptable methods as listed in the *NEC®*. Then, from properly supported outlet boxes placed above the ceiling and near the intended locations of the lay-in fluorescent fixtures, a flexible connection, commonly called a fixture whip, is made between the outlet box and the fixture (*Figure 43*). As appropriate, the fixture whip is:

- An 18" (minimum) to 6' (maximum) length of ⅜" flexible metal conduit using conductors suitable for the temperature requirement stated on the fixture label
- An 18" to 6' length of armored cable containing conductors suitable for the temperature as stated on the fixture label
- An 18" to 6' length of ⅜" flexible nonmetallic conduit with conductors and grounding conductor suitable for at least 90°C

7.3.0 Mechanical Installation of Suspended Lighting Fixtures

Suspended lighting fixtures used in commercial and industrial buildings can be supported in numerous ways, depending on the construction of the building and the particular lighting application. Because the methods of support used can vary so widely, it would be impossible to describe them all here. *Figure 44* shows some types of hangers commonly used to attach supporting rods, cables, chains, etc. to the building ceiling or beam structure.

Figure 45 shows some examples of the different kinds of supports used with suspended fluorescent fixtures.

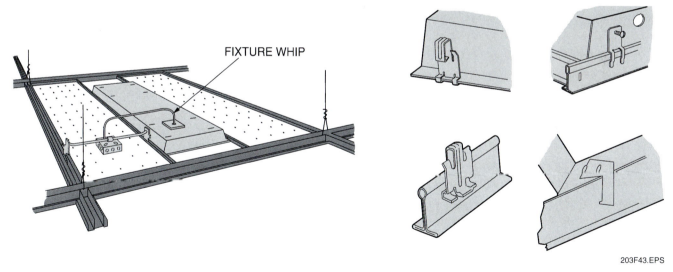

Figure 43 ◆ Typical clips used to fasten lay-in troffers to suspended ceiling grid systems.

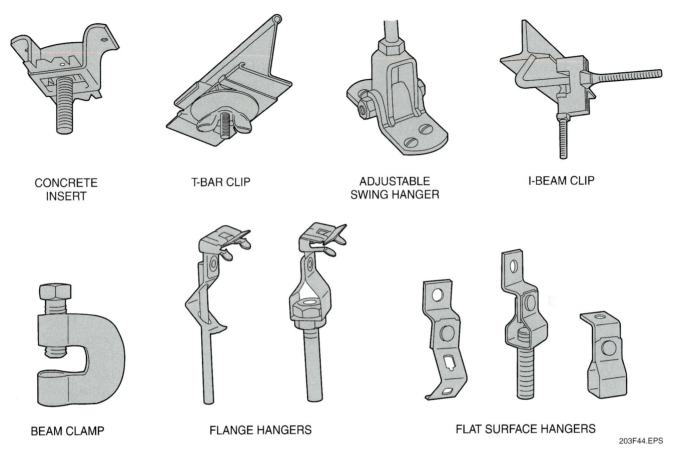

Figure 44 ◆ Examples of hangers used to attach fixtures to building structural members.

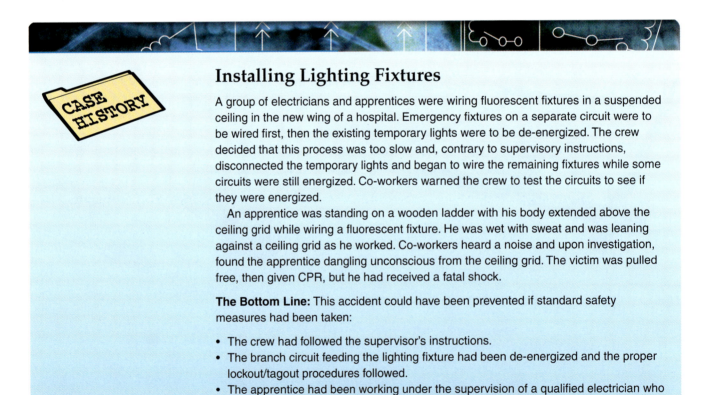

Installing Lighting Fixtures

A group of electricians and apprentices were wiring fluorescent fixtures in a suspended ceiling in the new wing of a hospital. Emergency fixtures on a separate circuit were to be wired first, then the existing temporary lights were to be de-energized. The crew decided that this process was too slow and, contrary to supervisory instructions, disconnected the temporary lights and began to wire the remaining fixtures while some circuits were still energized. Co-workers warned the crew to test the circuits to see if they were energized.

An apprentice was standing on a wooden ladder with his body extended above the ceiling grid while wiring a fluorescent fixture. He was wet with sweat and was leaning against a ceiling grid as he worked. Co-workers heard a noise and upon investigation, found the apprentice dangling unconscious from the ceiling grid. The victim was pulled free, then given CPR, but he had received a fatal shock.

The Bottom Line: This accident could have been prevented if standard safety measures had been taken:

- The crew had followed the supervisor's instructions.
- The branch circuit feeding the lighting fixture had been de-energized and the proper lockout/tagout procedures followed.
- The apprentice had been working under the supervision of a qualified electrician who had first checked the branch circuit wires feeding the fixture with a voltage tester to make sure that the wires were not energized.

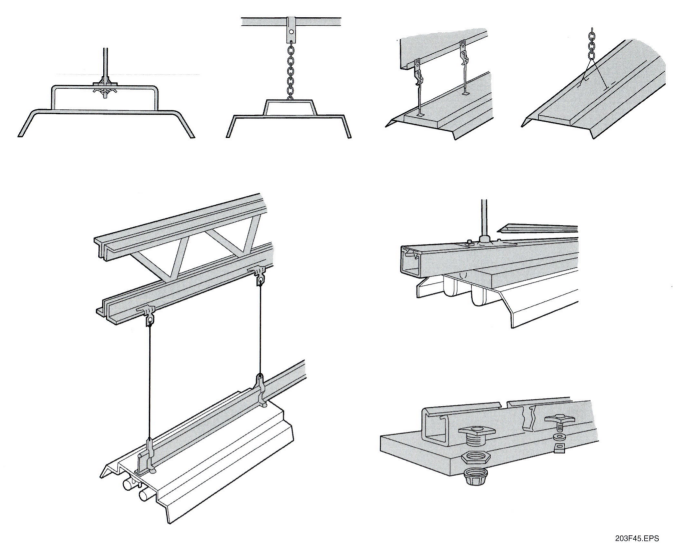

Figure 45 ◆ Typical methods of support for suspended fluorescent fixtures.

Figure 46 also shows fixture support methods typical of those found in commercial and industrial sites. Note that the method of presentation used in this figure is representative of what you will see on electrical installation drawings and blueprints.

7.4.0 Mechanical Installation of Track Lighting Fixtures

Track-mounted lighting systems include the track (rail), an electrical feed box, two or more fixtures that can be positioned along the track, and various types of accessories and other components that allow assembly in any pattern (*Figure 47*). The track lighting wiring is in the metal track rails. It can be either single-circuit or multiple-circuit wiring to allow two or more sets of lights to be independently controlled. The track is an extruded aluminum channel that holds two or more electrical connectors on opposite sides of the channel. The inside surfaces of the conductors are bare. When a track lighting fixture is snapped into the track, two terminals on the fixture contact the track conductors, supplying power to the fixture.

Track lighting systems are made so that the proper electrical polarity is always maintained. The polarity lines inside the track must be aligned when the sections of track are installed. The lighting fixtures will snap into the track in only one way.

NOTE
The *NEC*® requires that lighting fixtures, accessories, etc. used in a track system be specifically designed for the track on which they are to be installed.

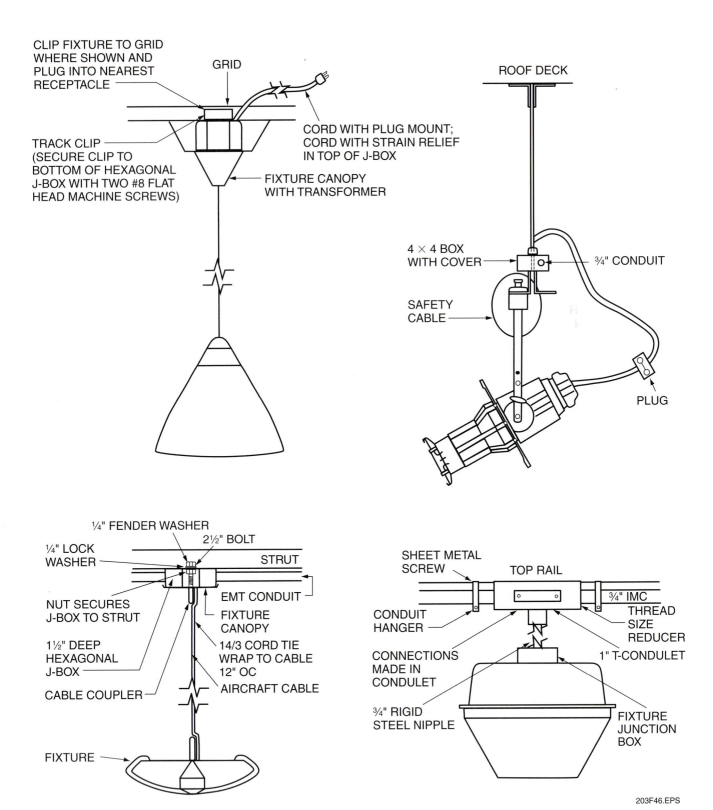

Figure 46 ◆ Examples of fixture supports for commercial/industrial applications.

3.42 ELECTRICAL LEVEL TWO ◆ TRAINEE GUIDE

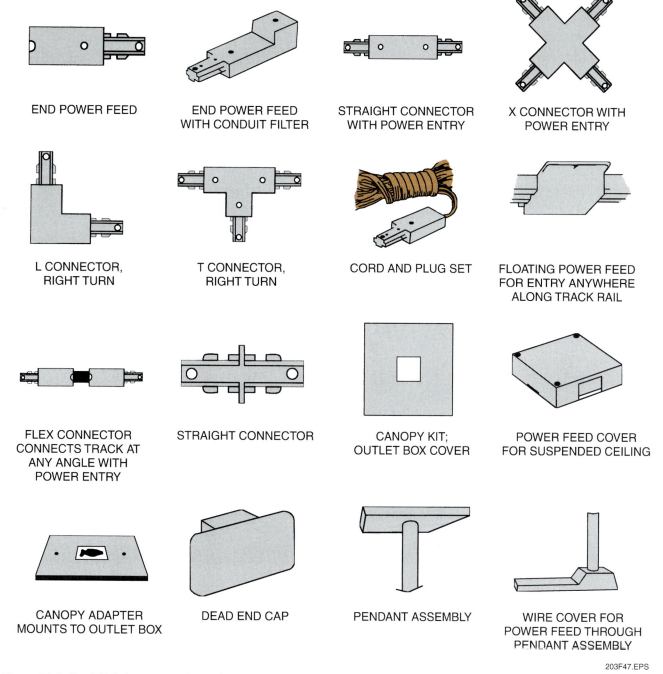

Figure 47 ♦ Track lighting accessories and components.

Track sections are made in lengths of 2', 4', and 8'. These sections are fastened together using straight, flexible, X, L, or T connectors. End caps provide a finished look. Track lighting can be connected to a branch circuit outlet box in many ways, including the use of a floating power feed, end power feed, or X, L, or T power feed connector. Some are connected to a receptacle outlet using a cord and plug connector.

NEC Sections 410.151 through 410.154 cover the installation requirements for track lighting. The track is mounted directly to the ceiling or wall using screws, toggle bolts, or other appropriate fasteners installed through holes in the track. When installed on a wall, the track must be at least 5' above the finished floor, except where protected from physical damage or when using low-voltage (less than 30V rms) track.

Inside Track

Methods of Hanging Suspended Lighting Fixtures

The mechanical methods used to hang and support specific suspended lighting fixtures can be quite different because of the wide variety of fixture types and the nature of the buildings or structures in which they are to be installed.

Generally, on large construction jobs, the specific method for hanging suspended fixtures is shown in the construction drawings. However, on many smaller jobs, the method used for suspending fixtures is frequently left up to the electrician. This can be a problem for less experienced electricians. Fortunately, technical help is readily available.

To aid customers, manufacturers of fasteners and metal framing devices produce numerous catalogs and technical product support bulletins that show example solutions to common hanging/support problems. Other good sources of technical advice are your supervisor, other qualified electricians, and fastener/framing company sales engineers or representatives.

203SA06.EPS

The *NEC®* requires that the track be fastened so that each fastener is suitable for supporting the maximum weight of the fixtures that can be installed. Unless identified for use with supports at greater intervals, a single section of track that is 4' or less in length must be supported in at least two places. If more track is being used, at least one additional support for each individual extension that is 4' long or less must be used.

A typical installation begins by first attaching the track electrical feed assembly to the designated outlet box. If using an end power feed for this purpose, the track adapter plate covers the outlet box and holds the track connector and electrical housing (*Figure 48*).

When using a wire-in connector, track clips that are spaced at the required distance apart are used to hold the track ¼" to ½" away from the ceiling or wall mounting surface. These clips are fastened in place using screws, toggle bolts, or other appropriate fasteners. It is a good idea to snap a chalkline from the center of the outlet box or

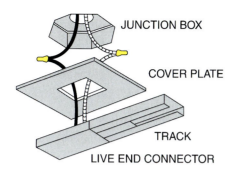
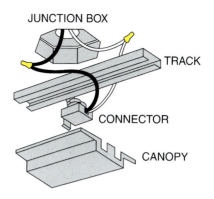

Figure 48 ♦ Typical end feed and floating canopy track connectors.

center slot on the track connector to the proposed end of the track run for use as a track installation guide. Once the clips are installed, the first section of track is fitted solidly into the track connector, and then the track is snapped into the clips. Following this, setscrews in the sides of the clips are tightened to hold the track firmly in place. The remaining sections of track are installed in the same manner using the appropriate connectors and accessories. When it is necessary to cut a section of track to length, it can be cut using a hacksaw. Make sure to follow the manufacturer's instructions as to where and how to make the cut and how to close off the cut with a proper end cap.

7.5.0 Electrical Connection of Lighting Fixtures

Making the electrical connections between the lighting fixture and related outlet box wires is done at the appropriate point during the mechanical installation of the fixture. The specific way in which the supply wires in the outlet box are connected to the lighting fixture wires varies depending on the design of the branch circuit and the way in which the branch circuit cable or raceway wiring is routed. Therefore, you must refer to the lighting plan to determine the exact connections.

7.5.1 Conventional Hard-Wired Installations

The connection of the lighting fixture wiring to the outlet box supply wires should be done in accordance with the manufacturer's instructions. The general procedure is basically the same, no matter what type of fixture is being connected.

> **WARNING!**
> Before attempting to connect lighting fixtures to a branch circuit outlet box, make sure that the power is turned off to the branch circuit involved at the circuit breaker panel. Then, make sure to follow the prevailing lockout and tagout procedures.

When cutting outlet box wires in preparation for connecting the fixture, do not cut the branch circuit wires too short or too long. *NEC Section 300.14* requires that at least 6" of free conductor length be left at each outlet or junction box where it emerges from its cable sheath or raceway. Furthermore, for splices or the connection of fixtures or devices, each conductor's length shall extend at least 3" outside any box opening that is less than 8" in any dimension. The 3" length is generally measured from the outer edge of the box. Note that leaving the wires too long will make it difficult to position the wires in the box and will overcrowd the box, possibly resulting in overheating.

The lighting fixture's black wire(s) are connected to the outlet box hot wires (black, red, or marked hot) and the white wire(s) to white wires (*Figure 49*). If dealing with nonmetallic cable and metal boxes, the bare grounding wires should be spliced with one end of a grounding jumper. The other end of the grounding jumper should be attached to the box using either a grounding clip or the box grounding screw. If dealing with nonmetallic cable and a nonmetallic outlet box, the bare cable wires should be spliced together. In some cases, a green wire from the fixture and the bare wires will be connected to a grounding screw

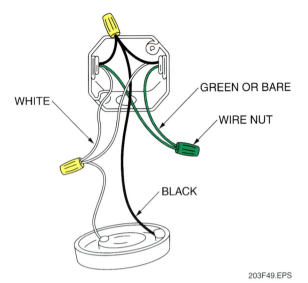

Figure 49 ◆ Example of basic lighting fixture connections.

nuts should cover all bare current-carrying conductors. Some wire nut manufacturers recommend twisting the wires together before installing the wire nut; others do not. Follow the manufacturer's directions for the wire nut you are using. The correct methods for stripping wires and using wire nuts were covered in the *Conductor Terminations and Splices* module.

The *NEC*® requires that all fixtures with exposed metal parts be grounded. If the fixture has a separate green grounding wire, it should be spliced together with the other grounding wires. Once the fixture box and/or metal mounting strap is grounded, the nipple or screws holding the fixture will ground the fixture.

7.5.2 Modular System Wiring

In some installations, lighting fixtures are connected using a modular wiring system (*Figure 50*). Modular wiring provides an alternative method to conventional hard wiring. Modular system components are interchangeable and can be reused. Component labels are color-coded by voltage, and the units are keyed to prevent mismatching voltages. Lighting branch circuit wiring is run from the building panel to an outlet or junction box that serves as the starting point for the modular wiring. There, a modular system feeder adapter is wired and mounted. From this point on, all the wiring to the lighting fixtures is modular using the appropriate cable adapters.

provided on the mounting bar supplied with the lighting fixture.

All wire splices should be made using the correct wire nuts sized for the number and gauges of wire you are splicing. If in doubt, most wire nut manufacturers mark their packages to show the maximum and minimum number of wires of different sizes that can be connected using a particular wire nut. Use this information as your guide. When properly installed and tightened, the wire

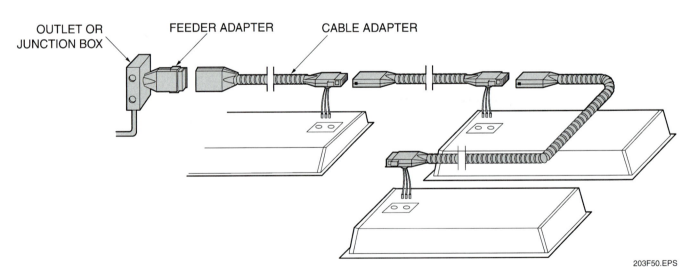

Figure 50 ◆ Basic modular wiring system.

8.0.0 ◆ CONTROLS FOR LIGHTING

Because of high energy costs, the use of lighting controls to manage the application of lighting is common. Several devices can be used to control lighting circuits in order to conserve energy. These include occupancy sensors, photosensors, timers, and similar devices.

8.1.0 Occupancy Sensors

Occupancy sensors (*Figure 51*) are devices that can be used to turn lights on and off automatically in an individual space such as a private office, restroom, or storage area. Occupancy sensors can be motion detecting (ultrasonic), heat sensing (infrared), or sound sensing. They can be recessed or surface-mounted on a wall or ceiling, they can replace wall switches, or they can plug into receptacles. The sensor turns the lights on when it senses someone coming into the room or area and turns the lights off some time after no longer sensing anyone present. Units come either with fixed, preset time delays and sensitivity levels, or with adjustable ones.

Ultrasonic sensors transmit ultrasound and receive a reflected signal to sense the presence of occupants in a space. They typically operate at a frequency between 25kHz and 40kHz. Passive infrared sensors detect the changes in infrared patterns across their segmented detection regions. The type of sensor used must be compatible with the application. For example, a motion detector or sound detector may not be the right choice if occupants of the space sit very quietly at desks. People in such situations have been known to complain that they must deliberately move or make noise from time to time to prevent the sensor from turning off the lights. On the other hand, infrared sensors must be placed so that no obstruction blocks their sensing field.

8.2.0 Photosensors

Electronic photosensors sense the level of visible light in the surrounding area and convert this level into an electrical signal. This signal can be used in one of two ways, depending on the type of system. In the first type, the signal can be used to activate a simple on/off switch or relay that functions to control the power to the related lighting fixtures. In the second type, a variable output signal is produced that can be sent to a controller that operates to continuously adjust the output of electric lighting in the area. The device shown in *Figure 52* includes a photoelectric sensor to automatically enable the device at dusk plus an infrared sensor to activate the lamps when motion is detected.

Photosensors can be an integral part of a lighting fixture, remote from the lighting fixture, or may control a circuit relay that operates several lighting fixtures. It is important that the area controlled by one photosensor have the same daylight illumination conditions (amount and direction) and that the area be contiguous with no high walls or partitions to divide it. When photosensors are used outdoors, the sensor should be aimed due north.

(A) INDOOR WALL SWITCH SENSOR

(B) OUTDOOR SENSOR

(C) ULTRASONIC CEILING SENSOR

(D) INFRARED CEILING SENSOR

Figure 51 ◆ Examples of infrared occupancy sensors.

Figure 52 ◆ Combination photoelectric/infrared sensor.

8.3.0 Timers

Timers (*Figure 53*) are used to turn lighting on or off in response to known or scheduled sequences of events. Timers can be very simple clock-like mechanisms, or they can be microprocessors that can program a sequence of events for years at a time. With a simple timer, the load is switched on and held energized for a preset period of time. Timer limits can range from a few minutes to 12 hours. Some models have a hold position for continuous service.

An electromechanical time clock/timer is driven by an electric motor, with contactors actuated by mechanical stops or arms attached to the clock face. Time clocks have periods ranging from 24 hours to seven days. They can initiate many operations. Some can actuate a momentary contact switch to provide on and off pulses for actuating relays or contactors. Electronic time clocks/timers provide programmable selection of many switching operations and can typically be controlled to the nearest minute over a seven-day period. Most can control multiple channels and have time-of-day scheduling, holiday programming, daylight savings time adjustment, leap year correction, annual override, and battery carryover for protection against power outages.

Time clocks are often used in conjunction with photosensors in order to turn off lighting when there is no longer a need for it. For example, an industrial building may use a photosensor signal to activate outdoor lighting at dusk, then a time clock to turn off that lighting after the last worker has left the facility. Another example is when a photosensor is used to signal when to turn the lights on, then a time clock is used to initiate dimming the lights when a high level of lighting is no longer needed.

(A) BASIC ELECTROMECHANICAL

(B) SIMPLE ELECTRONIC

(C) MULTI-FUNCTION PROGRAMMABLE

203F53.EPS

Figure 53 ◆ Typical timer controls.

9.0.0 ♦ ENERGY MANAGEMENT SYSTEMS

Modern buildings normally use some form of an energy management system (EMS) to economically control the amount of energy consumed by the building's lighting circuits and HVAC equipment. An EMS can be a fairly simple stand-alone unit connected to one or more pieces of equipment (e.g., room lighting, a heat pump or rooftop HVAC unit, etc.), or it can be more extensive and control all the lights and equipment throughout an entire large facility.

Whether large or small, an EMS typically consists of a computer or control processor, energy management and scheduling software, sensors and controls located where needed, and, in large systems, a communications network. When programmed, an EMS can automatically control lighting to:

- Turn off lights in unoccupied areas
- Maintain partial lighting before and after working hours or public-use hours
- Schedule lighting operation by hour of day and time of year

You will recognize the tasks performed by an EMS as being similar to those performed by the sensors and timers that were discussed earlier. The EMS simply receives signal inputs applied from these devices and processes them so as to perform tasks more reliably and precisely than can be performed by manual methods. Lighting control can be implemented in a building by a local approach, a central system, or both. The method used is determined both by the size of the controlled areas and by how the control inputs are integrated into the system. A local lighting system is divided into small, independently controllable zones based on size, environment, etc., or according to functional need. The inputs from the sensors located in the zone are wired directly to a control that is also located in the zone (*Figure 54*). In central systems, the sensor inputs from the individual zone sensors are all wired to a central control.

An EMS commonly turns lights on and off and/or initiates partial lighting in an area via relays that are activated or deactivated by the EMS control unit. *Figure 55* shows a typical lighting switching scheme involving the use of split-wired, multi-ballasted lighting fixtures. By split-wiring three-lamp and four-lamp lighting fixtures, multiple light intensities can be provided in a single zone. The relay-based control system provides full lighting for specific times of the day, while allowing a reduced lighting level and reduced power for those times when less lighting is needed.

Figure 55(A) shows a relay-controlled split-wiring system connected to two three-lamp lighting fixtures. This arrangement allows for four levels of lighting: 0%, 33⅓%, 66⅔%, and 100%. As shown, the relay to the inboard ballasts is closed, allowing two of the six lamps to be turned on, thus providing light at the 33⅓% level.

Figure 55(B) shows a similar relay-controlled split-wiring system connected to two four-lamp lighting fixtures. This arrangement allows for three levels of lighting: 0%, 50%, and 100%. As shown, the relay to the inboard ballasts is closed, allowing four of the eight lamps to be turned on, thus providing light at the 50% level.

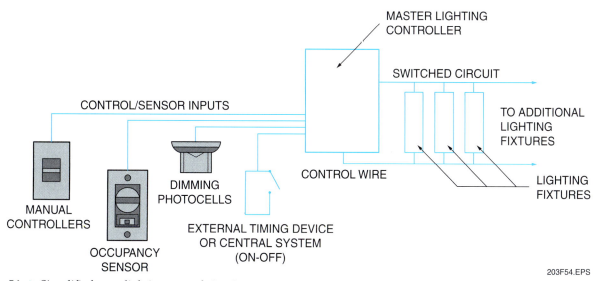

Figure 54 ♦ Simplified zone lighting control circuit.

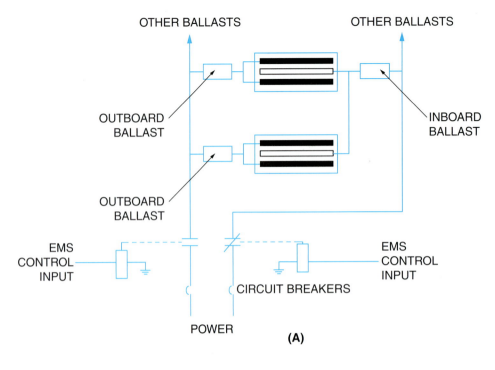

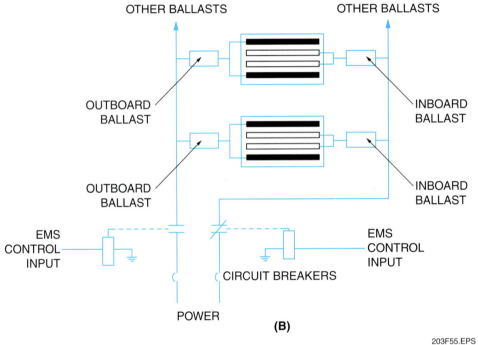

Figure 55 ◆ Simplified wiring diagrams of relay-controlled lighting fixture circuits.

GOING GREEN

LED Lighting Systems

LED lighting systems offer significant energy savings due to their long life and low power use. They are now being produced for a wide variety of applications, from task lighting to recessed troffers. The fixture shown here is used in commercial and industrial lowbay lighting systems.

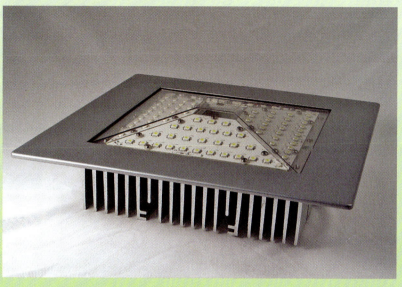

Putting It All Together

Examine the lighting system in your kitchen. What changes, if any, would help to maximize the lighting in the different work areas? What types of lighting could you use to make the other rooms in your home appear warmer and more inviting?

Review Questions

1. Light impulses are transferred from the eye to the brain by the part of the eye called the _____.
   a. rods
   b. cones
   c. pupil
   d. optic nerve

2. Light energy that is *not* absorbed by an object can be _____.
   a. reflected from the object
   b. transmitted through the object
   c. both reflected from and transmitted through the object
   d. dissipated as heat in the object

3. When all the wavelengths of the light spectrum are presented to the eye in nearly equal proportions, _____ light is seen.
   a. red
   b. white
   c. blue
   d. green

4. In an incandescent lamp, the filament resistance is _____.
   a. much higher when the lamp is turned on
   b. lower when the lamp is turned on
   c. somewhat higher when the lamp is turned on
   d. the same whether the lamp is turned on or off

5. True or False? The filament of a tungsten halogen lamp is encased in a capsule containing iodine or bromine gas.
   a. True
   b. False

6. The diameter of a 4', T-12 fluorescent lamp is _____.
   a. 1"
   b. 1⅛"
   c. 1¼"
   d. 1½"

7. During operation, a high-output (HO) fluorescent lamp typically has a current draw of _____.
   a. 400mA to 600mA
   b. 800mA to 1,000mA
   c. 1,100mA to 1,300mA
   d. 1,500mA to 1,700mA

8. The type of high-intensity discharge (HID) lamp that uses the oldest HID technology is the _____ lamp.
   a. low-pressure sodium
   b. high-pressure sodium
   c. mercury vapor
   d. metal halide

9. Lamps with color temperatures of _____ and lower produce a light that is considered warm in tone.
   a. 3,000K
   b. 3,500K
   c. 4,100K
   d. 5,000K

10. An electronic ballast typically operates the connected fluorescent lamps at a frequency of _____ to improve both ballast efficiency and lamp efficacy.
    a. 50Hz to 60Hz
    b. 10kHz to 20kHz
    c. 20kHz to 50kHz
    d. 60kHz to 100kHz

11. Most code requirements governing the installation of lighting fixtures, lampholders, and lamps are covered in _____.
    a. *NEC Article 411*
    b. *NEC Article 370*
    c. *NEC Article 410*
    d. *NEC Article 422*

12. Fixture whips used to connect from a junction box mounted above the ceiling to a fixture installed in a suspended ceiling cannot exceed _____ in length.
    a. 18"
    b. 2'
    c. 4'
    d. 6'

Review Questions

13. When cutting the branch circuit wires in an outlet box in preparation for connecting a light fixture, you should leave at least _____ of free conductor length after it emerges from the cable or raceway.
    a. 3"
    b. 5"
    c. 6"
    d. 8"

14. All of the following sensors are normally used as occupancy sensors *except* a(n) _____.
    a. motion detector
    b. infrared sensor
    c. photosensor
    d. sound sensor

15. If the contacts of both relays are closed in the circuit shown in *Figure 55(A)*, what percentage of lighting will be produced by the lighting fixtures?
    a. 0%
    b. 33⅓%
    c. 66⅔%
    d. 100%

Summary

Electric lighting serves to provide illumination for the performance of visual tasks with a maximum of comfort and a minimum of eyestrain and fatigue, allowing individuals to perform their daily living and work-related tasks.

The proper installation of lighting fixtures requires that you become familiar with the different types of lighting fixtures and the lamps used with them. You must acquire the skills needed to mechanically and electrically install lighting fixtures for different applications according to the prevailing building and electrical codes.

Notes

Trade Terms Introduced in This Module

Ballast: A circuit component in fluorescent and HID lighting fixtures that provides the required voltage surge at startup and then controls the subsequent flow of current through the lamp during operation.

Color rendering index (CRI): A measurement of the way a light source reproduces color. The higher the index number (0–100), the closer colors are to how an object appears in full sunlight or incandescent light.

Dip tolerance: The ability of an HID lamp or lighting fixture circuit to ride through voltage variations without the lamp extinguishing and cooling down.

Efficacy: The light output of a light source divided by the total power input to that source. It is expressed in lumens per watt (LPW).

Incandescence: The self-emission of radiant energy in the visible light spectrum resulting from thermal excitation of atoms or molecules such as occurs when an electric current is passed through the filament in an incandescent lamp.

Incident light: The light emitted from a self-luminous object such as the sun or an electric source.

Lumen (lm): The basic measurement of light. One lumen is defined as the amount of light cast upon one square foot of the inner surface of a hollow sphere with a one-foot radius with a light source of one candela in its center.

Lumen maintenance: A measure of how a lamp maintains its light output over time. It may be expressed either numerically or as a graph of light output versus time.

Lumens per watt (LPW): A measure of the efficiency, or, more properly, the efficacy of a light source. The efficacy is calculated by taking the lumen output of a lamp and dividing by the lamp wattage. For example, a 100W lamp producing 1,750 lumens has an efficacy of 17.5 lumens per watt.

Luminaire (fixture): A complete lighting unit consisting of a lamp (or lamps) and ballasts (where applicable), together with the parts designed to distribute the light, position and protect lamps, and connect them to the power supply.

Luminance: The luminous intensity of any surface in a given direction per unit area of that surface as viewed from that direction. Measured in candela/m^2. The term luminance is commonly used to express brightness.

Reflected: The bouncing back of light waves or rays by a surface.

Reflected light: Light reflected from an object that does not have self-luminating properties but reflects light provided from another source.

Refracted/refraction: The bending of a light wave or ray as it passes obliquely from one medium to another of different density, or through layers of different density in the same medium.

Starter: A circuit component of certain fluorescent lighting fixtures that is used to heat the lamp electrodes before the lamps are lighted.

Troffer: A recessed lighting fixture installed with the opening flush with the ceiling.

Appendix A

Summary of Lighting Fixture Symbols

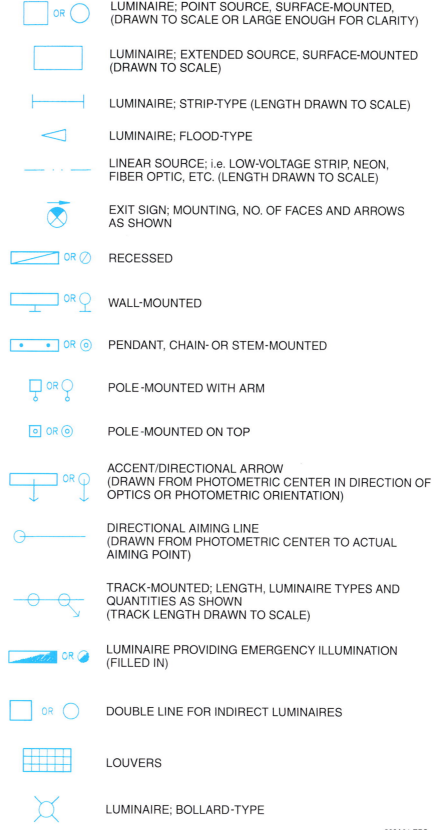

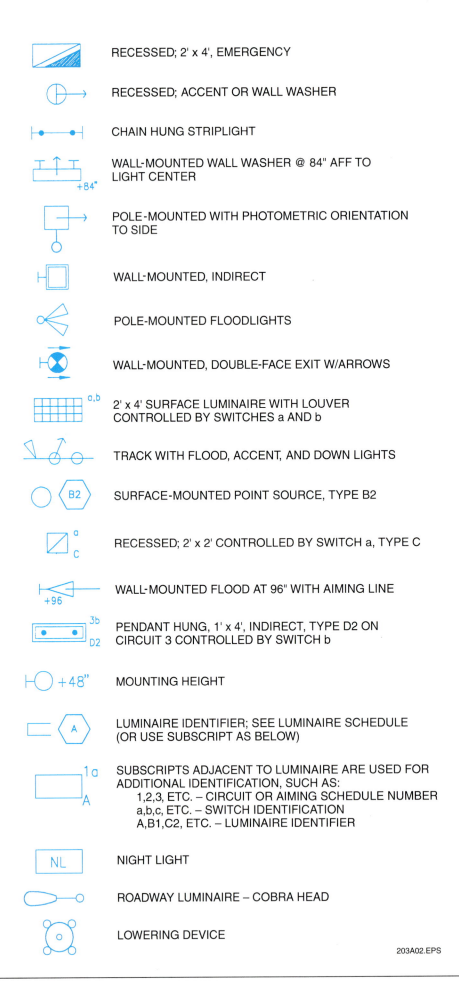

Appendix B

Testing Procedure and Checklist

Electrical Testing Procedures

Caution: High voltages, currents, and temperature are required to operate lamps. Therefore, shock and burn hazards exist, and testing or evaluating fixtures or components should be done only by qualified individuals.

A. Testing Lamps

The easiest method of troubleshooting a fixture is to try a **known** good lamp in the inoperative fixture. If the lamp being replaced exhibits any of the following conditions, replace with a new lamp.

1. **Sodium Leaker Lamp** - will have a brown/golden coating on the inside of the lamp envelope other than at the base of the lamp.

2. **Amalgam Leaker Lamp** - the lamp envelope will have a smoked bronze appearance on the inside of the envelope.

3. **Faulty Base to Lamp Envelope Seal** - a white powdery substance will appear at the base of the lamp where oxygen has leaked inside the lamp.

4. **End of Lamp Life** - the arc tube will be black on both ends or the entire length of the arc tube will be black.

5. **Broken Welds or Arc Tube Support Brackets** - mechanical breaks occasionally occur due to rough handling or internal thermal stresses. Broken welds in evacuated (HPS) lamps can also create a problem known as vacuum switching. Extremely high voltage surges occur in the lamp circuit if the weld opens while the lamp is operating. Secondary coil burnout, ignitor arcing and socket arcing can occur. Rewelding may occur and the lamp may **appear** to be satisfactory; however, if left in operation, failure of the ballast and/or ignitor is likely.

B. Ignitors/Starters

The starter provides the necessary voltage and energy required to initiate the arc in the lamp. The easiest way to check the ignitor on 35W to 150W HPS units is to install a 120V incandescent lamp in the fixture. If the incandescent lamp operates but a known good HPS lamp will not ignite, replace starter. In 200W to 1000W HPS fixtures, install a mercury lamp of similar wattage. If the mercury lamp lights and the HPS lamp will not, replace the starter.

Do not operate Incandescents or mercury lamps used to check the starter for extended periods of time (more than ½ hour).

C. Capacitors

Testing Capacitors may be accomplished by:

1. **Visual Inspection for swollen capacitors.** If the capacitor is swollen or bulged on the sides or top where the terminals are located, remove and replace with a new one.

2. **Verify the correct microfarad rating** as specified on the ballast I.D. label.

3. **Using an ohmmeter to check capacitors:**
   - discharge capacitor by shorting between the terminals
   - disconnect capacitor from circuit
   - remove bleed resistor

4. **Set ohmmeter to highest resistance scale** and connect leads to capacitor terminals.
   - if resistance starts low and gradually increases, the capacitor is good.
   - if resistance starts low and doesn't increase, the capacitor is shorted and should be replaced.
   - if resistance is high and remains relatively the same, the capacitor is open and should be replaced.

D. Ballast

Visual inspection of the coil for burned or charred windings is the easiest method for checking the ballast. If lamps, ignitors and capacitors test good, replace the ballast. Testing the voltage at the socket is another method of checking the ballast. However, to use this procedure you must know specific ballast/lamp voltage and amperage requirements. The starting aid (if present) should be disconnected prior to testing the voltage at the socket. Failure to remove the starting aid could damage the test equipment

Fluorescent Service Checklist

(Identify problem and test for cause in numerical sequence.)

PROBLEM	End of Lp. Life	Incorr. Lamp	Def. Blst.	Incorr. Blst.	Incorrect Supply Voltage	No Supply Voltage	Incomp. Lamp Seating	(B) Dirty Lamps	Incorr. Fixt. Wiring	Loose Socket Connect.	(C) Low Amb. Temp.	High Amb. Temp.
Failure to Start (A)	1	5	4	10	6	7	8	2	9	11	3	
Slow Starting		3			4			2			1	
Blinking Off and On - Very Short Duration Cycles	1				2					4	3	
Long Duration Cycles - Several Minutes to Hours (D)		3	2		4				5			1
Short Lamp Life		2	1	3	4		5		6		7	
Lamp End Blackening - One End Only (A)			4				1		2	3		
Lamp End Blackening - Both Ends (A)	1	2	3	6	7		4		5			
Lamp Ends Only Lighted		1	2	3					4			

A. Generally indicates lamp cathode heat is missing. Problem source must be identified and corrected.
B. In humid weather or when air conditioning systems are operated only during working hours, dust on the lamps may gather condensation and prevent reliable starting. Lamps and fixtures must be cleaned and the lamp waxed with a good silicone wax.
C. Energy saving lamp, ballast combinations will not reliably start near or below 60°F; all others 50°F unless low temperature systems are used.
D. Indicates ballast thermal protector is cycling. May be the result of high ambient temperature, fixture misapplication or restricted air circulation. With recessed fixtures, check to be sure insulation has not been placed directly on the recessed fixture body. With surface mounted fixtures mounted against insulated ceiling, make sure they are rated for such applications.

All service checklists detail the more probable problem causes. It is important to keep in mind that other elements may be involved or more than one cause present. All electrical service work must begin with a careful and detailed evaluation of the problem and thorough inspection of all components, paying special attention to all wiring connections.

The ability to field test ballasts is very limited. The normal verification is to substitute a known good ballast in the problem fixture.

On rapid start systems only, cathode heater voltage may be verified by a voltage reading between **the contacts on each socket.** (A voltage of 3.5 to 5 is considered normal.)

Caution - Voltages in excess of 500V may be present between the lamp socket parts and ground.

203A04.EPS

Mercury Vapor and Metal Halide Service Checklist

Identify problem and test for cause in numerical sequence.

PROBLEM	A End of Lamp Life	B Incorrect Lamp	C Open Cap.	D Def. Starter (if app.)	E Incorrect Supply Voltage	F Photo-Cont.	G New Lamp Repl.	H Line Voltage Dips	I Shorted Cap.	J Burned Ballast Windings	K Incorrect Ballast
FAILURE TO OPERATE	1	5	3	4	2						6
LAMP CYCLING		1			2	3					
COLOR SHIFT							1	2			
LOW LIGHT OUTPUT		2			1						
TRIPPED BREAKER OR BLOWN FUSE									1	2	3

A. End of Lamp Life
At a lamp's end of life, the voltage requirements of the lamp exceeds the output ability of the ballast. The usual failure mode for mercury is low light output followed by failure to operate, and for metal halide low light levels, color shifts, and lamp operating instability (cycling). Replace end of life lamp as dictated by lamp testing procedure.

B. Incorrect Lamp
Lamp wattage, voltage, burning position, and type must be checked against the fixture label to be sure the proper lamp has been installed.

C. Open Capacitor
This is the usual result of electrical failure or mechanical damage. Very often the capacitor can will be bulged or distorted. Where the capacitor is used in series with the lamp, the lamp will not operate. See tests for capacitors or replace with a known good capacitor.

D. Defective Starter
The function of the starter is to provide a high voltage pulse to ignite the lamp. To test, replace with a known good lamp. (Also see Ignitor/Starter testing procedure)

E. Incorrect Supply Voltage
When investigating problems of low light output or cycling, voltage readings first must be taken at the fixture to properly identify power distribution problems. In the case of multiple supply type ballasts, verification that the supply voltage is connected to the appropriate input lead is advised.

F. Photocontrol
Problems may result from electrical failure, incorrect wiring, or from an incorrect amount of light reaching the cell. First cover the eye of the cell with electrical tape to verify fixture and cell operation. If the fixture fails to operate, the cell must be bypassed electrically to identify the problem source. Problems of incorrect amount of light usually can be resolved by repositioning the fixture or by using cell caps to regulate the light level.

G. New Lamp Replacement
New lamps, when installed, go through a period of burn-in or seasoning which may extend for a period of 100 hours or more. The usual result is noticeable color variation between lamps. While metal halide lamps are noted for this, the system will stabilize as the burn-in period ends. It is important to understand that some variation in color may be noted between lamp manufacturers, or between old and new lamps.

H. Line Voltage Dips
When investigating voltage dips, it is important to identify distribution system loading. The usual cause is the starting of large motors or the use of electric welding equipment. Line voltage recorders will usually identify the problem. It is important to understand that mild dips will cause a color shift, while severe dips will cause the lamp to go out. Dip tolerance depends on lamp type, lamp age and ballast type.

I. Shorted Capacitor
This is the direct result of electrical failure or mechanical damage. The most common result of a shorted capacitor is ballast failure. In all cases of shorted capacitors, both capacitor and transformer should be replaced. See tests for capacitors or replace with a known good capacitor.

J. Burned Ballast Coils

1. Burned Primary Coils
This is the usual result of fixture connection to incorrect supply voltage. Repeated failure often in conjunction with capacitor failure may indicate short duration high voltage spikes on the distribution system. The use of a scope along with power company assistance is generally required to identify these spikes. Equally important as a cause of failure is a shorted capacitor (See "I" above - Shorted Capacitor). See tests for ballasts on preceding page and replace if necessary.

2. Burned Secondary Coils
This may be caused by a short circuit in the lamp circuit wiring or by mechanical failure in the lamp. Carefully inspect all lamp circuit wiring. See tests for ballasts on preceding page and replace it if necessary. **In all cases, replace the lamp.**

K. Incorrect Ballast
The requires checking only a new fixture or if a recurring problem is encountered. Carefully compare details on the transformer to the fixture label and lamp used in the circuit. Change components as required.

High-Pressure Sodium Service Checklist

Identify problem and test for cause in numerical sequence.

PROBLEM	A End of Lamp Life	B Incorrect Supp. Voltage	C Incorrect Lamp	D Shorted Capacitor	E Photo-Control	F Line Volt. Dip	G Defective Lamp	H Defective Starter	I Open Capacitor	J Burned Ballast Windings
LAMP CYCLING SINGLE FIXTURE	1	3	2		4					
LAMP CYCLING GROUP OF FIXTURES					2	1				
FAILURE TO START	2	3					4	1	5	6
LOW LIGHT OUTPUT	2	3	1				4			

A. End of Lamp Life

At a lamp's end of life, the operating voltage requirements of the lamp exceed the output ability of the ballast. This results in the lamp cycling off and on. It is important to understand that in the early stages of failure, the lamp may operate for several hours before cycling off. As the lamp nears total failure, the on time will decrease until the lamp fails to ignite at all. Cycling lamps should immediately be replaced to avoid starting aid damage. See lamp tests or replace with known good lamp.

B. Incorrect Supply Voltage

When investigating problems of low light output, cycling, or failure to start, voltage readings must be taken *at the fixture* to properly identify distribution system problems. For multiple supply type ballasts, proper lead connection must be verified.

C. Incorrect Lamp

Lamp wattage, voltage, burning position, and type must be checked against fixture label to be sure the proper lamp has been installed in the fixture.

D. Shorted Capacitor

This is the direct result of electrical failure or mechanical damage. The most common result is low light output; cycling may also occur.

E. Photocontrol

Problems may result from electrical failure or from an incorrect amount of light reaching the cell. First cover the eye of the cell with electrical tape, to verify fixture and cell operation. If the fixture fails to operate, the cell must be bypassed electrically to identify the problem source. Problems of incorrect amount of light usually can be resolved by repositioning the fixture or by using cell caps to regulate the light level. Occasionally, the cell will see light from the fixture reflected off a nearby object and turn itself off. Repositioning the cell or reflecting object may be required.

F. Line Voltage Dip

When investigation voltage dips, it is important to identify distribution system loading. The usual cause is the starting of large motors or the use of electric welding equipment. Line voltage recorders will usually identify the problem. It is important to understand that lamps nearing end of life will be more susceptible to voltage dips than new lamps, and lamp operating on reactor ballasts are more sensitive to voltage dips than those on regulating ballasts.

G. Defective Lamp

This is normally the result of some mechanical failure in the lamp. This can often be determined by brown or silver coating on the lamp outer jacket, or by deposits at the base of the lamp. See lamp testing procedures or replace with a known good lamp.

H. Defective Starter

The function of the starter is to provide a high voltage pulse to ignite the lamp. Failure to operate generally results from electrical failure in the starter. See tests for starters or replace with a known good starter.

I. Open Capacitor

This generally results from electrical failure or mechanical damage. Very often the capacitor can will be bulged or distorted. When the capacitor is used in the secondary (lamp) circuit, the fixture will not operate. See tests for capacitors or replace with a known good capacitor.

J. Burned Ballast Windings

This is often the result of fixture being connected to incorrect supply voltage. Repeated primary winding failures often in conjunction with capacitor failures may indicate short duration high voltage spikes on the distribution system. The use of a scope in conjunction with power company assistance is generally required to identify these spikes. See tests for ballasts and replace if necessary.

Additional Resources

This module is intended to present thorough resources for task training. The following reference works are suggested for further study. These are optional materials for continued education rather than for task training.

Lighting Handbook, Latest Edition. New York: Illuminating Engineering Society of North America (IESNA).

National Electrical Code® Handbook, Latest Edition. Quincy, MA: National Fire Protection Association.

CONTREN® LEARNING SERIES – USER UPDATE

NCCER makes every effort to keep these textbooks up-to-date and free of technical errors. We appreciate your help in this process. If you have an idea for improving this textbook, or if you find an error, a typographical mistake, or an inaccuracy in NCCER's Contren® textbooks, please write us, using this form or a photocopy. Be sure to include the exact module number, page number, a detailed description, and the correction, if applicable. Your input will be brought to the attention of the Technical Review Committee. Thank you for your assistance.

Instructors – If you found that additional materials were necessary in order to teach this module effectively, please let us know so that we may include them in the Equipment/Materials list in the Annotated Instructor's Guide.

Write: Product Development and Revision
National Center for Construction Education and Research
3600 NW 43rd St., Bldg. G, Gainesville, FL 32606

Fax: 352-334-0932

E-mail: curriculum@nccer.org

Craft _____ Module Name _____

Copyright Date _____ Module Number _____ Page Number(s) _____

Description
_____
_____
_____
_____

(Optional) Correction
_____
_____
_____

(Optional) Your Name and Address
_____
_____
_____

Conduit Bending

John Paul Jones Arena

The Barton Malow Company won an Aon Build America Award for New Construction Management for its work with the University of Virginia's sports and entertainment complex in Charlottesville, Va.

26204-08

26204-08
Conduit Bending

Topics to be presented in this module include:

1.0.0	Introduction	4.2
2.0.0	*NEC*® Requirements	4.3
3.0.0	Types of Bends	4.4
4.0.0	The Geometry of Bending Conduit	4.6
5.0.0	Mechanical Benders	4.12
6.0.0	Mechanical Stub-Ups	4.15
7.0.0	Mechanical Offsets	4.16
8.0.0	Electric and Hydraulic Conduit Benders	4.17
9.0.0	Segment Bending Techniques	4.23
10.0.0	Tricks of the Trade	4.34
11.0.0	PVC Conduit Installations	4.35
12.0.0	Bending PVC Conduit	4.36

Overview

Simple one-shot bends in small-diameter conduit can be made using a hand bender, with little or no calculations required. However, when complete conduit systems must be installed on exposed surfaces or pipe racks, many calculated bends may be required using hand- or power-bending equipment.

The *National Electrical Code*® regulates the number of degrees that can be bent into a single run of conduit. A run of conduit is considered to be any conduit installed between outlet or junction boxes, cabinets, or panels. The *NEC*® also states that all rigid conduit bends must be made so that the conduit is not damaged and the internal diameter of the conduit is not reduced. Additional *NEC*® regulations address the bending radius.

Electricians must know how to create offsets, saddles, kicks, and other bends in conduit without relying on the trial-and-error method, which results in wasted conduit and time. Precision bends can be made the first time by knowing and applying formulas such as those associated with the right triangle.

Note: NFPF 70®, *National Electrical Code*®, and *NEC*® are registered trademarks of the National Fire Protection Association, Inc., Quincy, MA 02269. All *National Electrical Code*® and *NEC*® references in this module refer to the 2008 edition of the *National Electrical Code*®.

Objectives

When you have completed this module, you will be able to do the following:

1. Describe the process of conduit bending using power tools.
2. Identify all parts of electric and hydraulic benders.
3. Bend offsets, kicks, saddles, segmented, and parallel bends.
4. Explain the requirements of the *National Electrical Code*® (*NEC*®) for bending conduit.
5. Compute the radius, degrees in bend, developed length, and gain for conduit up to six inches.

Trade Terms

Approximate ram travel
Back-to-back bend
Bending protractor
Bending shot
Concentric bending
Conduit
Degree indicator
Developed length
Elbow
Gain
Inside diameter (ID)
Kicks
Leg length
Ninety-degree bend
Offsets
One-shot shoe
Outside diameter (OD)
Radius
Rise
Segment bend
Segmented bending shoe
Springback
Stub-up
Sweep bend
Take-up (comeback)

Required Trainee Materials

1. Pencil and paper
2. Appropriate personal protective equipment
3. Copy of the latest edition of the *National Electrical Code*®

Prerequisites

Before you begin this module, it is recommended that you successfully complete *Core Curriculum; Electrical Level One* and *Electrical Level Two*, Modules 26201-08 through 26203-08. You should also read *NEC Articles 342, 344, 352, and 358*.

This course map shows all of the modules in *Electrical Level Two*. The suggested training order begins at the bottom and proceeds up. Skill levels increase as you advance on the course map. The local Training Program Sponsor may adjust the training order.

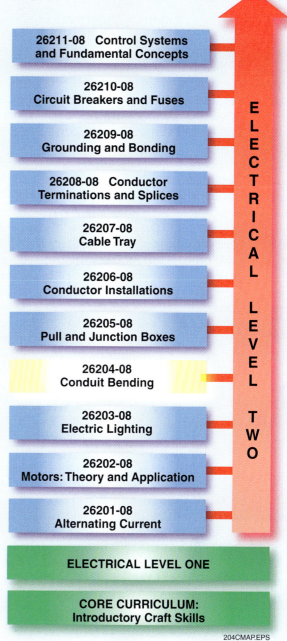

1.0.0 ◆ INTRODUCTION

The normal installation of intermediate metal conduit (IMC), rigid metal conduit (RMC), and electrical metallic tubing (EMT) requires many changes of direction in the conduit runs, ranging from simple offsets at the point of termination at outlet boxes and cabinets to complicated angular offsets at columns, beams, cornices, and so forth. This module will deal mainly with conduit bending machines.

Unless the contract specifications dictate otherwise, such changes in direction, particularly in the case of smaller sizes, are made by bending the conduit or tubing as is required. In the case of 1¼" and larger sizes, right angle changes of direction are sometimes accomplished with the use of factory elbows or conduit bodies. In most cases, however, such changes in direction are accomplished more economically by making conduit bends in the field.

On-the-job conduit bends are also performed when multiple runs of the larger conduit sizes are installed. Truer parallel alignment of multiple runs is maintained by using on-the-job conduit bends rather than factory elbows. Such bends can all be made from the same center, using the bends of the largest conduit in the run as the pattern for all other bends. This is just one of the useful techniques covered in this module.

Exposed conduit work is one area of electricians' work that puts their skills on display. Exposed conduit directly reflects on the ability of the installer. With these thoughts in mind, it will benefit you to learn several methods of bending conduit that will ensure accurate and precisely-bent conduit—conduit that you can step back and look at with pride and the knowledge that it was bent right the first time.

1.1.0 Safety Considerations

Keep the following in mind when bending conduit:

- Read and understand the operating instructions before using any tool. Always use the right tool for the job; never use bending tools for any other purpose.
- Never operate tools with damaged or missing parts.
- Watch for pinch points on power bending tools.
- Use tools in a well-lit, uncluttered area, with enough space for the conduit to move while bending.
- Make sure all hydraulic connections are clean and tight. Replace damaged or worn hoses before using.
- Be prepared for the unexpected.
- Make sure tools are complete and properly assembled before operating.
- Bending shoes and follow bars should be treated as precision instruments. Never toss them into a tool chest or allow them to become damaged or bent out of shape. The quality of the final job will depend on using the proper shoes and follow bars in good condition.
- When working with PVC conduit, make sure adequate ventilation is provided to carry off fumes from joint cement or glue.

Take Pride in Your Work

Even though the conduit in this installation will be hidden by the flooring and the final wall finish, notice how the installing electrician made a special effort to keep the bends perfectly spaced and aligned. Doing a good job even when it might never be seen or admired is the mark of a true professional.

2.0.0 ◆ NEC® REQUIREMENTS

NEC Section 344.24 requires that all rigid metal conduit bends be made so that the conduit will not be damaged and the internal diameter of the conduit will not be effectively reduced. To accomplish this, the *NEC®* further specifies that the minimum radius (*Figure 1*) to the centerline of the conduit shall not be less than that listed in *Table 1*. There is a good reason for this rule. When bends are too tight, pulling becomes extremely difficult and the insulation on the conductors may be damaged.

The *NEC®* requirements for other bends are shown in *Table 2*.

2.1.0 Number of Bends per Run

Every change of direction in a conduit run adds to the difficulty of the pull. The *NEC®* specifically states that no more than four quarter bends (360° total) may be made in any one conduit run between outlet boxes, cabinets, panels, or junction boxes; that is, between pull points.

Some electricians believe that offsets, kicks, and saddles are not bends, especially in areas where the electrical inspectors are lax. These electricians count only those bends that are actually a quarter circle (90°). The misconception of this is quickly apparent when wires are pulled. Offsets and saddles add just as much resistance to pulling conductors as any 90° elbow. A 45° offset, for example, takes two 45° bends, which equal one 90° bend. A saddle may be as low as 60° or as high as 180°, depending on the types of bends used.

Table 1   *NEC®* Minimum Requirements for Radius of Conduit Bends – One-Shot and Full-Shoe Benders (Data from *NEC Chapter 9, Table 2*)

Trade Size (Inches)	Radius to Center of Conduit (Inches)
½	4
¾	4½
1	5¾
1¼	7¼
1½	8¼
2	9½
2½	10½
3	13
3½	15
4	16
5	24
6	30

Reprinted with permission from NFPA 70, the *National Electrical Code®*. Copyright © 2007, National Fire Protection Association, Quincy, MA 02269. This reprinted material is not the complete and official position of the National Fire Protection Association on the referenced subject, which is represented only by the standard in its entirety.

Table 2   *NEC®* Minimum Requirements for Radius of Other Conduit Bends (Data from *NEC Chapter 9, Table 2*)

Trade Size (Inches)	Other Bends (Inches)
½	4
¾	5
1	6
1¼	8
1½	10
2	12
2½	15
3	18
3½	21
4	24
5	30
6	36

Reprinted with permission from NFPA 70, the *National Electrical Code®*. Copyright © 2007, National Fire Protection Association, Quincy, MA 02269. This reprinted material is not the complete and official position of the National Fire Protection Association on the referenced subject, which is represented only by the standard in its entirety.

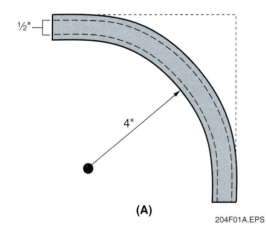

(A)

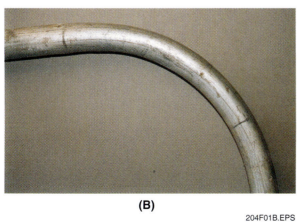

(B)

Figure 1 ◆ Inside radius requirements.

International Differences

Everywhere except in the United States, steel conduit is measured using the metric system, so it might be helpful to familiarize yourself with common metric conversion calculations, especially if you live near the Canadian or Mexican border. For example, to convert inches to centimeters (cm), multiply inches by 2.54; to convert yards to meters (m), multiply yards by 0.9.

Don't Ignore the Outside Diameter

Although conduit measurements are based on the inside diameter, don't forget that the outside diameter determines the size of the opening that the conduit must penetrate.

A 15° kick in a conduit run may seem insignificant, but after several of these are incorporated into the run, the difficulty of pulling wire becomes apparent. The number of degrees in each kick should be included in the total count, and in no case should the total number (number of bends × number of degrees in each bend) exceed 360°. This is the maximum number of degrees allowed. For example, you could have two 45° offsets, a 90° saddle, and two elbows. Many electricians prefer to install pull boxes at closer intervals to reduce the number of bends, especially when the larger conductor sizes are being pulled. The additional cost of the pull boxes and the labor to install them is often offset by the labor saved in pulling the conductors.

3.0.0 ◆ TYPES OF BENDS

There are various types of conduit bends: elbows, offset bends, back-to-back bends, saddles, etc. A brief review of each follows:

- *Elbow* – An elbow, or ell, is a 90° bend that is used when a conduit must turn at a 90° angle. In single conduit runs when the larger sizes of conduit are being installed, factory elbows are frequently used to save labor on setting up a power bending machine, calculating and marking the conduit for bending, and finally, making the bend. However, in multiple conduit runs, a neater job will result if on-the-job sweep bends or concentric bends are properly calculated and installed. See *Figure 2*.

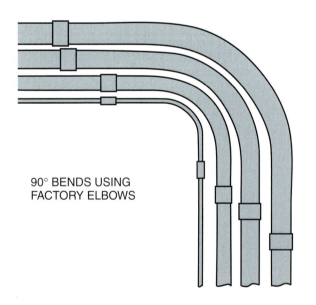
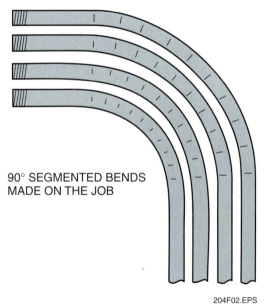

90° BENDS USING FACTORY ELBOWS

90° SEGMENTED BENDS MADE ON THE JOB

204F02.EPS

Figure 2 ◆ Typical 90° bends.

4.4 ELECTRICAL LEVEL TWO ◆ TRAINEE GUIDE

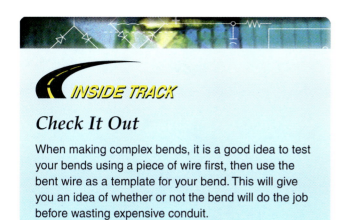

INSIDE TRACK

Check It Out

When making complex bends, it is a good idea to test your bends using a piece of wire first, then use the bent wire as a template for your bend. This will give you an idea of whether or not the bend will do the job before wasting expensive conduit.

- *Offset* – An offset consists of two equal bends and is used when the conduit run must go over, under, or around an obstacle. An offset is also used at outlet boxes, cabinets, panelboards, and pull boxes. See *Figure 3*.
- *Saddle* – A saddle is used to cross a small obstruction or other runs of conduit. A saddle is made by marking the conduit at a point where the saddle is required and placing a bender a few inches ahead of this point. Bends are made as shown in *Figure 4*. Both three-bend and four-bend saddles can be used, depending on the type of obstruction.
- *Kick* – A kick is a single change in direction of a conduit run of less than 90°. It is used mostly where the conduit run will be concealed in deck work. The first bend in an offset, for example, is really a kick, as shown in *Figure 5*; another kick in the opposite direction transforms the bend into an offset.

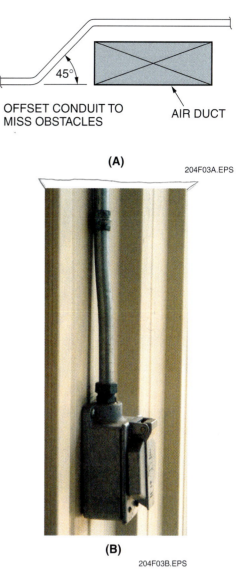

Figure 3 ♦ Applications of conduit offsets.

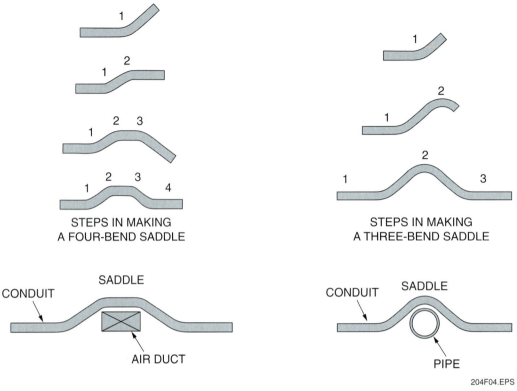

Figure 4 ◆ Practical application of a saddle bend.

Figure 5 ◆ Kick.

4.0.0 ◆ THE GEOMETRY OF BENDING CONDUIT

Learning to bend conduit involves a thorough knowledge of some basic geometry (the science that deals with the properties of lines, angles, surfaces, and solids). The basic bends discussed previously will handle the majority of the electrician's needs, as these simple bends are merely combined to form more complex bends. When these basic skills are mastered, you will be able to calculate and bend conduit to fit most situations.

The 90° or right angle bend is probably the most basic of all and is used much of the time, regardless of the type of conduit being installed. All other bends are typically made with angles less than 90°.

4.1.0 Right Triangle

A right triangle, as shown in *Figure 6(A)*, is defined as any triangle with one 90° angle. The side directly opposite the 90° angle is called the hypotenuse and the side on which the triangle sits is the base. The vertical side is called the height or altitude. For offset bends, right triangle characteristics can also be applied because the offset forms the hypotenuse of a right triangle, as shown in *Figure 6(B)*. There are reference tables available for sizing offset bends based on these characteristics.

4.1.1 Trigonometry Fundamentals

Right triangles are used to develop trigonometric equations (*Figure 7*). These equations can be used to determine the **rise** or **stub-up** distance between bend points, the distance between bend joints, the distance a kick needs to be above the surface, or the amount of additional conduit required. There are six basic trigonometric functions that will be required:

- Sine
- Cosine
- Tangent
- Cotangent
- Secant
- Cosecant

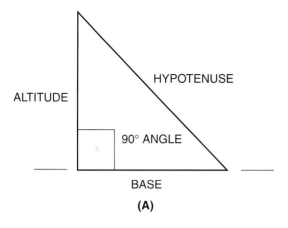
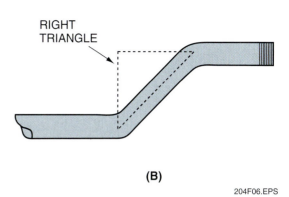

Figure 6 ♦ A right triangle and its relationship to a conduit offset.

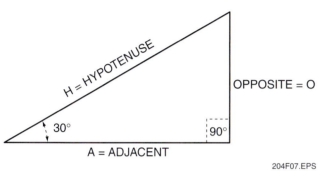

Figure 7 ♦ Trigonometry fundamentals of a right triangle.

The sine function can be computed by dividing the hypotenuse by the side opposite the angle being considered. This function can be represented by the following equation:

$$\text{Sine} = \frac{\text{opposite}}{\text{hypotenuse}} = \frac{O}{H}$$

The equations for the other functions are as follows:

$$\text{Cosine} = \frac{\text{adjacent}}{\text{hypotenuse}} = \frac{A}{H}$$

$$\text{Tangent} = \frac{\text{opposite}}{\text{adjacent}} = \frac{O}{A}$$

$$\text{Cotangent} = \frac{\text{adjacent}}{\text{opposite}} = \frac{A}{O}$$

$$\text{Secant} = \frac{\text{hypotenuse}}{\text{adjacent}} = \frac{H}{A}$$

$$\text{Cosecant} = \frac{\text{hypotenuse}}{\text{opposite}} = \frac{H}{O}$$

Where:

H = hypotenuse, side facing the right (90°) angle
O = side opposite the angle you are working with
A = side adjacent to the angle you are working with, but not the hypotenuse

Therefore, in *Figure 6(B)*, the hypotenuse or distance between bends could be computed by using the following cosecant trigonometric function:

$$\text{Cosecant} = \frac{\text{hypotenuse}}{\text{opposite}} = \frac{H}{O}$$

Let's say that the angle of the first bend is 30° and the rise or side opposite is 12". Therefore:

$$\text{Cosecant } 30° = \frac{H}{12"}$$

The cosecant of 30° is 2 (check using your calculator).

$$2 = \frac{H}{12"}$$

Cross multiply:

$$H = 2 \times 12"$$
$$H = 24"$$

Therefore, the hypotenuse or distance between bends would be 24".

Now let us complete another example. In this case, a kick is to be made on the end of a piece of conduit (*Figure 8*). The angle of the kick is to be 30°. The hypotenuse of the kick is 20". Determine how far off the surface the end of the kick needs to be.

When calculating the hypotenuse of a bend, the cosecant of the angle is multiplied by the side opposite. If the angle and the hypotenuse are known, then the inverse can be used to determine the side opposite. Therefore, divide 20" by the cosecant of 30°, or:

$$20" \div 2 = 10"$$

The end of the conduit needs to be brought 10" off the surface to acquire a 30° bend. This process eliminates the need for a protractor level.

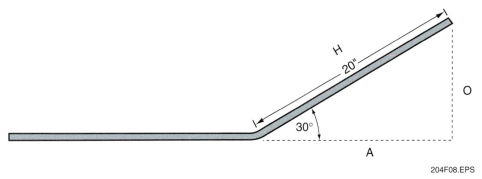

Figure 8 ♦ Kick example.

4.2.0 Circle

A circle is defined as a closed curved line whose points are all the same distance from its center, as shown in *Figure 9(A)*. The distance from the center point to the edge of the circle is called the radius and the length of a straight line from one edge through the center to the other edge is called the diameter. The distance around the circle is called the circumference. A circle can be divided into four equal quadrants, as shown in *Figure 9(B)*. Each quadrant accounts for 90°, making a total of 360°. When making 90° bends, you will be interested in ¼ of a circle, or one quadrant.

Concentric circles, shown in *Figure 9(C)*, are several circles that have a common center but different radii. The concept of concentric circles can be applied to concentric 90° bends in conduit. Such bends have the same center point, but the radius of each is different. *Figure 10* shows how parts of a circle relate to a 90° conduit bend.

For bending conduit, it is necessary to understand the dynamics of the unit circle (*Figure 11*). The unit circle is a circle with a given radius of 1.

In calculating the circumference of a circle, the following formula is used:

$$C = 2\pi R$$

Where:

$$C = \text{circumference}$$
$$\pi = 3.14 \text{ (pi)}$$
$$R = \text{radius}$$

By taking the radius of the unit circle and substituting it into the formula, the result is:

$$C = 2\pi \times 1$$

Any number multiplied by the number 1 is equal to that number, or:

$$1 \times 2 = 2$$

Therefore, $2\pi \times 1 = 2\pi$. This means that in terms of pi, the circumference of the circle is 2π; 360° is 2π and 180° = π. See *Figure 12*.

If you look at 90° in terms of π, 90° is half of 180° or half of π, so 90° is equal to ½ π or $\pi \div 2$. Again, π is a symbol for the numerical value 3.14

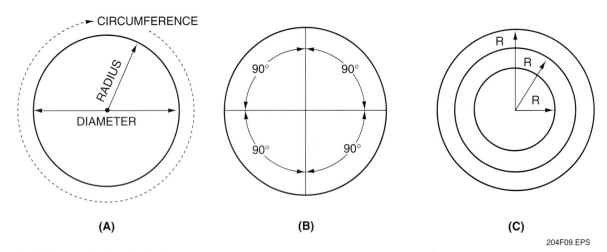

Figure 9 ♦ Characteristics of a circle.

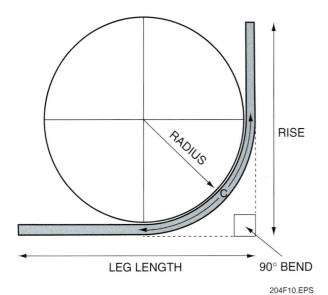

Figure 10 ♦ Parts of a circle related to conduit bending.

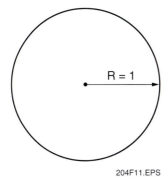

Figure 11 ♦ Unit circle.

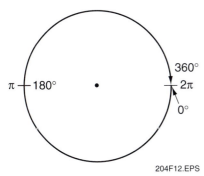

Figure 12 ♦ π and 2π.

(rounded). π ÷ 2 is 3.14 ÷ 2 or 1.57. Therefore, 90° is represented by the numerical value 1.57. Looking back at the circumference formula shows that multiplying 2 × π or 6.28 times the given radius of the circle gives the linear distance around the circle. With that in mind, if we multiply 1.57 times the given radius, it will provide the linear distance from 0° to 90° on the circle. This linear distance is

known as the **developed length** in regard to the amount of conduit that is required to make a 90° sweep.

4.2.1 Types of Angular Representations

There are three basic ways that angles can be represented or specified. Angles can be stated in degrees, as already mentioned; radians; or gradients.

Angles are measured in degrees from 0° to 360°. They may also be measured in radians from 0 to 6.28 or 0 to 2π. The third way is to specify angles in gradients, from 0 to 400 gradients (grads). These measurements can be equated as follows:

$$90° = \pi \div 2 = 1.57 \text{ radians} = 100 \text{ grads}$$
$$180° = \pi = 3.14 \text{ radians} = 200 \text{ grads}$$
$$270° = 3\pi \div 2 = 4.71 \text{ radians} = 300 \text{ grads}$$
$$360° = 2\pi = 6.28 \text{ radians} = 400 \text{ grads}$$

4.3.0 Equations

To calculate conduit bends accurately, you need to use some basic equations. The equations given here are the most relevant ones for the electrician. Examples are provided for clarification. To calculate the circumference of a circle, use the following equation:

$$C = \pi D$$

Where:

C = circumference
D = diameter
π = 3.14

As discussed earlier, another equation for finding the circumference of a circle is as follows:

$$C = 2\pi R$$

Where:

C = circumference
R = radius (½ the diameter)
π = 3.14

If the radius in a circle (and conduit bend) measures two feet, the circumference of the circle may be found as follows:

$$C = 2\pi R$$
$$C = 2 \times 3.14 \times 2' = 12.56'$$

The arc of a quadrant equals ¼ the circumference of the circle, and its length is found as follows:

Length of arc = $(0.25)(2\pi R)$
Length of arc = $(0.25)(2 \times 3.14 \times R)$
Length of arc = $1.57R$

Therefore, the length of the arc in the quadrant, if the radius is 2', may be found as follows:

Length of arc = 1.57R
Length of arc = 1.57 × 2' = 3.14'

If the radius of the bend and the outside diameter of the conduit are known, then the distance C (the length of the arc) can be calculated. This length, as related to conduit bending, is known as the developed length; it is the actual length of the bend.

Conduit bends with the circumference of the circle (*Figure 13*) and not at right angles. Therefore, the length of the conduit needed for a bend will not equal the right angle distances A and B. **Gain** is the difference between the right angle distances A and B and the shorter distance C, the length of conduit actually needed for the bend.

The gain for a 90° bend is arrived at by multiplying the radius of the bend × 0.43 (rounded up from the value of 0.4292 in *Table 3*). Therefore, if the radius of the bend in *Figure 13* is 2', the gain may be found as follows:

Gain = 2' × 0.43 = 0.86' = 10.32"

4.3.1 Making a 90° Bend

The 90° stub bend is probably the most basic bend of all. The stub bend is used much of the time, regardless of the type of conduit being installed. Before beginning to make the bend, you need to know two measurements:

- The desired rise or stub-up
- The **take-up (comeback)** distance of the bender

The desired rise is the height of the stub-up. The take-up is the amount of conduit the bender will use to form the bend. Take-up distances are usually listed in the manufacturer's instruction manual. Once the take-up has been determined, subtract it from the stub-up height. Mark that distance on the conduit (all the way around) at that distance from the end. The mark will indicate the point at which you will begin to bend the conduit. Line up the starting point on the conduit with the starting point on the bender. Most benders have a mark, like an arrow, to indicate the starting point.

NOTE

When bending conduit, the conduit is placed in the bender and the bend is made facing the end of the conduit from which the measurements were taken.

As discussed above, if the radius of the bend in *Figure 13* is 2', the gain is 10.32".

Table 4 shows the equivalent fractions for many decimals. For 10.32", search the decimal column

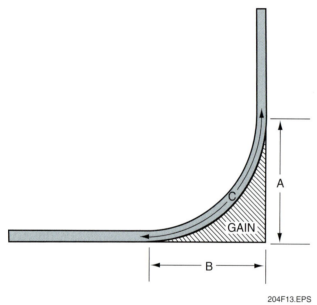

Figure 13 ♦ Gain.

Table 3 Gain Factors

	1°	2°	3°	4°	5°	6°	7°	8°	9°
0°	0	0	0	0	0	0	0.0001	0.0001	0.0003	0.0003
10°	0.0005	0.0006	0.0008	0.001	0.0013	0.0015	0.0018	0.0022	0.0026	0.0031
20°	0.0036	0.0042	0.0048	0.0055	0.0062	0.0071	0.0079	0.009	0.01	0.0111
30°	0.0126	0.0136	0.015	0.0165	0.0181	0.0197	0.0215	0.0234	0.0254	0.0276
40°	0.0298	0.0322	0.0347	0.0373	0.04	0.043	0.0461	0.0493	0.0527	0.0562
50°	0.06	0.0637	0.0679	0.0721	0.0766	0.0812	0.086	0.0911	0.0963	0.1018
60°	0.1075	0.1134	0.1196	0.126	0.1327	0.1397	0.1469	0.1544	0.1622	0.1703
70°	0.1787	0.1874	0.1964	0.2058	0.2156	0.2257	0.2361	0.247	0.2582	0.2699
80°	0.2819	0.2944	0.3074	0.3208	0.3347	0.3491	0.364	0.3795	0.3955	0.4121
90°	0.4292	—	—	—	—	—	—	—	—	—

Table 4 Decimal Equivalents of Some Common Fractions

Fraction	Decimal	MM	Fraction	Decimal	MM
1/64	0.015625	0.397	33/64	0.515625	13.097
1/32	0.03125	0.794	17/32	0.53125	13.494
3/64	0.046875	1.191	35/64	0.546875	13.891
1/16	0.0625	1.588	9/16	0.5625	14.288
5/64	0.078125	1.984	37/64	0.578125	14.684
3/32	0.09375	2.381	19/32	0.59375	15.081
7/64	0.109375	2.778	39/64	0.609375	15.478
1/8	0.125	3.175	5/8	0.625	15.875
9/64	0.140625	3.572	41/64	0.640625	16.272
5/32	0.15625	3.969	21/32	0.65625	16.669
11/64	0.171875	4.366	43/64	0.671875	17.066
3/16	0.1875	4.763	11/16	0.6875	17.463
13/64	0.203125	5.159	45/64	0.703125	17.859
7/32	0.21875	5.556	23/32	0.71875	18.256
15/64	0.234375	5.953	47/64	0.734375	18.653
1/4	0.25	6.35	3/4	0.75	19.05
17/64	0.265625	6.747	49/64	0.765625	19.447
9/32	0.28125	7.144	25/32	0.78125	19.844
19/64	0.296875	7.54	51/64	0.796875	20.241
5/16	0.3125	7.938	13/16	0.8125	20.638
21/64	0.32812	8.334	53/64	0.828125	21.034
11/32	0.34375	8.731	27/32	0.84375	21.431
23/64	0.359375	9.128	55/64	0.859375	21.828
3/8	0.375	9.525	7/8	0.875	22.225
25/64	0.390625	9.922	57/64	0.890625	22.622
13/32	0.40625	10.319	29/32	0.90625	23.019
27/64	0.421875	10.716	59/64	0.921875	23.416
7/16	0.4375	11.113	15/16	0.9375	23.813
29/64	0.453125	11.509	61/64	0.953125	24.209
15/32	0.46875	11.906	31/32	0.96875	24.606
31/64	0.484375	12.303	63/64	0.984375	25.003
1/2	0.5	12.7	1	1	25.400

for 0.32. The closest decimal in the table is 0.3125, equivalent to 5/16. Therefore, the number in question becomes 10 5/16".

A decimal may also be mathematically converted to a fraction. A decimal whose denominator is contained in the numerator without a remainder can easily be converted to a fraction by removing the decimal point from the numeral, which then becomes the numerator (the top numeral of the fraction). The denominator is always one plus as many zeros as there are decimal places in the decimal. For example:

$$0.75 = 75/100 = 3/4$$

or:

$$0.375 = 375/1000 = 3/8$$

NOTE

Decimals may also be converted to fractions by using an electronic calculator, provided the calculator has a fraction key. The exact procedure will vary with the different models of calculators.

Gain factors for 0° to 90° bends are shown in *Table 3*. To demonstrate the use of this table, assume that it is desired to find the gain on a 45° conduit bend with a 15" centerline radius. Referring to *Table 3*, look in the left-hand column; glance down the column until the number 40° is found. Since the bend is 45°, read to the right in this row until the column titled 5° is found; note the figure, 0.043. Therefore, the gain factor for a 45° bend is 0.043.

Multiply the gain factor by the centerline radius (15") to obtain the full gain of a 45° bend.

$$0.043 \times 15" = 0.645"$$

To convert this figure to a readable figure on the foot rule, convert the decimal to a common fraction: $^{645}/_{1000}$ = approximately ⅝. Thus, the full gain of the 45° bend is ⅝".

For example, the gain for a given bend is 2½". In *Figure 14*, there are two back-to-back 90° bends. The rise for the first 90° bend is 2' and the rise for the second 90° bend is 3'. The back-to-back measurement for these bends is 4'. The task is to determine the amount of straight conduit required to accomplish this task. The conduit is to be cut to length and threaded before bending.

The gain is the amount of conduit that is saved by the radius of the bend. The amount of conduit saved is 2½" per 90° bend. In this situation, there are two 90° bends, or a savings of 5". The total straight lengths add up to 9' of conduit. Therefore:

$$9' - 5" = 8'\text{-}7"$$

5.0.0 ◆ MECHANICAL BENDERS

Conduit bends are normally made in the smaller sizes of conduit and tubing by hand with the use of hickeys or EMT bending tools. However, on many projects, an advantage can be gained by the use of mechanical bending equipment with suitable adjustable stops and guides. With the use of such equipment, the exact bend can be duplicated in quantity with a minimum of effort. The angle of the bend and the location of the bend in relation to the end of the length of conduit are preset.

A popular mechanical bender is shown in *Figure 15*. This type of bender was originally called the Chicago bender, as it was made by the Chicago Equipment and Manufacturing Company. Today, however, this type of bender is manufactured by several different companies and the correct name is portable mechanical conduit bender. In any event, you may still hear the term Chicago bender on many jobs.

This type of mechanical bender is very popular with many electrical contractors. To use it, a length of conduit is placed in position and secured in place; a long bending handle is then pulled around and the bend completed. This type of bender may be used as a one-shot bender for the smaller sizes of conduit (bypassing the ratchet

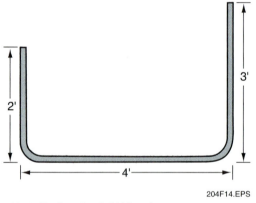

Figure 14 ◆ Back-to-back 90° bends.

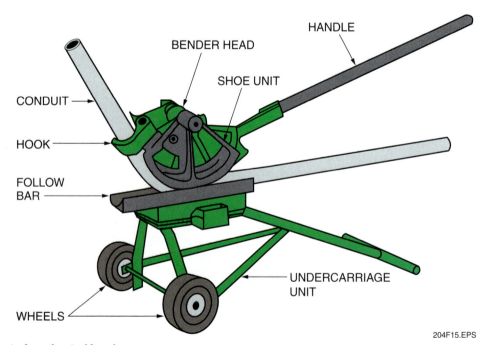

Figure 15 ◆ Typical mechanical bender.

mechanism). The ratchet mechanism, however, is usually activated when bending the larger sizes of conduit to make the work easier. It is suitable for making bends in conduit sizes up to 2" EMT, 1¼" IMC, and 1½" rigid conduit, provided the proper bending accessories are used (e.g., bending shoes, follow bars, etc.).

Bending shoes and follow bars are designed to form a particular radius bend for a certain type and size of conduit, that is, EMT, IMC, or rigid conduit. These accessories should be treated as precision instruments; any damage will result in inaccurate bends, kinks, and so on. The first consideration is to use only the proper shoe and follow bar for the type and size of conduit being bent. For example, never use an EMT shoe for bending rigid conduit or a 3" shoe for bending 2½" conduit. In general, make certain that the bending shoes and follow bar are compatible with the type and size of conduit to be bent; to do otherwise may damage the tool and result in inaccurate bends.

The ratchet feature is normally engaged for the larger sizes of conduit, whereas a spring-loaded pawl engages the ratchet for easier bending in segments. For the smaller sizes of conduit, however, the ratchet may be bypassed so that the bend can be made in one shot.

A bending gauge with an adjustable pointer on the bender is helpful when making multiple bends at the same angle. This pointer is set at the desired angle and then the setscrew is tightened. As the bend is being made, the handle is operated until the pointer reaches the index mark. To ensure the correct angle of bend, the first bend should be checked with a **bending protractor** (*Figure 16*) and any necessary adjustments made to the bending gauge pointer before continuing. All successive bends will be exactly the same as the first.

For example, follow this process for charting a bender. This chart will contain minimum size, gain, and centerline distances from the arrow on the bender to the center of the bend for 15°, 30°, 45°, and 60° bends. This is to be accomplished with one scrap piece of conduit.

For this discussion, the scrap piece will be ½" conduit and will be 3' in length (*Figure 17*).

Place a mark on the conduit at a given distance from the end of the conduit. For this discussion, the mark will be 10" in from the end of the conduit (*Figure 18*).

Take a ½" conduit bender and place the arrow, which represents the take-up or minimum rise of the bender, on the mark 10" in from the end of the bender.

Place a protractor level on the conduit in front of the bender. This will be on the piece of conduit marked for 10" back from the end. Bend the 10" portion until the protractor level reads 15°. Remove the bender from the conduit (*Figure 19*).

Now take a straightedge, such as a ruler or torpedo level, and lay it on the inside of the bend so the straightedge lies across the bend and rests against the straight portion of the conduit (*Figure 20*).

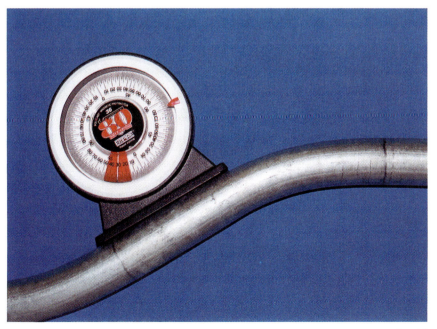

Figure 16 ◆ Bending protractor.

Figure 17 ◆ Conduit.

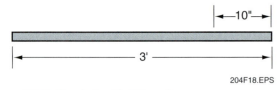

Figure 18 ◆ Conduit with 10" mark.

Figure 19 ◆ Kick of 15°.

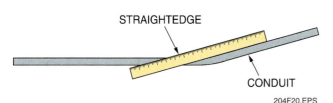

Figure 20 ◆ Conduit and straightedge.

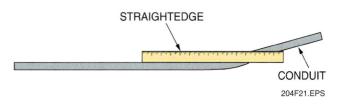

Figure 21 ◆ Conduit and horizontal straightedge.

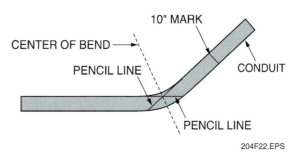

Figure 22 ◆ Center of bend.

With a sharp pencil, scribe a line across the bend of the conduit.

Now take the straightedge and lay it across the conduit so the straightedge is against the side adjacent to the previous position. It should extend across the bend once again (*Figure 21*).

Now scribe a line across the bend of the conduit. The two pencil lines should cross to form an X on the conduit. This represents the center of the bend (*Figure 22*).

Now take a tape measure and measure from the 10" mark to the centerline of the bend. Record this distance as follows:

15° – 1" (assuming 1" is the measured distance)

Now place the bender back on the conduit so that the take-up arrow is on the 10" mark on the conduit. Place the protractor level back on the conduit. Bend the conduit until the protractor level reads 30°. Take the bender off the conduit and repeat the line crossing process that was discussed for 15°, using the same straightedge. Once again, measure from the 10" mark to the point where the pencil lines cross in the center of the bend. Record this measurement.

15° – 1"

30° – 1½" (assuming this is the measured distance)

Repeat the previous process for 45°. Record this measurement.

15° – 1"

30° – 1½"

45° – 2" (assuming this is the measured distance)

Repeat the previous process for 60°. Record this measurement.

15° – 1"

30° – 1½"

45° – 2"

60° – 2¼" (assuming this is the measured distance)

The centerline distances for the different bends from the take-up mark of the bend have now been recorded. This portion of the chart is now completed.

The next step is to determine the minimum rise of the bender. Place the bender back on the conduit so the take-up mark of the bender is on the 10" mark on the conduit. Place the protractor level back on the conduit and bend the conduit until the protractor level reads 90°. Take the bender off the conduit. Measure from the back of the conduit to the 10" mark. The reading on the tape is 5". This is the minimum 90° stub length (*Figure 23*). Record this measurement.

15° – 1"
30° – 1½"
45° – 2"
60° – 2½"

Minimum rise – 5" (assuming this is the measured distance)

The next step is to determine the gain of the bender for 90°. Measure the length of both sides of the scrap piece of conduit (*Figure 24*).

Add the two measured stub lengths together.

15" + 1'-11⅝" = 3'-2⅝"

Subtract the original length of the conduit, which was 3', from 3'-2⅝".

3'-2⅝" – 3' = 2⅝"

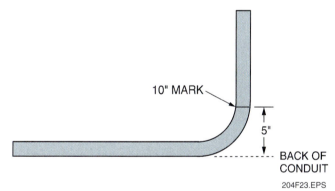

Figure 23 ◆ 90° stub-up.

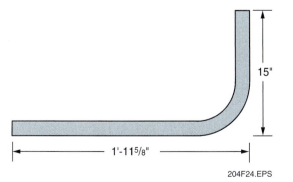

Figure 24 ◆ 90° elbow.

This is the gain for this particular ½" conduit bender. Record this information.

15° – 1"
30° – 1½"
45° – 2"
60° – 2½"

Minimum rise – 5"

Gain – 2⅝" (assuming this is the measured distance)

The ½" bender has now been charted. This same process would need to be repeated for a ¾" bender, 1" bender, 1¼" bender, 1½" bender, etc.

This charting process should help you to understand how the manufacturer of a bender comes up with the marks that determine the ability to bend exact 90° bends of any length, which is the minimum rise mark. The star mark seen on many benders is used for back-to-back bends. The star mark is 2⅝" back from the minimum rise mark (take-up). In other words, it is the measured gain distance back from the minimum rise mark. The centerline marks are used in lining up the centers of bends for various sizes of conduit. The centerline marks can also be used in lining up saddles on the centerlines of I-beams and process piping. This information is good for charting any type of wrap-around bender.

6.0.0 ◆ MECHANICAL STUB-UPS

Stub-ups are quickly and easily made with mechanical benders. A deduct decal is provided on many benders, but sometimes these decals become damaged, making them difficult to read. Therefore, backup charts should be provided on all jobs. With this deduct chart on hand, the following is an example of making a 90° stub-up to a given height.

Assume that you are working on a deck job and need a number of 1" rigid stub-ups with a rise of 15" each. When you check the deduct chart on the bender for a 15" stub-up using 1" rigid conduit, you note that 11" should be deducted from the total rise of 15". Since 15" – 11" = 4", measure back from the end of the conduit by 4" and make a mark. Encircle the entire conduit at this point so you will not lose the mark once the conduit is placed in the bender. Many electricians like to use a black felt-tip marker for marking conduit.

Load the conduit into the bender with the mark lined up with the front of the bender hook. Engage the ratchet and start pumping the bender handle until the bender pointer reaches the preset index mark for 90°. Move the bender handle forward, then remove the conduit from the bender and check its height. It should be exactly 15". If the height of the bend is slightly off, make the necessary adjustments before continuing. Once the correct height is reached, the remaining bends will also be correct.

7.0.0 ◆ MECHANICAL OFFSETS

The decal chart on the bender that provides deduct information for the stub-ups also contains data for making offsets that require 20°, 30°, and 45° bends. This chart is necessary to make perfect offsets every time using the mechanical bender.

Assume that you are running a raceway system with ½" rigid conduit and an air duct must be bypassed, requiring the conduit run to be offset. After taking measurements on the job, we find that an offset of 12" is needed to clear the air duct.

Measure the distance from the end of the conduit to the start of the first bend; mark the conduit as before. Referring to the chart on the bender with offset information, we decide to make the offset with two 45° bends. The chart indicates that the distance between bends is 16⁵⁄₁₆". Therefore, measure and mark this distance back from the first mark.

Insert the conduit into the bender and line up the first mark with the front of the bender hook. The ratchet may be used, but for ½" conduit, the ratchet override on the front of the bender shoe is normally employed. Make the first 45° bend. Move the bender handle forward to release the conduit. Now slide the conduit forward through the bender hook until the second mark lines up with the front of the bender hook, and then turn the bend over so the end of the conduit is pointing downward toward the deck. Also, make sure that the first bend lines up with the next bend to be made to prevent a crooked bend (commonly referred to as a dog leg) in the conduit. Once everything is aligned, engage the bender handle and make another 45° bend. The height of the offset should be exactly 12" (see *Figure 25*).

The distances between marks for offset bends will vary depending upon the size and type of conduit being bent, so always check the offset information on the bender.

7.1.0 General Tips for Bending with the Chicago Bender

The following are some general tips to keep in mind when bending with the portable mechanical conduit bender:

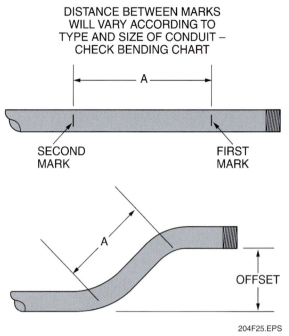

Figure 25 ◆ Bending offsets in conduit.

- An engineer's rule marked in hundredths will simplify formula bending by eliminating the need to convert to fractions.
- When minimum length stubs are being bent, the shoe tends to creep and deforms the end of the conduit and threads. Screwing a coupling onto the pipe stops the shoe from creeping forward and protects the threads.
- When bending offsets, the front of the bender can be temporarily elevated for clearance requirements.
- Most bender shoes are made of cast aluminum and are easily pitted and gouged if foreign material gets between the shoe and the pipe. For longer shoe life, keep the pipe clean and the shoes wiped down.
- When the remaining pipe length is not long enough to reach the roller or pipe support, a larger diameter conduit can be slid over the pipe being bent to complete the bend; or, if the pipe has threads, screw on a coupling and a short piece of scrap pipe.
- Segment and concentric bending of smaller sizes of pipe can be performed with this bender. Bend a scrap piece of pipe and measure from the center of the bend to the front of the bending shoe. Use this measurement to adjust the start mark using the segment and concentric bending procedures.
- To make matching bends in two different sizes of conduit when using a mechanical bender, make both bends using the larger shoe.

Making Accurate Offsets

Be sure to use the same angle for both bends of an offset. To avoid dog legs, make sure your first bend lines up evenly with the rest of the conduit, the handle, and the bender. Take your time, and be sure of proper alignment before making the second bend. A No-Dog® is a simple, pocket-sized device that may be used to prevent crooked bends. It is screwed onto the end of the conduit and has a built-in level to ensure straight bends.

NOTE

For matching bends in sizes ½", ¾", and 1" rigid conduit and IMC, bend all pipes with the 1" shoe.

8.0.0 ◆ ELECTRIC AND HYDRAULIC CONDUIT BENDERS

Bends in larger sizes of conduit (over 2" EMT, 1¼" IMC, or 1½" rigid conduit) are normally made using hydraulic benders. In some instances, bending tables have been developed for use with bending tools or hydraulic benders to simplify making bends to certain dimensions. Various adaptations of benders have been developed to serve certain specific purposes. For example, one type of hydraulic bender is designed for making bends in a section of conduit that has been installed in a raceway system.

8.1.0 Electric Conduit Benders

Electric conduit benders operate on the same basic principle as the mechanical benders described previously except that the bending is accomplished by a gear motor rather than manual power. There are several types of electric benders on the market. A typical electric unit is shown in *Figure 26*. This bender will make stub-ups, offsets, and saddles quickly and easily at any location where a 120V receptacle is available. Furthermore, this bender will make 90° one-shot bends in ½" through 2" EMT, IMC, or rigid conduit.

Bending charts are often included in the form of decals attached to the bender for quick reference on the job. These include deduct and springback figures along with information for making offsets.

Many electric benders are sold as a power unit and the bending shoes for various types of conduit are sold as accessories. These shoes quickly snap into place on the power unit, ready for use in seconds.

To describe the operation of the example bender, assume that you need a 16" stub-up in a piece of 1¼" rigid conduit. Measure and mark a piece of conduit 16" from one end. When you check the deduct chart on the bender for this size conduit, the figure is 12¾". Therefore, 12¾" must be deducted from 16", which leaves 3¼". So a mark is

Figure 26 ◆ Typical electric bender.

made at this distance from the end of the conduit. Encircle the conduit with this mark so it will be plainly visible during the bend.

Insert the conduit in the bender with the mark lined up with the front of the bender hook. For this particular bend, the mark is 3¼" from the end of the conduit.

The three-position operating switch on the example bender is attached to a flexible cord. The center position de-energizes the machine, the up position (forward) is for bending, and the down position (reverse) is for unloading the conduit after the bend has been made. When released, the switch automatically springs to the center or off position. A pointer on the shoe indicates the degree of bend as the bend is being made.

The machine is jogged or inched up until the shoe pointer lines up with the zero mark on the degree scale. The switch is then pressed upward and held in this position to start the bend. As the pointer approaches the 90° mark, refer to the springback chart for the size and type of conduit being bent. In this case, 1¼" rigid conduit, the chart indicates 95°. Therefore, the pointer should pass the 90° mark and stop at 95° to allow for springback when the conduit is removed from the bender; the stub-up is ready for installation.

An offset is made in the electric bender similar to the method described for the mechanical bender; that is, the first mark is located on the conduit and then offset information is obtained from the bender chart. In the case of a 16" offset in a length of 1¼" rigid conduit, the chart indicates a distance of 22⅝" between marks for a 16" offset using 45° bends.

Mark the conduit and insert the conduit into the bending shoe, positioning the mark so that it is even with the front of the bender hook. Start the bend as discussed previously until the pointer reaches the 45° mark on the scale; allow for any springback as indicated in the chart. Reverse the motor until the conduit is loose, then turn the conduit upside down. Position the second mark at the front of the bender hook, making certain the conduit is aligned to prevent a dog leg in the bend. Start the second bend until the pointer reaches the 45° mark, reverse the bender, and unload the conduit. A magnetic torpedo level placed on the side of the conduit will help align bends for offsets. An anti-dog device can also be used for this.

Finally, check the bends for accuracy. Although rebending is possible, it is not a good practice. Rebending puts considerable strain on the conduit, and while it may not break, the coating may crack and cause corrosion.

8.1.1 *Speed Benders*

A speed bender operates basically the same as the standard bender except the speed bender utilizes remote digital control with easy-to-use bending charts and instructions to ensure fast, accurate, and consistent bends in conduit sizes from ½" to 2".

This bender can be operated upright or laid on its back for large offsets and saddles. In operation, rather than holding the switch up in the bend position as with the standard bender, the operator sets the bend wanted via digital controls and the bender automatically bends to that degree once the conduit is placed in the bending shoe and the start control is activated. Otherwise, the conduit is marked and bent in the same way as described for the standard bender.

8.2.0 Hydraulic Conduit Benders

The hydraulic bender is indispensable for bending large conduit. Only by the addition of the hydraulic pump and cylinder to a bender frame is the necessary power available for bending rigid conduit up to 6" trade size.

The typical hydraulic bender (*Figure 27*) provides extended bending capability, but also requires additional responsibility. Hydraulic bending equipment represents a large initial investment, and the bender must be properly cared for if it is to last and give the service necessary to justify the cost.

WARNING!
Dangerous pressures are used in hydraulic benders. Exercise care when working with or around hydraulic benders.

The bender frame, shoes, follow bars, and saddles will require little more than occasional wiping down to remove the dirt and oil film. The pump, cylinder, and hydraulic hoses, however, will demand more attention.

The pump should have its fluid level checked periodically. Low fluid pressure will keep the pump from developing its rated hydraulic pressure. When the level is low and fluid must be added, use only approved hydraulic oil. Ordinary oil will cause damage to the pump assembly.

Hydraulic benders have two types of bending shoes, one-shot and segment.

The one-shot shoe has a full 90° radius. The conduit can be formed around the shoe to a full 90° bend without collapsing the walls of the pipe.

Hydraulic Bender Safety

A hydraulic bender generates a tremendous amount of power and operates under dangerous pressures. Do not attempt to use a hydraulic bender unless you have been properly trained and are thoroughly familiar with the unit's operating and safety instructions. Even then, exercise extreme caution and always work in conjunction with a bending partner.

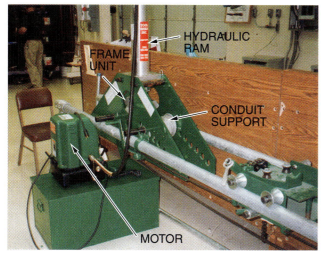

Figure 27 ♦ Hydraulic conduit bender.

One-shot benders require no new bending techniques. The take-up is figured to the center of the bend and the bender shoe will have an indicating mark at its center. If take-up values are not listed on the bender frame or bender storage case, scrap pipe can be bent and take-up figures found. All the methods and layout techniques discussed for EMT and rigid metal conduit can be used with a hydraulic bender and one-shot shoes. Offsets will have to be adjusted (less angle of bend) so the spacing between the bend marks is far enough apart to allow the first bend to be rolled 180° and advanced enough to clear the shoe.

Segment shoes are shorter and have a radius that is far less than 90°. A **ninety-degree bend** cannot be made in one operation, as the conduit walls would collapse. Bends, then, must be made in several steps (as few as four and as many as 30) to form a smooth radius. The **segmented bending shoe** allows pipe to be bent to larger size radii.

Segmented shoes are used for concentric bending (bending several conduits with increasing or decreasing radii). One-shot shoes can be used for a **segment bend**, but are not as convenient. Concentric bending is covered in detail later in this module.

Accurate bending of large conduit is possible but requires practice, patience, and ability. With few exceptions, all formulas and bending techniques discussed to this point will apply.

8.2.1 Bending Tips for Rigid Aluminum

Rigid aluminum conduit is available in all trade sizes from ½" through 6". It is lightweight, corrosion resistant, and has low ground impedance. It is, however, difficult to bend consistently and accurately. Two pipes out of the same bundle will act differently when bent. Even if two pipes are bent using the same layout, they do not always come out the same. Do not be discouraged by this; it is the nature of the metal and it cannot be helped.

Another disadvantage of aluminum is that a one-shot bending shoe will dig in and score the pipe. Also, where the pipe rides on the shoe, it is prone to wrinkling and scoring. Applying petroleum jelly or a lubricant such as WD-40® to the shoe will allow the pipe to slide without the shoe digging in. Petroleum jelly will also make it easier to remove the conduit when the bend is complete.

8.2.2 One-Shot Bending

Accurate one-shot stub-ups are easily made on hydraulic benders by applying a little basic geometry in making calculations, and then knowing the operating principles of the bender. *Figure 28* shows the reference points of a common 90° bend. To make one-shot 90° bends, first determine the **leg length** and rise, the gain, the radius of the bend, and the half-gain.

Use the following procedure for laying out accurate stub-ups:

Step 1 Determine lengths A and B.

Step 2 Add lengths A and B. Subtract X for the length of pipe required.

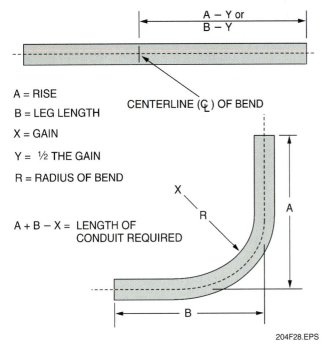

Figure 28 ♦ Laying out stub-ups.

8.2.3 90° Segment Bends

When bending conduit in segments with a hydraulic bender, the following factors must be determined:

- The size of conduit to be bent
- The radius of the bend
- The total number of degrees in the bend
- The developed length
- The gain of the bend

To determine the developed length for a 90° bend, multiply the radius by 1.57. The next step is to locate the center of the bend. Most benders have the center mark indicated on the bending shoes. Once the center mark on the conduit is found, it is easy to locate the other bend marks (see *Figure 29*).

You must now determine the number of **bending shots** that will make the bend to suit the requirements, preferably an odd number so that there are an equal number of bends on each side of the center mark. Next, calculate the width of the spaces for each segment bend and make the layout on the conduit. Make an equal number of spaces on each side of the center mark. The gain need only be determined if the bend is being fitted between two existing conduit runs or junction boxes.

To determine the bending data for a 90° bend using 3" conduit with a rise of 48", a leg length of 46", and a centerline radius of 30", proceed as follows (see *Figure 30*).

Step 1 Multiply the radius by 1.57 to determine the developed length. Therefore:

30" × 1.57 = 47.10" (47⅛") developed length

Step 2 Determine the gain for a 90° bend.

Gain = (2 × R) − developed length
Gain = (2 × 30") − 47⅛" = 12⅞"

Step 3 To calculate the overall length (OL) of the conduit, add the leg and stub-up lengths and subtract the gain. See *Figure 30*.

OL = leg length + rise − gain
OL = 46" + 48" − 12⅞"
OL = 81⅛"

Step 3 Subtract Y from length A or B to get the center of the bend.

Step 4 Calculate the developed length and the length of conduit required.

Step 5 Determine the center of the bend. This can be done by taking the half-gain value from the distance A or B. See *Table 5*.

Table 5	Dimensions of Stub-Ups for Various Sizes of Conduit			
Pipe and Conduit Size	**Radius of Bend R**	**Minimum Developed Length 90°**	**Gain X**	**½ Gain Y**
½"	4"	6⁵⁄₁₆"	1¹¹⁄₁₆"	²⁷⁄₃₂"
¾"	4½"	7¹⁄₁₆"	1¹⁵⁄₁₆"	³¹⁄₃₂"
1"	5¾"	9"	2½"	1¼"
1¼"	7¼"	11⅜"	3⅛"	1⁹⁄₁₆"
1½"	8¼"	13"	3½"	1¾"
2"	9½"	14¹⁵⁄₁₆"	4¹⁄₁₆"	2¹⁄₃₂"

Developed Length

Why is the radius multiplied by a factor of 1.57 to determine the developed length for a 90° bend?

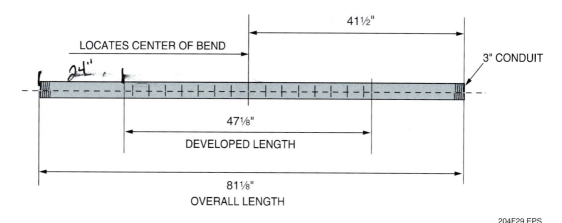

Figure 29 ◆ Laying out segment bends.

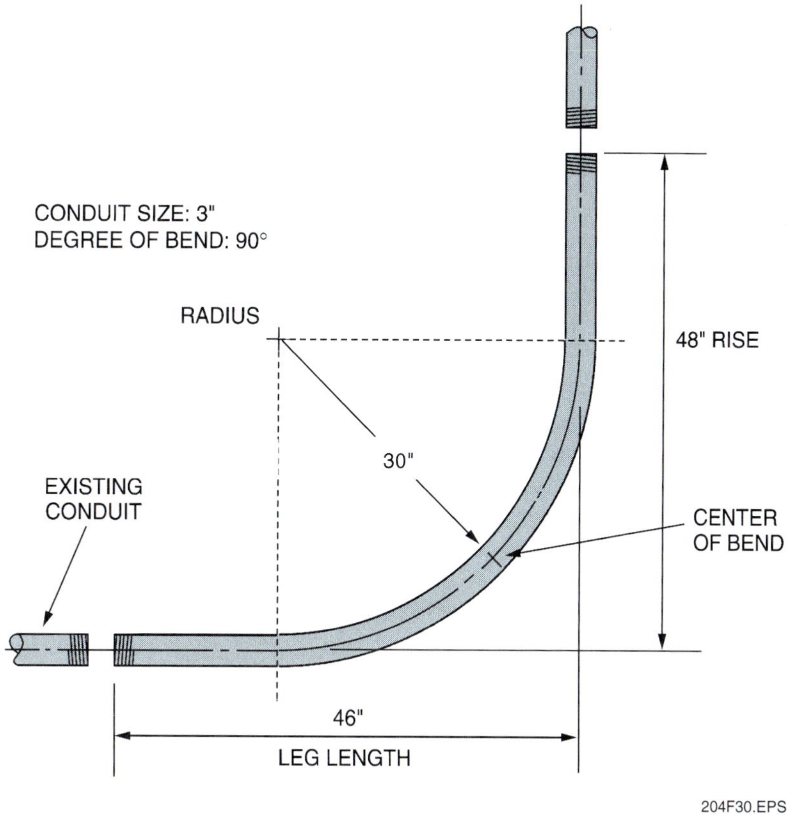

Figure 30 ◆ Specifications for sample bend.

Step 4 Now locate the center of the required bend. First, determine one-half of the developed length:

$$\tfrac{1}{2}\,(47\tfrac{1}{8}") = 23.56" = 23\tfrac{1}{2}"$$

Use the rise or stub-up dimension of 48". Subtract the radius (30") and add one-half of the developed length:

$$48" - 30" + 23\tfrac{1}{2}" = 41\tfrac{1}{2}"$$

Step 5 As a rule, 6° or less per bend will produce a good bend for a 30" radius. In this case, 6° per bend will be used, making 15 segment bends (90 ÷ 6 = 15). An odd number of segment bends is easy to lay out after finding the center mark because there will be an equal number of spaces on each side of the center mark.

MODULE 26204-08 ◆ CONDUIT BENDING 4.21

Step 6 To determine the space between the segment marks, divide the developed length by the total number of segments:

$$47\frac{1}{8}" = 47.125"$$
$$47.125" \div 15 = 3.14"$$
$$3.14" = 3\frac{1}{8}"$$

Step 7 Position the conduit in the pipe holders, making sure to clamp them securely.

Step 8 Place the center mark 41½" from one end of the conduit. Next, mark seven points on each side of the center point, 3⅛" apart, for a total of 15 marks. These are the centers of the segment bends.

Step 9 It is a good idea to check the distance between the first and last bend marks to be sure the layout is correct before starting the first bend. The distance from the first mark to the last is the developed length minus the length of one bend. (Actually, you are subtracting one-half of a segment bend from each end of the conduit.)

$$47\frac{1}{8}" - 3\frac{1}{8}" = 44"$$

Step 10 After positioning the conduit in the bender (*Figure 31*), attach the pipe bending **degree indicator** in a convenient location.

Step 11 Attach the pipe supports with the proper face toward the conduit and insert the pipe support pins. Lock them in position by turning the small lock pin. Now, proceed to make the series of bends.

Step 12 Begin by bending 6° on the first mark. When this is done, the indicator will read 6°. Release the pressure and check the springback; if any is found, overbend by the same amount.

Step 13 When using a bender with a rigid frame, move the pipe support one hole position in (toward the ram) on the side that you have bent the conduit.

Step 14 Continue to bend to 12° on the second mark. Check for springback. When the first bend in the conduit is moved past the one pipe support, the **approximate ram travel** for the remaining bends will be exactly the same.

Step 15 Follow this procedure until you get to the last mark, where you will be bending to 90°. Stop at exactly 90°, release the pressure, and check for springback, correcting if necessary. The result will be a 90° bend without any bows or twists.

For example, suppose the task is to bend a 90° sweep with a given radius of 25". This particular sweep has no definite height. This is to be done on a hydraulic bender using a segmented bending process.

The first step is to determine the linear length or developed length of the conduit required to be used for the bend. This can be done by multiplying the radius of 25" by $\pi \div 2$ or $3.14 \div 2 = 1.57$. The answer is 39.25, as shown below:

$$1.57 \times 25" = 39.25"$$

Therefore, it takes 39.25" of conduit to accomplish a 90° bend with a 25" radius.

The next step is to decide how to lay out the number of segments to be bent to attain the 90° sweep. A rule of thumb is that there will be 20 segments or 21 shots. A shot is the actual bending process. There are 21 shots because the sweep is always laid out from the center of the developed length (*Figure 32*).

Once the center of the developed length is established, half of the developed length is measured out in each direction from the centerline.

To establish the linear distance between segments, divide the developed length by the number of shots (39.25 ÷ 20) or 1.9625" between segments (*Figure 33*).

Figure 31 ◆ Conduit placed in hydraulic bender for segment bends.

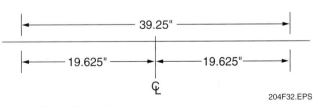

Figure 32 ◆ Conduit center.

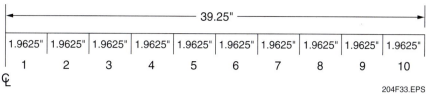

Figure 33 ◆ Conduit segments.

Once the 20 segments are laid out, the number of degrees per shot (bend) needs to be determined. There are 21 shots, so dividing 90° by 21 will give you the number of degrees per shot:

$$90° \div 21 = 4.29°$$

You are now ready to do the bending. Starting at one end of the developed length, place the centerline of the bending shoe on the first shot mark and bend to 4.29°. Repeat the process for the remaining 20 bend marks. If the bend is not quite 90°, make the final adjustment in the last bend.

There is another way to lay out the segment marks for the developed length of a bend. Cut a piece of white elastic band to the length of the outside radius of the innermost bend and lay it out on a table. Make sure that it is not stretched. From one end of the elastic, measure 15 marks spaced evenly apart and mark these points with a fine-tip ink pen. Now place the end of the elastic tape on the centerline of the developed bend length. Stretch out the elastic until the 15th mark is on the end of the developed length. Place a mark on the conduit beside each mark on the elastic tape. Repeat this process on the other half of the developed length. You are now ready to do the segmented bends.

9.0.0 ◆ SEGMENT BENDING TECHNIQUES

Now that the procedure for laying out the conduit with bending marks is understood, the next step is bending. Because segment bending requires several small angle bends to complete a 90° stub, some method to measure the amount of bend will be required. This can be done in four ways:

- Bend degree protractor
- Magnetic angle finder
- Amount of travel method
- Number of pumps method

Bend degree protractor – This is a device that hooks onto the pipe being bent. The circular face is divided into four sections (18, 20, 21, and 30 shots) and is capable of being rotated to whichever scale is to be used. The indicating pointer is weighted and swings free. To use this device, proceed as follows:

Step 1 Level the conduit and secure it using a pipe vise or other means.

Step 2 Rotate the face to the desired scale that corresponds to the number of shots being used.

Step 3 Adjust the scale so the pointer is on 0.

Step 4 Bend the pipe until the pointer reaches the first mark. (Bend a little past to compensate for springback, release the pressure, and then check the pointer. Only a few bends will be needed to find out how much you must bend past the mark to account for springback.)

Step 5 Move the pipe forward in the bender to the second bend mark, and bend until the pointer reaches the second mark on the protractor face. (Again, bend past for springback.)

Step 6 Move the conduit to the third mark and bend the pipe so the pointer is at the third mark (after allowing for springback). Follow this procedure at each bend mark until the 90° stub is achieved, the pipe is level, and the bender is in a vertical position.

NOTE

It is a good idea to check the developing stub length before the last few bends are made. Make spacing corrections as required (e.g., shorten the spacing if the stub is coming up short). If the stub length is reached before the stub is plumb, do not bend at any of the remaining marks. Instead, move the pipe in the bender and bend at the start mark. This will make the stub plumb without adding to the stub length.

Magnetic angle finder – With the magnetic angle finder, the pipe must be kept level and bent vertically (the bender is in vertical position). When a magnetic angle finder is used, care must be taken with each bend as a very small error may become multiplied by 15 to 30 times, becoming a large error. To use the angle finder:

Step 1 Level the conduit and place the angle finder on the stub end. The angle finder will indicate the number of degrees in each bend as determined by the number of shots (e.g., for 20 shots, each bend is about 4.5°).

Step 2 Bend the pipe until the angle finder indicates that the bend is just past the desired degree of bend. This allows for springback. Release hydraulic pressure. If the right amount of overbend was made to allow for springback, the angle finder should read the desired angle of bend. If you bent too much or too little, make the adjustment on the next bend. In two or three bends, you will find the right amount to overbend at each mark to allow for springback.

Step 3 Move the pipe to the second bend mark. Bend at this mark until the next setting on the angle finder (allowing for springback) is indicated by the angle finder pointer. (For example, first bend 4½°, second bend 9°, third bend 13½°, etc.)

Step 4 Bend the pipe at each successive bend mark using the angle finder to indicate the proper degree of bend.

NOTE
Again, check the developing stub length before bending the last few bends. Adjust the spacing or amount of bend as required.

Amount of travel method – The pipe may be bent in any position. To find the amount of theoretical travel for a 90° bend, proceed as follows:

Step 1 Set up the bender with the pivot shoes in the proper holes for the conduit to be bent.

Step 2 Measure the distance (D) center-to-center between the pivot shoe pins (i.e., from center of pin A to center of pin B). The plunger (also called the ram) will have to travel half this distance to bend a full 90° stub.

Step 3 The travel per shot is one-half the distance from pin to pin divided by the number of shots. For example, the distance from the center of the pins is 24" and the pipe is to be bent in 18 shots. The travel per shot equals one-half the distance from the center of the pins (12") divided by the number of shots (18), which equals 0.666 or approximately ⅔" travel per shot.

Step 4 Bending with this method will follow a slightly different procedure.

Place the center of the bending shoe on the first bending mark (not the start mark) and activate the pump until ⅔" of ram travel is measured. Do not allow for springback with this method.

Move the pipe to the next mark and activate the pump until another ⅔" of ram travel has been measured. Continue to move the pipe and measure ram travel.

As you approach the last few marks (4 or 5), check both the developing stub length and the angle of the bend. The spacing and/or amount of travel can be adjusted on these last marks, as required.

NOTE
Once you have found the amount of travel for 90°, you can also find the amount of travel for any other angle for offsets, kicks, etc.

Number of pumps method – The pipe may be bent in any position. This method depends on the fact that a given hydraulic pump will produce the same amount of bend for a given number of pumps of the handle; however, it does not take into account that the number of pumps will change with a change in fluid level, the condition of the pump, and the condition of the O-rings.

For example, if it takes 40 pumps to bend a 90° stub, it should only take 20 pumps to achieve 45°, 10 pumps for 22½°, 2 pumps for 4½°, etc. As you can see, this method should work very well, but it will require additional time to determine pump/degree values. However, this can be offset by not having to measure ram travel at each bend as required by the amount of travel method.

Use the same procedure for bending as outlined in the amount of travel method. Check the developing stub length and degree of bend prior to bending the last few shots. Use these remaining shots for final adjustments in stub length and for checking the 90° bend.

NOTE

Use whichever one of the four methods is the most convenient, but bear in mind that extreme care and attention to detail are required in all methods. A small amount of error at each bend will compound itself, and accuracy in bending will be impossible to achieve.

Using a bending table – To make hydraulic bending easier, a bending table, either a commercial model or one constructed on the job, is a necessity. The table will hold the conduit and the bender, make leveling and plumbing easier, and produce more accurate bends. The table will also eliminate the need to continually wrestle with the conduit and bender.

Hydraulic bending example – For example, suppose that the task is to bend an offset containing two sweeping bends. The radius for the two sweeping bends is 30". This is to be done on a hydraulic bender using a segmented bending process. The angles of the sweeping bends are to be 30° each. The height of the offset is to be 30" (*Figure 34*).

This problem continues to explore the concept of the unit circle. This information is brought together with the cosecant trigonometric function.

In continuing with the concept of the unit circle and the linear distance on the circumference of the circle in terms of pi, it is necessary to understand the development of the pi relationship from 0° to 90°. All of these relationships from 0° to 90° and other interval angles in the circle are always in reference to 180°. For example: 360° = 2π or 2 × 180°; 180° = π or 1 × 180°; and 90° = π ÷ 2 or ½ × 180°.

The same process holds true for any angle from 0° through 90°. For example: 45° = π ÷ 4 or ¼ × 180°; 60° = ⅓π or ⅓ × 180°; 30° = ⅙π or ⅙ × 180°; and 15° = ¹⁄₁₂π or ¹⁄₁₂ × 180° (*Figure 35*).

To figure the developed length of the conduit to be used by one swing in the 30" radius, simply multiply π ÷ 6, which represents pi at 30°, times the 30" radius. It is nothing more than a converted circumference formula.

$C = 2\pi R$ (linear distance around a complete circle)

$C = \dfrac{\pi}{6} R$ (linear distance from 0° to 30°)

$C = \dfrac{\pi}{6} R$ or $C = 0.523 \times R$ or $C = 0.523 \times 30" = 15.7"$

The developed length for the first radius is 15.7". That means that the developed length for the second sweeping radius is also 15.7". The total amount of conduit used to develop the two sweeps is 31.4".

The next step is to figure out the distance between the centerline of each of these bends on the hypotenuse of the imaginary triangle. To do this, simply use the cosecant of the angle trigonometric function or Cosecant = H ÷ O. Converting

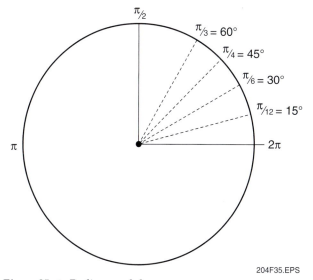

Figure 35 ◆ Radians and degrees.

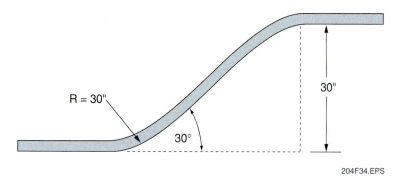

Figure 34 ◆ Two 30° sweeping bends.

the formula gives H = O × Cosecant or H = 30" × 2 or 60". The layout of these measurements is shown in *Figure 36*.

Again, laying out the segments for the bending process is done from the centerline of the developed length of each sweeping bend (*Figure 37*). Therefore:

15.7" ÷ 2 = 7.85"

When building a 90° sweep, the rule of thumb is 21 shots or 20 segments. This gives the conduit the appearance of being a natural sweep and not a line of straight segments. (The same odd shot/even segment process will appear again.) If you choose to have seven shots or six segments within the 15.7" of developed length, each segment will be 15.7" ÷ 6 or 2.62" long. Because there are seven shots (bends), each shot will be 30 ÷ 7 or 4.29°. The layout of the bend is shown in *Figure 38*.

Once you have both sweeps laid out as described previously and the centerlines of these sweeps are 60" apart, you are ready to start the bending process. Again, start at one end of the offset and swing the bender segmented shoe down on the conduit. Because you are working an offset, an anti-dog device should be attached to one end of the conduit. This device maintains the centerline for each bend. Begin bending the first sweep, and once this is accomplished, roll the conduit in the bender and re-level the anti-dog device. Bend the second sweep and your offset is built. The final measurements should be as follows:

Developed length of each sweep = 15.7"
Distance between centerline of each sweep = 60"
Distance between shot marks in each sweep = 2.62"
Degrees per shot in each sweep = 4.29°

9.1.0 Concentric Bending

If two or more parallel runs of conduit must be bent in the same direction, as shown in *Figure 39*, the best results will be obtained by using concentric bends. When laying out concentric bends, the bend for the innermost conduit is calculated first. In the example in *Figure 39*, the first bend has a radius of 20". If this dimension is multiplied by

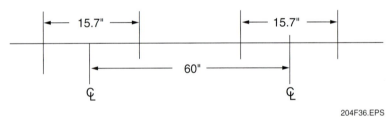

Figure 36 ◆ Bend centerline distance.

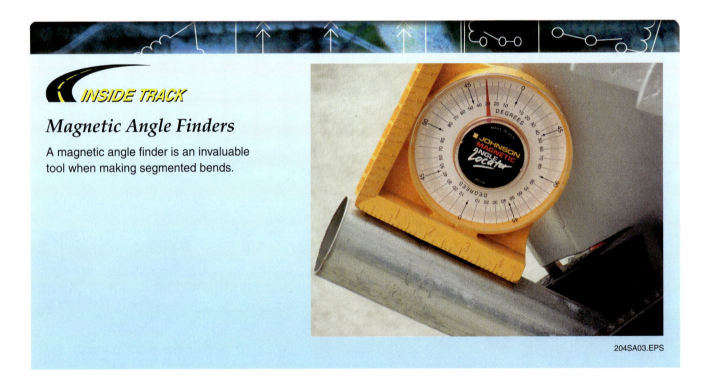

INSIDE TRACK

Magnetic Angle Finders

A magnetic angle finder is an invaluable tool when making segmented bends.

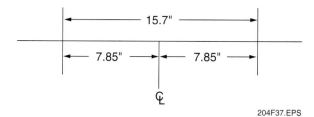

Figure 37 ◆ Bend centerline.

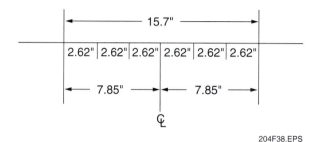

Figure 38 ◆ Bend segments.

1.57, it yields a value of 31¹³⁄₃₂". This is the developed length of the shortest radius bend.

The second radius is found by adding the **outside diameter (OD)** size of the first pipe to the radius of the first pipe and the desired spacing between pipes. In this example, the radius of the second bend would be equal to 24⅜" (see *Figure 39*).

Radius of first pipe = 20"
OD size of first pipe = 2⅜"
Spacing between pipes = 2"
Radius of second pipe:
20" + 2⅜" + 2" = 24⅜" or 24.375"

Once the developed length of the second radius is found, the spacing between marks can be determined. The start mark must be the same distance from the end of the second pipe as it was for the first pipe (*Figure 39*). This will be true for each additional pipe as well. The spacing will increase

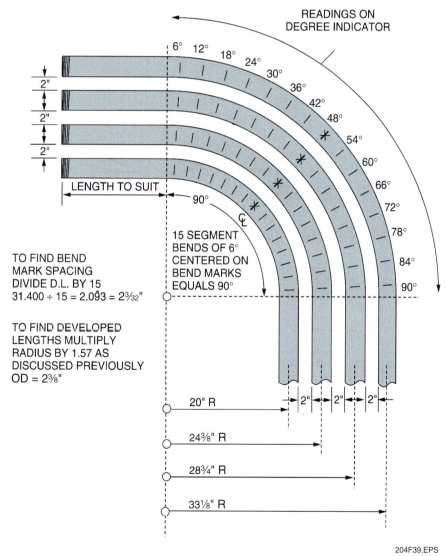

Figure 39 ◆ Principles of concentric bending.

between marks as the radius of the pipes increases. This is due to increasing developed length. The pipes in *Figure 39* are all laid out for 15 shots.

The equation for the developed length (DL) is 1.57 × R (radius). The developed length for the second bend is found by multiplying 1.57 × 24.375" = 38.27".

The radius and developed length of each successive bend are found in a similar manner. After determining the developed length, you must establish the number of segment bends needed to form each 90° bend. In concentric bending, every bend must receive the same number of segment bends to maintain concentricity. As illustrated, 15 segment bends of 6° each will total 90°.

NOTE

The radius change from one bend to another affects the spacing of segment bends. To find the segment bend spacing, divide the developed length of each bend by 15. For the first bend illustrated, the spacing for each segment bend is 2³⁄₃₂"; for the second bend, it is 2⁹⁄₁₆", and so on, as shown in *Figure 39*.

If the legs of the bends have to be a certain length, the gain must be considered just as in any segment bending procedure. When the runs of conduit are not the same size, the radius of each successive bend can be found as follows:

Step 1 Determine the radius of the innermost bend.

Step 2 Calculate one-half the outside diameter of the innermost conduit and of the next adjacent conduit.

Step 3 Note the distance between the two runs of conduit.

Step 4 Add these quantities.

9.2.0 Offset Bends

Many situations require a conduit to be bent so that it can pass by or over objects such as beams and other conduit or to enter panelboards and junction boxes. The bends used for this purpose are called offsets. To produce an offset, two equal bends of less than 90° are required at a specified distance apart. This distance is determined by the angle of the two bends and can be calculated by using the following procedure and the table in *Figure 40*.

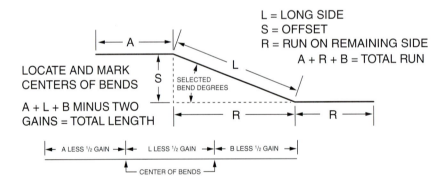

TO FIND UNKNOWN	KNOWN	TIMES CORRESPONDING MULTIPLIER EQUALS								UNKNOWN
		TABLE OF MULTIPLIER FOR SELECTED DEGREES OF BEND								
		5⅝°	11¼°	15°	22½°	30°	37½°	45°	60°	
L	S	10.207	5.126	3.864	2.613	2.00	1.643	1.414	1.155	L
S	L	0.098	0.195	0.259	0.383	0.50	0.609	0.707	0.866	S
R	S	10.158	5.027	3.732	2.414	1.732	1.303	1.00	0.577	R
S	R	0.098	0.199	0.268	0.414	0.577	0.767	1.00	1.732	S
L	R	1.005	1.02	1.035	1.082	1.155	1.260	1.414	2.00	L
R	L	0.995	0.981	0.966	0.933	0.866	0.793	0.707	0.50	R
GAIN PER BEND	RADIUS OR SHOE	0.0002	0.0006	0.0015	0.0051	0.0124	0.0212	0.0430	0.1076	GAIN PER BEND

204F40.EPS

Figure 40 ♦ Conduit offset bending table.

Leave Room for System Expansion

It's a good idea to make your first concentric bend large enough to fit another bend or two inside it in case you have to add additional conduit runs later on. Another method is to add any additional conduit runs on the outside of the original bend. Both methods produce a neat and professional installation.

First, determine the offset needed, then find the degree of bend. Next, multiply the offset measurements by the figure directly under the degree of bend (see *Figure 40*). This applies to all sizes of conduit. For example, to form an 18" offset with two 45° bends, first make the following calculation to determine the distance between bends:

$$18" \times 1.414 = 25\tfrac{1}{2}"$$

This is the distance between bends and is labeled side *L* (see *Figure 40*). To connect the two ends of an offset to two pieces of conduit that are already in place, it is necessary to know the total or overall length (OL) of the offset from end to end before bending.

Note the following equation:

$$OL = (A + L + B) - (2 \times gain)$$

Where:

$A = 36"$
$L = 25\tfrac{1}{2}"$
$B = 48"$
$2 \times gain = 1\tfrac{9}{32}"$

The gain is calculated by multiplying the shoe radius by the decimal figures shown on the last line under DEGREES OF BEND in *Figure 40*. For example, to calculate how an offset might be made in a length of 3" conduit using 45° bends:

$$OL = (A + L + B) - (2 \times gain)$$

NOTE
This offset uses a 15" radius.

$$OL = 109\tfrac{1}{2}" - 1\tfrac{9}{32}" = 108\tfrac{7}{32}"$$
$15" \times 0.0430$ (from the table in *Figure 40*) $= 0.645°$

For two gains, you have:

$$1.290 = 1\tfrac{9}{32}$$

Since you already know the long dimension of the offset (25"), to find the amount of offset, refer to the table in *Figure 40*, second line down (to find

Concentric Bends

What does concentric mean? Why can't you just bend all your conduit sections to the same radius and lay them side by side?

S), then move across the row until you come to the 45° column. Note that the multiplier is 0.707.

Therefore:

25" × 0.707 = 17.675"

This is the height of the offset. To make the offset, the conduit is positioned in the hydraulic bender, as shown in *Figure 41*. The bending shoe makes one 45° bend on the first center mark, the conduit is reversed in the bender, and the next 45° bend is made at the second mark. When making bends, always refer to the bending charts that accompany the bender being used for the operation.

For example, an offset is to be built to go over an I-beam that is 24" in height. The distance from the end of the conduit run to the J box on the opposite side of the I-beam is 8' (see *Figure 42*).

The task is to determine the distance between bend points where a 45° bend is being used. Determine the exact length of conduit needed so threading can occur before bending. The conduit will have no clearance over the top of the I-beam.

To find the distance between bend points, multiply the amount of offset required by the cosecant of the bending angle (in this case, 45°):

24" × 1.41 = 33.84"

The straight distance from the end of the conduit run to the J box is 8'. To find the amount of conduit to add to this because of the 45° bend, use the cosine formula (C = A ÷ H). The hypotenuse or the distance between bend points is 33.84". The side adjacent is found by converting C = A ÷ H to A = C × H, or:

A = 0.707 × 33.84" = 23.92"

Subtract 23.92" from 33.84" and this gives the amount of extra conduit needed when making this 45° bend:

33.84" − 23.92" = 9.92" or $9^{92}/_{100}$"

Add the 9.92" to 8' and the amount of straight conduit required has been determined:

8' + 9.92" = 8'-9.92" of straight conduit

The final measurements should be as follows:

Distance between bend points = $33^{21}/_{25}$" or 33.84"

Length of straight conduit = 8'-9.92" or 8'-$9^{92}/_{100}$"

Now let us complete another example. In this case, an offset is to be made without the use of a protractor level. The offset is to be 20" in height (*Figure 43*). The bend angle is 45°. Explain the process required to accomplish this.

The cosecant for 45° is 1.41. The distance between bend points is 1.41 × 20". Therefore:

1.41 × 20" = 28.2"

The first mark on the conduit is placed 14.1" back from the end of the conduit. The end of the conduit needs to be brought up 10" to accomplish a 45° bend. The second mark on the conduit is placed 28.2" down the conduit from the first mark.

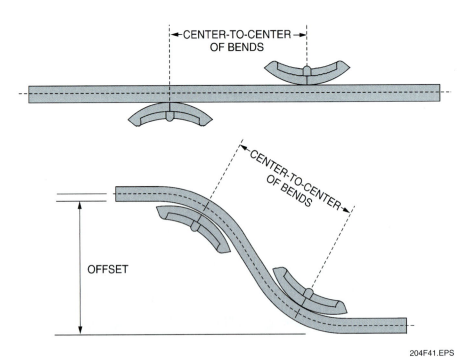

Figure 41 ◆ Position of conduit in bender for making offsets.

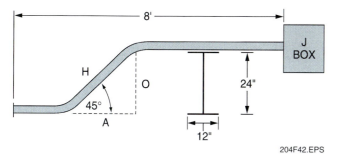

Figure 42 ◆ 24" offset.

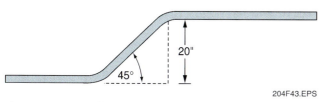

Figure 43 ◆ 20" offset.

The bender is placed on the second mark. The conduit is bent until the two surfaces are parallel. At this point, a 20" offset has been created.

9.3.0 Concentric Offsets

To make concentric offsets, use the same procedure as for 90° stubs. Increase the radius of the second pipe by an amount equal to the radius of the first pipe plus the OD size of the first pipe plus the spacing desired between pipes. Although the examples used to explain offset bending had several shots per bend, fewer shots can be used. Successful offsets can be made with only two or three shots per bend. However, with offsets, 90° bends, or any other segment bends, the extra time spent bending smoother radius bends (more shots) is returned when the conductors are pulled into the completed raceway.

9.3.1 Saddle Bends

As mentioned previously, saddle bends are used to cross small obstructions or other runs of conduit. Saddle bends in conduit up to four inches **inside diameter (ID)** may be made in hydraulic conduit benders. Always refer to the charts that accompany the bender. For example, to make a saddle bend on a length of 2" rigid conduit so that it can pass over a 3" water pipe with ¼" clearance, three bending operations are required, as shown in *Figure 44*.

The bend is calculated by referring to appropriate bending charts. The one shown in *Table 6* gives measurements for 2" conduit with ¼" clearance. Therefore, look in the left column under straight-run conduit and go down the column until the 3" row is reached. Moving to the right in this row, it can be seen that the spacing between bends is 15⅝" on center, and bends No. 1 and No. 2 will be 15°, while the third bend is 30°.

Using the information found in the bending table, mark the conduit accordingly. Insert the conduit in the bender and make bends No. 1 and No. 2 first, both at 15°. Back off the pump pressure, reverse the conduit in the bender, and then make the third bend at 30°. Release the pump pressure, remove the conduit from the bender, and check the saddle for accuracy.

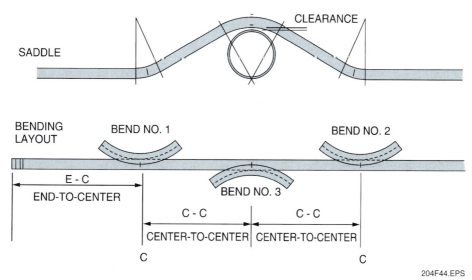

Figure 44 ◆ Principles of saddle bending.

Table 6 Saddle Table

Straight-Run Conduit	Minimum Length	Bend Spacing	Bend Degrees	Bend Degrees	Bend Degrees
—	E-C	C-C	No. 1	No. 2	No. 3
1"	20"	16"	6	6	12
1¼"	20"	16"	7	7	14
1½"	20"	16"	8	8	16
2"	20"	15⅞"	10	10	20
2½"	20"	15¾"	12½	12½	25
3"	20"	15⅝"	15	15	30
3½"	20"	15½"	18	18	36
4"	20"	15½"	20	20	40

For another example, suppose a saddle is to be built to go over an I-beam that is 36" in height (*Figure 45*). The distance from the end of the conduit to the center of the I-beam is 5'.

The task is to determine the distance between bend points when a 45° bend is being used. Determine the exact length of conduit needed so threading can be completed before bending. The conduit is to clear the top of the I-beam by 1". Determine the centerline layout of the saddle (*Figure 46*).

The 16" between bend points in this problem is an arbitrary measurement. It is determined by the person doing the bending. Enough conduit should be used to extend past the outside dimensions of the 12" I-beam so the radius of each bend does not run into the I-beam.

The distance between bend points on the hypotenuse of the imaginary triangle is calculated by multiplying the total height of the I-beam (36" + 1" clearance = 37") by the cosecant of the 45° angle of the bend or 1.41 × 37":

1.41 × 37" = 52¹⁷⁄₁₀₀" or 52.17"

The centerline layout mark for the saddle would be 5' if the conduit were to be installed on a straight line from the end of the existing conduit run to the center of the I-beam. However, the conduit is not

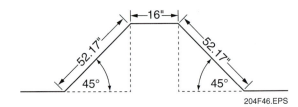

Figure 46 ◆ Saddle.

being installed on a straight line. Therefore, the measurement in question is the amount of additional conduit needed because of the 45° angle of deviation (*Figure 47*). The amount of additional conduit is equal to the side adjacent subtracted from the hypotenuse.

The side adjacent (A) first needs to be calculated, then that measurement can be subtracted from the hypotenuse (52.17"). To calculate the side adjacent, use the cosine formula (C = A ÷ H). Because we are looking for A, convert the formula to A = C × H, or:

A = 0.707 × 52.17"

A = 36.88"

Subtract 36.88 from 57.17, or:

52.17" − 36.88" = 15.29"

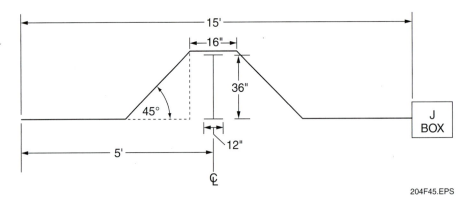

Figure 45 ◆ 36" saddle.

Inside Track

Saddle Bends

Whenever possible, a three-bend saddle should not be sharper than 30°. If the object you are trying to saddle is more than 6" in diameter, use a four-bend saddle rather than a three-bend saddle. In this installation, two three-bend saddles were required.

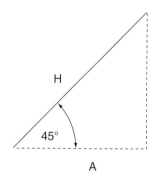

Figure 47 ◆ Cosine function.

H − A = AMOUNT OF ADDITIONAL CONDUIT

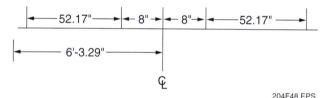

Figure 48 ◆ Conduit layout.

That is how much extra conduit is needed to get the center of the saddle to the center of the I-beam. Therefore, the layout mark for the saddle is:

$$5' + 15.29" = 6'\text{-}3.29"$$

The layout of the conduit is shown in *Figure 48*.

To find the total amount of conduit to be used from the end of the existing conduit to the J box, simply take the straight line distance of 15' and add the difference between the hypotenuse and side adjacent to 15' two times, or:

$$15' + 15.29" + 15.29" = 15' + 30.58" = 17'\text{-}6.58"$$

The final measurements should be as follows:

Distance between bend points on the top portion of the saddle = 16"

Distance between bend points that form each hypotenuse = $52^{17}/_{100}"$ or 52.17"

Distance from the end of the conduit to the centerline layout mark of the saddle = 6'-3.29" or 6'-3$^{29}/_{100}"$

Overall length of the conduit required to reach the J box = 17'-6$^{58}/_{100}"$ or 17'-6.58"

10.0.0 ♦ TRICKS OF THE TRADE

There are many tricks of the trade that you will learn during your career as an electrician. Some of these will be handed down to you by experienced workers; others will be learned through experience. Most professionals are constantly seeking new methods to improve their efficiency and to make the work go smoother and faster without sacrificing workmanship.

One of the handiest personal tools for bending conduit is the small, magnetic torpedo level with a 35° and 45° bubble. This tool should be in the tool pouch of every electrician. In many cases, it can make the difference between a good job and a poor job, that is, obtaining level and plumb conduit runs or not.

Workers have also designed their own custom tools to help in conduit installations. One tip is to take a pair of vice grips and weld a small, flat piece of iron on the top of the jaws for use with the magnetic level. During multiple bends, the vice grips are positioned at the desired point on the end of a piece of conduit, locked in place, and then the magnetic level is placed on the flat plate so the bubble may be watched during the bend. There are also some new commercial tools that are designed for leveling conduit bends; keep up with these developments by reading trade journals.

10.1.0 Eliminating Dog Legs

One of the problems that exists when bending an offset or saddle is the creation of a dog leg. A dog leg occurs when the centerline that runs the length of the conduit is not maintained for both bends of an offset and for the four bends of a saddle. Dog legs can be eliminated using an anti-dog device, as discussed earlier, or by following the procedure described here.

See *Figure 49*. When the conduit is snugged up by the bender, take a piece of Unistrut® and clamp it to the end of the conduit farthest from the bending shoe. Before tightening the conduit strap, level the Unistrut®. Once this is done, make your first bend. Release the conduit from the bending shoe and move the conduit forward to the next bending mark. Roll the conduit 180° and re-level the Unistrut®. Tighten the conduit in the bender and again make sure that the Unistrut® is level. Bend the second bend and you will have an offset with no dog leg.

Inside Track

Trapeze Hangers

Trapeze hangers are often used to support multiple overhead conduit runs. This reduces the number of individual supports and also reduces the number of bends as the conduit can be run below obstacles, as shown here.

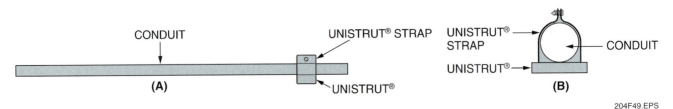

Figure 49 ♦ Dog leg elimination.

10.2.0 Using a Table Corner or Plywood Sheet for Calculating Added Length

The task is to use a table corner or the corner of a sheet of plywood and demonstrate how to come up with the amount of additional conduit needed to go from one point to another once a bend has been made. The distance between points is 8'. The amount of offset to be made is 25" (*Figure 50*). The degree of bend is 30°.

See *Figure 51*. Measure 25" up one side of the sheet of plywood or the corner of a table. The hypotenuse of the triangle will be 2 × 25" or 50" (2 is the cosecant for 30°). Measure from the 25" mark to the surface that is tangent to the side that is 25" long, until the 50" mark on the tape measure crosses the tangent side. Mark this 50" point. Then, place the end of the ruler on the 50" mark and let the 50" point on the ruler extend beyond the end of the plywood or table. The amount of tape that extends beyond the plywood or table edge is the amount of extra conduit that is needed to accomplish this bend. This is 6⅝" more conduit than if going on a straight line for 8' or 8'-6⅝" of conduit required to accomplish the task.

11.0.0 ◆ PVC CONDUIT INSTALLATIONS

Rigid nonmetallic conduit and fittings (PVC electric conduit) may be used where the potential is 600V or less in direct earth burial; in walls, floors, and ceilings of buildings; in cinder fill; and in damp and dry locations. Its use is prohibited by the *NEC®* in certain hazardous locations, for support of fixtures or other equipment, where subject to physical damage, and under certain other conditions. See *NEC Article 352*.

PVC conduit can be cut easily at the job site without special tools, although PVC cutters help in cutting square ends. Sizes from ½" through 1½" can be cut with a fine-tooth saw. For sizes 2" through 6", a miter box or similar saw guide should be used to keep the conduit steady and ensure a square cut. To ensure satisfactory joining, care should be taken not to distort the end of the conduit when cutting.

After cutting, deburr the pipe ends and wipe clean of dust, dirt, and plastic shavings. Deburring is accomplished easily using a pocket knife or file.

One of the important advantages of PVC conduit, in comparison with other rigid conduit materials, is the ease and speed with which solvent-cemented joints can be made. The following steps are required for a proper joint:

Step 1 The conduit should be wiped clean and dry.

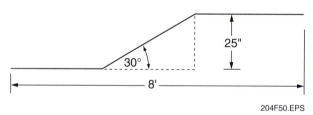

Figure 50 ◆ 25" offset.

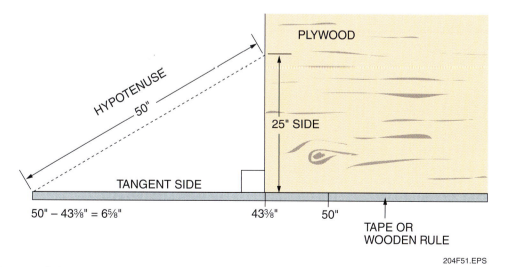

Figure 51 ◆ Extra conduit.

Inside Track

Magnetic Level

A magnetic torpedo level can be attached directly to the conduit. It will remain where you put it, even if the conduit is turned over.

204SA07.EPS

 WARNING!
The cement used with PVC can be hazardous. Provide adequate ventilation, avoid skin contact, and always refer to the MSDS and follow the manufacturer's usage and safety instructions.

Step 2 Apply a full, even coat of PVC cement to the end of the conduit and the fitting. The cement should cover the area that will be inserted in the fitting.

Step 3 Push the conduit and fitting firmly together with a slight twisting action until it bottoms and then rotate the conduit in the fitting (about a half turn) to distribute the cement evenly. Avoid cement buildup inside the conduit. The cementing and joining operation should not exceed more than 20 seconds.

Step 4 When the proper amount of cement has been applied, a bead of cement will form at the joint. Wipe the joint with a brush to remove any excess cement. The joint should not be disturbed for 10 minutes.

12.0.0 ◆ BENDING PVC CONDUIT

Most manufacturers of PVC conduit offer various radius bends in a number of segments. Where special bends are required, PVC conduit is easy to bend on the job. Stub-ups, saddles, concentric bends, offsets, and kicks are all possible with PVC conduit, just as with metallic conduit.

PVC conduit is bent with the aid of a heating unit. The PVC must be heated evenly over the entire length of the curve. Heating units are available from various sources that are designed specifically for the purpose in sizes to accommodate all conduit diameters (*Figure 52*). While some heaters use gas for the heat source, most employ infrared heat energy, which is more quickly absorbed in the conduit. Small sizes are ready to bend after a few seconds in the hotbox. Larger diameters require two or three minutes, depending on the conditions. Other methods of heating PVC conduit for bending include heating blankets and hot air blowers. The use of torches or other flame-type devices is not recommended. PVC conduit exposed to excessively high temperatures may take on a brownish color. Sections showing evidence of such scorching should be discarded.

 WARNING!
Always wear gloves when working with heat.

Figure 52 ◆ PVC heating units.

204F52.EPS

If a number of identical bends are required, a template can be helpful (*Figure 53*). A simple template can be made by sawing a sheet of plywood to match the desired bend. Nail to a second sheet of plywood. The heated conduit section is placed in the template, sponged with water to cool, and is then ready to install. Care should be taken to fully maintain the ID of the conduit when bending.

If only a few bends are needed, scribe a chalk line on the floor or workbench. Then, match the heated conduit to the chalk line and cool. The conduit must be held in the desired position until it is relatively cool since the PVC material will tend to revert to its original shape. Templates are also available for many bends.

Another method is to take the heated conduit section to the point of installation and form it to fit the actual installation (*Figure 54*). This method is especially effective for making blind bends or compound bends using smaller sizes of PVC. After bending, wipe a wet rag over the bend (*Figure 55*) to cool it.

Bends in small-diameter PVC conduit (½" to 1½") require no filling for code-approved radii. When bending PVC of 2" or larger diameter, there is a risk of wrinkling or flattening the bend. To help eliminate this problem, a plug set is used. A plug is inserted in each end of the piece of PVC being bent, and then a hand pump is used to pressurize the conduit before bending it. The pressure is about three to five pounds per square inch.

Place airtight plugs (*Figure 56*) in each end of the conduit section before heating. The retained air will expand during the heating process and hold the conduit open during bending. Do not remove the plugs until the conduit has cooled.

In applications where the conduit installation is subject to constantly changing temperatures and the runs are long, precautions should be taken to allow for expansion and contraction of PVC conduit. When expansion and contraction are factors, an O-ring expansion coupling should be installed near the fixed end of the run, or fixture, to take up any expansion or contraction that

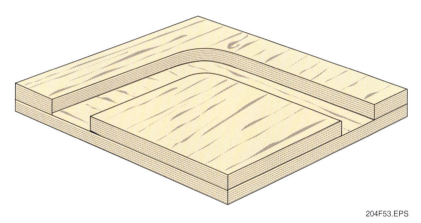

Figure 53 ◆ Plywood template.

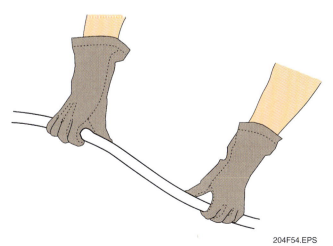

Figure 54 ◆ Some PVC bends may be formed by hand.

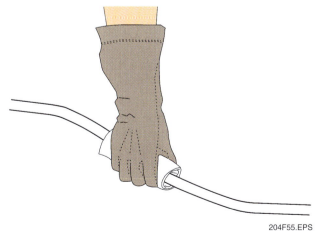

Figure 55 ◆ After the bend is formed, wipe a wet rag over the bend to cool it.

Figure 56 ◆ Typical plug set.

may occur. Confirm the expansion and contraction lengths available in these fittings as they may vary by manufacturer. Charts are available that indicate what expansion can be expected at various temperature levels. The coefficient of linear expansion of PVC conduit is 0.0034" per 10' per °F. *Table 7* lists the expansion rates for various temperatures.

Expansion couplings are seldom required in underground or slab applications. Expansion and contraction may generally be controlled by bowing the conduit slightly or burying the conduit immediately. After the conduit is buried, expansion and contraction cease to be factors. Care should be taken, however, in constructing a buried installation. If the conduit should be left exposed for an extended period of time during widely-variable temperature conditions, an allowance should be made for expansion and contraction.

In above-ground installations, care should be taken to provide proper support of PVC conduit because of its semi-rigidity. This is particularly important at high temperatures. The distance between supports should be based on temperatures encountered at the specific installation. Charts are available that clearly outline at what intervals support is required for PVC conduit at various temperature levels.

Table 7	PVC Expansion Rates		
Temperature Change in °F	Length Change in Inches per 100 Ft of PVC Conduit	Temperature Change in °F	Length Change in Inches per 100 Ft of PVC Conduit
5	0.2	105	4.2
10	0.4	110	4.5
15	0.6	115	4.7
20	0.8	120	4.9
25	1.0	125	5.1
30	1.2	130	5.3
35	1.4	135	5.5
40	1.6	140	5.7
45	1.8	145	5.9
50	2.0	150	6.1
55	2.2	155	6.3
60	2.4	160	6.5
65	2.6	165	6.7
70	2.8	170	6.9
75	3.0	175	7.1
80	3.2	180	7.3
85	3.4	185	7.5
90	3.6	190	7.7
95	3.8	195	7.9
100	4.1	200	8.1

PVC Conduit Bending

PVC conduit has a memory. As soon as you have finished bending the PVC, cool it down with cold, wet towels or hose it down with cold water. Otherwise, it will begin to return to its original position.

Putting It All Together

Examine the visible conduit bends in your home or workplace. Are they neat and professionally made? If not, what might you have done differently to improve the installation?

Review Questions

1. The minimum bending radius for 3" rigid conduit is ____.
   a. 8"
   b. 10"
   c. 11"
   d. 13"

2. The minimum bending radius for ½" rigid conduit is ____.
   a. 4"
   b. 6"
   c. 8"
   d. 10"

3. The maximum number of 90° bends allowed between pull points in a conduit system is ____.
   a. one
   b. two
   c. three
   d. four

4. A saddle bend is counted as ____.
   a. 90°
   b. 120°
   c. 180°
   d. It depends on the type of bends used.

5. When referring to the two bends required to make an offset in a length of conduit, which of the following is always true?
   a. The degree of bend for each must be exactly 45°.
   b. The degree of bend for each must be exactly 30°.
   c. The degree of bend for each must be equal.
   d. The degree of bend for each must be unequal.

6. The reason for making a saddle bend in a run of conduit is to ____.
   a. change direction in the height of a conduit run
   b. make a 90° bend
   c. cross a small obstacle
   d. make a conduit termination in an outlet box

7. The equation used to calculate the circumference of a circle is ____.
   a. $C = \pi D$
   b. $C = 1\pi R$
   c. $C = \pi R^2$
   d. $C = \pi^2$

8. The fractional equivalent of the decimal 0.015625 is ____.
   a. $\frac{1}{64}$
   b. $\frac{1}{32}$
   c. $\frac{3}{64}$
   d. $\frac{1}{16}$

9. The rise in a conduit bend is best described as the ____.
   a. horizontal distance that runs from the point of bend and parallel with the deck
   b. leg length
   c. radius
   d. height of the stub-up

10. The best type of bending shoe for making concentric bends is the ____ shoe.
    a. concentric
    b. one-shot
    c. segmented
    d. long

11. Rigid aluminum conduit is available in standard sizes from ____.
    a. ½" to 4"
    b. ½" to 6"
    c. ¾" to 6"
    d. ¾" to 4"

12. Two or more parallel bends that are bent in the same direction are known as ____.
    a. stubs
    b. perpendicular bends
    c. concentric bends
    d. axial bends

13. Which of the following formulas should be used to find the side adjacent?
    a. $S = O \times H$
    b. $S = O \div H$
    c. $C = A \div H$
    d. $C = A \times H$

Review Questions

14. An important safety precaution to use when working with PVC conduit is to _____.
    a. provide proper ventilation to carry off fumes from the joint cement
    b. use lengths of PVC of not more than 8'
    c. use inside diameters of 4" or less
    d. never wear gloves that may adhere to the joint cement

15. When bending larger diameter PVC pipe, _____ is/are used to prevent flattening.
    a. plugs and air pressure
    b. sand fill
    c. water fill
    d. higher temperature

Summary

Many conduit installations are visible; that is, they run exposed. Consequently, electricians must take special care to ensure that all exposed conduit runs are parallel, level, and plumb. Nothing else will do. This is one phase of the electrical construction industry where electricians have an opportunity to show off. In fact, an expert installation of a conduit system is similar to a work of art.

Learn the basics of conduit bending and installations, put your knowledge to practical use, and take pride in your abilities to perform a professional conduit installation. Make your work second to none. Of course, contractors and clients want speed, but if you have a good basic knowledge of conduit bending and then put this knowledge to use, your craftsmanship will let you bend conduit smarter and faster, giving a good-looking installation along with speed to satisfy your employer, the building owners, and all concerned.

Notes

Trade Terms Introduced in This Module

Approximate ram travel: The distance the ram of a hydraulic bender travels to make a bend. To simplify and speed bending operations, many benders are equipped with a scale that shows ram travel. Using a simple table (supplied with many benders), the degree of bend can easily be converted to inches of ram travel. This measurement, however, can only be approximated because of the variation in springback of the conduit being bent.

Back-to-back bend: Any bend formed by two 90° bends with a straight section of conduit between the bends.

Bending protractor: Made for use with benders mounted on a bending table and used to measure degrees; also has a scale for 18, 20, 21, and 22 shots when using it to make a large sweep bend.

Bending shot: The number of shots needed to produce a specific bend.

Concentric bending: The process of making 90° bends in parallel runs of conduit. This requires increasing the radius of each conduit from the inside of the bend toward the outside.

Conduit: Piping designed especially for pulling electrical conductors. Types include rigid, IMC, EMT, PVC, aluminum, and other materials.

Degree indicator: An instrument designed to indicate the exact degree of bend while it is being made.

Developed length: The amount of straight pipe needed to bend a given radius. Also, the actual length of the conduit that will be bent.

Elbow: A 90° bend.

Gain: The amount of pipe saved by bending on a radius as opposed to right angles. Because conduit bends in a radius and not at right angles, the length of conduit needed for a bend will not equal the total determined length. Gain is the difference between the right angle distances A and B and the shorter distance C—the length of conduit actually needed for the bend.

Inside diameter (ID): The inside diameter of conduit. All electrical conduit is measured in this manner. The outside dimensions, however, will vary with the type of conduit used.

Kicks: Bends in a piece of conduit, usually less than 90°, made to change the direction of the conduit.

Leg length: The distance from the end of the straight section of conduit to the bend, measured to the centerline or to the inside or outside of the bend or rise.

Ninety-degree bend: A bend in a piece of conduit that changes its direction by 90°.

Offsets: Two equal bends made to avoid an obstruction blocking the run of the conduit.

One-shot shoe: A large bending shoe that is designed to make 90° bends in conduit.

Outside diameter (OD): The size of any piece of conduit measured on the outside diameter.

Radius: The relative size of the bent portion of a pipe.

Rise: The length of the bent section of conduit measured from the bottom, centerline, or top of the straight section to the end of the bent section.

Segment bend: Any bend formed by a series of bends of a few degrees each, rather than a single one-shot bend.

Segmented bending shoe: A smaller type of shoe designed for bending segmented bends only (always less than 15°).

Springback: The amount, measured in degrees, that a bent conduit tends to straighten after pressure is released on the bender ram. For example, a 90° bend, after pressure is released, will pull back about 2° to 88°.

Stub-up: Another name for the rise in a section of conduit.

Sweep bend: A 90° bend with a radius larger than that produced by a standard one-shot shoe.

Take-up (comeback): The amount that must be subtracted from the desired stub length to make the bend come out correctly using a point of reference on the bender or bending shoe.

Additional Resources

This module is intended to present thorough resources for task training. The following reference works are suggested for further study. These are optional materials for continued education rather than for task training.

Benfield Conduit Bending Manual, 2nd Edition. KS: EC&M Books.

National Electrical Code® Handbook, Latest Edition. Quincy, MA: National Fire Protection Association.

Tom Henry's Conduit Bending Package (includes video, book, and bending chart). Winter Park, FL: Code Electrical Classes, Inc.

CONTREN® LEARNING SERIES – USER UPDATE

NCCER makes every effort to keep these textbooks up-to-date and free of technical errors. We appreciate your help in this process. If you have an idea for improving this textbook, or if you find an error, a typographical mistake, or an inaccuracy in NCCER's Contren® textbooks, please write us, using this form or a photocopy. Be sure to include the exact module number, page number, a detailed description, and the correction, if applicable. Your input will be brought to the attention of the Technical Review Committee. Thank you for your assistance.

Instructors – If you found that additional materials were necessary in order to teach this module effectively, please let us know so that we may include them in the Equipment/Materials list in the Annotated Instructor's Guide.

Write: Product Development and Revision
National Center for Construction Education and Research
3600 NW 43rd St., Bldg. G, Gainesville, FL 32606

Fax: 352-334-0932

E-mail: curriculum@nccer.org

Craft _____ Module Name _____

Copyright Date _____ Module Number _____ Page Number(s) _____

Description

_____

_____

_____

_____

(Optional) Correction

_____

_____

_____

(Optional) Your Name and Address

_____

_____

_____

Pull and Junction Boxes

Zachry Corporate Headquarters

Zachry Construction Corporation won Aon Build America Award for its work on its own Corporate Headquarters in San Antonio, Texas. Zachary constructed a new conference and employment center using the U.S. Green Building Council (USGBC) Leadership in Energy and Environmental Design (LEED) Programs.

26205-08

26205-08
Pull and Junction Boxes

Topics to be presented in this module include:

1.0.0	Introduction	.5.2
2.0.0	Sizing Pull and Junction Boxes	.5.5
3.0.0	Conduit Bodies	.5.7
4.0.0	Handholes	.5.10
5.0.0	Fittings	.5.12

Overview

The pull and junction boxes used in an electrical system are selected according to their volume capacity. This volume capacity, called box fill, is measured in cubic inches or centimeters and is regulated by the *National Electrical Code®*.

There are many types of pull and junction boxes to choose from. It is essential to select the right box for the job in order to make conductor installation easier, and to comply with *NEC®* regulations that govern the types and sizes of boxes. Some of the factors affecting box selection include environmental conditions, mounting surfaces or materials, and raceway design, as well as the size, type, and number of conductors.

Conduit bodies, or condulets as they are often called, also play an important role in conductor installation. The *NEC®* regulates the type of conduit and conduit bodies that may be installed in hazardous locations.

Note: *NFPF 70®*, *National Electrical Code®*, and *NEC®* are registered trademarks of the National Fire Protection Association, Inc., Quincy, MA 02269. All *National Electrical Code®* and *NEC®* references in this module refer to the 2008 edition of the *National Electrical Code®*.

Objectives

When you have completed this module, you will be able to do the following:

1. Describe the different types of nonmetallic and metallic pull and junction boxes.
2. Properly select, install, and support pull and junction boxes and their associated fittings.
3. Describe the *National Electrical Code® (NEC®)* regulations governing pull and junction boxes.
4. Size pull and junction boxes for various applications.
5. Understand the NEMA and IP classifications for pull and junction boxes.
6. Describe the purpose of conduit bodies and Type FS boxes.

Trade Terms

Conduit body
Explosion-proof
Handhole
Junction box
Mogul
Pull box
Raintight
Waterproof
Watertight
Weatherproof

Required Trainee Materials

1. Pencil and paper
2. Copy of the latest edition of the *National Electrical Code®*
3. Appropriate personal protective equipment

Prerequisites

Before you begin this module, it is recommended that you successfully complete *Core Curriculum; Electrical Level One;* and *Electrical Level Two*, Modules 26201-08 through 26204-08.

This course map shows all of the modules in *Electrical Level Two*. The suggested training order begins at the bottom and proceeds up. Skill levels increase as you advance on the course map. The local Training Program Sponsor may adjust the training order.

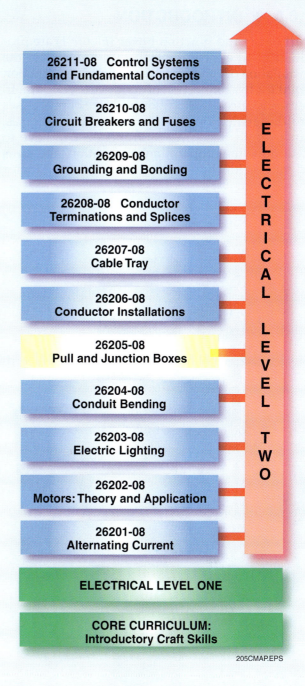

MODULE 26205-08 ◆ PULL AND JUNCTION BOXES 5.1

1.0.0 ♦ INTRODUCTION

Pull boxes and **junction boxes** (*Figure 1*) are provided in an electrical installation to facilitate the installation of conductors, or to provide a junction point for the connection of conductors, or both. In some instances, the location and size of pull boxes are designated on the drawings. In most cases, however, the electricians on the job will have to determine the proper number, location, and sizes of pull or junction boxes to facilitate conductor installation.

> **NOTE**
> Smaller conductors may use a regular octagon or square device box with a blank cover as a junction box.

Pull and junction boxes must be sized, installed, and supported to meet the requirements of *NEC Article 314*. Since the *NEC*® limits the number of conductors allowed in each box according to its volume, you must install boxes that are large enough to accommodate the number of conductors that must be spliced in the box or fed through it. Therefore, a knowledge of the various types of boxes and their volumes is essential.

Besides being able to calculate the required box sizes, you must also know how to select the proper type of box for any given application. For example, metallic boxes used in concrete deck pours are different from those used as pull and junction boxes in residential or commercial buildings. Boxes for use in hazardous locations must be rated as **explosion-proof**.

You must also know what fittings are available for terminating the various wiring methods in these boxes.

1.1.0 Boxes for Damp and Wet Locations

In damp or wet locations, boxes and fittings must be placed or equipped to prevent moisture or water from entering and accumulating within the box or fitting. It is recommended that approved boxes of nonconductive material be used with nonmetallic sheathed cable or approved nonmetallic conduit when the cable or conduit is used in moisture-prone locations. Boxes installed in wet locations must be approved for the purpose per *NEC Section 314.15*.

A wet location is any location subject to saturation with water or other liquids, such as locations exposed to weather or water, washrooms, garages, and interiors that might be hosed down.

Figure 1 ♦ Pull and junction boxes.

INSIDE TRACK

Floor Boxes

A floor box that is listed specifically for installation in a floor is required when junction boxes are installed in a floor. Listed floor boxes are provided with covers and gaskets to exclude surface water and cleaning compounds.

Outdoor Boxes

Outdoor wiring must be able to resist the entry of water. Outdoor boxes are either driptight, which means sealed against falling water from above, or watertight, which means sealed against water from any direction. Driptight boxes simply have lids that deflect rain; they are not waterproof. Watertight boxes are sealed with gaskets to prevent the entry of water from any angle.

Underground installations or those in concrete slabs or masonry in direct contact with the earth must be considered to be wet locations. *Raintight*, *waterproof*, or *watertight* equipment (including fittings) may satisfy the requirements for *weatherproof* equipment. Boxes with threaded conduit hubs and gasketed covers will normally prevent water from entering the box except for condensation within the box.

A damp location is a location subject to some degree of moisture. Such locations include partially protected outdoor locations—such as under canopies, marquees, and roofed open porches. It also includes interior locations subject to moderate degrees of moisture—such as some basements, some barns, and cold storage warehouses.

1.2.0 NEMA and IP Enclosure Classifications

Like switch enclosures, pull and junction boxes are also classified according to the degree of protection they provide from the elements. The NEMA enclosures are as follows:

- *NEMA Type 1: General Purpose* – This enclosure is primarily intended to prevent accidental contact with the enclosed apparatus. It is suitable for general-purpose applications indoors where it is not exposed to unusual service conditions. A NEMA Type 1 enclosure serves as protection against dust and light and indirect splashing, but is not dust-tight.
- *NEMA Type 3: Dust-Tight, Raintight* – This enclosure is intended to provide suitable protection against specified weather hazards. A NEMA Type 3 enclosure is suitable for outdoor applications, such as construction work. It is also sleet-resistant.
- *NEMA Type 3R: Rainproof, Sleet Resistant* – This enclosure protects against interference in operation of the contained equipment due to rain and resists damage from exposure to sleet. It is designed with conduit hubs and external mounting as well as drainage provisions.
- *NEMA Type 4: Watertight* – A watertight enclosure is designed to meet a hose test, which consists of a stream of water from a hose with a 1" nozzle, delivering at least 65 gallons per minute. The water is directed on the enclosure from a distance of not less than 10' for a period of five minutes. During this period, it may be pointed in one or more directions, as desired. There should be no leakage of water into the enclosure under these conditions.
- *NEMA Type 4X: Watertight, Corrosion-Resistant* – These enclosures are generally constructed along the lines of NEMA Type 4 enclosures except that they are made of a material that is highly resistant to corrosion. For this reason, they are ideal in applications such as meat packing and chemical plants, where contaminants would ordinarily destroy a steel enclosure over a period of time.
- *NEMA Type 7: Hazardous Locations, Class I* – These enclosures are designed to meet the application requirements of the *NEC®* for Class I hazardous locations. Class I locations are those in which flammable gases or vapors are or may be present in the air in sufficient quantities to produce explosive or ignitible mixtures. In this type of equipment, the circuit interruption occurs in the air.
- *NEMA Type 9: Hazardous Locations, Class II* – These enclosures are designed to meet the application requirements of the *NEC®* for Class II hazardous locations. Class II locations are those that are hazardous because of the presence of combustible dust. The letter or letters following the type number indicate the particular group or groups of hazardous locations (as defined in the *NEC®*) for which the enclosure is designed. The designation is incomplete without a suffix letter or letters.

- *NEMA Type 12: Industrial Use* – This type of enclosure is designed for use in those industries where it is desired to exclude such materials as dust, lint, fibers and flyings, oil seepage, or coolant seepage. There are no conduit openings or knockouts in the enclosure, and mounting is by means of flanges or mounting feet.
- *NEMA Type 13: Oil-Tight, Dust-Tight* – NEMA Type 13 enclosures are generally made of cast iron, gasketed, or permit use in the same environments as NEMA Type 12 devices. The essential difference is that due to its cast housing, a conduit entry is provided as an integral part of the NEMA Type 13 enclosure, and mounting is by means of blind holes rather than mounting brackets.

Table 1 provides an explanation of the ingress protection (IP) classification system. Each enclosure is designated IP followed by a two-digit number. For example, a dust-tight enclosure that is suitable for continuous immersion would be an IP68. *Table 2* is a cross reference of NEMA and IP enclosure classifications.

Table 2 NEMA and IP Enclosures

NEMA Enclosure Type	IP Classification
1	IP20
2	IP21
3	IP54
3R	IP24
3S	IP54
4,4X	IP56
5	IP52
6,6P	IP67
12,12K	IP52
13	IP54

Table 1 IP Classification System

First Number — Degree of protection against solid objects	Second Number — Degree of protection against water
0. Not protected.	0. Not protected.
1. Protected against a solid object greater than 50 mm, such as a hand.	1. Protected against water dripping vertically, such as condensation.
2. Protected against a solid object greater than 12 mm, such as a finger.	2. Protected against dripping water when tilted up to 15°.
3. Protected against a solid object greater than 2.5 mm, such as a wire or tool.	3. Protected against water when spraying at an angle of up to 60°.
4. Protected against a solid object greater than 1.0 mm, such as wire or thin strips of metal.	4. Protected against water splashing from any direction.
5. Dust-protected. Prevents ingress of dust sufficient to cause harm.	5. Protected against jets of water from any direction.
6. Dust-tight. No dust ingress.	6. Protected against heavy seas or powerful jets of water. Prevents ingress sufficient to cause harm.
	7. Protected against harmful ingress of water when immersed between a depth of 150 mm to 1 m.
	8. Protected against submersion. Suitable for continuous immersion in water.

2.0.0 ◆ SIZING PULL AND JUNCTION BOXES

Pull boxes should be as large as possible. Workers need space within the box for both hands and in the case of the larger wire sizes, workers will need room for their arms to feed the wire. *NEC Section 314.16* specifies that pull and junction boxes must provide adequate space and dimensions for the installation of conductors.

Long runs of conductors should not be made in one pull. Pull boxes, installed at convenient intervals, will relieve much of the strain on the conductors. The length of the pull, in many cases, is left to the judgment of the workers or their supervisor, and the conditions under which the work is installed.

The installation of pull boxes may seem to cause a great deal of extra work and trouble, but they save a considerable amount of time and hard work when pulling conductors. Properly placed, they eliminate many bends and elbows and do away with the necessity of fishing from both ends of a conduit run.

If possible, pull boxes should be installed in a location that allows electricians to work easily and conveniently. For example, in an installation where the conduit comes up a corner of a wall and changes direction at the ceiling, a pull box that is installed too high will force the electrician to stand on a ladder when feeding conductors, and will allow no room for supporting the weight of the wire loop or for the cable-pulling tools.

Unless the contract drawings or project engineer state otherwise, it is just as easy for the pull boxes to be placed at a convenient height that allows workers to stand on the floor with sufficient room for both wire loop and tools.

In some electrical installations, a number of junction boxes must be installed to route the conduit in the shortest, most economical way. The *NEC*® requires all junction boxes to be accessible. This means that a person must be able to get to the conductors inside the box without removing plaster, wall covering, or any other part of the building.

Junction boxes detract from the decorative scheme of a building. Therefore, where such boxes will be used in areas open to the public or in other areas where the boxes will be unattractive, they should be installed above suspended ceilings, in closets, or at least in corners of the room or area.

Junction boxes or pull boxes must be securely fastened in place on walls or ceilings or adequately suspended.

While certain sizes of factory-constructed boxes are available with knockouts, it may be necessary to have them custom built to meet the job requirements. When it is not possible to accurately anticipate the raceway entrance requirements, it will be necessary to cut the required knockouts on the job. In the case of large pull boxes and troughs, shop drawings should be prepared prior to the construction of these items with all required knockouts accurately indicated in relation to the conduit run requirements.

2.1.0 Sizing Pull and Junction Boxes for Systems Under 600V

For raceways containing conductors of No. 4 or larger, and for cables containing conductors of No. 4 or larger but carrying less than 600V, the minimum dimensions of pull boxes, junction boxes, auxiliary gutters, and wireways installed in a raceway or cable run shall comply with the following:

- In straight pulls, the length of the box shall not be less than eight times the trade diameter of the largest raceway.
- Where angle or U pulls are made, the distance between each raceway entry inside the box and the opposite wall of the box shall not be less than six times the trade diameter of the largest raceway in a row. This distance shall be increased for additional entries by the amount of the sum of the diameter of all other raceway entries in the same row on the same wall of the box. Each row shall be calculated individually, and the single row that provides the maximum distance shall be used.
- The distance between raceway entries enclosing the same conductor shall not be less than six times the trade diameter of the larger raceway.
- When transposing cable size into raceway size, the minimum trade size raceway required for the number and size of conductors in the cable shall be used.

Figure 2 shows a junction box with several runs of conduit. Since this is a straight pull, and 4" conduit is the largest size in the group, the minimum length required for the box can be determined by the following calculation:

Trade size of conduit × 8 [per *NEC Section 314.28(A)(1)*] = minimum length of box

or:

$$4" \times 8 = 32"$$

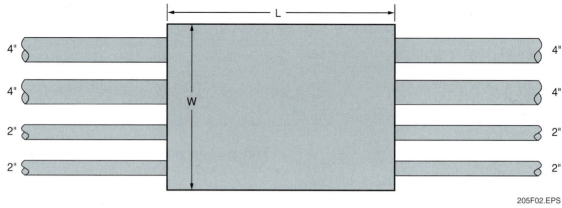

Figure 2 ◆ Pull box with two 4" and two 2" conduit runs.

Therefore, this particular pull box must be at least 32" in length. The width of the box, however, need only be of sufficient size to enable locknuts and bushings to be installed on all the conduits or connectors entering the enclosure.

Junction or pull boxes in which the conductors are pulled at an angle (*Figure 3*) must have a distance of not less than six times the trade diameter of the largest conduit [*NEC Section 314.28(A)(2)*]. The distance must be increased for additional conduit entries by the amount of the sum of the diameters of all other conduits entering the box on the same side. The distance between raceway entries enclosing the same conductors must not be less than six times the trade diameter of the largest conduit.

Since the 4" conduit is the largest size in this case:

$$L_1 = 6 \times 4" + (3 + 2) = 29"$$

Since the same conduit runs are located on the adjacent wall of the box, L_2 is calculated in the same way; therefore, $L_2 = 29"$.

The distance (D) = $6 \times 4"$ or 24". This is the minimum distance permitted between conduit entries enclosing the same conductor.

The depth of the box need only be of sufficient size to permit locknuts and bushings to be properly installed. In this case, a 6" deep box would suffice.

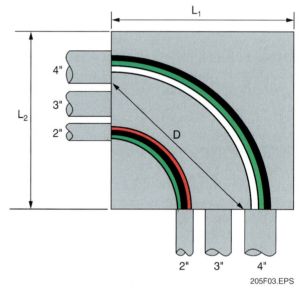

Figure 3 ◆ Pull box with conduit runs entering at right angles.

2.2.0 Sizing Pull and Junction Boxes for Systems over 600V

NEC Article 314, Part IV covers requirements for pull and junction boxes used on systems over 600V. Because the conductors used with these systems are larger and heavier than those used on lower-voltage systems, the *NEC*® requires additional

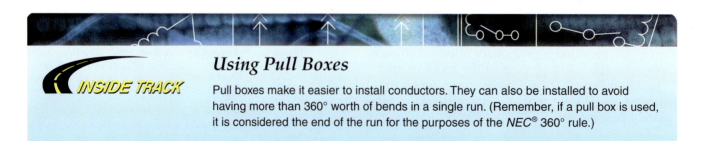

Using Pull Boxes

Pull boxes make it easier to install conductors. They can also be installed to avoid having more than 360° worth of bends in a single run. (Remember, if a pull box is used, it is considered the end of the run for the purposes of the *NEC*® 360° rule.)

Label Junction Boxes

It's a good idea to label every junction box plate with the circuit number, the panel it came from, and its destination. The next person to service the installation will be grateful for this extra help.

---

pulling and bending space. The *NEC®* requirements for these systems include the following:

- For straight pulls, the length of the box must be at least 48 times the outside diameter of the largest shielded or lead-covered conductor entering the box. See *NEC Section 314.71(A)*. When the cable is unshielded, the length may be reduced to 32 times the outside diameter of the largest nonshielded conductor entering the box.
- For angle or U pulls, the distance between each cable or conductor entry inside the box and the opposite wall of the box must be at least 36 times the diameter, over sheath, of the largest cable or conductor. See *NEC Section 314.71(B)(1)*. The distance must be increased by the sum of the outside diameters, over sheath, of all other cables or conductors entering through the same wall of the box. When the cable is unshielded and not lead covered, the length may be reduced to 24 times the outside diameter of the largest nonshielded conductor entering the box.
- The distance between a cable or conductor entry and its exit must be at least 36 times the outside diameter, over sheath, of that cable or conductor. See *NEC Section 314.71(B)(2)*. When the cable is both unshielded and not lead covered, the distance may be reduced to 24 times the outside diameter.
- In addition to a suitable cover, pull boxes containing conductors over 600V must be provided with at least one removable side.
- Per *NEC Section 314.72(E)*, boxes housing conductors over 600V must be completely enclosed and permanently marked DANGER – HIGH VOLTAGE – KEEP OUT.

3.0.0 ◆ CONDUIT BODIES

Conduit bodies, also called condulets, are defined in *NEC Article 100* as a separate portion of a conduit or tubing system that provides access through a removable cover to the interior of the system at a junction of two or more sections of the system or at a terminal point of the system. Conduit bodies are usually used with RMC and IMC. The cost of conduit bodies, because they are cast, is significantly higher than the stamped steel boxes. Splicing in conduit bodies is typically not recommended; however, it is permitted under certain conditions as specified in *NEC Section 314.16(C)(2)*.

As an electrical trainee, you will hear such terms as LL, LR, as well as other letters to distinguish between the various types of conduit bodies. To identify certain conduit bodies, an old trick of the trade is to hold the conduit body like a pistol (*Figure 4*). When doing so, if the oval-shaped opening of the conduit body is to your left, it is called an LL—the first L stands for elbow and the second L stands for left. If the opening is on your right, it is called an LR—the R stands for right. If the opening is facing upward, this type of conduit body is called an LB—the B stands for back. If there is an opening on both sides, it is called an LRL—for both left and right. The other popular shapes are named for their letter look-alikes, that is, T and X. The only exception is the C conduit body. Let us take a closer look at each of these.

3.1.0 Type C Conduit Body

A Type C conduit body (*Figure 5*) may be used to provide a pull point in a long conduit run or a conduit run that has bends totaling more than 360° (see *NEC Sections 342.26, 344.26, and 358.26*). In this application, the Type C conduit body is used as a pull point.

3.2.0 Type L Conduit Bodies

A Type L conduit body (*Figure 6*) is used as a pulling point for conduit that requires a 90° change in direction. (Again, the letter L is short for elbow.) To use a Type L conduit body, the cover is removed, the wire is pulled out and coiled on the ground (or floor), and then it is reinserted into the other opening and pulled. Type L conduit bodies are available with the cover on the back (Type LB), on the sides (Type LL and Type LR), or on both sides (Type LRL). The cover and gasket for conduit body fittings must be ordered separately; do not assume that these parts come with the conduit body when it is ordered.

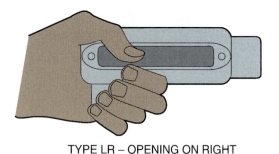
TYPE LR – OPENING ON RIGHT

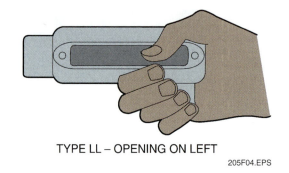
TYPE LL – OPENING ON LEFT

Figure 4 ◆ Identifying conduit bodies.

Figure 5 ◆ Type C conduit body.

Figure 6 ◆ Type L conduit bodies.

3.3.0 Type T Conduit Body

A Type T conduit body, also known as a tee, is used to provide a junction point for three intersecting conduits (*Figure 7*). Tees are used extensively in rigid conduit systems. The cost of a tee conduit body is more than twice that of a standard 4" square box with a cover. Therefore, the use of Type T conduit bodies with EMT is limited. According to *NEC Section 314.16(C)(2)*, conductor splicing is permitted in a Type T conduit body if the necessary conditions are satisfied.

Figure 7 ◆ Type T conduit body.

3.4.0 Type X Conduit Body

A Type X conduit body is used to provide a junction point for four intersecting conduits. The removable cover provides access to the interior of the X so that wire pulling and splicing may be performed (see *Figure 8*).

3.5.0 FS and FD Boxes

FS boxes are cast boxes available in single-gang, two-gang, and three-gang configurations. They are sized to permit the installation of switches and receptacles. Covers for switches and receptacles are available for FS boxes that have formed openings much like switch and receptacle plates. FD boxes are similar to FS boxes. The letter D in the FD box indicates it is a deeper box (2½" deep versus 1⅝" deep for an FS box). Neither FS nor FD boxes are considered by the *NEC®* to be conduit bodies. FS and FD boxes may be used in environments defined by NEMA 1 (dry, clean environments); NEMA 3R (outdoor, wet environments); and NEMA 12 (dusty, oily environments). NEMA stands for the National Electrical Manufacturers Association.

Engineers specify and electricians install FS and FD boxes for a reason. Never alter these boxes

5.8 ELECTRICAL LEVEL TWO ◆ TRAINEE GUIDE

Figure 8 ◆ Type X conduit body.

by drilling mounting holes in them. Most are provided with cast-in mounting eyes for this purpose. Mount these boxes only as recommended by the manufacturer.

3.6.0 Pulling Elbows

Pulling elbows are used exclusively for pulling wire at a corner point of a conduit run. The volume of a pulling elbow is too low to permit splicing wire. See *Figure 9(A)*.

3.7.0 Entrance Ells (SLBs)

An entrance ell, or SLB, is built with an offset so that it may be attached directly to the surface that is to have a conduit penetration. A cover on the back of the SLB permits wire to be pulled out and reinserted into the conduit that penetrates the support surface. See *Figure 9(B)*.

> **CAUTION**
> Never make splices in an entrance ell (SLB). Splices can be made in a conduit body that has the cubic inches marked, but never in an SLB.

3.8.0 Moguls

Mogul conduit bodies (*Figure 10*) are available in the same types as standard conduit bodies (Type L, Type T, and so on). They have raised covers to provide better access to large conductors during conductor installation or maintenance. Moguls also allow right angle bends where splices, pulls, and taps are needed in a weatherproof chamber. Like regular boxes, they must comply with the bending space requirements of NEC Section 314.28. Larger moguls may contain built-in cable-pulling rollers to facilitate installation.

INSIDE TRACK

All-in-One Conduit Bodies

All-in-one conduit bodies are also available. The conduit body shown here easily converts between LB, LL, LR, T, and C types by removing the appropriate covers. It is threaded for use with rigid conduit but also includes setscrews for use with EMT.

INSIDE TRACK

FS and FD Boxes

FS and FD boxes are precision-molded boxes with a cover and gasket designed to provide a tight seal. Because they cost significantly more than standard boxes, they are typically used only in industrial environments where the boxes are subjected to moisture, dirt, dust, and corrosion.

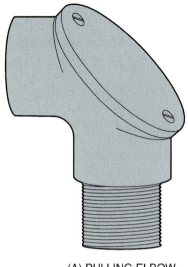
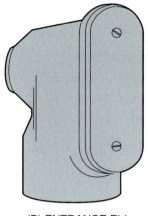

(A) PULLING ELBOW

(B) ENTRANCE ELL

Figure 9 ◆ Elbows.

Figure 10 ◆ Mogul.

> **NOTE**
>
> When installing conduit bodies, remember that just because the body matches the conduit size does not mean it matches the conductor size. For example, if the conductor fill is near the limit for a four-inch conduit, it may not be acceptable for a four-inch conduit body per *NEC Section 314.28*. If this is the case, a six-inch conduit body would be installed, along with the appropriate reducing fitting.

4.0.0 ◆ HANDHOLES

According to *NEC Article 100*, a handhole (*Figure 11*) is "an enclosure identified for use in underground systems, provided with an open or closed bottom, and sized to allow personnel to reach into, but not enter, for the purpose of installing, operating, or maintaining equipment or wiring or both." Handholes are often used with underground PVC conduit for the installation of landscape lighting, light poles, traffic lights, and other applications. Because handholes are exposed to various levels of foot traffic and/or vehicular loading, it is critical to select the correct enclosure for the application. *NEC Section 314.30, FPN* refers to *ANSI/SCTE 77-2002, Specification for Underground Enclosure Integrity,* for additional information on loading for underground enclosures. Some of the *NEC®* requirements for handholes are as follows:

- Like boxes and conduit bodies, handholes must be installed in such a way that the enclosed wiring remains accessible without removing any portion of the building, such as wall coverings, or without requiring excavation, such as for enclosures serving underground wiring installations. See *NEC Section 314.29*.
- Per *NEC Section 314.30(A)*, handhole enclosures must be sized in accordance with *NEC Section 314.28(A)* for conductors operating at 600 volts or below, and in accordance with *NEC Section 314.71* for conductors operating at over 600 volts.

Figure 12 ◆ Handhole containing traffic signal wiring.

Figure 11 ◆ Handhole.

- Underground raceways and cable assemblies entering a handhole enclosure must extend into the enclosure, but they are not required to be mechanically connected to the enclosure per *NEC Section 314.30(B)*.
- Where handhole enclosures without bottoms are installed, all enclosed conductors and any splices or terminations, if present, must be listed as suitable for wet locations per *NEC Section 314.30(C)*.
- Handhole enclosure covers shall have an identifying mark or logo that prominently identifies the function of the enclosure, such as ELECTRIC (*Figure 11*) or TRAFFIC SIGNALS (*Figure 12*). To discourage unauthorized entry, handhole enclosure covers must either weigh over 100 pounds or require the use of tools to open.
- Per *NEC Section 314.30(D)*, metal covers and other exposed conductive surfaces must be bonded in accordance with *NEC Article 250*.

4.1.0 Handhole Construction

Handhole enclosures may be constructed of a variety of materials, including precast Portland cement concrete, precast polymer concrete, thermoplastic, fiberglass-reinforced resin, steel, and aluminum. The type of material selected depends on the environment and the expected load.

 CAUTION

Some enclosures use dissimilar materials for the body and cover of the enclosure. Dissimilar materials are likely to have different responses to temperature and pressure variations, which may result in cracking or other performance failures.

4.2.0 ANSI/SCTE Requirements

According to *ANSI/SCTE 77-2002*, handhole enclosures must be constructed in such a way that they withstand all loads likely to be imposed and remain safe and reliable for the intended application. To meet the ANSI standard, enclosures must pass various physical, environmental, and internal equipment protection tests. The ANSI standard defines loading requirements for enclosures based upon anticipated loads and separates these requirements into various tiers defined by the application. Some manufacturers mark tier designations on the enclosure cover.

Lighting Poles

While the *NEC®* states that handholes are used in underground systems, the access holes on lighting poles are also typically referred to as handholes.

205SA02.EPS

Handhole Failure

In July 2003, an eight-year-old girl was electrocuted when she stepped on the cover of an in-ground pull box (handhole) in a ballpark in Ohio. Over the years, the unsupported box sank, bringing the lid in contact with the conductors, one of which contained a damaged splice. The exposed conductor energized the lid of the handhole. When the box was installed, there were no performance standards for the construction of underground enclosures and the *NEC®* did not require bonding.

The Bottom Line: Because of this and similar incidents involving foot traffic over energized handholes, the *NEC®* now requires that handholes be effectively bonded and designed to support the expected load.

While the *NEC®* does not explicitly require third party listing, some enclosure manufacturers have submitted products to UL for testing and enclosures that meet the testing requirements may now be UL Listed to the ANSI standard.

WARNING!
Without a solid footing, a handhole can sink and result in structural failure. A solid footing is essential and must be designed by a qualified individual in accordance with good engineering practices. Refer to the manufacturer's instructions for the proper installation of underground enclosures.

5.0.0 ♦ FITTINGS

Certain fittings are required in every raceway system for joining runs of conduit and also when the raceway terminates in an outlet box or other enclosure. Most metallic raceways qualify as an equipment grounding conductor provided they are tightly connected at each joint and termination point to provide a continuous grounding path.

5.1.0 EMT Fittings

Because EMT or thinwall is too thin for threads, special fittings must be used. For wet or damp locations, compression fittings such as those shown in *Figure 13* are used. This type of fitting contains compression rings made of plastic or other soft material that forms a watertight seal.

> **NOTE**
> To be used outdoors, an EMT compression fitting must be listed as raintight.

EMT fittings for dry locations can be either the setscrew type or the compression type. To use the setscrew type, the reamed ends of the EMT are inserted into the sleeve and the setscrews are tightened with a screwdriver to secure them and the conduit in place. Various types of setscrew couplings are shown in *Figure 14*.

EMT also requires connectors at each termination point, and EMT connectors are available to match the couplings described previously, that is, compression and setscrew types. They are similar to the couplings except that one end of the connector is threaded to accept a locknut and bushing.

5.2.0 Rigid, Aluminum, and IMC Fittings

Rigid metal conduit, aluminum conduit, and intermediate metal conduit all have sufficient wall thicknesses to permit threading. Consequently, all three types may be joined with threaded couplings (*Figure 15*) and when any of these types terminate into an outlet box or other enclosure, double locknuts are used to secure the conduit to the box opening. Running threads are not permitted for connection at couplings.

Sometimes rigid conduit or tubing must be connected to flexible metal conduit for connection to electric motors and other machinery that may

Figure 14 ◆ Setscrew fittings.

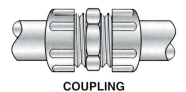
COUPLING

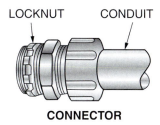
CONNECTOR

Figure 13 ◆ EMT compression fittings.

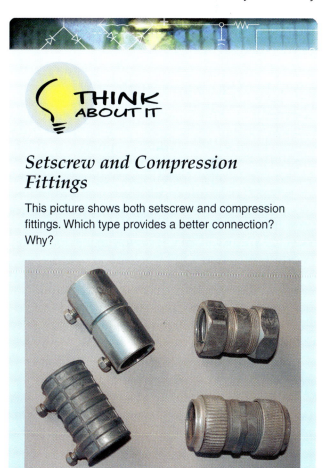

Setscrew and Compression Fittings

This picture shows both setscrew and compression fittings. Which type provides a better connection? Why?

Figure 15 ◆ Rigid metal conduit with coupling.

vibrate during operation. Combination couplings (*Figure 16*) are used to make the transition. When using combination couplings, be sure the flexible conduit is pushed as far as possible into the coupling. This covers the sharp edges of the conduit to protect the conductors from damage.

Threadless couplings and connectors may also be used under certain conditions with rigid, IMC, and aluminum conduit. When used, they must be made up wrench-tight and where buried in masonry or concrete, they must be concrete-tight. Where installed in wet locations, they must be rainproof. This type of coupling is not permitted in most hazardous locations.

Other types of fittings used with raceway systems are shown in *Figure 17*.

5.3.0 Locknuts and Bushings

In general, locknuts are used on the inside and outside walls of outlet boxes or other enclosures to which threaded conduit is connected. When conduit connectors are used, such as EMT connectors, only one locknut is required on the threads that protrude inside the box. A grounding locknut may be needed if bonding jumpers are used inside the box or enclosure. Special sealing locknuts are also available for use in wet locations. Locknuts are shown in *Figure 18*.

Bushings protect the wires from the sharp edges of the conduit or connector. Bushings are

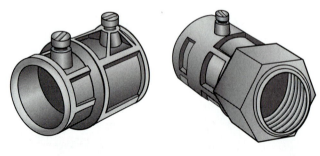

Figure 16 ◆ Combination couplings.

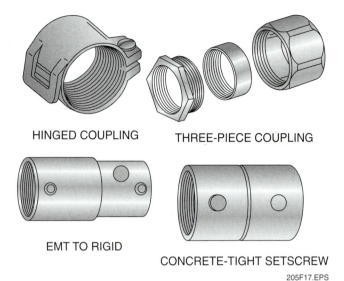

Figure 17 ◆ Metal conduit couplings.

Figure 18 ◆ Common types of locknuts.

usually made of plastic, fiber, or metal. Some metal bushings have a grounding screw to permit an equipment or bonding jumper wire to be installed. Several types of bushings are shown in *Figure 19*.

An insulating bushing is installed on the threaded end of conduit that enters a sheet metal enclosure. The purpose of the bushing is to protect the conductors from being damaged by the sharp edges of the threaded conduit end. Any ungrounded conductor, No. 4 AWG or larger, that enters a raceway, box, or enclosure must be protected with an insulating bushing, as required in *NEC Sections 300.4(G), 312.6(C), and 314.17(D)*.

A grounding insulated bushing has provisions for protecting conductors and also has provisions for the connection of a ground wire. The ground wire, once connected to the grounding bushing, may be connected to the box to which the conduit is connected. See *Figure 20*.

According to *NEC Section 314.15*, boxes, conduit bodies, and fittings installed in damp or wet locations must provide protection against the entry of moisture. This can be accomplished using either a sealing locknut or a gasketed fitting. Myers hubs are gasketed fittings with an integral O-ring to provide a watertight seal for enclosures

Figure 19 ◆ Typical bushings used at termination points.

Stainless Steel Gasketed Hubs

Stainless steel gasketed hubs are available for use with stainless steel conduit systems. Stainless steel is commonly used in the food industry.

installed in wet locations (*Figure 21*). They are often used in place of sealing locknuts because they provide a more reliable seal in areas where water tends to accumulate. For example, in wet locations where the knockout entry is at the side of the box, a sealing locknut might be used because the water tends to shed itself. However, if there is a knockout on top of the box where the water tends to settle, a Myers hub should be used

(A)

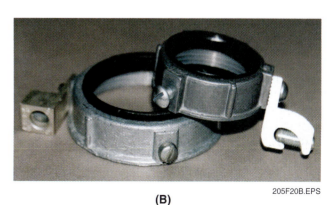

(B)

Figure 20 ◆ (A) Regular insulating bushings and (B) grounding insulating bushings.

MODULE 26205-08 ◆ PULL AND JUNCTION BOXES 5.15

Figure 21 ◆ Myers-type (gasketed) hub.

because it provides a better seal. Myers hubs are also listed for use in certain hazardous locations.

NOTE

Many manufacturers will void the warranty on a watertight enclosure if a Myers hub or other gasketed fitting is not used. Always read and follow the manufacturer's instructions for any product you are installing.

Figure 22 ◆ Knockout punch kit.

An opening must be provided in the outlet box or enclosure for the entrance of conduit and connectors when raceway systems terminate. Most boxes and enclosures are provided with an adequate number of concentric knockouts. However, some may not have precut knockouts or the ones that are available may not be in the required location. In these cases, a knockout punch must be used to make a hole for the conduit connection. A hand-operated knockout punch is shown in *Figure 22*.

To use the knockout punch, the center of the hole is located in the box or enclosure and marked with a center punch. A pilot hole is then drilled to accept the threaded drive bolt of the knockout punch. The punch is separated from the drive screw, the screw is then placed through the pilot hole with the die on one side of the box wall, and the punch is screwed onto the drive screw on the opposite side of the wall. The punch is then aligned and screwed onto the drive screw hand-tight, in which case the punch should lightly bite into the wall of the enclosure. Finally, a wrench is then used to tighten the drive nut until the punch is drawn through the enclosure wall, making a neat circular opening.

INSIDE TRACK

Step Bits

Step bits can be used to quickly create holes for smaller size conduit.

5.16 ELECTRICAL LEVEL TWO ◆ TRAINEE GUIDE

Where many such openings must be made, or when holes for the larger sizes of conduit must be cut, power knockout tools are often used to speed up the process. The battery-powered knockout kit shown in *Figure 23* can be used to punch up to 2" conduit, while the hydraulic kit in *Figure 24* can be used to punch up to 4" conduit.

Figure 24 ◆ Hydraulic knockout kit.

Figure 23 ◆ Battery-powered knockout kit.

Review Questions

1. Where in the *NEC®* will you find the most information on pull and junction boxes?
   a. NEC Article 240
   b. NEC Article 250
   c. NEC Article 314
   d. NEC Article 320

2. When calculating the pull box size for straight pulls on systems operating at less than 600V, the length of the box must not be less than _____ times the trade diameter of the largest raceway.
   a. two
   b. four
   c. six
   d. eight

3. If the largest trade diameter of a raceway entering a pull box is 3", and it is a straight pull for a system operating at less than 600V, the minimum size box allowed is _____.
   a. 20"
   b. 24"
   c. 30"
   d. 36"

4. For a straight pull, the length of a pull box housing unshielded conductors carrying over 600V must be _____ times the outside diameter of the largest unshielded conductor entering the box.
   a. 24
   b. 32
   c. 48
   d. 60

5. For a straight pull, the length of a pull box housing shielded conductors carrying over 600V must be _____ times the outside diameter of the largest shielded conductor entering the box.
   a. 24
   b. 32
   c. 48
   d. 60

6. True or False? In addition to a suitable cover, pull boxes containing conductors over 600V must be provided with at least one removable side.
   a. True
   b. False

7. A _____ conduit body has openings on four different sides plus an access opening.
   a. Type X
   b. Type C
   c. Type T
   d. Type LL

8. Which of the following best describes how Type FS boxes should be installed?
   a. Holes should be drilled in back of the box for mounting screws.
   b. Holes should be drilled on the sides of the box only for mounting screws.
   c. No holes should be drilled in the box for mounting.
   d. Holes may be drilled only in existing installations.

Review Questions

9. Which of the following enclosures is best suited for splicing conductors?
   a. Pulling ell
   b. SLB
   c. Conduit body
   d. Mogul

10. Item B in *Figure 1* is used to connect _____.
    a. two sections of flexible conduit
    b. flexible conduit to EMT
    c. EMT to rigid conduit
    d. flexible conduit to rigid conduit

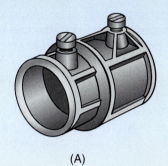

(A)

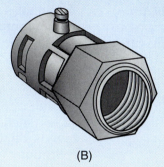

(B)

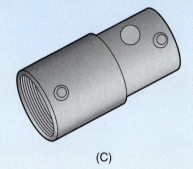

(C)

Summary

Pull and junction boxes play an important role in conductor installation and maintenance. They are installed at various locations in a conduit run to provide additional pull points to facilitate conductor installation or to provide access for system service and maintenance. *NEC Article 314* provides details on the proper sizing and selection of pull and junction boxes. The rules for systems carrying over 600V are much more stringent than those for systems operating at lower voltages. Always refer to the latest edition of the *NEC*® as well as the manufacturer's instructions for the proper application and installation of pull and junction boxes.

Conduit bodies provide access to the electrical system at junction and termination points. Moguls are a special type of conduit body with a raised cover to provide extra space when using larger conductors.

Handholes are underground enclosures that typically serve outdoor lighting systems and traffic signal wiring. Because they are subject to varying load conditions and may become energized if they fail, it is essential to select and install them for the specific environment to which they will be exposed.

Fittings are required to connect a raceway to a pull or junction box. The type of fitting selected depends on the type of conduit used and the application.

Notes

Trade Terms Introduced in This Module

Conduit body: A separate portion of a conduit or tubing system that provides access through a removable cover (or covers) to the interior of the system at a junction of two or more sections of the system or at a terminal point of the system. Boxes such as FS and FD or larger cast or sheet metal boxes are not classified as conduit bodies.

Explosion-proof: Designed and constructed to withstand an internal explosion without creating an external explosion or fire.

Handhole: An enclosure used with underground systems to provide access for installation and maintenance.

Junction box: An enclosure where one or more raceways or cables enter, and in which electrical conductors can be, or are, spliced.

Mogul: A type of conduit body with a raised cover to provide additional space for large conductors.

Pull box: A sheet metal box-like enclosure used in conduit runs to facilitate the pulling of cables from point to point in long runs, or to provide for the installation of conduit support bushings needed to support the weight of long riser cables, or to provide for turns in multiple-conduit runs.

Raintight: Constructed or protected so that exposure to a beating rain will not result in the entrance of water under specified test conditions.

Waterproof: Constructed so that moisture will not interfere with successful operation.

Watertight: Constructed so that water will not enter the enclosure under specified test conditions.

Weatherproof: Constructed or protected so that exposure to the weather will not interfere with successful operation.

Additional Resources

This module is intended to be a thorough resource for task training. The following reference work is suggested for further study. This is optional material for continued education rather than for task training.

National Electrical Code® Handbook, Latest Edition. Quincy, MA: National Fire Protection Association.

CONTREN® LEARNING SERIES – USER UPDATE

NCCER makes every effort to keep these textbooks up-to-date and free of technical errors. We appreciate your help in this process. If you have an idea for improving this textbook, or if you find an error, a typographical mistake, or an inaccuracy in NCCER's Contren® textbooks, please write us, using this form or a photocopy. Be sure to include the exact module number, page number, a detailed description, and the correction, if applicable. Your input will be brought to the attention of the Technical Review Committee. Thank you for your assistance.

Instructors – If you found that additional materials were necessary in order to teach this module effectively, please let us know so that we may include them in the Equipment/Materials list in the Annotated Instructor's Guide.

Write: Product Development and Revision
National Center for Construction Education and Research
3600 NW 43rd St., Bldg. G, Gainesville, FL 32606

Fax: 352-334-0932

E-mail: curriculum@nccer.org

Craft _____ Module Name _____

Copyright Date _____ Module Number _____ Page Number(s) _____

Description
_____
_____
_____
_____

(Optional) Correction
_____
_____
_____

(Optional) Your Name and Address
_____
_____
_____

Conductor Installations

U.S. Department of Transportation Headquarters

Clark Construction Group built the new headquarters for the U.S. Department of Transportation (DOT) in Washington, D.C. on an 11-acre parcel. The 2,000,000-square-foot office complex spans two city blocks.

26206-08

26206-08
Conductor Installations

Topics to be presented in this module include:

1.0.0	Introduction	.6.2
2.0.0	Planning the Installation	.6.3
3.0.0	Setting Up for Wire Pulling	.6.7
4.0.0	Cable-Pulling Equipment	.6.17
5.0.0	High-Force Cable Pulling	.6.22
6.0.0	Supporting Conductors	.6.26
7.0.0	Pulling Cable in Cable Trays	.6.27
8.0.0	Physical Limitations of Cable	.6.28
9.0.0	Cable-Pulling Instruments	.6.32

Overview

The first step in any conductor installation is planning. Most small conductors can be pulled into a conduit using manual techniques such as the two-person method in which one person pulls the conductors and another feeds the conductors at the supply end. A common manual pulling tool is the fish tape, which may be made of spring steel or fiberglass. The stripped ends of the conductors are fastened to the end of the fish tape and secured in place with electrical tape. As the person on the pulling end pulls, the person on the feeding end feeds the conductors into the conduit in unison with the pull, protecting the conductors from damage.

Wire-pulling ropes are often used in place of fish tapes because of their flexibility and non-conductive properties. To initially install a pulling rope in a conduit system, either a vacuum or blower system is used to install a lightweight string into the system. The pulling rope is then attached to the string and pulled through the conduit system. The conductors to be installed are attached to the pulling rope and either manually pulled through the raceway system or installed using a power pulling device.

There are several types of power pulling devices available. Most of them operate on the same principle as a power winch. Only properly trained personnel may install conductors using a power winch system, as high torques can be developed that can cause personal injury, damage to conductors, or both.

Note: *NFPF 70®*, *National Electrical Code®*, and *NEC®* are registered trademarks of the National Fire Protection Association, Inc., Quincy, MA 02269. All *National Electrical Code®* and *NEC®* references in this module refer to the 2008 edition of the *National Electrical Code®*.

Objectives

When you have completed this module, you will be able to do the following:

1. Explain the importance of communication during a cable-pulling operation.
2. Plan and set up for a cable pull.
3. Set up reel stands and spindles for a wire-pulling installation.
4. Explain how mandrels, swabs, and brushes are used to prepare conduit for conductors.
5. Properly install a pull line for a cable-pulling operation.
6. Explain how and when to support conductors in vertical conduit runs.
7. Describe the installation of cables in cable trays.
8. Calculate the probable stress or tension in cable pulls.

Trade Terms

Basket grip
Cable grip
Capstan
Clevis
Conductor support
Conduit piston
Fish line
Fish tape
Setscrew grip
Sheave
Soap

Required Trainee Materials

1. Paper and pencil
2. Appropriate personal protective equipment
3. Copy of the latest edition of the *National Electrical Code®*

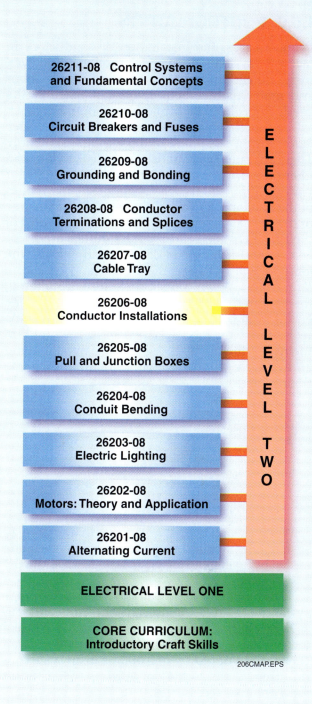

Prerequisites

Before you begin this module, it is recommended that you successfully complete *Core Curriculum; Electrical Level One;* and *Electrical Level Two,* Modules 26201-08 through 26205-08.

This course map shows all of the modules in *Electrical Level Two.* The suggested training order begins at the bottom and proceeds up. Skill levels increase as you advance on the course map. The local Training Program Sponsor may adjust the training order.

1.0.0 ♦ INTRODUCTION

In most cases, the installation of conductors in raceway systems is merely routine. However, there are certain practices that can reduce labor and materials and help prevent damage to the conductors. The use of modern equipment, such as vacuum fish tape systems, is one way to reduce labor during this phase of the wiring installation. There are three types of fish tape: steel, nylon, and fiberglass. They also come in different weights for various applications. The proper size and length of the fish tape, as well as the type, should be one of the first considerations. For example, if most of the runs between branch circuit outlets are 20' or less, a short fish tape of 25' will easily handle the job and will not have the weight and bulk of a larger tape. When longer runs are encountered, the required length of the fish tape should be enclosed in one of the metal or plastic fish tape reels. This way, the fish tape can be rewound on the reel as the pull is being made to avoid having an excessive length of tape lying around on the floor or deck.

WARNING!
Never fish a steel tape through or into enclosures or raceways containing an energized conductor.

When several bends are present in the raceway system, the insertion of the fish tape may be made easier by using flexible fish tape leaders on the end of the fish tape.

The combination blower and vacuum fish tape systems are ideal for use on long runs and can save much time. Basically, the system consists of a tank and air pump with accessories. An electrician can vacuum or blow a line or tape in any size conduit from ½" through 4", or even up to 6" conduit with optional accessories.

After the fish tape is inserted in the raceway system, the conductors must be firmly attached by some approved means. On short runs, where only a few conductors are involved, all that is necessary is to strip the insulation from the ends of the wires, bend these ends around the hook in the fish tape, and securely tape them in place. Where several wires are to be pulled together, the wires should be staggered and the fish tape securely taped at the point of attachment so that the overall diameter is not increased any more than is absolutely necessary.

Basket grips (*Figure 1*) are available in many sizes for almost any size and combination of conductors. They are designed to hold the conductors firmly to the fish tape and can save much time and trouble that would be required when taping wires.

In all but very short runs, the wires should be lubricated with wire lubricant prior to attempting the pull, as well as during the pull. Some of this lubricant should also be applied to the inside of the conduit.

Wire dispensers are great aids in keeping the conductors straight and facilitating the pull. Many different types of wire dispensers are available to handle virtually any size spool of wire or cable. Some dispensers are stationary, while others have casters that make it easy to move heavy spools to the pulling location (*Figure 2*). Wheeled dispensers are sometimes called wire caddies.

Figure 1 ♦ Basket grip.

Nonconductive Fish Tape

When fishing conductors in a conduit or raceway that already contains existing energized (live) conductors, the safest method is to turn off and lock out/tag the power sources for all the live conductors. However, in some rare instances, fishing a conductor through a conduit or raceway containing other live conductors may be unavoidable. In this case, always request your supervisor's approval before proceeding and always use a nonconductive fish tape made of nylon or fiberglass.

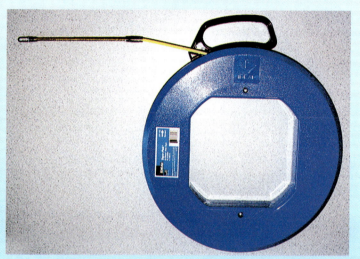

2.0.0 ♦ PLANNING THE INSTALLATION

The importance of planning any wire-pulling installation cannot be over-stressed. Proper planning will make the work go easier and much labor will be saved.

Large sizes of conductors are usually shipped on reels, involving considerable weight and bulk. Consequently, setting up these reels for the pull, measuring cable run lengths, and similar preliminary steps will often involve a relatively large amount of the total cable installation time. Therefore, consideration must be given to reel setup space, proper equipment, and moving the cable reels into place.

Whenever possible, the conductors should be pulled directly from the shipping reels without pre-handling them. This can usually be done through proper coordination of the ordering of the conductors with the job requirements. While doing so requires extremely close checking of the drawings and on-the-job measurements (allowing for adequate lengths of conductors in pull boxes, elbows, troughs, connections, and splices), the extra effort is well worth the time to all involved.

When the lengths of cable have been established, the length of cable per reel can be ordered so that the total length per reel will be equal to the total of a given number of raceway lengths, and the reel so identified.

(A) CABLE DISPENSER

(B) WIRE CADDY

Figure 2 ♦ Wire dispensers.

In most cases, the individual cables of the proper length for a given number of runs are reeled separately onto two or more reels at the factory, depending on the number of conductors in the runs.

When individual conductors are shipped on separate reels, it is necessary to set up for the same number of reels as the number of conductors to be pulled into a given run, as shown in *Figure 3*.

As an extra precaution against error in calculating the lengths of conductors involved, it is a good idea to actually measure all runs with a fish tape before starting the cable pull, adding for makeup to reach the terminations, and accounting for discarding the cable underneath the pulling sleeve or pulling wrap, as it will have been excessively stressed during the pull. Check these totals against the totals indicated on the reels. Under normal cable delivery schedules, when the feeder raceways have been installed at a relatively early stage of the overall building construction, it may not delay the final completion of the electrical installation to delay ordering the cables until the raceways can actually be measured.

When pulling conductors directly from the reels, care must be taken that each given run be cut off from the reel so that there is a minimum amount of waste. In other words, preclude the possibility of the final run of cable taken from the reel being too short for that run.

2.1.0 Pulling Location

Each job will have to be judged separately as to the best location of pulling setups; the number of setups should be reduced to a minimum in line with the best direction of the pull. The pulling location for a particular job is determined by the weight of the cable, height of the pull, practicality of moving the equipment to the pulling location, number of setups required, as well as the location of any bends. Also, a separate setup might have to be made at the top of each rise, whereas a single setup might be made at a ground floor pull box location from which several feeders are served with the same size and number of conductors.

The location of the pulling equipment determines the number of workers required for the job. A piece of equipment that can be moved in and set up on the first floor by four workers in an hour's time may require six workers working two hours when set up in basements or parking levels of buildings. Therefore, when planning a cable pull, make certain enough workers are on hand to adequately handle the installation.

It is a simple operation for a few workers to roll cable reels from a loading platform to a first floor setup, whereas moving them to upper floors involves much handling and usually requires a crane or other hoists. In addition, the reel jacks have to be moved to the setup point when a downward pull is made, and after the pulling operation is completed, they must be taken back to the first floor.

Figure 3 ◆ Multi-cable pull.

Multiple Reels

Why are multiple reel pulling setups used?

2.2.0 Cable Pull Operations

These operations are performed to a lesser or greater degree in almost all cable pulls with larger sizes of conductors:

Step 1   Measure or re-check runs and establish communications between both ends of the pull.

Step 2   Provide pulling equipment.

Step 3   Receive and unload pulling equipment.

Step 4   Move pulling equipment to the pulling location.

Step 5   Set up and anchor pulling equipment.

Step 6   Remove to another location or move to a loading platform and load on a truck or forklift.

Step 7   Receive and store cable; may be moved directly to setup location if job conditions permit.

Step 8   Move to setup point.

Step 9   Move reel jacks and mandrel to setup point.

Step 10   Set up reels, then identify and tag the cables.

Step 11   Prepare cable ends.

Step 12   Install fish tape.

Step 13   Install pulling line or cable.

Step 14   Connect pulling line.

Step 15   Lubricate with proper lubricants.

Step 16   Pull cable.

Step 17   Disconnect pulling line.

Step 18   Remove reels.

Step 19   Permanently identify both ends of the cables/conductors.

Step 20   Rack cables in pull boxes and troughs.

Step 21   Splice or connect cables.

Step 22   Check and test.

INSIDE TRACK

Measuring Tape

A waterproof polyester tape with permanent markings every foot from 0' to 3,000' is available on reels for use in measuring conduit/raceway runs. It can be fished through the run manually or using a power fishing system.

In some instances, the following additional operations are involved, depending upon the exact details of the project. Other items of importance will be discussed later in this module.

Step 23   Remove lagging or other protective covering from reels.

Step 24   Unreel cable and cut it to length.

Step 25   Re-reel cable for pulling.

Step 26   Replace lagging on reels.

Step 27   Operate such auxiliary equipment as guide-through cabinets or pull boxes and signal systems between reel and pulling setups.

Each of these pulling operations is discussed in detail in this module (also see *Figure 4*).

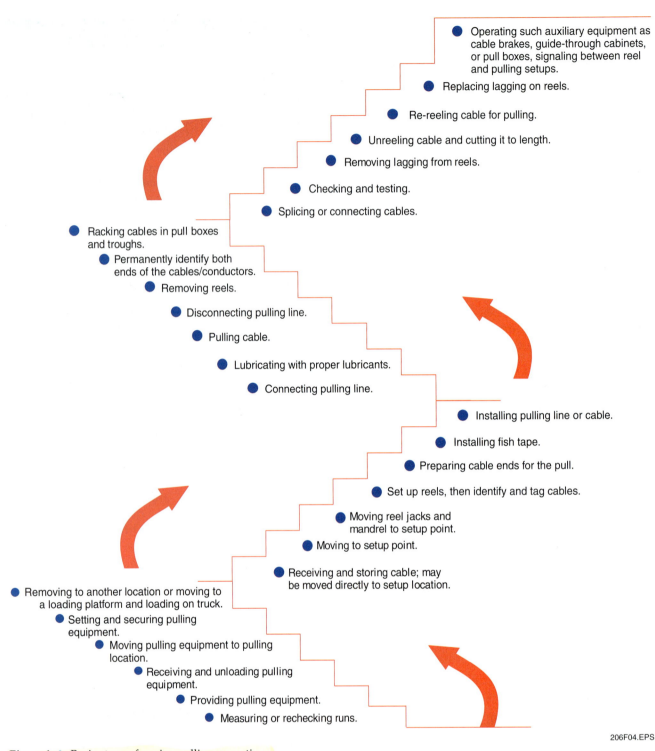

Figure 4 ♦ Basic steps of a wire-pulling operation.

3.0.0 ◆ SETTING UP FOR WIRE PULLING

As mentioned previously, much planning is required for pulling the larger sizes of conductors in raceways. There are several preliminary steps required before the actual pull begins.

The proper use of appropriate equipment is crucial to a successful cable installation. The equipment needed for most installations is shown in the checklist in *Figure 5*. Some projects may require all of these items, while others may require only some of them. Each cable-pulling project must be taken on an individual basis and analyzed accordingly. Seldom will two pulls require identical procedures.

NOTE

Think of everything that can go wrong and take every precaution.

3.1.0 Setting Up the Cable Reels

When reels of cable arrive at the job site, it is best to move them directly to the setup location if at all possible. This prevents having to handle the reels more than necessary. However, if this is not practical, arrangements must be made for storage until the cable is needed.

The exact method of handling reels of cable depends on their size and the available tools and equipment. In many cases, the reels may be rolled to the pulling location by one or more workers. For reels up to 24" wide × 40" in diameter, a cable reel transporter can be used to transport the cable reel; it also acts as a dispenser during the pulling operation. When available, a forklift is ideal for lifting and transporting cable reels. See *Figure 6*.

However, for very large reels (48" or more in diameter), a crane or similar hoisting apparatus is usually necessary for lifting the reels onto reel jacks supported by jack stands to acquire the necessary height. *Figure 7* shows a summary of proper and improper ways to transport reels of wire or cable on the job site.

Figure 8 shows several types of reel jacks, including the spindle. For a complete setup, two stands and a spindle are required for each reel. The reel jacks or stands are available in various sizes from 13" to 54" high to accommodate reel diameters up to 96". Extension stands used in conjunction with reel stands can accommodate larger reels.

Reel-stand spindles are commonly available in diameters from 2⅜" to 3½" and from 59" to 100" in length for carrying reel loads up to 7,500 pounds. However, some heavy-duty spindles are rated for loads up to 15,000 pounds.

EQUIPMENT CHECKLIST

- ❏ PORTABLE ELECTRIC GENERATOR
- ❏ EXTENSION CORDS AND GFCI
- ❏ PUMP, DIAPHRAGM
- ❏ MAKEUP BLOWER AND HOSE
- ❏ MANHOLE COVER HOOKS
- ❏ WARNING FLAGS, SIGNS
- ❏ ELECTROSTATIC kV TESTER
- ❏ ELECTRIC SAFETY BLANKETS AND CLAMPS
- ❏ RADIOS OR TELEPHONES
- ❏ GLOVES
- ❏ FLOOD LAMPS
- ❏ FISH TAPE OR STRING BLOWER/VACUUM
- ❏ HAND LINE
- ❏ DUCT-CLEANING MANDRELS
- ❏ DUCT-TESTING MANDRELS
- ❏ CAPSTAN-TYPE PULLER
- ❏ SNATCH BLOCKS
- ❏ SHORT ROPES FOR TEMPORARY TIE-OFFS
- ❏ GUIDE-IN FLEXIBLE TUBING (ELEPHANT TRUNKS)
- ❏ SEVERAL WIRE ROPE SLINGS OF VARIOUS LENGTHS
- ❏ SHACKLES/ROPE CLEVIS
- ❏ GANG ROLLERS WITH AT LEAST 4' EFFECTIVE RADIUS
- ❏ HAND WINCHES
- ❏ MANHOLE EDGE SHEAVE
- ❏ PULLING ROPE
- ❏ SWIVELS
- ❏ BASKET GRIP PULLERS
- ❏ 0-1/5/10 KIP DYNAMOMETER
- ❏ REEL ANCHOR
- ❏ REEL JACKS
- ❏ CABLE CUTTERS
- ❏ LINT-FREE RAGS
- ❏ CABLE-PULLING LUBRICANTS
- ❏ PRELUBING DEVICES
- ❏ PLYWOOD SHEETS
- ❏ DIAMETER TAPE
- ❏ 50' MEASURING TAPE
- ❏ SILICONE CAULKING (TO SEAL CABLE ENDS)

Figure 5 ◆ Wire-pulling equipment checklist.

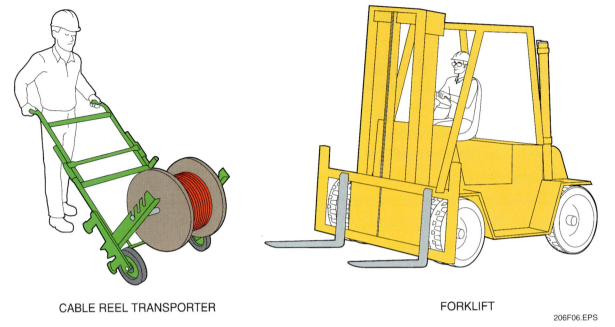

Figure 6 ♦ Two methods of transporting cable reels.

3.2.0 Preparing Raceways for Conductors

Another preliminary step prior to pulling conductors in raceway systems is to inspect the raceway itself. Few things are more frustrating than to pull four 1,000 kcmil conductors through a conduit and find out when the pull is almost done that the conduit is blocked or damaged. Such a situation usually requires pulling the conductors back out, repairing the fault, and starting all over again.

These problems are not too serious with exposed banks of conduit, as the conduit can usually be separated at the fault, the fault corrected, and another piece of conduit installed, using unions if necessary. However, in underground conduit runs—especially those encased in concrete—the situation can be both time-consuming and costly.

A test pull will detect any hidden obstructions in the conduit prior to the pull (*Figure 9*). Go and no-go steel and aluminum mandrels are available for pulling through runs of conduit before the cable installation. Mandrels should be approximately 80% of the conduit size (twice the 40% fill factor). If any obstructions are found, they can be corrected before wasting time on an installation that might result in conductor damage and the possibility of having to re-pull the conductors.

Figure 10 shows several devices used to inspect raceway systems, as well as to prepare raceways for easier and safer conductor pulls. The conduit swab in *Figure 10* may be used ahead of the conductors during a pull. Its main purpose is to swab out debris from the raceway and spread a uniform film of pulling compound inside the conduit for easier pulling. The flexible mandrel is also used to clean the conduit and check the run for obstructions. The flexible discs make it easier to pull around tight bends.

The conduit brush in *Figure 10* helps clean and polish the interior of conduit before pulling the cable. Such brushes remove sand and other light obstructions. Note that this brush has a pulling eye on one end and a twisted eye on the opposite end, enabling it to be pushed or pulled through the conduit.

One final step before starting the pull is to measure the length of the raceway, including all turns in junction boxes and the like. A fish tape may be pushed through the raceway system and a piece of tape used to mark the end. When it is pulled back out, a tape measure may be used to measure the exact length. An easier way, however, is to use a power fishing system to push or pull a measuring tape through the conduit run. Details of this operation are explained in the next section.

When measuring the conductor length, be sure to allow sufficient room where measurements are made through a pull box. Conductors should enter and leave pull boxes in such a manner as to allow the greatest possible sweep for the conductors. Large conductors are especially difficult to bend but with proper planning, you can simplify the feeding of these conductors from one conduit to another.

ON THE RIMS OF THE SPOOL (MOVING EQUIPMENT DOES NOT COME INTO CONTACT WITH CABLE)

ON THE FLAT SIDE OF THE SPOOL OR ON THE CABLE (MOVING EQUIPMENT COMPRESSES INSULATION AND MAY DAMAGE CABLE)

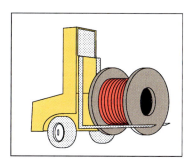
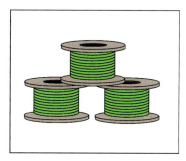
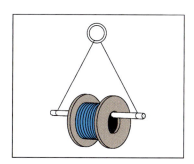
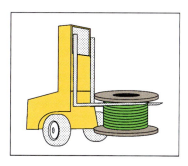
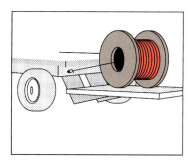
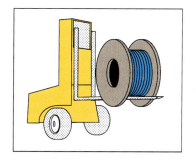
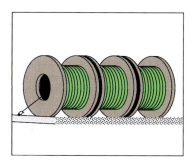
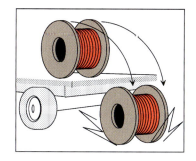

206F07.EPS

Figure 7 ◆ Proper and improper ways of transporting cable reels.

Figure 8 ◆ Typical reel stands.

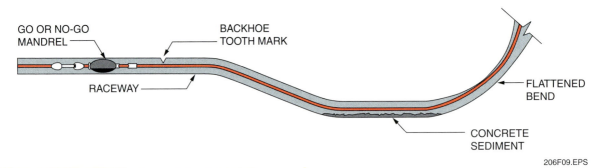

Figure 9 ◆ Faults that may be detected with a conduit mandrel.

Figure 10 ◆ Devices used to inspect, clean, and lubricate raceway systems.

For example, if a conduit run makes a right-angle bend through a pull box, the conduit for a given feeder should come into the box at the lower left-hand corner and leave diagonally opposite at the upper right-hand corner, as shown in *Figure 11*. This gives the conductors the greatest possible sweep with the box, eliminating sharp bends and consequent damage to the conductor insulation.

Runs should also be calculated to allow for splices and terminations in junction boxes, panelboards, motor-control centers, the cable discarded under the sleeve or pulling wrap, etc.

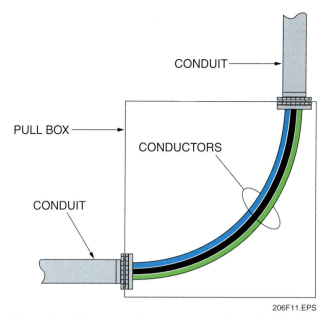

Figure 11 ♦ Obtaining the greatest possible conductor sweep in a pull box.

3.3.0 Installing the Pull Line

At one time, pull-in wires were frequently placed in conduit runs as the raceways were installed. However, in recent times, with modern cable-fishing equipment, this practice is seldom used.

Pull lines are sometimes manually fished in with a steel fish tape, but much time can be saved by using a blower/vacuum fish tape system. In general, a conduit piston—sometimes referred to as a mouse or missile—is blown with air pressure or vacuumed through the run. The foam piston is sized to the conduit and has a loop on both ends. In most cases, fish line or measuring tape is attached to the piston as it is blown or vacuumed through the conduit run. The measuring tape serves two purposes: it provides an accurate measurement of the conduit run, and the tape is used to pull the cable-pulling rope into the conduit run. In some cases, if the run is suitable, the pulling rope is attached directly to the piston and vacuumed into the run. *Figure 12* shows a blower/vacuum fish tape system being used to vacuum a pull line in a conduit while *Figure 13* shows the same apparatus blowing the piston through. Most of these units provide enough pressure to clean dirt or water from conduit during the fishing operation. *Figure 14* shows two types of pistons used with this system. The one on the right utilizes air-guide vanes to prevent the piston from tumbling inside the larger sizes of conduit.

3.4.0 Preparing Cable Ends for Pulling

The pulling-in line or cable must be attached to the cable or conductors in such a manner that it cannot part from the cable during the pull. Two common methods include direct connection with the cable conductors themselves and connection by means of pulling grips or baskets placed over the cable or group of conductors. The use of the proper type of grip or basket will facilitate the pull, but in many cases—especially on long pulls—workers prefer to use three- or four-hole cable grips with setscrews that secure each conductor to the pulling block. *Figure 15* shows several types of pulling grips.

Most pulling blocks have a rope clevis as an integral part of the block. However, when using pulling grips or baskets in 2" or larger conduit, a rope clevis is normally used to facilitate connecting the pulling rope to the wire grip. Two types are currently used: the straight clevis and the swivel clevis.

3.4.1 Stripping the Cable Ends

When the type of grip being used requires that the ends of the cable insulation be stripped from the conductors, conventional methods are used—the same as for terminating conductors for splices or connections to terminal lugs in panelboards, switchgear, etc.

Fish Poles

Rigid fish poles can be used in areas such as over drop-in ceilings where traditional fish tape might get caught up on ductwork or the ceiling grid. Glow-in-the-dark fish poles are also available.

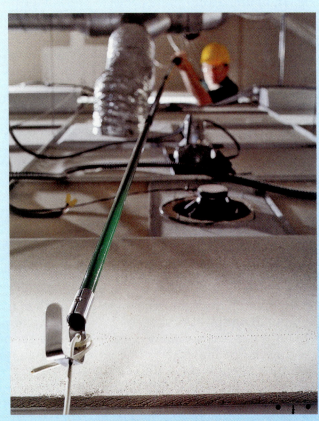

(A) FISH POLE IN USE

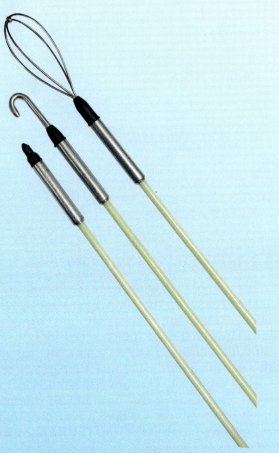

(B) GLOW-IN-THE-DARK FISH POLES

206SA03.EPS

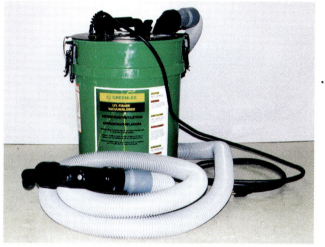

(A)

(B)

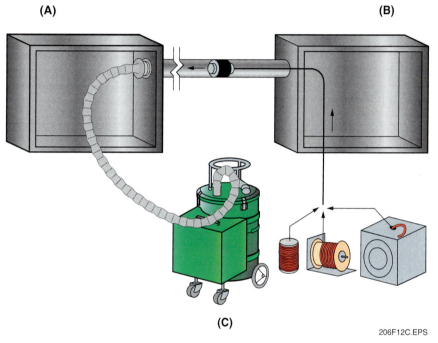

(C)

Figure 12 ♦ Power fishing system.

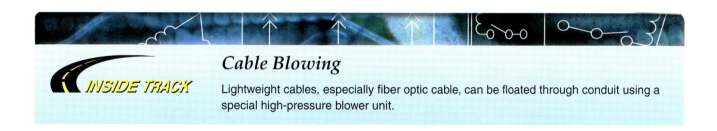

Cable Blowing

Lightweight cables, especially fiber optic cable, can be floated through conduit using a special high-pressure blower unit.

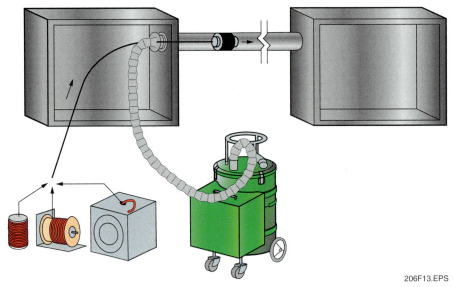

Figure 13 ◆ Blower/vacuum fish tape system used to blow a pull line in conduit.

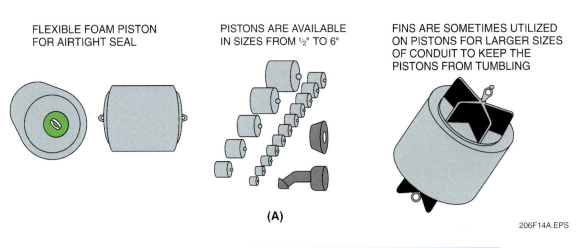

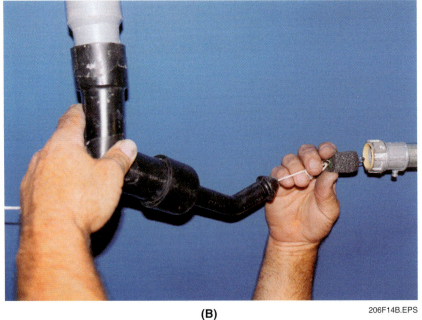

Figure 14 ◆ Types of pistons in common use.

6.14 ELECTRICAL LEVEL TWO ◆ TRAINEE GUIDE

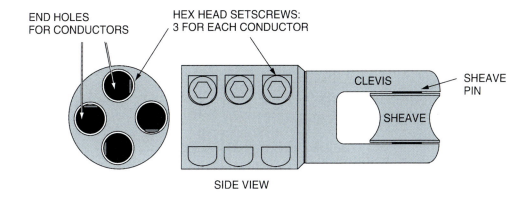
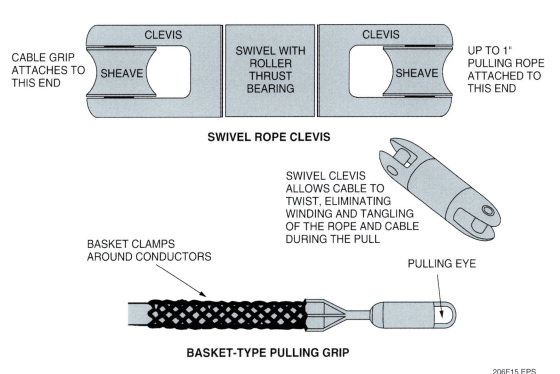

Figure 15 ♦ Various types of pulling grips used during conductor installation.

In general, the ends of conductors should first be trimmed. Cable cutters capable of cutting conductors through 1,000 kcmil save workers much time over using a hacksaw. There are also cable strippers, adjustable from 1/0 AWG through 1,000 kcmil, that handle midspan and termination stripping of THHN, THWN, XHHW, and similar insulation. These tools are excellent for stripping conductors for use in setscrew clamp-type pulling grips.

To use a stripping tool, first mark the required distance from the ends of the conductors, using the pulling grip as a gauge. Close the jaws of the stripping tool on the cable and twist. These self-feeding devices ensure positive progression down the cable to any position desired. To stop stripping, apply back pressure to the stripper until a full circle has been completed.

Once the conductor ends have been stripped, insert one conductor at a time into the setscrew grip. Make sure the end of the bare conductor is firmly in place, and then tighten the setscrews with a hex wrench. Continue on to the next conductor, and so on, until all conductors are secured in the pulling grip.

INSIDE TRACK

Grips and Swivels

Conductors can be attached to a pulling line using various methods, including setscrew grips or basket grips. No matter what method is used, always insert a swivel of some sort between the pulling line or fish tape and the conductors to alleviate twisting of the conductors. Breakaway swivels are available that release at a specified tension to avoid damage to the conductors if excess force is applied or if the conductors get hung up during the pull.

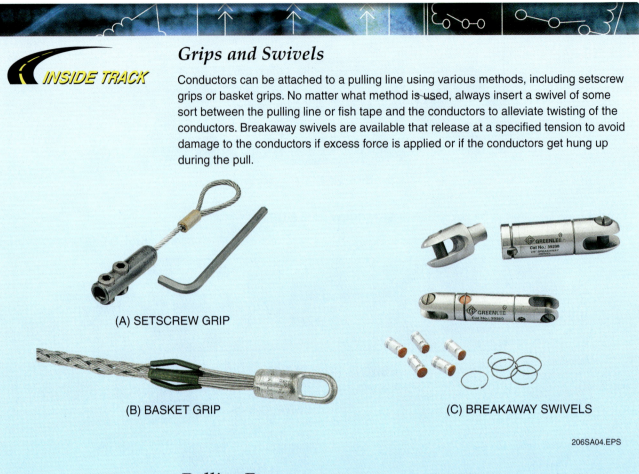

(A) SETSCREW GRIP

(B) BASKET GRIP

(C) BREAKAWAY SWIVELS

206SA04.EPS

Pulling Eyes

Smaller conductors may require only a pulling eye attached to the fish tape or pulling line. To attach the pulling eye, the conductors are first stripped, exposing a length of bare wire. These bare wires are inserted through the eye and twisted back onto themselves. The exposed twisted wires of the conductors can then be wrapped with smaller copper wire to prevent them from untwisting. They are then completely taped with three layers of electrical tape, starting from the conductor insulation, to prevent snagging as they are drawn through the conduit run.

3.4.2 Safety with Cutters

When using a stripping tool for the first time, make sure that you read and understand all instructions and warnings before using the tool. The following should also be observed when working with cable terminators:

- Wear eye protection.
- Inspect tools before using, and replace damaged, missing, or worn parts.
- Be prepared for the unexpected. Make sure your footing and body position are such that you will not lose your balance.
- Use only the type and size material in the stated capacity.
- Do not use the tool on or near live circuits.

3.5.0 Types of Pulling Lines

Wire-pulling ropes have come a long way from the hemp ropes used by electricians a couple of decades ago. The most common wire-pulling ropes on the market include the following:

- Nylon
- Polypropylene
- Multiplex polyester
- Double-braided polyester composite

The type of rope selected will depend mainly on the pulling load; that is, the weight of the cable, the length of the pull, and the total resistance to the pull. For example, Greenlee's Multiplex cable-pulling rope is designed for low-force cable pullers. It has a low stretch characteristic that makes it suitable for pulls up to 2,000 pounds. Lengths are available from 100 to 1,200 feet. Greenlee's double-braided composite rope for high-force cable pullers is designed for pulls up to 6,500 pounds.

CAUTION

Any equipment associated with the pull must have a working load rating in excess of the force applied during the pull. All equipment must be used and mounted in strict accordance with the manufacturer's instructions.

Care must be used in selecting the proper rope for the pull, and then every precaution must be taken to make sure that the cable-pulling force does not exceed the rope capacity. There are several reasons for this, but the main one is safety. For example, think of a 100' length of nylon rope with a 10,000-pound breaking strength. Such ropes can stretch 40' before breaking, releasing 200,000 foot-pounds of energy in the process. Think of the damage this amount of energy could do to a raceway system and to nearby workers if the rope broke under this amount of force.

In some cases, a power blower/vacuum fish tape system is used to vacuum the piston through the raceway with the pulling rope attached. In other cases, a line is first blown or vacuumed through the conduit and then the pulling rope is attached to the line for pulling it through the raceway system. In either case, there are certain precautions that should be taken when using a power fish tape system:

- Read and understand all instructions and warnings before using the tool.
- Never fish in runs that might contain live power.
- Be prepared for the unexpected. Make sure your footing and body position are such that you will not lose your balance in any unexpected event.
- Use blower/vacuum systems only for specified light fishing and exploring the raceway system.
- Never use pliers or other devices that are not designed to pull a fish tape. They can kink or nick the tape, creating a weak spot.

4.0.0 ♦ CABLE-PULLING EQUIPMENT

Except for short cable pulls, hand-operated or power-operated cable pullers or winches are used to furnish the pulling power. In general, cable reels are set in place at one end of the raceway system, and the cable puller is set up at the opposite end. One end of the previously-installed pulling rope is attached to a clevis, basket, or other cable grip to which the conductors are attached. The other end of the rope (at the cable puller) is wrapped around the rotating drum capstan on the cable puller (*Figure 16*).

Be Sure to Check the Pulling Rope Rating

Many pulling ropes available today are made of synthetic materials designed for pulling by hand or with a winch-type puller. However, if you are using certain synthetic pulling ropes on friction-type capstan power pullers, make sure the rope is rated by the manufacturer for this use so that it will be able to withstand the heat generated by any extended slippage of the rope on the capstan. During high-force pulling operations, melting damage and possible failure of unrated synthetic rope can occur quickly during periods of capstan slippage.

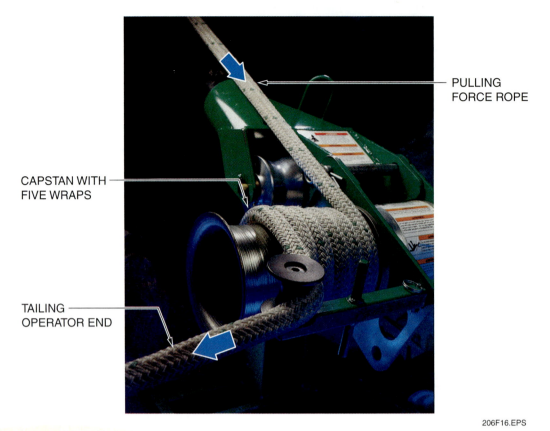

Figure 16 ♦ Power cable puller capstan.

 WARNING!
Always use a wire-pulling lubricant that is compatible with the type of cable being pulled. Check with the cable manufacturer for their recommendations, and always contact the lubricant manufacturer about the compatibility of their products with specific cables. Also check the product's MSDS for any applicable safety requirements.

Wire-pulling lubricant—sometimes referred to as **soap**—is inserted into the empty conduit as well as wiped thoroughly onto the front of the cable. One or more operators are on hand to feed the cable, while one worker is usually all that is required on the pulling end.

The number of wraps on the puller drum decides the amount of force applied to the pull. For example, the operator needs to apply only 10 pounds of force to the pulling rope in all cases. With this amount of force applied by the operator, and with one wrap around the rotating drum, 21 pounds of pulling force will be applied to the pulling rope; 2 wraps, 48 pounds; 3 wraps, 106 pounds, etc. This principle is known as the capstan theory and is the same principle applied to block-and-tackle hoists or the lone cowboy who is able to rope and hold a 2,000-pound bronco by wrapping his lariat around the center post in a corral. *Table 1* gives the amount of pulling force with various numbers of wraps when the operator applies only 10 pounds of tailing force for a particular model puller.

Table 1	Example Pulling Forces for Various Wraps	
Number of Wraps	Operator Force (Lbs)	Pulling Force (Lbs)
1	10	21
2	10	48
3	10	106
4	10	233
5	10	512
6	10	1,127
7	10	2,478

4.1.0 Pulling Safety

Adhere to the following precautions when using power cable-pulling equipment:

- Read and understand all instructions and warnings before using the tool.
- Use compatible equipment; that is, use the properly-rated cable puller for the job, along with the proper rope and accessories.
- Always be prepared for the unexpected. Make sure your footing and body position are such that you will not lose your balance. Keep out of the direct line of force.
- Make sure all cable-pulling systems, accessories, and rope have the proper rating for the pull.
- Inspect tools, rope, and accessories before using; replace damaged, missing, or worn parts.
- Personally inspect the cable-pulling setup, rope, and accessories before beginning the cable pull. Make sure that all equipment is properly and securely rigged.
- Make sure all electrical connections are properly grounded and adequate for the load.
- Use cable-pulling equipment only in uncluttered areas.

NOTE

The strain placed on the wrapped cable during the pull may weaken this part of the cable. Be sure to discard the wrapped cable after making the pull.

CAUTION

Make absolutely certain that all communications equipment is in working order prior to the pull. Place personnel at strategic points with operable communications equipment to stop and start the pull as conditions warrant. Anyone involved with the pull has the authority to stop the pull at the first sign of danger to personnel or equipment.

4.2.0 Types of Cable Pullers

There are several types of cable pullers on the market. Most, however, operate on the same principle. The self-contained hand-crank wire puller in *Figure 17(A)* is designed to pull up to 1,500 pounds with only 30 pounds of handle force. It is used on projects where only a few cable runs need to be pulled.

Guiding and Lubricating Conductors

When guiding conductors into a conduit/raceway during a pull, the conductors may tend to twist, overlap, or become crossed during the pulling operation, especially if fed from boxes instead of reels. Excessive twisting, overlaps, crossovers, etc. can cause binding of the conductors in conduit/raceway turns, create bunching obstructions in the conduit/raceway, and can contribute to insulation burns. Operators at the feeding end of the pull must attempt to keep the conductors as straight as possible during the pulling operation and lubricants should be applied liberally during the pull to allow the conductors to slide against each other and the sides of the run.

206SA05.EPS

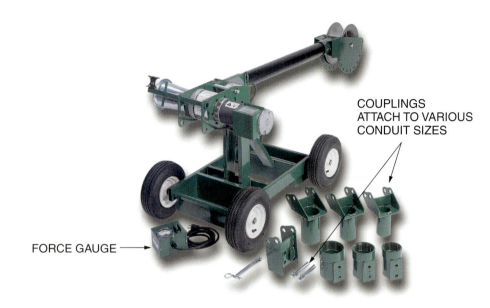

Figure 17 ◆ Cable pullers.

Power cable pullers are available in various sizes and pulling capacities, from lightweight units that can be set up and used by a single person to heavy-duty units for high-force cable pulling. *Figure 17(B)* shows two typical units.

 WARNING!
Make absolutely certain that all cable-pulling equipment is anchored properly. Follow the manufacturer's recommendations for the type of puller being used.

4.2.1 Setting Up Cable Pullers

Figure 18 shows the basic setup for a down pull using the portable 2,000-pound puller shown in *Figure 17*. To set up for the pull, first adjust the elbow and boom to the correct angle using the attached pins. The elbow attaches to the conduit using the locking knob. This unit has a universal conduit latch so it will attach to all conduit sizes without having to change couplings.

Figure 19 shows a setup for an up pull. The setup is similar to the one described for the down pull except the elbow is attached to the bottom conduit rather than the top conduit.

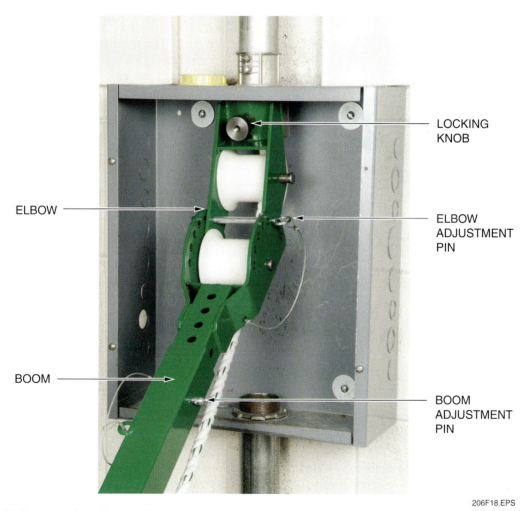

Figure 18 ◆ Puller setup for a down pull.

Figure 19 ◆ Puller setup for an up pull.

5.0.0 ◆ HIGH-FORCE CABLE PULLING

High-force cable pulling is not done every day. It is actually a small part of any electrical raceway installation. Workers may take days, weeks, or even months to install the complete raceway system; the cable-pulling operation typically takes less time.

There are several items to consider during high-force cable pulling:

- The design of the raceway system must be studied by consulting the working drawings and by examining the installed system to ensure that it is properly installed for the type of conductors that will be pulled through the system. Items to consider include conduit sizes, number and size of conductors, number of bends, sufficient pull boxes, and adequate supports.
- The conductors need to be matched to the proper pulling equipment.
- The proper pulling rope must be selected. Choose a rope that has at least four times the strength of the required pulling force; that is, if the estimated pulling force is 1,200 pounds, the rope should be rated for no less than 4,800 pounds. Also check the rope carefully for wear or damage prior to the pull. Remember that a rope is only as strong as its weakest point.
- A decision must be made as to the best end of the raceway for pulling. The reel setup must also be carefully placed. In general, conductors that are to be installed downward should be fed off the top of the reel and where conductors are to be fed upward, the best method is to feed from the bottom of the reel. This eliminates sharp kinks or bends in the conductors.
- The appropriate pulling equipment must be used. The equipment must be of the proper capacity for the job. Space consideration for the equipment is also important, as is the particular type of mounting.
- Having enough workers is critical. Never be caught short. This is where experienced workers earn their pay. In high-force wire pulling, experience is the best teacher.
- Safety must be foremost in everyone's mind.

When planning the cable pull, make sure that enough cable is on hand for the run. Cable may be verified while still on the reel by using a cable-length meter. Of course, the conduit run length should have already been checked, as discussed previously. Do not proceed further until everyone is assured that enough cable is on hand for all bends, sweeps in junction boxes, etc. If the cable falls short, take steps to correct the situation before continuing.

Communications

What type of communications equipment would most likely be used during cable-pulling operations?

Monitored Cable Pullers

Many power cable pullers include force gauges with an automatic over-tension shutoff device. These units prevent you from accidentally exceeding the maximum recommended pulling tension for the conductor(s) being pulled.

Equipment with feeding sheaves will help provide a smooth guide for the cable. Also, have sufficient lubricant on hand and use it both before and during the pull.

Select and install proper cable grips on the cable ends. Gripping must be adequate to handle the imposed force.

When pulling cables in a horizontal run, the worker simply has to reduce the amount of operator pull in order to slow down the cable pull. Releasing all operator pull stops the pulling force entirely. However, when pulling vertical runs, the rope must be tied off after stopping the pull to keep the cable from reversing in the raceway.

The entire operation should be supervised from start to finish. Therefore, communications equipment must be utilized on both ends of the pull at all times.

During the pull, make sure that the pulling rope remains free and is not wrapped around any part of the body.

A typical cable-pulling setup is shown in *Figure 20*, including cable reels, reel stands, a power cable puller, conduit system, and other accessories necessary for the pull.

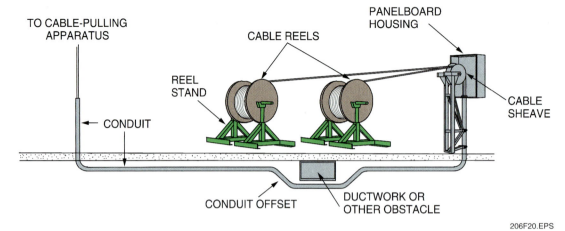

Figure 20 ◆ Typical cable-pulling setup.

5.1.0 The Feeding End—Sheaves and Rollers

The feed-in setup should unreel the cable along its natural curvature, as shown in *Figure 21(A)*, as opposed to a reverse S curvature, as shown in *Figure 21(B)*. Feed-in setups are shown in *Figure 22* for manhole, underfloor duct, and overhead cable tray. Note the use of auxiliary equipment in some of these drawings, that is, cable reels, guide-in tubes, sheaves, and rollers.

Single sheaves, such as those shown in *Figure 23(A)*, may be used only for guiding cables. Multiple blocks should be arranged to hold the cable-bending radius wherever the cable is deflected. For pulling around bends, use a conveyor sheave assembly of the appropriate radius series, such as the one shown in *Figure 23(B)*.

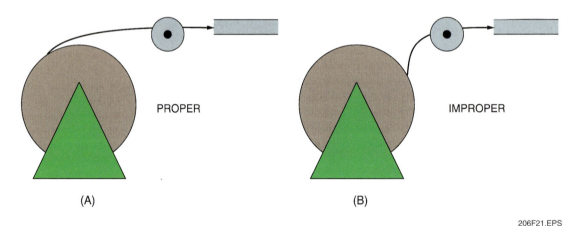

Figure 21 ◆ Unreel the cable along its natural curvature.

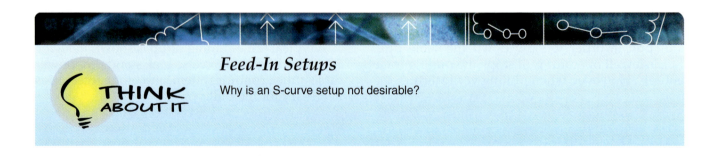

Feed-In Setups

Why is an S-curve setup not desirable?

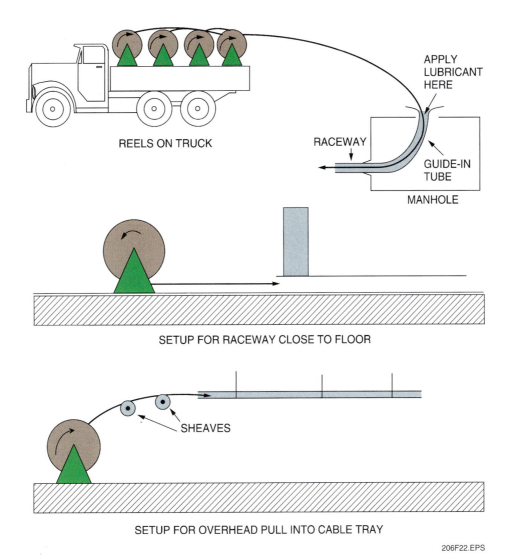

Figure 22 ◆ Cable feed-in setups.

Figure 23 ◆ Cable sheaves.

MODULE 26206-08 ◆ CONDUCTOR INSTALLATIONS 6.25

Sheaves and pulleys must be positioned to ensure the effective curvature is smooth and deflected evenly at each pulley. Never allow a polygon curvature to occur, as shown in *Figure 24*.

CAUTION

Use the radius of the surface over which the cable is bent, not the outside flange diameter of the pulley. For example, a 10" cable sheave typically has an inside (bending) radius of 3".

6.0.0 ◆ SUPPORTING CONDUCTORS

Conductors in vertical raceways must be supported in accordance with *NEC Section 300.19* if the vertical rise exceeds the values in *Table 2*. In general, one conductor support must be provided at the top of the vertical raceway or as close to the top as practical. Intermediate supports must also be provided as necessary to limit supported conductor lengths to not greater than those values specified in *NEC Table 300.19(A)*.

The *NEC®* allows several different methods of supporting conductors in vertical raceways; the following are typical:

- Conductors may be supported by clamping devices constructed of or employing insulating wedges inserted in the ends of the conduit, as shown in *Figure 25*. Where clamping of insulation does not adequately support the cable, the conductor itself must also be clamped.
- Conductors may be supported by inserting boxes at the required intervals in which insulating supports are installed and secured in a satisfactory manner to withstand the weight of the conductors attached to them. The boxes must be provided with covers.
- Cables may be supported in junction boxes by deflecting the cables not less than 90° and carrying them horizontally to a distance not less than twice the diameter of the cable. The cables are carried on two or more insulating supports and additionally secured to these supports by tie wires, if desired. Where this method is used, cables must be supported at intervals not greater than 20% of those mentioned in *NEC Table 300.19(A)*.
- Cables may be supported by a method of equal effectiveness.

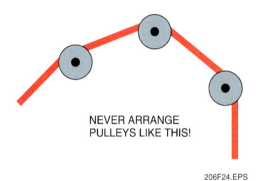

Figure 24 ◆ Never allow a polygon curvature to occur in a cable-pulling operation.

Table 2 Spacings for Conductor Supports in Vertical Raceways

Size of Wire	Support (Aluminum or Copper-Clad Aluminum Conductor)	Support (Copper Conductor)
18 AWG through 8 AWG	≤ 100'	≤ 100'
6 AWG through 1/0 AWG	≤ 200'	≤ 100'
2/0 AWG through 4/0 AWG	≤ 180'	≤ 80'
Over 4/0 through 350 kcmil	≤ 135'	≤ 60'
Over 350 kcmil through 500 kcmil	≤ 120'	≤ 50'
Over 500 kcmil through 750 kcmil	≤ 95'	≤ 40'
Over 750 kcmil	≤ 85'	≤ 35'

Figure 25 ◆ Conductors supported with wedges.

CAUTION

When installing conductors in vertical raceways, install proper supporting devices to hold the conductors before removing the pulling equipment or cutting the conductors.

A variety of supports are manufactured specifically for supporting cable in vertical conduit runs, and many ideas can be obtained from the manufacturers' catalogs. Therefore, make an effort to obtain such catalogs from electrical suppliers or manufacturer's representatives. In fact, manufacturers' catalogs of electrical tools and equipment are excellent study guides for any phase of the electrical industry. Manufacturers of electrical equipment want their equipment used, and they have found that one of the best ways to accomplish this is to provide easy-to-understand instructions and examples of practical applications. Most manufacturers go to great expense to provide this information, but it is usually free of charge for those working in the industry.

There are several precautionary measures that must be taken when pulling cables in vertical runs. The worst danger is runaways; that is, the weight of the cable combined with gravity exceeds the speed of the pull and falls at a rapid rate down the raceway run. Such a situation can cause injury to workers on both ends of the pull. Consequently, braking systems should be used on all long vertical cable pulls. To do otherwise is asking for trouble. Get specific instructions from your supervisor before beginning a vertical pull—either up or down.

WARNING!

Guard against runaways on all vertical pulls. Make certain that proper braking equipment is used to stop a conductor fall should a runaway occur.

7.0.0 ◆ PULLING CABLE IN CABLE TRAYS

When long lengths of cable are to be installed in raceways or cable trays, problems are frequently encountered, particularly when the cable has to be pulled directly into the tray and changes in direction of the tray sections are involved. An entire cable-pulling system has to be planned and set up so that the cable may be pulled into the trays without scuffing or cutting the sheathing and insulation, and also to avoid damaging the cable trays or the tray hangers. To accomplish a successful cable tray pull, a complete line of installation tools is available for pulling lengths of cable up to 1,000 feet or longer. These tools consist mainly of conveyor sheaves and cable rollers. *Figure 26* shows a partial cable tray system with a sheave and roller in place.

Short lengths of cable can be laid in place without tools or pulled with a basket grip. Long lengths of small cable (2" or less in diameter) can also be pulled with a basket grip. Larger cables, however, should be pulled by the conductor and the braid, sheath, or armor. This is done with a pulling eye applied at the cable factory or by tying the conductor to the eye of a basket grip and taping the tail end of the grip to the outside of the cable.

In general, the pull exerted on the cables pulled with a basket grip that is not attached to the conductor should not exceed 1,000 pounds. For heavier pulls, care should be taken not to stretch the insulation, jacket, or armor beyond the end of the conductor nor bend the ladder, trough, or channel out of shape.

The bending radius of the cable should not be less than the values recommended by the cable manufacturer, which range from four times the diameter for a rubber-insulated cable with a 1" maximum outside diameter without lead, shield, or armor, to eight times the diameter for interlocked armor cable. Cables or special construction such as wire armor and high-voltage cables require a larger radius bend.

Best results are obtained in installing long lengths of cable up to 1,000 feet with as many as a dozen bends by pulling the cable in one continuous operation at a speed of 20 to 25 feet per minute. It may be necessary to brake the reel to reduce sagging of the cables between rollers and sheaves.

The pulling line diameter and length will, of course, depend on the pull to be made and the tools and equipment available. Winch and power units must be of adequate size for the job and capable of developing the high pulling speeds required for the best and most economical results.

In general, single or multi-cable rollers are placed in the bottom of trays to protect the cable as it is pulled along. Sheaves are placed at each change of direction—either horizontally or vertically. The bottom rollers may be secured to the tray bottoms except at vertical changes in direction. Extra support is necessary at these locations to prevent damage or movement of the tray system.

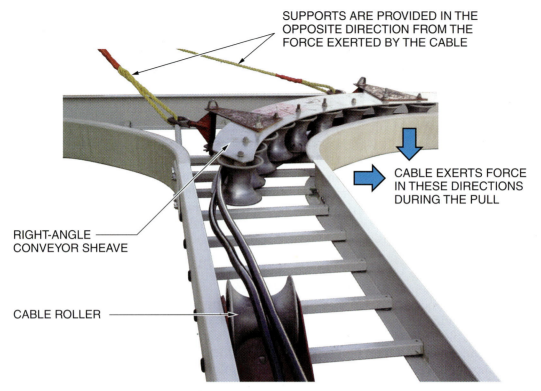

Figure 26 ◆ Typical cable tray cable-pulling arrangement.

NOTE
If single cables are to be installed, always place them on the outside of a bend to allow room on the inside of the bend for pulling other cables.

Sheaves must be supported in the opposite direction of the pull. For example, all right-angle conveyor sheaves should be supported at two locations, as shown in *Figure 26*, to compensate for the pull of the cable.

NOTE
Power cable pulls should not be stopped unless absolutely necessary. However, anyone associated with the pull—upon evidence of danger to either the cable or the workers—may stop a cable-pulling operation. Communication is the most important factor in these cases.

CAUTION
Workers feeding a cable pull must carefully inspect the cable as it is paid off the cable reel. Any visible defects in the cable at the feeding end warrants stopping the pull.

WARNING!
At the first sign of any type of malfunctioning equipment, broken sheaves, or other events that could present a danger to either the workers or the cable, the pull should be stopped. Make certain that all communications equipment is in proper order before starting a pull.

8.0.0 ◆ PHYSICAL LIMITATIONS OF CABLE

Consideration must be given to the physical limitations of a cable as it is being pulled into position. Pulling subjects cable to extreme stress and, if done improperly, can displace a cable's components. Thus, it is important that the following guidelines be observed:

- While reels are in storage—either before or after a pull—the conductor ends must be sealed to prevent moisture from entering or creeping into the cable ends.
- The minimum ambient working temperatures for cable-pulling operations depend on the cable jackets. In cold weather, cable reels should be stored in a warm area overnight so that the cable jackets will be at the proper temperature for pulling the next day.

- Calculate and stay within the cable's maximum pulling tension, maximum sidewall load, and minimum bending radii.
- Ensure that the raceway joints are aligned and that the wiring space is sufficient.
- Train the cable to avoid dragging on the edge of the raceway; also avoid laying or dragging cable on the ground.
- If using a basket grip, secure it to the cable with steel strapping and cut well behind the areas it covers once the cable is in place. The portion of the cable under the grip should be discarded.
- Ensure that the elongation of the pull rope minimizes jerking.
- Pull with a capstan and no faster than 40 feet per minute.
- Do not stop a pull unless absolutely necessary.
- Never pull the middle of the cable.
- Seal the ends with appropriate putty or silicone caulking and overwrap with tape until the conductors are terminated.

8.1.0 Maximum Pulling Tension

Maximum pulling tension should not exceed the smaller of the following values:

- Allowable tension on pulling device
- Allowable tension on conductor
- Allowable sidewall load

Installation Aids

Other tools available to simplify the pulling of cables into cable tray systems include triple pulleys, bull wheels, and both wide and narrow rollers. Be sure to position these pulleys and rollers at the proper locations to prevent damage to the cables during the installation and also to help the installation proceed as quickly as possible. Various rollers are shown here.

(A) SNAP-IN SPINDLE AND CABLE ROLLER

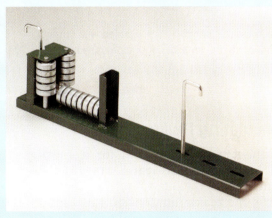

(B) CABLE ROLLER WITH ADJUSTABLE BRACKETS

(C) ELBOW ROLLER

8.1.1 Allowable Tension on Pulling Device

Do not exceed the working load stated by the manufacturer of the pulling devices (pulling eyes, ropes, anchors, basket grips, etc.). If catalog information is not available, work at 10% of the rated braking tensile strength.

The allowable tension with a basket grip must not exceed the lbs/cmil value (as shown in *Table 3*) or 1,000 pounds, whichever is smaller. Exceptions to this rule, however, do occur, but seldom will this figure rise to over 1,250 pounds.

8.1.2 Allowable Tension on Conductors

The metallic phase conductors are the tensile members of the cables and should bear all of the pulling force. Never use shielding drain wires or braids for pulling. *Table 3* provides the allowable pulling tensions of various types of conductors. The listed values should never be exceeded.

Reduce the maximum pulling tension by 20% to 40% if several conductors are being pulled simultaneously since the tension is not always evenly distributed among the conductors.

CAUTION

When smaller conductors are pulled with large conductors, the smaller conductors may be damaged.

8.2.0 Calculating Pulling Tension

Normally, the maximum tension for a specific type of cable can be found using data from the cable manufacturer.

In general, the maximum tension for a single-conductor cable should not exceed 6,000 pounds. The maximum tension for two or more conductors should not exceed 10,000 pounds.

The maximum stress for leaded cables must not exceed 1,500 pounds per square inch of lead sheath area when pulled with a basket grip.

The maximum tension must not exceed 1,000 pounds per square inch of insulation area for non-leaded cables when pulled with a basket grip.

The maximum tension at a bend must not exceed 500 times the radius of curvature of the conduit or duct expressed in feet.

8.2.1 Tension in Horizontal Pulls

The pulling tension in a given horizontal raceway section may be calculated as follows:

Table 3 Physical Limitations of Cable

Material	Cable Type	Temper	Lbs/Cmil
Copper	All	Soft	0.008
Aluminum	Power	Hard	0.008
Aluminum	Power	¾ Hard	0.006
Aluminum	Power	AWM	0.005
Aluminum	URD (solid)	Soft (½ hard)	0.003
All	Thermocouple	—	0.008

For a straight section, the pulling tension is equal to the length of the duct run multiplied by the weight per foot of the cable and the coefficient of friction, which will vary depending on the type and amount of lubrication used. Therefore, the equation is as follows:

$$T = L \times w \times f$$

Where:

T = total pulling tension
L = length of raceway run in feet
w = weight of cable in pounds per foot
f = coefficient of friction

For ducts having curved sections, the following equation applies:

$$T_{OUT} = T_{in} \, e^{fa}$$

Where:

T_{OUT} = tension of bend
T_{in} = tension into bend
f = coefficient of friction
e = Napierian logarithm base 2.718
a = angle of bends in radians

INSIDE TRACK

Software Programs

Software programs are now available to take the math out of calculating pulling tensions. These programs typically calculate pulling tension and sidewall pressures based on conductor size using their own data on friction coefficients. These programs also determine conduit fill, conductor configuration, jam ratios, and the amount of lubricant required for the pull.

To aid in solving the above equation, values of e^{fa} for specific angles of bend and coefficients of friction are listed in *Table 4*. For more precise values, tables are available from cable manufacturers.

8.3.0 Sidewall Loading

Before calculating cable-pulling tension, sidewall loading or sidewall bearing pressure must be considered.

The sidewall load is the radial force exerted on a cable being pulled around a conduit bend or sheave. Excessive sidewall loading can crush a cable and is, therefore, one of the most restrictive factors in installations having bends or high tensions. *Figure 27* shows a section taken across a conduit run in a 90° bend. Note that pulling tension is exerted parallel with the walls of the conduit. However, due to the 90° bend, pressure is also exerted downward against the wall of the conduit. Once again, this is known as the sidewall load. Sidewall loading is reduced by increasing the radius of bends.

In general, the sidewall load on any raceway run should not exceed 500 pounds/foot of bend radius. This pressure, however, must be reduced on some types of cables. For example, *Table 5* shows one manufacturer's recommendations for the maximum sidewall pressures permitted for various types of cables. Always refer to the cable manufacturer's instructions for the type of cable being used.

Table 5 Maximum Sidewall Pressures for Various Types of Cable

Cable Type	Sidewall Pressure in Pounds/Foot of Bend Radius
600V nonshielded control	300
600V and a kV nonshielded EP power	500
5kV and 15kV EP power	500
25kV and 35kV power	300
Interlocked armored cable (all voltages)	300

8.4.0 Practical Applications

Figure 28 shows a typical raceway system containing three 500 kcmil lead sheath copper conductors. Note the straight 300' run from A to B, a 45° kick, and then another 100' straight run from C to D.

To find the calculated pulling tension from D to A, refer to cable data in the manufacturer's catalogs. Suppose this cable weighs 8 pounds per foot and has a 0.141" lead sheath. The outside diameter of the three-conductor cable assembly is 3". Use 0.5 as the coefficient of friction and calculate the pulling tension from A to B (*Figure 28*) as follows:

Step 1 Find the tension between points A and B.

$$\text{Tension at B} = T_1 = L_1 \times w \times f$$

L_1 is the length between points A and B while w is the weight of the cable per foot, and f is the coefficient of friction, which is 0.5. Substituting the known values in the equation gives:

$$T_1 = 300 \times 8 \times 0.5 = 1{,}200 \text{ pounds}$$

Table 4 Angle of Bend vs. Coefficients of Friction

Angle of Bend (degrees/radians)	Values of e^{fa} for Coefficients of Friction		
	f = 0.75	f = 0.50	f = 0.35
15/0.2618	1.22	1.14	1.10
30/0.5236	1.48	1.30	1.20
45/0.7854	1.80	1.48	1.32
60/1.0472	2.19	1.68	1.44
90/1.5708	3.25	2.20	1.73

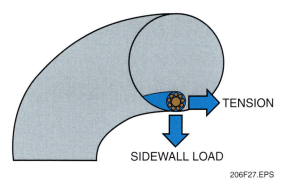

Figure 27 ◆ Sidewall loading.

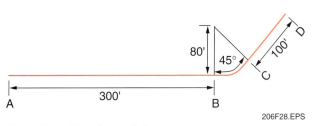

Figure 28 ◆ Sample conduit run.

Step 2 Find the tension between points B and C.

$$\text{Tension at C} = T_1 e^{fa}$$

Since the distance between B and C involves an angle, refer to *Table 4* of this module. Looking in the left column of the table, we see that 45° equals 0.7854 radians; this figure multiplied by the coefficient of friction (0.5) equals 0.3927. Radians of angles may also be found with electronic pocket calculators if they have scientific functions. Follow the instructions in the manual accompanying the calculator. The exact key strokes will vary with the brand of calculator, but most require pressing the degree key, entering the numeral for degrees, then pressing the convert key, and finally pressing the radian or RAD key. The radians of the entered angle will be displayed.

Once again, refer to *Table 4*. Find the 45° angle of bend in the left column; read across the row to the column under 0.50—the coefficient of friction figure. The number is 1.48, the value of e^{fa}. Substituting these values in the equation gives the following:

$$A = 45° = 0.7854 \text{ radians}$$
$$fa = 0.7854 \times 0.5 = 0.3927$$
$$e^{fa} = 1.48$$
$$1{,}200 \times 1.48 = 1{,}776 \text{ pounds}$$

Therefore, the tension at C is 1,776 pounds.

Step 3 Find the tension from points C to D.

$$\text{Tension from C to D} = T_2 = L_2 \times w \times f$$
$$T_2 = 100 \times 8 \times 0.5 = 400 \text{ pounds}$$

Step 4 Find the total pulling tension by adding the figures obtained previously.

$$T = T_2 + T_1 e^{fa} = 400 + 1{,}776$$
$$= 2{,}176 \text{ pounds}$$

The maximum pulling force using a basket grip for this size cable should not exceed 1,900 pounds. Therefore, if the pull is made from point A to point D, a pulling eye will have to be used since the total pulling tension exceeds 1,900 pounds. However, if the pull is reversed—pulling from point D to point A—the total pulling tension will be reduced since the distance from point D to the 45° angle (point C) is ⅓ the distance from point A to the 45° angle (point B).

$$\text{Tension at C} = 400 \text{ pounds}$$
$$\text{Tension at B} = 400 \times 1.48 = 592 \text{ pounds}$$
$$\text{Total tension at A} = 1{,}200 + 592$$
$$= 1{,}792 \text{ pounds}$$

Therefore, if the cable is pulled from point D to point A, either a pulling eye or basket grip may be used.

NOTE
A lower tension is obtained by feeding the pull from the end nearest the bend.

9.0.0 ♦ CABLE-PULLING INSTRUMENTS

There are several instruments used in conjunction with cable-pulling operations. Since details of operation vary with the manufacturer, these instruments will only be briefly discussed in this module. Study the operation manuals for all instruments before using them.

- *Cable length meter* – Cable length meters (*Figure 29*) are available for direct reading in lengths from 2,000' to 20,000'. Most are calibrated for different wire sizes whereas the sizes are selected with a selector switch on the instrument.

Cable Pulling

At a large commercial job site recently, workers began a complex pull shortly before the end of the day. At 5 P.M., they left for the day, having completed only a portion of the pulling operation. The next day, they resumed work promptly at 8 A.M., but the pulling lubricant had already dried in the conduit. As a result of the excess friction on the pull, the rope broke, delaying the job and causing extensive rework.

The Bottom Line: Don't stop a pull in the middle of a job.

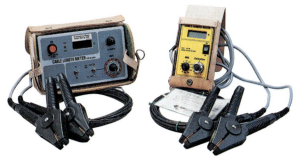

Figure 29 ♦ Cable length meters.

Figure 30 ♦ Circuit tester.

Controls may also be set for either copper or aluminum conductors. These instruments are ideal for determining the exact length of conductors on reels prior to making a pull.

- *Circuit tester/wire sorter* – This instrument is used to trace conductors on unenergized circuits. See *Figure 30*. One lead of the transmitter is attached to ground while the other lead is attached to the wire being traced. The receiver is then taken to the opposite end of the circuit, where it will show a strong signal on the traced conductor. That wire is marked and other wires are traced in the same way.
- *Dynamometer (force gauge)* – This type of meter is designed to read dynamic friction (pulling force) during a cable pull. Many are designed for a specific cable-pulling tool and are shipped as an integral or optional part of the cable puller. For example, one common cable puller electronically displays both actual speed and actual force while running. It automatically shuts down when maximum preset force is reached to prevent damage to the cable being pulled. Others are portable units for use with any type of cable-pulling equipment. Such instruments are invaluable for use during high-force cable pulls to avoid damage to conductors during the pull. When the instrument indicates that the maximum pulling force has been reached, the pull can be stopped before damage occurs to the cable or conductors.

Cable Pulling

In June 2006, an electrician was tagging instrumentation cable on a work platform about 30 feet off the ground. At the same time, a crew was pulling an assembly in an adjacent cable tray but failed to install caution signs or barrier controls. The inexperienced crew did not monitor the tension on the pulling assembly and it failed during the pull, whipping back along the cable tray and picking up a loose cable roller, which propelled into the victim at high speed. It amputated his arm, broke his ribs and femur, and ruptured several internal organs. He died from his injuries.

The Bottom Line: Only qualified individuals may conduct a cable pull. All pulling equipment must be tested and certified for the expected load, and the area must be secured with caution/barrier controls during the pull.

Complex Installations

Complex cable-pulling installations are best handled by highly experienced contractors. This is one small portion of an award-winning installation that took over 70,000 hours to complete.

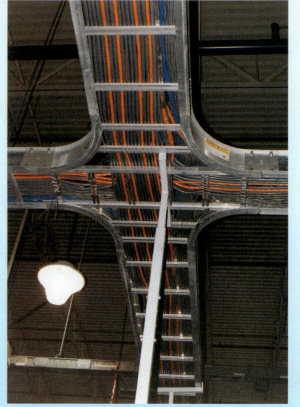

Review Questions

1. Who is allowed to stop the pull at the first sign of danger to the cable, raceway system, or personnel?
   a. Only the project supervisor
   b. Only workers on the feeding end of the run
   c. Only workers on the pulling end of the run
   d. Anyone involved with the pull

2. After some length of conductor has been used from a cable reel, what should be done before placing the reel in storage?
   a. A stress test should be made to see if the cable has been damaged.
   b. An insulation test should be made using a megger.
   c. The ends of the conductors should be sealed.
   d. The conductor jacket should be warmed before storing.

3. Which of the following precautions should be taken when making a cable pull in extremely cold weather?
   a. The cable should be stored in a warm area overnight.
   b. The cable should be stored outside overnight.
   c. Antifreeze should be mixed with the pulling lubricant.
   d. Electric warming blankets should be used during the pull.

4. Which of the following pulling methods should be used when the pulling tension exceeds 1,000 pounds?
   a. A basket grip by itself
   b. A pulling eye
   c. Conductors bent around a hook and taped
   d. Conductors bent around a snake hook and left untaped

5. All the following are acceptable methods of transporting or storing cable reels *except*:
   a. Reels should be stored in an upright position.
   b. Reels should be laid flat on their sides.
   c. Reels should be lifted by a rigged crane with a spindle through the reel.
   d. Reels should be carried with a forklift so that the forks do not touch the cable.

6. Where is the best place to seek information about a wire-pulling lubricant that is compatible with a particular type of cable insulation?
   a. The lubricant manufacturer
   b. The cable manufacturer
   c. Both the cable and lubricant manufacturers
   d. The *NEC*®

7. How should the feed-in setup be handled during a cable installation?
   a. The cable should unreel with a reverse S curvature.
   b. It makes no difference.
   c. Only the pulling end is important when unreeling the cable.
   d. The cable should unreel along its natural curvature.

8. All of the following may be used to guide a cable during a pull *except*:
   a. A conveyor sheave
   b. A polygon setup
   c. A guide-in tube
   d. A cable reel

9. Which of the following is the most important consideration when pulling cables in vertical pulls?
   a. Calculate the cable weight before installation.
   b. Install the cable from the bottom up.
   c. Guard against runaways.
   d. Use less wire lubricant than in horizontal applications.

10. Which of the following best describes the location of sheave supports for pulling cable in cable trays?
    a. They should be supported in the opposite direction of the pull.
    b. They should be supported in the same direction as the pull.
    c. They should be supported only by the tray assembly itself.
    d. No support is necessary when sheaves are used in cable trays.

Summary

A cable-pulling operation involves careful planning. Furthermore, the proper use of appropriate equipment is crucial to a successful cable installation. Think of everything that can go wrong, and then plan accordingly.

Communications equipment is another crucial part of a cable installation. Workers on both ends must be in constant contact with each other in case something goes wrong. The workers feeding the cable must carefully inspect the cable as it is paid off the reels and stop the pull immediately if the cable appears damaged. In fact, anyone involved with the pull should be able to stop the pull whenever the safety of the cable or personnel is threatened. Again, good communication is crucial.

Safety precautions must be followed exactly during any cable installation. Use the correct size and type of equipment for the job. Never exceed the maximum force of the weakest component. Do not position yourself in a direct line with a cable pull.

Notes

Trade Terms Introduced in This Module

Basket grip: A flexible steel mesh grip that is used on the ends of cable and conductors for attaching the pulling rope. The more force exerted on the pull, the tighter the grip wraps around the cable.

Cable grip: A device used to secure ends of cables to a pulling rope during cable pulls.

Capstan: The turning drum of a cable puller on which the rope is wrapped and pulled. An increase in the number of wraps increases the pulling force of the cable puller.

Clevis: A device used in cable pulls to facilitate connecting the pulling rope to the cable grip.

Conductor support: The act of providing support in vertical conduit runs to support the cables or conductors. The *NEC*® gives several methods in which cables may be supported, including wedges in the tops of conduits, supports to change the direction of cable in pull boxes, etc.

Conduit piston: A cylinder of foam rubber that fits inside the conduit and is then propelled by compressed air or vacuumed through the conduit run to pull a line, rope, or measuring tape. Also called a mouse.

Fish line: Light cord used in conjunction with vacuum/blower power fishing systems that attaches to the conduit piston to be pushed or pulled through the conduit. Once through, a pulling rope is attached to one end and pulled back through the conduit for use in pulling conductors.

Fish tape: A flat iron wire or fiber cord used to pull conductors or a pulling rope through conduit.

Setscrew grip: A cable grip, usually with built-in clevis, in which the cable ends are inserted in holes and secured with one or more setscrews.

Sheave: A pulley-like device used in cable pulls in both conduit and cable tray systems.

Soap: Slang for wire-pulling lubricant.

Additional Resources

This module is intended to present thorough resources for task training. The following reference works are suggested for further study. These are optional materials for continued education rather than for task training.

Cable Installation Manual, Latest Edition. New York: Cablec Corp.

National Electrical Code® Handbook, Latest Edition. Quincy, MA: National Fire Protection Association.

CONTREN® LEARNING SERIES – USER UPDATE

NCCER makes every effort to keep these textbooks up-to-date and free of technical errors. We appreciate your help in this process. If you have an idea for improving this textbook, or if you find an error, a typographical mistake, or an inaccuracy in NCCER's Contren® textbooks, please write us, using this form or a photocopy. Be sure to include the exact module number, page number, a detailed description, and the correction, if applicable. Your input will be brought to the attention of the Technical Review Committee. Thank you for your assistance.

Instructors – If you found that additional materials were necessary in order to teach this module effectively, please let us know so that we may include them in the Equipment/Materials list in the Annotated Instructor's Guide.

Write: Product Development and Revision
National Center for Construction Education and Research
3600 NW 43rd St., Bldg. G, Gainesville, FL 32606

Fax: 352-334-0932

E-mail: curriculum@nccer.org

Craft _____ Module Name _____

Copyright Date _____ Module Number _____ Page Number(s) _____

Description
_____
_____
_____
_____

(Optional) Correction
_____
_____
_____

(Optional) Your Name and Address
_____
_____
_____

Cable Tray

Dallas/Fort Worth International Airport

Hensel Phelps Construction Co. won an ABC Excellence in Construction award in the mega-projects division for its work on the automated people mover stations at Dallas/Fort Worth International Airport.

26207-08

26207-08
Cable Tray

Topics to be presented in this module include:

1.0.0	Introduction	7.2
2.0.0	Cable Tray Loading	7.7
3.0.0	Cable Tray Support	7.13
4.0.0	Center Rail Cable Tray Systems	7.16
5.0.0	*NEC*® Requirements	7.17
6.0.0	Cable Installation	7.19
7.0.0	Cable Tray Drawings	7.22
8.0.0	Pulling Cable in Tray Systems	7.24
9.0.0	Safety	7.25

Overview

Electricians who regularly work on commercial and industrial installations projects may not only be called upon to install conduit systems, but may also find themselves installing elevated or suspended cable trays to support cables or conductors, especially in industrial locations.

Cable tray systems can be accessorized with fittings that either snap or bolt into place to provide tees, wyes, crosses, reducers, barriers, covers, and even box connectors. Cable tray installations, like any raceway system, are regulated by the *National Electrical Code*®.

The support of cable tray is an important consideration and is determined by the number and size of conductors or cables to be installed in the tray. If the support is not adequate for the load at various stress points, overloading can cause rung and/or rail failure. To prevent these failures, load stress points must always be identified before making any splices in straight cable tray sections. Methods used to hang and support cable tray assembles include trapeze, direct rod, and wall mounting.

Note: *NFPF 70*®, *National Electrical Code*®, and *NEC*® are registered trademarks of the National Fire Protection Association, Inc., Quincy, MA 02269. All *National Electrical Code*® and *NEC*® references in this module refer to the 2008 edition of the *National Electrical Code*®.

Objectives

When you have completed this module, you will be able to do the following:

1. Describe the components that make up a cable tray assembly.
2. Explain the methods used to hang and secure cable tray.
3. Describe how cable enters and exits cable tray.
4. Select the proper cable tray fitting for the situation.
5. Explain the *National Electrical Code®* (*NEC®*) requirements for cable tray installations.
6. Select the required fittings to ensure equipment grounding continuity in cable tray systems.
7. Interpret electrical working drawings showing cable tray fittings.
8. Size cable tray for the number and type of conductors contained in the system.

Trade Terms

Barrier strip
Cable pulley
Cross
Direct rod suspension
Dropout
Dropout plate
Elbow
Expansion joints
Fittings
Interlocked armor cable
Ladder tray
Pipe racks
Swivel plates
Tee
Trapeze mounting
Tray cover
Trough
Unistrut®
Wall mounting
Wye

Required Trainee Materials

1. Paper and pencil
2. Appropriate personal protective equipment
3. Copy of the latest edition of the *National Electrical Code®*

Prerequisites

Before you begin this module, it is recommended that you successfully complete *Core Curriculum; Electrical Level One;* and *Electrical Level Two*, Modules 26201-08 through 26206-08.

This course map shows all of the modules in *Electrical Level Two*. The suggested training order begins at the bottom and proceeds up. Skill levels increase as you advance on the course map. The local Training Program Sponsor may adjust the training order.

1.0.0 ◆ INTRODUCTION

NEC Section 392.2 defines a cable tray system as a unit or assembly of units or sections and associated **fittings** forming a structural system used to securely fasten or support cables and raceways. *NEC Article 392* covers cable tray installations, along with the types of conductors to be used in various systems. Whenever a question arises concerning cable tray installations, this is the *NEC®* article to use.

Cable trays are the usual means of supporting cable systems in industrial applications. The trays themselves are usually made up into a system of assembled, interconnected sections and associated fittings, all of which are made of metal or noncombustible units. The finished system forms into a continuous rigid assembly for supporting and carrying single, multiconductor, or other electrical cables and raceways from their origin to their point of termination, frequently over considerable distances.

Cable tray is fabricated from both aluminum and steel. Some manufacturers provide an aluminum cable tray that is coated with PVC for installation in caustic environments. Relatively new all-nonmetallic trays are also available; this type of tray is ideally suited for use in corrosive areas and in areas requiring voltage isolation. Cable tray is available in various forms, including ladder, **trough**, center rail, and solid bottom, and can be supported by either side mounts or center mounts.

Ladder tray, as the name implies, consists of two parallel channels connected by rungs, similar in appearance to a conventional straight or extension ladder. Trough types consist of two parallel channels (side rails) having a corrugated, ventilated bottom. The solid bottom cable tray is similar to the trough. All of these types are shown in *Figure 1*. Ladder, trough, and solid bottom trays are completely interchangeable; that is, all three types can be used in the same run when needed.

Cable tray is manufactured in 12' and 24' lengths. Common widths range from 6" to 36". All sizes are provided in either 3", 4", 5", or 6" depths.

Cable tray sections are interconnected with various types of fittings. Fittings are also used to provide a means of changing the direction or dimension of the cable tray system. Some of the more common fittings include:

- Horizontal and vertical **tees**
- **Wyes**
- Horizontal and vertical bends
- Horizontal **crosses**
- Reducers
- **Barrier strips**
- Covers
- Splice plates
- Box connectors

The area of a cable tray cross section that is usable for cables is defined by width (W) × depth (D), as shown in *Figure 2*. The overall dimensions of a cable tray, however, are greater than W times D because of the side flanges and seams. Therefore, overall dimensions vary according to the tray design. Cables rest on the bottom of the tray and are held within the tray area by two longitudinal side rails, as shown in *Figure 3*.

A channel is used to carry one or more cables from the main tray system to the vicinity of the cable termination (see *Figure 4*). Conduit is then used to finish the run from the channel to the actual termination.

Certain *NEC®* regulations and National Electrical Manufacturers Association (NEMA) standards should be followed when designing or installing cable tray. Consequently, practically all projects of any great size will have detailed drawings and specifications for the workers to follow. Shop drawings may also be provided.

Cable Tray Installation

Cable trays are widely used in many commercial and industrial installations.

Special-Application Cable Management Systems

This lightweight, flexible nonmetallic cable management system is used to route wiring to the large motors in a manufacturing facility.

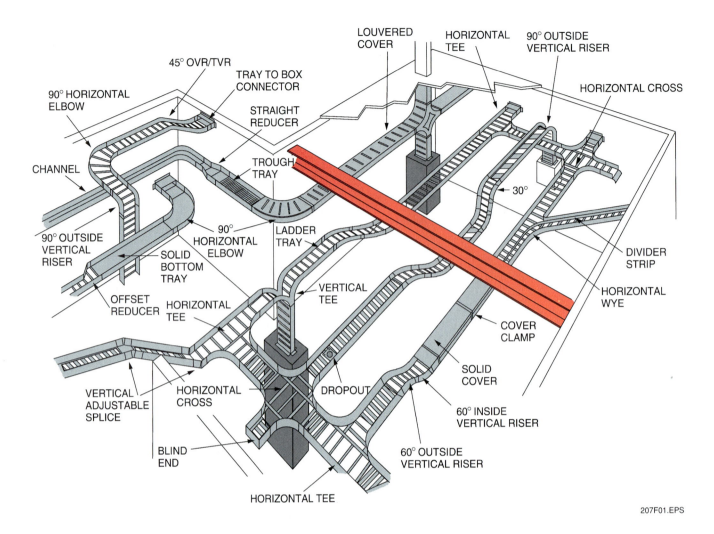

Figure 1 ◆ Typical cable tray system.

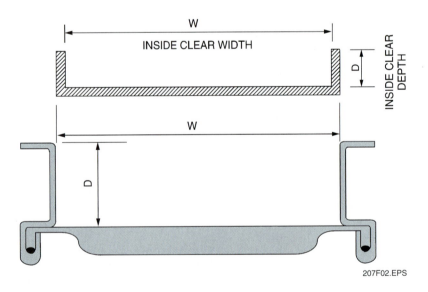

Figure 2 ◆ Cross section of cable tray comparing usable dimensions to overall dimensions.

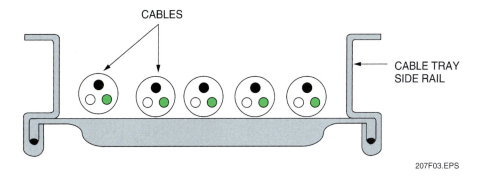

Figure 3 ◆ Cables rest on the bottom of the tray and are held in place by the longitudinal side rails.

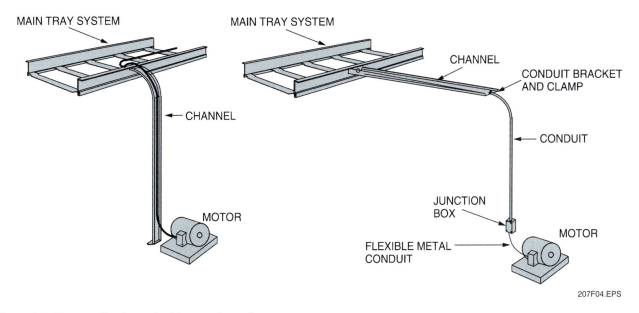

Figure 4 ◆ Two applications of cable tray channel.

INSIDE TRACK

Ladder Tray

Ladder tray is currently used for a majority of today's installations. It offers numerous advantages:

- Ladder tray has greater air circulation than other trays, which helps to dissipate heat.
- The ladder rungs are convenient for tying down conductors.
- Ladder tray does not collect moisture.
- Conductors can be dropped through the bottom of the tray at any point.

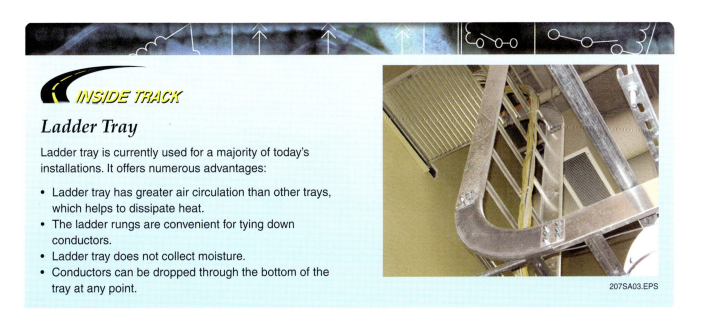

MODULE 26207-08 ◆ CABLE TRAY 7.5

Basket Tray

Basket tray offers many of the same advantages as ladder tray. It is lightweight and easy to install using special clips that attach to threaded rods.

Basket tray can be easily cut using bolt cutters or a basket tray cutter, such as that shown here.

One additional advantage of basket tray is that it can be cut, bent, and connected to create tees, crosses, and elbows without the need for separate fittings.

SPLICE PLATE

2.0.0 ◆ CABLE TRAY LOADING

The *NEC*® requires that cable trays have sufficient strength to support all contained wiring, and that they be supported at the intervals provided in the manufacturer's installation instructions. The following sections describe some additional load considerations in the design of a cable tray system.

2.1.0 Load Factors

The load capacity of cable tray varies with each manufacturer and depends on the shape and thickness of the side rails, the shape and thickness of the bottom members, spacing of rungs (if any), material used, safety factor used, method used to determine the allowable load, method of supporting the tray, and volume capacity. Consequently, each manufacturer publishes load data.

2.2.0 Determining Fill

The density of fill can only be determined by personnel laying out the system. In doing so, however, be aware that cables packed closely together can impair each other's efficiency (see *Figure 5*).

2.3.0 Determining the Load on Supports

Each support should be capable of safely supporting approximately 1.25 times the full weight of the cable and tray on a typical span, as shown in *Figure 6*.

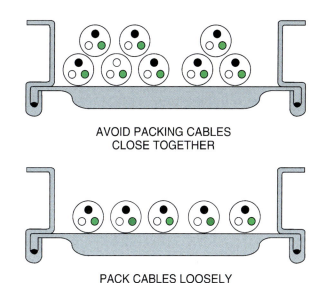

Figure 5 ◆ Cables packed closely together can impair efficiency.

2.4.0 Deflection Under Load

Consider the case of a cable tray spanning only two supports (simple span). As the tray is loaded, the side rails take the deflected form. Simultaneously, in cross section, the side rails rotate inward and the tray bottom deflects. The amount of inward or outward movement of the side rails is a critical factor in the ability of the tray to carry a load.

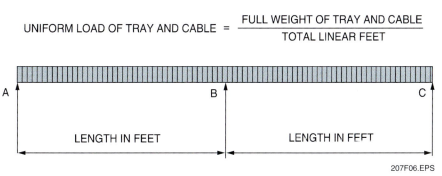

Figure 6 ◆ Each tray support should be capable of safely supporting 1.25 times the weight of the entire tray and cable assembly span.

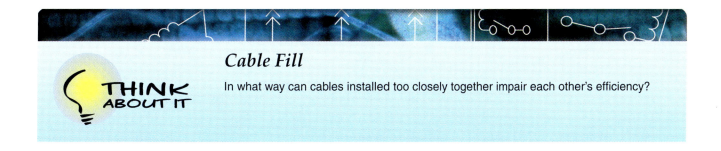

Cable Fill

In what way can cables installed too closely together impair each other's efficiency?

2.5.0 Failure Under Load

There are two types of failures that can occur with a loaded cable tray. These are longitudinal (side rail) failures and transverse (rung) failures. Transverse (rung) failures occur when the load applied causes the fibers on the tension side (bottom edge) of the rung to stretch and permanently deform. Simultaneously, fibers on the compression side (top edge) are permanently crushed. Longitudinal (side rail) failures occur either as bending failures when the tray is supported on a larger span, or, on longer spans, buckling failures may occur because the side rails of the tray have little resistance to inward or outward movement. As the tray deflects, the side rails rotate and the top (compression) flanges of the tray buckle. Bending failures occur on short spans because the side rails of the tray have greater resistance to rotation and remain reasonably upright. The tray does not fail until the load is such that it causes the fibers on the tension side (bottom edge) of the side rail to stretch and permanently deform. Simultaneously, fibers on the compression side (top edge) are permanently crushed. For any tray on an intermediate span, it is difficult to anticipate whether a bending or buckling failure would occur (see *Figure 7*).

2.6.0 Splicing Straight Sections

In *Figure 8*, the load on the cable tray creates bending moments along the spans. The stress in the

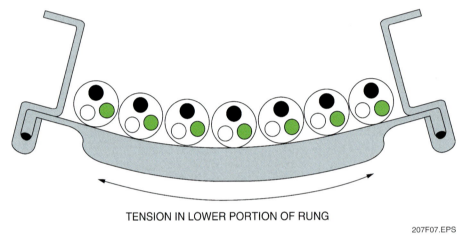

Figure 7 ◆ Bending of loaded tray.

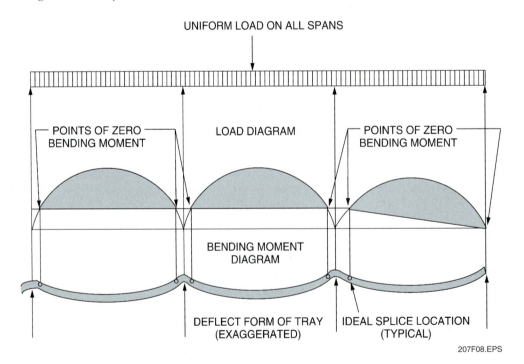

Figure 8 ◆ Load of cable creates bending moments along the span.

side rails of the tray is directly related to the bending moments at all points along the tray. The magnitude of the bending moment at any point is determined by measuring the vertical height of the shaded portion. In any cable tray system, a splice is a point of weakness. Consequently, splices should be located at the points of least stress. Ideally, splices would be located at the points of zero bending moment, and the strength of the tray system would be at a maximum. In actual practice, if the splice is located within one-fourth of the span's distance from the support, the result will be close to ideal.

Locating the splice within the one-fourth points of the span requires extra labor on the part of the installer. When a splice occurs within the central length of a span between the one-fourth points, the tray will support the load for which it was designed, but the safety factor will be greatly reduced.

2.7.0 Cable Placement in Tray

Cables are placed in the tray either by being pulled along the tray or by being laid in over the side. *Figure 9* shows a cable pulley being used to facilitate a cable pull in a tray.

2.8.0 Cable Tray Cover

A cable tray cover is used primarily for two reasons:

- To protect the insulation of the cables against damage that might be caused if an object were to fall into the tray. Prime hazards are tools, discarded cigarettes, and weld splatter.
- To protect certain types of cable insulation against the damaging effects of direct sunlight.

2.8.1 Cover Selection

When maximum protection is desired, solid covers should be used. However, if accumulation of heat from the cables is expected, caution should be used. Ventilated covers should be used if some protection of the cable is desired and provisions must be made to allow the escape of heat developed by the cables.

2.9.0 Cable Exit from Tray

Several different ways in which cables may exit from a cable tray are shown in *Figure 10*. While all of these methods are *NEC*®-approved and endorsed by most cable tray manufacturers, the engineering specifications on some projects may

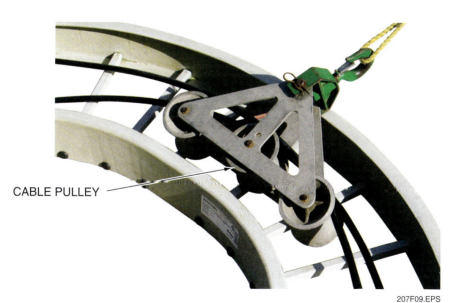

Figure 9 ◆ Cable pulley used to facilitate a cable pull in a tray.

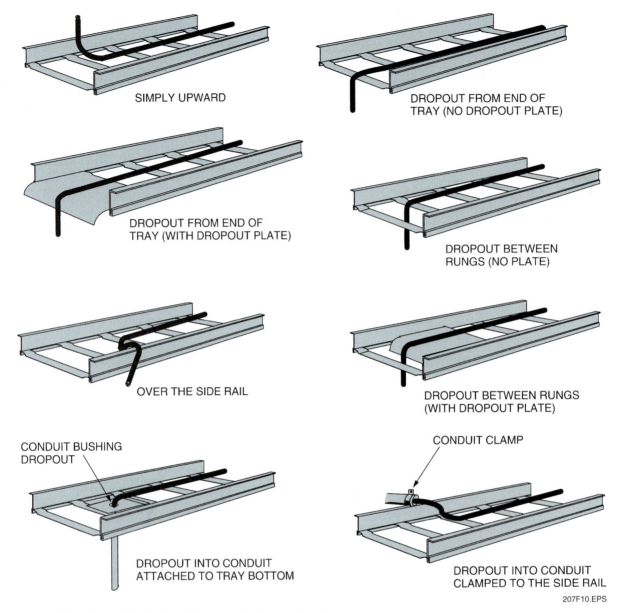

Figure 10 ♦ Several ways in which cables may exit from a cable tray.

prohibit the use of some of these methods. Most notable are the dropout between rungs method and the dropout from the end of the tray method. The cable radius may be too short with either of these methods. Also, since no dropout plates are used, the cable or conductors are not protected. Although *NEC Section 392.5(B)* specifically requires that cable trays have smooth edges to ensure that cable will not be damaged, accidents do occur. For example, a tool might be dropped on a cable tray rung during the installation, which may cause a burr or other sharp edge on the rung. Then, after the cable is installed and the system is in use, vibrating machinery may cause this burr to cut into the cable insulation, resulting in a ground fault and possible power outage. Always review the project specifications carefully and/or check with the project supervisor before using either of these methods.

2.9.1 Dropout Plates

A dropout plate provides a curved surface for the cable to follow as it passes from the tray, as shown in *Figure 11(A)*. Without a dropout plate, cables can be bent sharply, causing damage to the insulation. See *Figure 11(B)*.

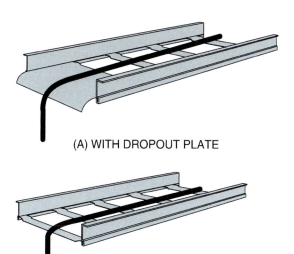

Figure 11 ◆ A dropout plate provides a curved surface for the cable to follow as it leaves the tray.

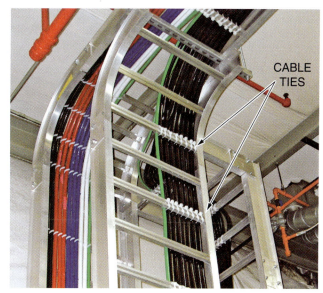

Figure 12 ◆ Typical application of supports in a vertical run.

2.10.0 Cable Supports in Vertical Trays

A cable hanger elbow is used to suspend cables in long vertical runs. Care should be taken to ensure that the weight of the suspended length of cable does not exceed the cable manufacturer's recommendation for the maximum allowable tension in the cable.

In short vertical runs, the weight of cables can be supported either by the outside vertical riser elbow or by the vertical straight section when the cables are tied to it. *Figure 12* shows a typical application of a cable hanger elbow in a vertical run.

2.11.0 Cable Edge Protection

The bottom of the solid bottom tray might be convex, concave, or flat. When two pieces of tray are butted together, the bottoms may be out of alignment. An alignment strip, also known as an H bar, is placed between the tray bottoms.

2.12.0 Splice Plates

There are several types of splice plates available, including vertical, horizontal, and expansion plates.

2.12.1 Vertical Adjustment Splice Plates

Vertical adjustment splice plates are used to change the elevation in a run of cable tray. They should not be used when it is important to maintain a cable bending radius. Vertical adjustment splice plates are useful when the change in elevation is so slight or the angle so unusual that it would not be possible to install a standard outside vertical riser elbow and an inside vertical riser elbow in the space available.

In general, four swivel plates are used to build an offset in a cable tray system. Once the proper angles have been calculated, proceed as follows:

Step 1 Bolt four swivel plates together at the proper angles, using the inner holes as the center or pivot hole (see *Figure 13*).

Step 2 Using a flat surface such as a bench or concrete deck, space two swivel plates at the proper center-to-center distance apart (refer to *Figure 13*).

Step 3 Measure and cut the amount of tray needed to complete the offset.

2.12.2 Horizontal Adjustment Splice Plates

These plates are sometimes used in place of horizontal elbows to change the direction in a run of cable tray. They are used primarily where there is insufficient space or an unusual angle that prevents the use of a standard elbow.

2.12.3 Expansion Splice Plates

These plates are used at intervals along a straight run of cable tray to allow space for thermal expansion or contraction of the tray to occur, or where offsets or expansion joints occur in the supporting structure.

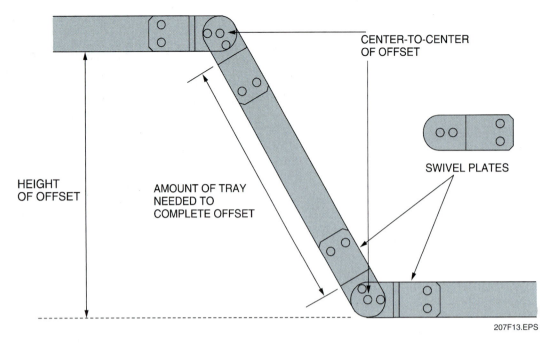

Figure 13 ◆ Fabricating a cable tray offset with swivel plates.

To enable the expansion joint to function properly, the cable tray must be allowed to slide freely on its supports. Any cable tray hold-down device used in an installation subject to expansion or contraction must give clearance to the tray. An expansion joint and splice plates are shown in *Figure 14*.

2.13.0 Barrier Strips

Barrier (divider) strips are used to separate certain types of cable as a result of the nature of the installation, types of circuits used, type of equipment used, local codes, or the *NEC*®. Some reasons for using divider strips are to:

- Separate or isolate electrical circuits
- Separate or isolate cables of different voltages
- Separate cable or wire runs from each other to prevent fire or ground fault damage from spreading to other cables or wires in the same tray
- Aid neatness in the arrangement of the cables
- Warn electricians of the difference between cables on each side of the divider strip

2.13.1 Barrier Strip Cable Protectors

Barrier strip cable protectors are used to bind any raw metallic edge over which a cable is to pass. Their purpose is to protect the cable insulation against damage.

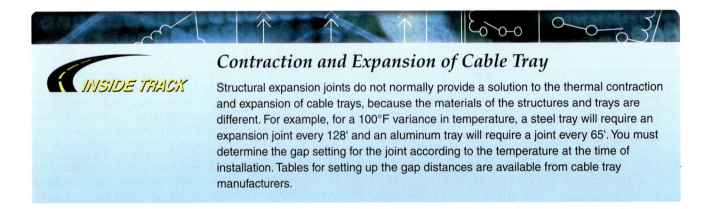

Contraction and Expansion of Cable Tray

Structural expansion joints do not normally provide a solution to the thermal contraction and expansion of cable trays, because the materials of the structures and trays are different. For example, for a 100°F variance in temperature, a steel tray will require an expansion joint every 128' and an aluminum tray will require a joint every 65'. You must determine the gap setting for the joint according to the temperature at the time of installation. Tables for setting up the gap distances are available from cable tray manufacturers.

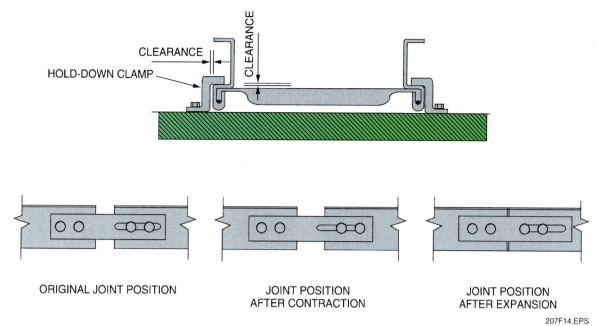

Figure 14 ◆ Expansion joint and splice plates.

3.0.0 ◆ CABLE TRAY SUPPORT

Proper supports for cable tray installations are very important in obtaining a good overall layout. Cable is usually supported in one or more of the following ways:

- Trapeze mounting
- Direct rod suspension
- Wall mounting
- Center hung support
- Pipe rack mounting

3.1.0 Trapeze Mounting

When trapeze mounting is used, a structural member—usually a steel channel or Unistrut®—is connected to the vertical supports to provide an appearance similar to a swing or trapeze (*Figure 15*). The cable tray is mounted to the structural member using bolts, anchor clips, or J-clamps. The underside of the channel or Unistrut® may also be used to support conduit.

3.2.0 Direct Rod Suspension

The direct rod suspension method of supporting cable tray uses threaded rods and hanger clamps. The threaded rod is connected to the ceiling or other overhead structure and is connected to the hanger clamps that are attached to the cable tray side rails, as shown in *Figure 16(A)*.

3.3.0 Wall Mounting

Wall mounting is accomplished by supporting the tray with structural members attached to the wall, as shown in *Figure 16(B)*. This method of support is often used in tunnels (mining operations) and other underground or sheltered installations where large numbers of conductors interconnect equipment that is separated by long distances. When using this or any other method of supporting cable tray, always examine the structure to which the hangers are attached, and make absolutely certain that the structure is of adequate strength to support the tray system.

3.4.0 Trapeze Mounting and Center Hung Support

Trapeze mounting of cable tray is similar to direct rod suspension mounting. The difference is in the method of attaching the cable tray to the threaded rods. A structural member, usually a steel channel or strut, is connected to the vertical supports to provide an appearance similar to a swing or trapeze. The cable tray is mounted to the structural member. Often, the underside of the channel or strut is used to support conduit. A trapeze mounting assembly is shown in *Figure 16(C)*.

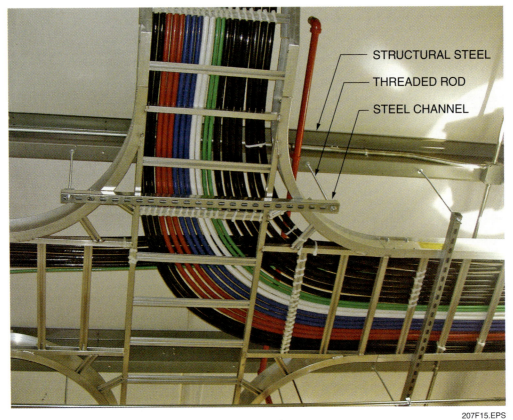

Figure 15 ◆ Channel support.

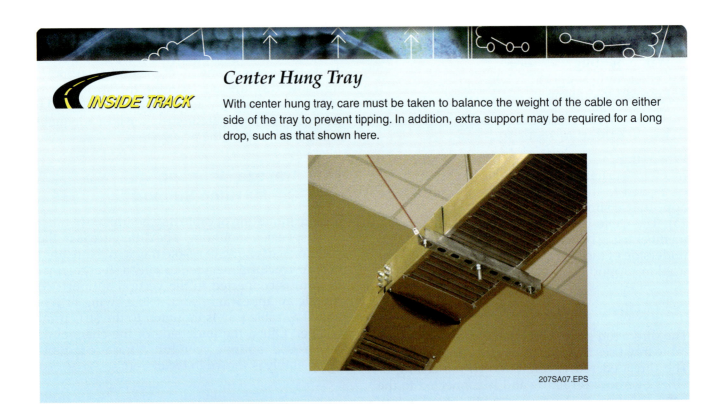

Center Hung Tray

With center hung tray, care must be taken to balance the weight of the cable on either side of the tray to prevent tipping. In addition, extra support may be required for a long drop, such as that shown here.

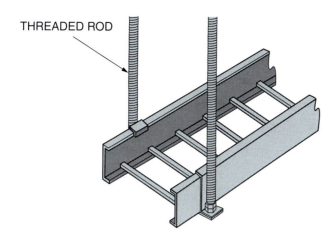
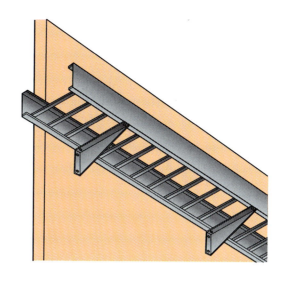
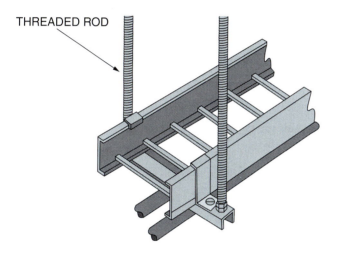
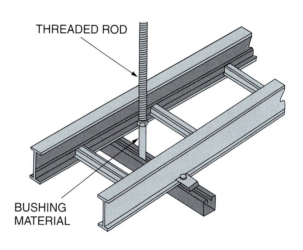

Figure 16 ♦ Alternate ways to hang cable tray.

A method that is similar to trapeze mounting is a center hung tray support, as shown in *Figure 16(D)*. In this case, only one rod is used and it is centered between the cable tray side rails. A bushing sleeve, such as a short piece of small-diameter PVC, may be used over the center rod to protect the conductors.

3.5.0 Pipe Rack Mounting

Pipe racks are structural frames used to support the piping that interconnects equipment in outdoor industrial facilities. Usually, some space on the rack is reserved for conduit and cable tray. Pipe rack mounting of cable tray is often seen in petrochemical plants where power distribution and instrumentation wiring is routed over a large area and for long distances.

4.0.0 ♦ CENTER RAIL CABLE TRAY SYSTEMS

Center rail or monorail cable tray systems (*Figure 17*) are light, easy to install, and provide open sides for ready access when changing or adding cables. Center rail cable tray is used in light-duty applications such as sound, telephone, and other communications systems. In addition to its weight limitations, it cannot be used with dividers so its use is restricted to systems where dividers are not necessary.

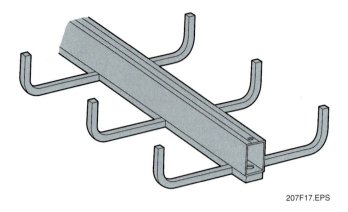

Figure 17 ♦ Center rail cable tray.

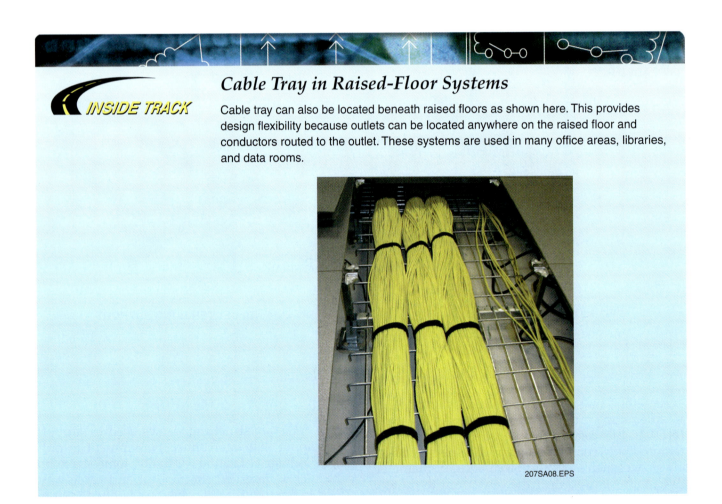

5.0.0 ◆ NEC® REQUIREMENTS

NEC Article 392 deals with cable tray systems along with related wire, cable, and raceway installations.

Figures 18, *19*, and *20* summarize the *NEC®* requirements for cable tray installations. For in-depth coverage, always refer to the *NEC®*. Cable tray manufacturers also have some excellent reference material available that details the installation of their products. Consult your local dealer for more information.

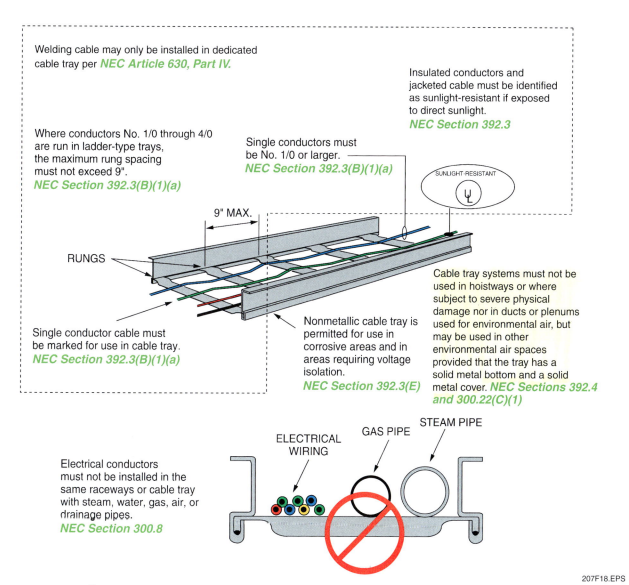

Figure 18 ◆ NEC® regulations governing the use of cable tray.

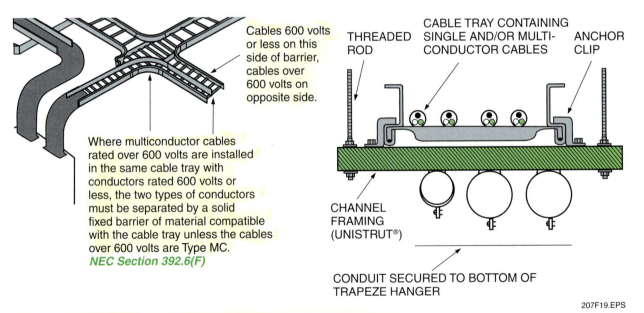

Figure 19 ◆ *NEC® regulations governing cable tray construction and installation.*

NEC Section 392.7, Grounding

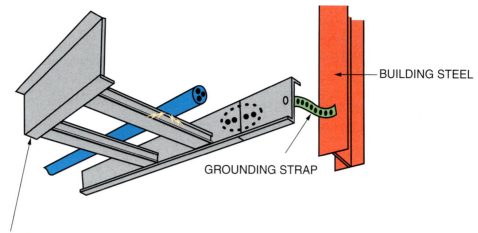

Steel or aluminum cable tray systems are permitted to be used as equipment grounding conductors if they meet all of the requirements in *NEC Section 392.7(B).*

Proper grounding lessens hazards due to ground faults. Therefore, the *NEC®* requires all metallic cable tray to be grounded as required for conductor enclosures in accordance with *NEC Section 250.96.*

Per *NEC Section 392.7(B)(2)*, the minimum cross-sectional area of cable tray must conform to the requirements in *NEC Table 392.7(B).*

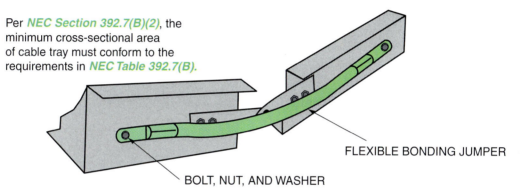

Where supervised by qualified personnel, grounded metallic cable tray may also be used as an equipment grounding conductor. *NEC Section 392.3(C).*

Cable tray sections and fittings must be bonded in accordance with *NEC Section 250.96.*

Figure 20 ♦ *NEC®* regulations governing cable tray grounding.

6.0.0 ♦ CABLE INSTALLATION

NEC Section 392.8 covers the general installation requirements for all conductors used in cable tray systems; that is, splicing, securing, and running conductors in parallel. For example, cable splices are permitted in cable trays provided they are made and insulated by *NEC®*-approved methods. Furthermore, any splices must be readily accessible and must not project above the side rails of the tray.

In most horizontal runs, the cables may be laid in the tray without securing them in place. However, on vertical runs or any runs other than horizontal, the cables must be secured to transverse members of the cable tray.

Cables may enter and leave a cable tray system in a number of different ways, as discussed previously. In general, no junction box is required where such cables are installed in bushed conduit or tubing. Where conduit or tubing is used, it must be secured to the tray with the proper fittings. Further precautions must be taken to ensure that the cable is not bent sharply as it enters or leaves the conduit or tubing.

6.1.0 Conductors Connected in Parallel

Where single-conductor cables comprising each phase, neutral, or grounded conductor of a circuit are connected in parallel as permitted in *NEC Section 310.4*, the conductors must be installed in groups consisting of not more than one conductor per phase, neutral, or grounded conductor to prevent a current imbalance in the paralleled conductors due to inductive reactance. This also prevents excessive movement due to fault current magnetic forces.

6.2.0 Number of Cables Allowed in Cable Tray (2,000V or Less)

The number of multiconductor cables rated at 2,000V or less that are permitted in a single cable tray must not exceed the requirements of *NEC Section 392.9*. This section applies to both copper and aluminum conductors.

6.2.1 All Conductors Size 4/0 or Larger

Per *NEC Section 392.9(A)(1)*, where all of the cables installed in ladder or ventilated trough tray are 4/0 or larger, the sum of the diameters of all cables shall not exceed the cable tray width, and the cables must be installed in a single layer. For example, if a cable tray installation is to contain three 4/0 multiconductor cables (1.5" in diameter), two 250 kcmil multiconductor cables (1.85"), and two 350 kcmil multiconductor cables (2.5"), the minimum width of the cable tray is determined as follows:

$$3(1.5) + 2(1.85) + 2(2.5) = 13.2"$$

The closest standard cable tray size that meets or exceeds 13.2" is 18". Therefore, this is the size to use.

6.2.2 All Conductors Smaller Than 4/0

Per *NEC Section 392.9(A)(2)*, where all of the cables are smaller than 4/0, the sum of the cross-sectional area of all cables smaller than 4/0 must not exceed the maximum allowable cable fill area as specified in Column 1 of *NEC Table 392.9;* this gives the appropriate cable tray width. To use this table, however, you must have the manufacturer's data for the cables being used. This will give the cross-sectional area of the cables.

The steps involved in determining the size of cable tray for multiconductors smaller than 4/0 AWG are as follows:

Step 1 Calculate the total cross-sectional area of all cables used in the tray. Obtain the area of each from the manufacturer's data.

Step 2 Look in Column 1 of *NEC Table 392.9* and find the smallest number that is at least as large as the calculated number.

Step 3 Look at the number to the left of the row selected in Step 2. This is the minimum width of cable tray that may be used.

For example, determine the minimum cable tray width required for the following multiple conductor cables—all less than 4/0 AWG:

- Four at 1.5" diameter
- Five at 1.75" diameter
- Three at 2.15" diameter

Step 1 Determine the cross-sectional area of the cables from the equation:

$$A = \frac{\pi \times D^2}{4}$$

Where:

A = area

D = diameter

The area of a 1.5" diameter cable is:

$$\frac{(3.14159)(1.5^2)}{4} = 1.7671 \text{ square inches}$$

The area of the four 1.5" cables is:

$$4 \times 1.7662 \text{ square inches} = 7.0648 \text{ square inches}$$

The area of 1.75" diameter cable is:

$$\frac{(3.14159)(1.75^2)}{4} = 2.4053 \text{ square inches}$$

The area of the five 1.75" cables is:

$$5 \times 2.4041 \text{ square inches} = 12.0205 \text{ square inches}$$

The area of the three 2.15" cables is:

$$\frac{(3.14159)(2.15^2)}{4} = 3.6305 \text{ square inches}$$

Therefore, the total area of the three cables is:

$$3 \times 3.6287 \text{ square inches} = 10.8861 \text{ square inches}$$

The total cross-sectional area is found by adding the above three totals to obtain:

$$7.0648 + 12.0205 + 10.8861 = 29.9714 \text{ square inches}$$

Step 2 Look in Column 1 of *NEC Table 392.9* and find the smallest number that is at least as large as 29.9714 square inches. The number is 35.

Step 3 Look to the left of 35 and you will see the inside tray width of 30". Therefore, the minimum tray width that can be used for the given group of conductors is 30".

Low-Voltage Cable

Increasingly, cable trays are being used to carry many low-voltage conductors in communication centers. In a large commercial building, thousands of telecommunications cables are distributed in bundles from the equipment room to the telecommunications closets on each floor. Cable trays offer a convenient means of running these cables through the building as well as a much easier method of allowing for system expansion. Instead of having to access conduit buried within the building walls, a new communication cable can simply be added to the cable tray system, which is normally accessible through a dropped ceiling.

6.2.3 Combination Cables

Per *NEC Section 392.9(A)(3)*, where 4/0 or larger cables are installed in the same ladder or ventilated trough cable tray with cables smaller than 4/0, the sum of the cross-sectional area of all cables smaller than 4/0 must not exceed the maximum allowable fill area from Column 2 of *NEC Table 392.9* for the appropriate cable tray width. The 4/0 and larger cables must be installed in a single layer, and no other cables can be placed on them.

To determine the tray size for a combination of cables as discussed in the above paragraph, proceed as follows:

Step 1 Repeat the steps from the procedure used previously to determine the minimum tray width required for the multiconductor cables having conductors sized 4/0 and larger.

Step 2 Repeat the steps from the procedure used previously to determine the cross-sectional area of all multiconductor cables having conductors smaller than 4/0 AWG.

Step 3 Multiply the result of Step 1 by the constant 1.2 and add this product to the result of Step 2. Call this sum A. Search Column 2 of *NEC Table 392.9* for the smallest number that is at least as large as A. Look to the left in that row to determine the minimum size cable tray required.

To illustrate these steps, we will assume that we need to find the minimum cable tray width of two multiconductor cables, each with a diameter of 2.54" (conductors size 4/0 or larger); three cables with a diameter of 3.30" (conductors size 4/0 or larger), plus eight cables with a diameter of 1.92" (conductors less than 4/0).

Step 1 The sum of all the diameters of cable having conductors 4/0 or larger is:

$$2(2.54) + 3(3.30) = 14.98"$$

Step 2 The sum of the cross-sectional areas of all cables having conductors smaller than 4/0 is:

$$\frac{(8)(3.14159)(1.92^2)}{4} = 23.1623 \text{ square inches}$$

Step 3 Multiply the result of Step 1 by 1.2 and add this product to the result of Step 2:

$$(1.2 \times 14.98) + 23.1506$$
$$= 41.1266 \text{ square inches}$$

The smallest number in Column 2 of *NEC Table 392.9* that is not larger than 41.1383 is 42. The tray width that corresponds to 42 is 36". Therefore, select a cable tray width of 36".

6.2.4 Solid Bottom Tray

Per *NEC Section 392.9(C)*, where solid bottom cable trays contain multiconductor power or lighting cables, or any mixture of multiconductor power, lighting, control, and signal cables, the maximum number of cables must conform to the following:

- Where all of the cables are 4/0 or larger, the sum of the diameters of all cables must not exceed 90% of the cable tray width, and the cables must be installed in a single layer.
- Where all of the cables are smaller than 4/0, the sum of the cross-sectional areas of all cables must not exceed the maximum allowable cable fill area in Column 3 of *NEC Table 392.9* for the appropriate cable tray width.
- Where 4/0 or larger cables are installed in the same cable tray with cables smaller than 4/0, the sum of the cross-sectional areas of all of the smaller cables must not exceed the maximum allowable fill area resulting from the computation in Column 4, *NEC Table 392.9* for the appropriate cable tray width. The 4/0 and larger cables must be installed in a single layer, and no other cables can be placed on them.

Where a solid bottom cable tray with a usable inside depth of 6" or less contains multiconductor control and/or signal cables only, the sum of the cross-sectional areas of all cables at any cross section must not exceed 40% of the interior cross-sectional area of the cable tray. A depth of 6" must be used to compute the allowable interior cross-sectional area of any cable tray that has a usable inside depth of more than 6".

In a previous example, we determined that the minimum tray size for multiconductor cables with all conductors size 4/0 or larger was 18". The sum of all cable diameters for this example was 13.2". To see if an 18" solid bottom tray can be used, multiply the tray width by 0.90 (90%):

$$18 \times 0.9 = 16.2"$$

Therefore, 16.2" is the minimum width allowed for solid bottom tray. Since we are using 18" tray, this meets the requirements of *NEC Section 392.9(C)(1)*.

When dealing with solid bottom trays and using *NEC Table 392.9*, use Columns 3 and 4 instead of Columns 1 and 2, as used for ladder and trough-type cable trays.

6.2.5 Single-Conductor Cables

Calculating cable tray widths for single-conductor cables (2,000V or under) is similar to the calculations used for multiconductor cables, with the following exceptions:

- Conductors that are 1,000 kcmil and larger are treated the same as multiconductor cables having conductors size 4/0 or larger.
- Conductors that are smaller than 1,000 kcmil are treated the same as multiconductor cables having conductors smaller than size 4/0.

NEC Section 392.10 covers the details of installing single-conductor cables with rated voltages of less than 2,000V in cable tray systems.

6.3.0 Ampacity of Cable Tray Conductors

NEC Section 392.11 gives the requirements for cables used in tray systems with rated voltages of 2,000 volts or less. Cables in cable tray use the same ampacity tables as other cable (see *NEC Article 310*), with additional derating applied for the following conditions:

- Cable tray construction
  – Open or covered
  – Solid or ventilated/ladder
- Type(s) of cable
  – Ampacity (less than 2,000V or over 2,001V)
  – Single conductor or multiconductor
- Number of cables and cable configuration in the tray

NEC Sections 392.12 and 392.13 cover the installation and ampacity of cables with voltages of 2,001 and over.

7.0.0 ◆ CABLE TRAY DRAWINGS

For an economical and satisfactory installation, working out the details of supports and hangers for a cable tray system is usually done beforehand by the engineering department or project engineer and is seldom left to the judgment of a field force that is not acquainted with the loads and forces to be encountered. As a result, drawings and specifications will usually be furnished to the work crew to provide details about the cable tray system. All workers involved with the installation should know how to interpret these drawings.

The exact method of showing cable tray systems on working drawings will vary, so always consult the symbol list or legend before beginning the installation. Also, study any shop or detail drawings that might accompany the construction documents.

If space permits, many engineers prefer to draw the cable tray system as close to scale as possible, using various symbols to show the different types of cable tray to be installed.

Look at the floor plan drawing in *Figure 21*. The cable tray system in this project originates at several power panels and motor control centers to feed and control motors in other parts of the building. The trays run from the motor control centers, are offset to miss beams and other runs of cable tray, and then branch off to various parts of the building.

Although experienced workers in the industrial electrical trade will have little trouble reading the information in this drawing, new electricians may have some difficulty in visualizing the system. However, if a supplemental drawing were provided with the floor plan drawing in *Figure 21*, it would provide a clearer picture of the system and leave little doubt as to how the cable tray system is to be installed. Even new workers in the trade would be able to see how the system should be installed, and this would take some of the load off experienced workers and give them more time to accomplish other tasks.

In actual practice, however, consulting engineering firms seldom furnish the isometric drawing; they merely show the layout in plan view. Consequently, plan views involving the construction details of cable tray systems must be studied carefully during the planning stage—before the work is begun.

> **NOTE**
> In some areas, cable tray components are ordered in metric units. A metric conversion chart is included at the back of this module.

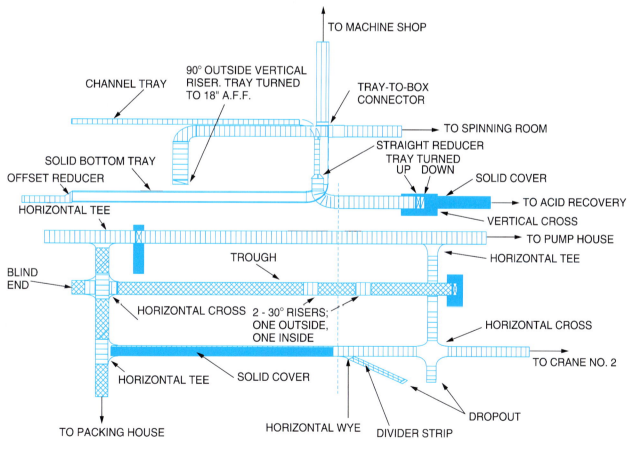

Figure 21 ◆ Sample floor plan of a cable tray system.

8.0.0 ◆ PULLING CABLE IN TRAY SYSTEMS

When installing cables in tray systems, proper precautions must be taken to avoid damaging the cables. A complete line of installation tools is available for pulling long lengths of cable up to 1,000' or longer. These tools save considerable installation time.

Short lengths of cable can be laid in place without power pulling tools, or the cable can be pulled manually using a basket grip and pulling rope. Long lengths of small cable, 2" or less in diameter, can also be pulled with a basket grip and pulling rope. Larger cables, however, should be pulled by the conductor and the braid, sheath, or armor. This is usually done with a pulling eye applied to the cable at the factory, or by tying the conductor to the eye of a basket grip and taping the tail end of the grip to the outside of the cable.

In general, the pull exerted on the cables pulled with a basket grip that is not attached to the conductor should not exceed 1,000 pounds. For heavier pulls, care should be taken not to stretch the insulation, jacket, or armor beyond the end of the conductor nor bend the ladder, trough, or channel out of shape.

The bending radius of the cable should not be less than the values recommended by the cable manufacturer, which range from four times the diameter for a rubber-insulated cable that has a 1" maximum outside diameter without lead, shield, or armor, to eight times the diameter for *interlocked armor cable*. Cables of special construction such as wire armor and high-voltage cables require a larger radius bend.

When installing long lengths of cable up to 1,000' with as many as a dozen bends, best results are obtained by pulling the cable in one continuous operation at a speed of 20' to 25' per minute. The pulling line diameter and length will depend on the pull to be made and construction equipment available. The winch and power unit must be of an adequate size for the job and capable of developing the high pulling speed required for the best and most economical results. A complete description of pulling equipment was covered in the *Conductor Installations* module.

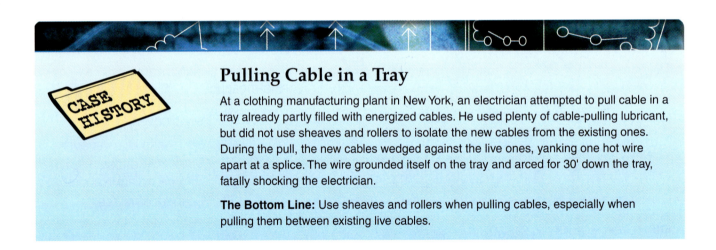

Pulling Cable in a Tray

At a clothing manufacturing plant in New York, an electrician attempted to pull cable in a tray already partly filled with energized cables. He used plenty of cable-pulling lubricant, but did not use sheaves and rollers to isolate the new cables from the existing ones. During the pull, the new cables wedged against the live ones, yanking one hot wire apart at a splice. The wire grounded itself on the tray and arced for 30' down the tray, fatally shocking the electrician.

The Bottom Line: Use sheaves and rollers when pulling cables, especially when pulling them between existing live cables.

Big Pulls

For very long pulls, some contractors use an auxiliary winch. Power winches may exceed their preset tension limits and abort the pull, so a second winch is installed in a high-tension area of the pull in a straight section of tray. The second winch is used to pull a loop of slack cable and reduce the tension on the main winch, which now pulls only the length of slack cable between itself and the auxiliary winch.

9.0.0 ◆ SAFETY

Installing cable tray means working at heights above the floor. Consequently, workers must take the necessary precautions. In general, workers installing cable tray will use ladders, scaffolds, lifts, work from the tray assembly itself, or a combination of all these.

Keep the tray assembly uncluttered during installation. Tools, tray fittings, and the like are ideal obstacles for tripping workers. They may also fall off the tray and injure workers below. To help prevent the latter, set up work barriers beneath the section of tray assembly being worked on. If you are working on the ground, never penetrate or move these barriers until the work above is complete.

Learn to secure your lifeline properly, and also make sure your full body harness is a proper fit. A harness that is too large may slip, resulting in a serious injury or even death.

What's wrong with this picture?

207SA11.EPS

Putting It All Together

If possible, examine the conduit system at your workplace or school. Did the installers make use of cable tray? If not, is the system large enough to warrant the use of it?

Review Questions

1. When a cable tray section branches off from a main section in two 90° turns, the section is called a(n) _____.
   a. wye
   b. tee
   c. divider strip
   d. ell

2. A section of cable tray that makes a single horizontal 90° turn is known as a(n) _____.
   a. tee
   b. wye
   c. elbow
   d. cross

3. Which of the following best describes the type of fitting that will be used with a solid cover cable tray?
   a. An Ericson coupling
   b. Vertical splices
   c. A horizontal cross
   d. A cover clamp

4. A(n) _____ branches in four different directions.
   a. horizontal wye
   b. horizontal cross
   c. inside vertical riser
   d. dropout

5. Which of the following best describes how closely packed cables will be affected when energized?
   a. Their efficiency will be increased.
   b. Their efficiency will be decreased.
   c. No change will be encountered.
   d. The operating temperature of each cable will be lower.

6. When a cable tray system must be protected from dropping objects that may damage the cable, a _____ is used.
   a. solid bottom tray
   b. trough-type tray
   c. tray cover
   d. blind end

7. When cable exits downward from a tray without a dropout plate, _____ could occur.
   a. current surges
   b. reduced amperage
   c. voltage surges
   d. insulation damage

8. The main purpose of vertical adjustment splice plates is to _____.
   a. increase the strength of a cable tray section
   b. support the cable tray system
   c. change the elevation in a run of cable tray
   d. secure the cable within the tray

9. A barrier strip is a _____.
   a. division strip installed in a raceway to separate certain types of cables
   b. strip used to provide the tray with added support
   c. glass strip used as a cable tray insulator
   d. metal strip used solely for grounding purposes

10. Which of the following is a violation of *NEC Section 300.8?*
    a. Placing an instrumentation air line in the same tray with electrical conductors
    b. Spacing tray rungs 9" or less apart
    c. Running single conductors in trays that are larger than size 1/0
    d. Using nonmetallic tray systems in corrosive areas

Summary

A cable tray system includes the assembly of units or sections and associated fittings that form a rigid structural system used to support cables and raceways. Cable tray systems are commonly used in industrial applications and are normally constructed of either aluminum or steel, although PVC-coated and all-nonmetallic trays are also available.

Cable tray is available in various forms, including ladder, trough, center rail, and solid bottom, and can be supported by either side mounts or center mounts. *NEC Article 392* covers cable tray installations, along with the types of conductors to be used in various cable tray systems.

Notes

Trade Terms Introduced in This Module

Barrier strip: A metal strip constructed to divide a section of cable tray so that certain kinds of cable may be separated from each other.

Cable pulley: A device used to facilitate pulling conductor in cable tray where the tray changes direction. Several types are available (single, triple, etc.) to accommodate almost all pulling situations.

Cross: A four-way section of cable tray used when the tray assembly must branch off in four different directions.

Direct rod suspension: A method used to support cable tray by means of threaded rods and hanger clamps. One end of the threaded rod is secured to an overhead structure, while the other end is connected to hanger clamps that are attached to the cable tray side rails.

Dropout: Cable leaving the tray assembly and travelling directly downward; that is, the cable is not routed into a conduit or channel.

Dropout plate: A metal plate used at the end of a cable tray section to ensure a greater cable bending radius as the cable leaves the tray assembly.

Elbow: A section of cable tray used to change the direction of the tray assembly a full quarter turn (90°). Both vertical and horizontal elbows are common.

Expansion joints: Plates used at intervals along a straight run of cable tray to allow space for thermal expansion or contraction of the tray.

Fittings: Devices used to assemble and/or change the direction of cable tray systems.

Interlocked armor cable: Mechanically protected cable; usually a helical winding of metal tape formed so that each convolution locks mechanically upon the previous one (armor interlock).

Ladder tray: A type of cable tray that consists of two parallel channels connected by rungs, similar in appearance to the common straight ladder.

Pipe racks: Structural frames used to support the piping that interconnects equipment in outdoor industrial facilities.

Swivel plates: Devices used to make vertical offsets in cable tray.

Tee: A section of cable tray that branches off the main section in two other directions.

Trapeze mounting: A method of supporting cable tray using metal channel, such as Unistrut®, Kindorf®, etc., supported by two threaded rods, and giving the appearance of a swing or trapeze.

Tray cover: A flat piece of metal, fiberglass, or plastic designed to provide a solid covering that is needed in some locations where conductors in the tray system may be damaged.

Trough: A type of cable tray consisting of two parallel channels (side rails) having a corrugated, ventilated bottom or a corrugated, solid bottom.

Unistrut®: A brand of metal channel used as the bottom bracket for hanging cable trays. Double Unistrut® adds strength and stability to the trays and also provides a means of securing future runs of conduit.

Wall mounting: A method of supporting cable tray systems using supports secured directly to the wall.

Wye: A section of cable tray that branches off the main section in one direction.

Appendix

Metric Conversion Chart

METRIC CONVERSION CHART

INCHES Fractional	Decimal	METRIC mm	INCHES Fractional	Decimal	METRIC mm	INCHES Fractional	Decimal	METRIC mm
.	0.0039	0.1000	.	0.5512	14.0000	.	1.8898	48.0000
.	0.0079	0.2000	9/16	0.5625	14.2875	.	1.9291	49.0000
.	0.0118	0.3000	.	0.5709	14.5000	.	1.9685	50.0000
1/64	0.0156	0.3969	37/64	0.5781	14.6844	2	2.0000	50.8000
.	0.0157	0.4000	.	0.5906	15.0000	.	2.0079	51.0000
.	0.0197	0.5000	19/32	0.5938	15.0813	.	2.0472	52.0000
.	0.0236	0.6000	39/64	0.6094	15.4781	.	2.0866	53.0000
.	0.0276	0.7000	.	0.6102	15.5000	.	2.1260	54.0000
1/32	0.0313	0.7938	5/8	0.6250	15.8750	.	2.1654	55.0000
.	0.0315	0.8000	.	0.6299	16.0000	.	2.2047	56.0000
.	0.0354	0.9000	41/64	0.6406	16.2719	.	2.2441	57.0000
.	0.0394	1.0000	.	0.6496	16.5000	2 1/4	2.2500	57.1500
.	0.0433	1.1000	21/32	0.6563	16.6688	.	2.2835	58.0000
3/64	0.0469	1.1906	.	0.6693	17.0000	.	2.3228	59.0000
.	0.0472	1.2000	43/64	0.6719	17.0656	.	2.3622	60.0000
.	0.0512	1.3000	11/16	0.6875	17.4625	.	2.4016	61.0000
.	0.0551	1.4000	.	0.6890	17.5000	.	2.4409	62.0000
.	0.0591	1.5000	45/64	0.7031	17.8594	.	2.4803	63.0000
1/16	0.0625	1.5875	.	0.7087	18.0000	2 1/2	2.5000	63.5000
.	0.0630	1.6000	23/32	0.7188	18.2563	.	2.5197	64.0000
.	0.0669	1.7000	.	0.7283	18.5000	.	2.5591	65.0000
.	0.0709	1.8000	47/64	0.7344	18.6531	.	2.5984	66.0000
.	0.0748	1.9000	.	0.7480	19.0000	.	2.6378	67.0000
5/64	0.0781	1.9844	3/4	0.7500	19.0500	.	2.6772	68.0000
.	0.0787	2.0000	49/64	0.7656	19.4469	.	2.7165	69.0000
.	0.0827	2.1000	.	0.7677	19.5000	2 3/4	2.7500	69.8500
.	0.0866	2.2000	25/32	0.7813	19.8438	.	2.7559	70.0000
.	0.0906	2.3000	.	0.7874	20.0000	.	2.7953	71.0000
3/32	0.0938	2.3813	51/64	0.7969	20.2406	.	2.8346	72.0000
.	0.0945	2.4000	.	0.8071	20.5000	.	2.8740	73.0000
.	0.0984	2.5000	13/16	0.8125	20.6375	.	2.9134	74.0000
7/64	0.1094	2.7781	.	0.8268	21.0000	.	2.9528	75.0000
.	0.1181	3.0000	53/64	0.8281	21.0344	.	2.9921	76.0000
1/8	0.1250	3.1750	27/32	0.8438	21.4313	3	3.0000	76.2000
.	0.1378	3.5000	.	0.8465	21.5000	.	3.0315	77.0000
9/64	0.1406	3.5719	55/64	0.8594	21.8281	.	3.0709	78.0000
5/32	0.1563	3.9688	.	0.8661	22.0000	.	3.1102	79.0000
.	0.1575	4.0000	7/8	0.8750	22.2250	.	3.1496	80.0000
11/64	0.1719	4.3656	.	.8858	22.5000	.	3.1890	81.0000
.	0.1772	4.5000	57/64	.89063	22.6219	.	3.2283	82.0000
3/16	0.1875	4.7625	.	.9055	23.0000	.	3.2677	83.0000
.	0.1969	5.0000	29/32	.90625	23.0188	.	3.3071	84.0000
13/64	0.2031	5.1594	59/64	.92188	23.4156	.	3.3465	85.0000
.	0.2165	5.5000	.	.9252	23.5000	.	3.3858	86.0000
7/32	0.2188	5.5563	15/16	.93750	23.8125	.	3.4252	87.0000
15/64	0.2344	5.9531	.	.9449	24.0000	.	3.4646	88.0000
.	0.2362	6.0000	61/64	.95313	24.2094	3 1/2	3.5000	88.9000
1/4	0.2500	6.3500	.	.9646	24.5000	.	3.5039	89.0000
.	0.2559	6.5000	31/32	.96875	24.6063	.	3.5433	90.0000
17/64	0.2656	6.7469	.	.9843	25.0000	.	3.5827	91.0000
.	0.2756	7.0000	63/64	.98438	25.0031	.	3.6220	92.0000
9/32	0.2813	7.1438	1	1.000	25.40	.	3.6614	93.0000
.	0.2953	7.5000	.	1.0039	25.5000	.	3.7008	94.0000
19/64	0.2969	7.5406	.	1.0236	26.0000	.	3.7402	95.0000
5/16	0.3125	7.9375	.	1.0433	26.5000	.	3.7795	96.0000
.	0.3150	8.0000	.	1.0630	27.0000	.	3.8189	97.0000
21/64	0.3281	8.3344	.	1.0827	27.5000	.	3.8583	98.0000
.	0.3346	8.5000	.	1.1024	28.0000	.	3.8976	99.0000
11/32	0.3438	8.7313	.	1.1220	28.5000	.	3.9370	100.0000
.	0.3543	9.0000	.	1.1417	29.0000	4	4.0000	101.6000
23/64	0.3594	9.1281	.	1.1614	29.5000	.	4.3307	110.0000
.	0.3740	9.5000	.	1.1811	30.0000	4 1/2	4.5000	114.3000
3/8	0.3750	9.5250	.	1.2205	31.0000	.	4.7244	120.0000
25/64	0.3906	9.9219	1 1/4	1.2500	31.7500	5	5.0000	127.0000
.	0.3937	10.0000	.	1.2598	32.0000	.	5.1181	130.0000
13/32	0.4063	10.3188	.	1.2992	33.0000	.	5.5118	140.0000
.	0.4134	10.5000	.	1.3386	34.0000	.	5.9055	150.0000
27/64	0.4219	10.7156	.	1.3780	35.0000	6	6.0000	152.4000
.	0.4331	11.0000	.	1.4173	36.0000	.	6.2992	160.0000
7/16	0.4375	11.1125	.	1.4567	37.0000	.	6.6929	170.0000
.	0.4528	11.5000	.	1.4961	38.0000	.	7.0866	180.0000
29/64	0.4531	11.5094	1 1/2	1.5000	38.1000	.	7.4803	190.0000
15/32	0.4688	11.9063	.	1.5354	39.0000	.	7.8740	200.0000
.	0.4724	12.0000	.	1.5748	40.0000	8	8.0000	203.2000
31/64	0.4844	12.3031	.	1.6142	41.0000	.	9.8425	250.0000
.	0.4921	12.5000	.	1.6535	42.0000	10	10.0000	254.0000
1/2	0.5000	12.7000	.	1.6929	43.0000	20	20.0000	508.0000
.	0.5118	13.0000	.	1.7323	44.0000	30	30.0000	762.0000
33/64	0.5156	13.0969	1 3/4	1.7500	44.4500	40	40.0000	1016.000
17/32	0.5313	13.4938	.	1.7717	45.0000	60	60.0000	1524.000
.	0.5315	13.5000	.	1.8110	46.0000	80	80.0000	2032.000
35/64	0.5469	13.8906	.	1.8504	47.0000	100	100.0000	2540.000

TO CONVERT TO MILLIMETERS; MULTIPLY INCHES X 25.4
TO CONVERT TO INCHES; MULTIPLY MILLIMETERS X 0.03937*
*FOR SLIGHTLY GREATER ACCURACY WHEN CONVERTING TO INCHES; DIVIDE MILLIMETERS BY 25.4

Additional Resources

This module is intended to present thorough resources for task training. The following reference works are suggested for further study. These are optional materials for continued education rather than for task training.

American Electrician's Handbook, Latest Edition. New York: Croft and Summers, McGraw-Hill.

National Electrical Code® Handbook, Latest Edition. Quincy, MA: National Fire Protection Association.

CONTREN® LEARNING SERIES – USER UPDATE

NCCER makes every effort to keep these textbooks up-to-date and free of technical errors. We appreciate your help in this process. If you have an idea for improving this textbook, or if you find an error, a typographical mistake, or an inaccuracy in NCCER's Contren® textbooks, please write us, using this form or a photocopy. Be sure to include the exact module number, page number, a detailed description, and the correction, if applicable. Your input will be brought to the attention of the Technical Review Committee. Thank you for your assistance.

Instructors – If you found that additional materials were necessary in order to teach this module effectively, please let us know so that we may include them in the Equipment/Materials list in the Annotated Instructor's Guide.

Write: Product Development and Revision
National Center for Construction Education and Research
3600 NW 43rd St., Bldg. G, Gainesville, FL 32606

Fax: 352-334-0932

E-mail: curriculum@nccer.org

Craft _____ Module Name _____

Copyright Date _____ Module Number _____ Page Number(s) _____

Description _____

_____

_____

_____

_____

(Optional) Correction _____

_____

_____

_____

(Optional) Your Name and Address _____

_____

_____

_____

Conductor Terminations and Splices

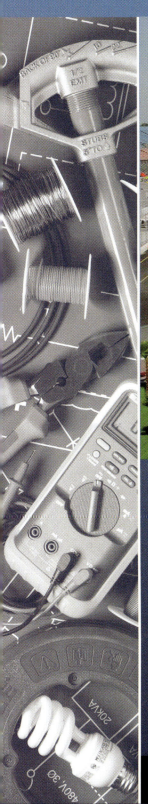

Daytona International Speedway Infield

QuestCom won an ABC Excellence in Construction award for its work on updating Daytona International Speedway's 45-year-old facilities with state-of-the-art telecommunications, fire alarm systems, and concert-quality sound.

26208-08

26208-08
Conductor Terminations and Splices

Topics to be presented in this module include:

1.0.0 Introduction 8.2
2.0.0 Stripping and Cleaning Conductors 8.2
3.0.0 Wire Connections Under 600 Volts 8.8
4.0.0 Control and Signal Cable 8.12
5.0.0 Low-Voltage Connectors and Terminals 8.13
6.0.0 Guidelines for Installing Connectors 8.14
7.0.0 Bending Cable and Training Conductors 8.24
8.0.0 *NEC*® Termination Requirements 8.26
9.0.0 Taping Electrical Joints 8.28
10.0.0 Motor Connection Kits 8.30

Overview

Specific procedures must be followed when splicing, terminating, and identifying conductors. First, the correct length of conductor insulation is removed to expose the conductor. Next, the conductor is cleaned to remove any remaining insulation residue or contamination, and then it is labeled and terminated. A properly prepared conductor end provides the maximum conduction and least resistance between the conductor and termination point or splice.

There are many methods used to prepare the conductor for termination or splicing. The tools available for preparing conductor ends range from simple handheld knives to specialized cable stripping tools used for large power cables and wires. Regardless of the tool or method used to prepare the wire or cable, the conductor must be left undamaged, with a cleanly exposed conductor end. The type of termination or splice used depends on the application. Common connectors include mechanical connectors and compression (crimp) connectors.

Note: *NFPF 70*®, *National Electrical Code*®, and *NEC*® are registered trademarks of the National Fire Protection Association, Inc., Quincy, MA 02269. All *National Electrical Code*® and *NEC*® references in this module refer to the 2008 edition of the *National Electrical Code*®.

Objectives

When you have completed this module, you will be able to do the following:

1. Describe how to make a good conductor termination.
2. Prepare cable ends for terminations and splices and connect using lugs or connectors.
3. Train cable at termination points.
4. Understand the *National Electrical Code®* (*NEC®*) requirements for making cable terminations and splices.
5. Demonstrate crimping techniques.
6. Select the proper lug or connector for the job.

Trade Terms

AL-CU
Amperage capacity
Connection
Connector
Drain wire
Grooming
Insulating tape
Lug
Mechanical advantage (MA)
Pressure connector
Reducing connector
Shielding
Splice
Strand
Terminal
Termination

Required Trainee Materials

1. Pencil and paper
2. Copy of the latest edition of the *National Electrical Code®*
3. Appropriate personal protective equipment

Prerequisites

Before you begin this module, it is recommended that you successfully complete *Core Curriculum*; *Electrical Level One*; and *Electrical Level Two*, Modules 26201-08 through 26207-08.

This course map shows all of the modules in *Electrical Level Two*. The suggested training order begins at the bottom and proceeds up. Skill levels increase as you advance on the course map. The local Training Program Sponsor may adjust the training order.

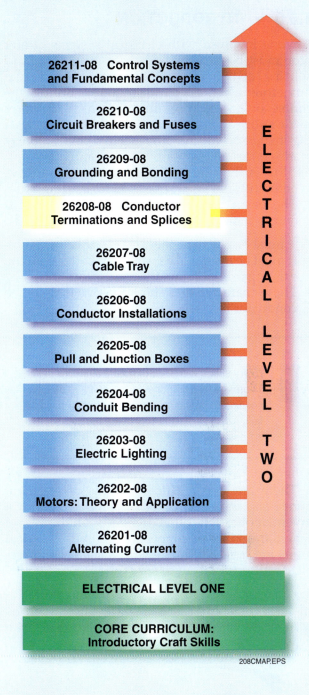

1.0.0 ♦ INTRODUCTION

Anyone involved with electrical systems of any type must be familiar with wire connectors and splicing, as they are both necessary to make the numerous electrical joints required during the course of an electrical installation. A properly made splice and connection should last as long as the insulation on the wire itself, while a poorly made connection will always be a source of trouble; that is, the joints will overheat under load and eventually fail with the potential for starting a fire.

The basic requirements for a good electrical connection include the following:

- It must be mechanically and electrically secure.
- It must be insulated as well as or better than the existing insulation on the conductors.
- These characteristics should last as long as the conductor is in service.

NOTE

Every splice is a point of potential failure. Therefore, splicing should not be used as a solution to providing shorter conductor pulls. In fact, some applications prohibit the use of splices in critical areas.

There are many different types of electrical joints, and the selection of the proper type for a given application often depends on how and where the splice or connection is used. Electrical joints are normally made with a solderless pressure connector or lug to save time.

Conductor Terminations and Splices

Poor electrical connections are responsible for a large percentage of equipment burnouts and fires. Many of these failures are a direct result of improper terminations, poor workmanship, and the use of improper splicing devices.

2.0.0 ♦ STRIPPING AND CLEANING CONDUCTORS

Before any connection or splice can be made, the ends of the conductors must be properly cleaned and stripped. To ensure a low-resistance connection and avoid contaminating the termination, clean the areas of the cable where it is to be cut and stripped. Remove any pulling compound, dirt, oil, grease, or water to avoid contaminating the exposed conductor.

Stripping is the removal of insulation from the end of the conductor or at the location of the splice. Conductors should only be stripped using the appropriate stripping tool. This will help to prevent cuts and nicks in the wire, which can reduce the conductor area and weaken the conductor.

Poorly stripped conductors can result in nicks, scrapes, or burnishes. Any of these can lead to a stress concentration in the damaged area. Heat, rapid temperature changes, mechanical vibration, and oscillatory motion can aggravate the damage, causing faults in the circuitry or even total failure. Lost strands are a problem in splice or crimp-type terminals, while exposed strands might be a safety hazard.

Faulty stripping can pierce, scuff, or split the insulation. This can cause changes in dielectric strength and lower the conductor's resistance to moisture and abrasion. Insulation particles often get trapped in solder and crimp joints. These form the basis for a defective termination. A variety of factors determine how precisely a conductor can be stripped, including the wire size and type of insulation.

It is a common misconception that a certain gauge of stranded conductor has the same diameter as a solid conductor. This is a very important consideration in selecting the proper blades for strippers. *Table 1* shows the nominal sizes referenced for the different wire gauges.

To eliminate nicking, cutting, and fraying, wires should only be stripped using the appropriate stripping tool. The specific tool used depends on the size and type of wire being stripped.

2.1.0 Stripping Small Conductors

There are many kinds of wire strippers available. *Figure 1* shows a common type of wire stripper for small conductors. It can be used to cut, strip, and crimp wires from No. 22 through No. 10 AWG. To use this tool, insert the conductor into the proper

Table 1 Dimensions of Common Wire Sizes

Size (AWG/kcmil)	Area (Circular Mils)	Overall Diameter in Inches	
		Solid	Stranded
18	1,620	0.040	0.046
16	2,580	0.051	0.058
14	4,130	0.064	0.073
12	6,530	0.081	0.092
10	10,380	0.102	0.116
8	16,510	0.128	0.146
6	26,240	—	0.184
4	41,740	—	0.232
3	52,620	—	0.260
2	66,360	—	0.292
1	83,690	—	0.332
1/0	105,600	—	0.373
2/0	133,100	—	0.419
3/0	167,800	—	0.470
4/0	211,600	—	0.528
250	—	—	0.575
300	—	—	0.630
350	—	—	0.681
400	—	—	0.728
500	—	—	0.813
600	—	—	0.893
700	—	—	0.964
750	—	—	0.998
800	—	—	1.03
900	—	—	1.09
1,000	—	—	1.15
1,250	—	—	1.29
1,500	—	—	1.41
1,750	—	—	1.52
2,000	—	—	1.63

Figure 1 ♦ Wire stripper/crimper.

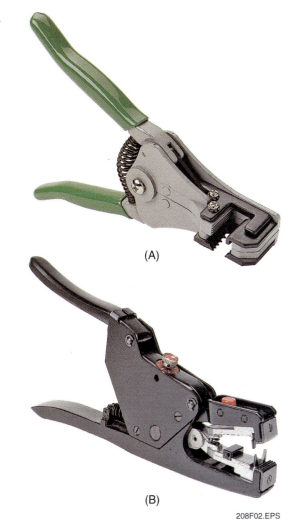

Figure 2 ♦ Wire strippers.

size knife groove, then squeeze the tool handles. The tool cuts the conductor insulation, allowing the conductor to be easily removed without crushing its stripped end. The length of the strip is regulated by the amount of wire extending beyond the blades when it is inserted in the knife groove.

Figure 2 shows production-grade stripping tools. *Figure 2(A)* can be used to strip conductors from No. 20 to No. 10 AWG, while *Figure 2(B)* strips wires from No. 18 to No. 6 AWG. Note that this tool has front entry jaws for use in tight spaces.

2.2.0 Stripping Power Cables and Large Conductors

Larger conductors can be cut using a ratchet-type cable cutter. The cable cutter shown in *Figure 3* can be used to cut cables up to 1,000 kcmil. Heavy-duty strippers are used to strip large power cables. *Figure 4* shows a heavy-duty stripper used to strip power cables with outside diameters ranging from 1/0 through 1,000 kcmil. Strippers can be used to remove insulation from the end of a cable or to make window cuts (*Figure 5*). All stripping tools should be operated according to the manufacturer's instructions. The procedures for using the tool shown in *Figure 4* to strip insulation from the end of a cable and to make a window cut are described here.

Cable Strippers

Open-blade (knife) stripping can damage the conductors and is prohibited on many job sites. Adjustable strippers such as the tool shown here are used instead.

Figure 3 ◆ Ratchet-type cable cutter.

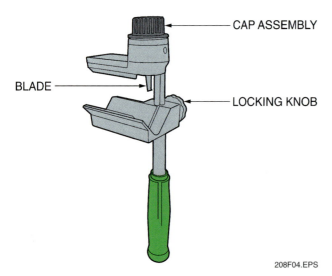

Figure 4 ◆ Heavy-duty cable stripper.

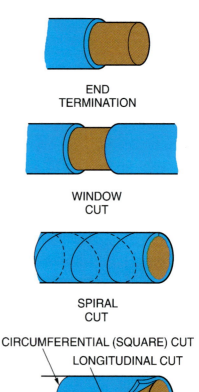

Figure 5 ◆ Types of cable stripping.

8.4 ELECTRICAL LEVEL TWO ◆ TRAINEE GUIDE

WARNING!

Keep fingers away from the blade when using any stripping tool.

To strip insulation from the end of a cable, proceed as follows:

Step 1 Loosen the locking knob to open the tool to the maximum position. Place the cable in the V-groove and close the tool firmly around the cable. Tighten the locking knob.

Step 2 Turn the cap assembly until the blade reaches the required depth.

CAUTION

Do not allow the blade to contact the conductor because damage to the conductor and/or the blade can result.

Step 3 Rotate the tool around the cable, advancing to the required strip length.

Step 4 Rotate the tool in the reverse direction to produce a square end cut.

Step 5 Loosen the locking knob to release the tool and remove it from the cable.

Step 6 Peel off the insulation.

To make a window cut, proceed as follows:

Step 1 With the tool opened to the maximum position, place the cable in the V-groove and close the tool firmly around the cable. Tighten the locking knob.

Step 2 Turn the cap assembly until the blade reaches the required depth.

Step 3 Rotate the tool to produce the first square cut.

Step 4 Rotate the tool in the reverse direction to cut the required window strip length.

Step 5 Rotate the tool in the original direction to produce the second square cut.

Step 6 Loosen the locking knob assembly to release the tool and remove it from the cable.

Step 7 Peel off the insulation.

Figure 6 shows a round cable slitting and ringing tool that can be used to strip single- or multi-conductor cables. This tool can be used to cut around the cable (square cut) or slit the length of the cable jacket (longitudinal cut) for easy removal. The tool blade is adjustable to accommodate different jacket thicknesses.

2.3.0 Stripping Control and Signal Cable/Conductors

A scissor-action type of cutting tool is preferable to cutting tools with jaws that butt against each other. These tools have a tendency to produce a flattened chisel end on the conductor, especially when the cutting edges become dull, as shown in *Figure 7*. This makes it difficult to insert the conductor into the barrel of a terminal.

Observe the following points when stripping conductors:

- Remove the cable jacket using strippers with an adjustable blade or a die designed for the particular wire size (*Figure 8*). The terminal manufacturer will recommend a stripping length. Be careful to avoid nicking or stretching the conductor or insulation in a multi-conductor cable. If this type of damage occurs, it is likely that the wrong groove of the stripping tool was used,

Figure 6 ♦ Round cable slitting and ringing tool.

Stripping Cables

When using any bladed-type cable stripper, make sure to replace the blade whenever it becomes dull. Remember, a dull blade is more dangerous than a sharp one because you are more likely to apply undue pressure when using a dull blade, causing it to slip.

Cutting Large Power Cable

When cutting large power cables prior to stripping, you can make the job easier by using a battery-powered cable cutter such as the one shown here. Electric and hydraulic cable cutters are also available.

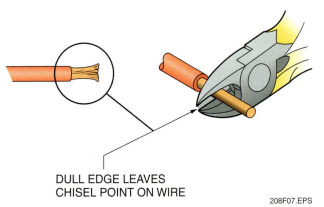

Figure 7 ◆ Chisel point on a conductor.

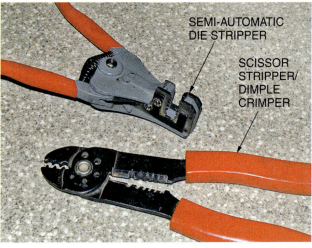

Figure 8 ◆ Cable and wire stripping tools.

8.6   ELECTRICAL LEVEL TWO ◆ TRAINEE GUIDE

the tool was improperly set or damaged, the wrong type of stripping blade was used, or the tool was used incorrectly.

- Cut the insulation so that no frayed pieces or threads extend past the point of cutoff. Frayed pieces or threads of insulation indicate the use of an improper tool or dull cutter blades.
- If possible, do not twist, spread, or disturb the wire strands from their original position in the cable. Retwisting or tightening the twist of the strands will eventually result in damage.
- Terminate stripped wire as soon as possible. The exposed strands will invariably become bent and spread, making termination difficult. A minimum amount of handling and storage after stripping will result in better terminations.

Figure 9 shows the positioning of the wire in the crimp barrel when stripped to the proper length. The conductor insulation must be in the belled mouth of the terminal. This relieves stress on the strands or wire and increases the strength of the connection. Allowing conductor strands to protrude out of the inspection hole more than $\frac{1}{32}$" will interfere with the terminal screw. Cutting the strands too short will reduce the contact surface area.

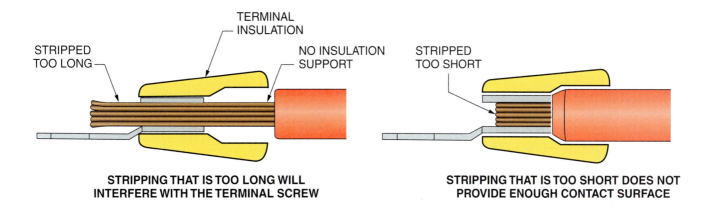

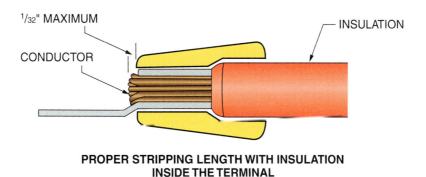

Figure 9 ◆ Proper stripping length.

3.0.0 ◆ WIRE CONNECTIONS UNDER 600 VOLTS

NEC Section 110.14 governs electrical connections, including terminations and splices. Wire connections are used to connect a wire or cable to electrically operated devices such as fan coil units, duct heaters, oil burners, motors, pumps, and control circuits of all types.

A variety of wire connectors for stranded wire are shown in *Figure 10*. These connectors are available in various sizes to accommodate wire sizes No. 22 AWG and larger. They can be installed with crimping tools having a single or double indenter. The range is normally stamped on the tongue of each terminal.

Mechanical compression-type terminators are also available to accommodate wires from No. 8 AWG through 1,000 kcmil. One-hole lugs, two-hole lugs, split-bolt connectors, and other types are shown in *Figure 11*.

Crimp-type connectors for wires smaller than No. 8 AWG are normally made to accept at least two wire sizes and are often color coded. For example, one manufacturer's color code is red for No. 18 or No. 20 wire, blue for No. 16 or No. 14 wire, and yellow for No. 12 or No. 10 wire. Crimp-type connectors for wire sizes No. 8 and larger,

INSIDE TRACK

Tightening Compression Connector Screws and Bolts

Mechanical compression connectors must be tightened to a specified torque using a torque screwdriver or torque wrench. Overtightening can cut the wires or break the fitting, while undertightening may lead to loose connections, resulting in overheating and failure.

commonly called lugs, are made to accept one specific conductor size. Crimp-type *reducing connectors* are used to connect two different size wires. Mechanical compression-type connectors and lugs are made to accommodate a range of different wire sizes.

The parallel-tap connector with an insulated cover shown in *Figure 11* is one example of a pre-insulated, molded mechanical compression connector. There are several kinds available. They come in setscrew/pressure plate and insulation-piercing configurations made for use in a variety of feeder tap and splice applications. Because they are equipped with an insulating cover, the requirement for taping the joint is eliminated.

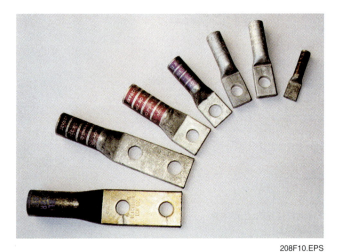

Figure 10 ◆ Crimp-on wire lugs.

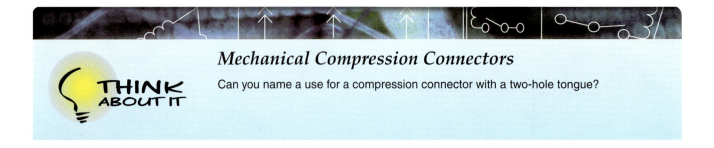

Mechanical Compression Connectors
Can you name a use for a compression connector with a two-hole tongue?

8.8 ELECTRICAL LEVEL TWO ◆ TRAINEE GUIDE

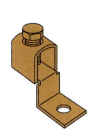
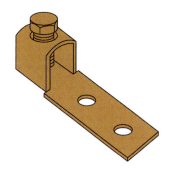
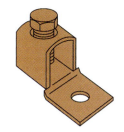

ONE BARREL, OFFSET TONGUE
ONE HOLE
NO. 14 AWG THROUGH 1,000 KCMIL

ONE BARREL, STRAIGHT TONGUE
TWO HOLE
NO. 14 AWG THROUGH 1,000 KCMIL

ONE BARREL, FIXED TONGUE
ONE HOLE
NO. 14 AWG THROUGH 500 KCMIL

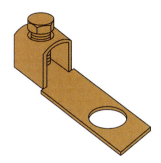
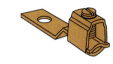

ONE BARREL, STRAIGHT TONGUE
ONE HOLE
NO. 14 AWG THROUGH 1,000 KCMIL

SINGLE HOLE
NO. 14 AWG THROUGH 4/0

ONE BARREL, OFFSET TONGUE
TWO HOLE
NO. 14 AWG THROUGH 1,000 KCMIL

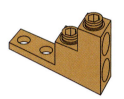

TWO HOLE, PANELBOARD CONNECTOR
NO. 2 AWG THROUGH 750 KCMIL

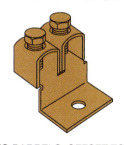
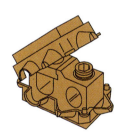

TWO BARRELS, OFFSET TONGUE
ONE HOLE
NO. 6 AWG THROUGH 500 KCMIL

PARALLEL-TAP CONNECTOR
WITH INSULATED COVER
(VARIOUS WIRE SIZE COMBINATIONS)

SPLIT BOLT CONNECTOR
(2) NO. 14 AWG THROUGH (2) 1,000 KCMIL
RUN AND TAP COMBINATIONS

Figure 11 ◆ Various mechanical compression connectors.

3.1.0 Aluminum Connections

Aluminum has certain properties that are different from copper and special precautions must be taken to ensure reliable connections. These properties are: cold flow (aluminum does not spring back in the same way as copper and has a tendency to deform), coefficient of thermal expansion, susceptibility to galvanic corrosion, and the formation of oxide film on the surface.

Because of the thermal expansion and cold flow of aluminum, standard copper connectors cannot be safely used on aluminum wire. Most manufacturers design their aluminum connectors with greater contact area to counteract this problem. Tongues and barrels of all aluminum connectors are larger or deeper than comparable copper connectors.

The electrolytic action between aluminum and copper can be controlled by plating the aluminum with a neutral metal (usually tin). The plating prevents electrolysis from taking place, and the joint remains tight. Connectors should also be tin-plated and prefilled with an oxide-inhibiting compound.

The insulating aluminum oxide film must be removed or penetrated before a reliable aluminum joint can be made. Aluminum connectors are designed to bite through this film as they are applied to conductors. The conductor should also be wire brushed and coated with a joint compound to ensure a reliable joint.

CAUTION

Never use connectors designed strictly for use on copper conductors on aluminum conductors. Connectors listed for use on both metals will be marked AL-CU. All connectors must be applied and installed in the manner for which they are listed and labeled.

NEC Section 110.14 prohibits conductors made of dissimilar metals (copper and aluminum, copper and copper-clad aluminum, or aluminum and copper-clad aluminum) from being intermixed in a terminal or splicing connector unless the device is identified for the purpose. As a general rule:

- Connectors marked with only the wire size should only be used with copper conductors.
- Connectors marked with AL and the wire size should only be used with aluminum conductors.
- Connectors marked with **AL-CU** and the wire size may be safely used with either copper or aluminum.

3.2.0 Heat-Shrink Insulators

Heat-shrink insulators for small connectors provide skintight insulation protection and are fast and easy to use. They are designed to slip over wires, taper pins, connectors, terminals, and splices. When heat is applied, the insulation becomes semi-rigid and provides positive strain relief at the flex point of the conductor. A vapor-proof band seals and protects the conductor against abrasion, chemicals, dust, gasoline, oil, and moisture. Extreme temperatures, both hot and cold, will not affect the performance of these insulators. The source of heat can be any number of types, but most manufacturers of these insulators also produce a heat gun especially designed for use on heat-shrink insulators. It is similar to a conventional hair dryer, as shown in *Figure 12*.

In general, a heat-shrink insulator may be thought of as tubing with a memory. After it is ini-

HEAT GUN

SLIP INSULATOR OVER OBJECT TO BE INSULATED, THEN APPLY HEAT FOR A FEW SECONDS

WHEN FINISHED, IT PROVIDES PERMANENT INSULATION PROTECTION

208F12.EPS

Figure 12 ♦ Method of installing heat-shrink insulators.

tially manufactured, it is heated and expanded to a predetermined diameter and then cooled. Upon a second application of heat, the tubing compound "remembers" its original size and shrinks to that smaller diameter. This property enables it to conform to the contours of any object. Various types of heat-shrink tubing are available, some of which include the following:

- *PVC* – This is a general-purpose, economical tubing widely used in the electronics industry. It provides good electrical and mechanical protection and resists cracking and splitting.
- *Polyolefin* – Polyolefin tubing has a wide range of uses for wire bundling, harnessing, strain relief, and other applications where cables and components require additional insulation. It is flame-retardant, flexible, and comes in a variety of colors.

- *Double wall* – This tubing is designed for outstanding protective characteristics. It is a semi-rigid tubing with an inner wall that melts and an outer wall that shrinks to conform to the melted area.
- *Teflon®* – This type is considered by many users to be the best overall heat-shrink tubing—physically, electrically, and chemically. Its high temperature rating of 250°C resists brittleness from extended exposure to heat and will not support combustion.
- *Neoprene* – This is a highly durable and flexible tubing that provides superior protection against abrasion.
- *Kynar®* – This is a thin-wall, semi-rigid tubing with outstanding resistance to abrasion. This transparent tubing enables easy inspection of components.

Tubing is available in a wide variety of colors and configurations. The manufacturer's tubing selector guide can help in the selection of the best tubing for any given application. A typical tubing selector guide appears in *Table 2*.

Table 2 Tubing Selector Guide

Type	Material	Temp. Range (°C)	Shrink Ratio	Max. Long. Shrinkage (%)	Tensile Strength (psi)	Colors	Dielectric Strength (V/mil)
Nonshrinkable	PVC	+105	—	—	2,700	White, red, clear, black	800
Shrinkable	PVC	−35 to +105	2:1	10	2,700	Clear, black	750
Nonshrinkable	Teflon®	−65 to +260	—	—	2,700	Clear	1,400
Shrinkable	Flexible polyolefin	−55 to +135	2:1	5	2,500	Black, white, red, yellow, blue, clear	1,300
Nonshrinkable	Teflon®	−65 to +260	—	—	7,500	Clear	1,400
Shrinkable	Polyolefin double wall	−55 to +110	6:1	5	2,500	Black	1,100
Shrinkable	Kynar®	−55 to +175	2:1	10	8,000	Clear	1,500
Shrinkable	Teflon®	+250	1.2:2	10	6,000	Clear	1,500
Shrinkable	Teflon®	+250	1½:1	10	6,000	Clear	1,500
Shrinkable	Neoprene	+120	2:1	10	1,500	Black	300

4.0.0 ◆ CONTROL AND SIGNAL CABLE

For the most part, electricians install all the associated power cables and many of the signal cables for the instrumentation systems in industrial facilities. It is necessary to select the proper type and rating of control and signal cables to meet *National Electrical Code®* (*NEC®*) requirements, local codes, and equipment manufacturer requirements for the system being installed. With the exception of optical fiber or communication cables that enter a structure and are terminated in an enclosure within 50' of the entrance, the *NEC®* requires that control and signal cables be listed and marked with an appropriate classification code.

4.1.0 *NEC®* Classifications and Ratings

Control and signal cables are type-classified, rated, and listed for use in various areas of a structure in accordance with the following:

- *NEC Article 725*, Remote Control, Signaling, and Power-Limited Circuits (Class 1, 2, and 3 Circuits)
- *NEC Article 727*, Instrumentation Tray Cable
- *NEC Article 760*, Fire Alarm Systems
- *NEC Article 770*, Optical Fiber Cables and Raceways
- *NEC Article 800*, Communications Circuits

All cable conforming to the requirements of the *NEC®* is normally marked by the cable manufacturer with the appropriate classification code. For more detail, refer to the latest edition of the *NEC®*.

4.2.0 Fire Alarm, Instrumentation, and Control Cable

Fire alarm, instrumentation, and control cable carry low-level signals that require less current than power cable. Conductor sizes range from No. 12 to No. 24 AWG, depending on the circuit requirements. Conductors can be tinned or bare, solid or stranded, and twisted or parallel. Shielding, if used, is usually an aluminum-coated Mylar® film with one or more drain wires (ground wires) inside a jacket. The jacket and conductor insulation are rated for the applicable usage. *Figure 13* shows typical fire alarm and instrumentation cable.

To be effective, the shielding drain wire(s) must be grounded. The installation loop diagrams or instructions for the system equipment will indicate how and where the shields are connected and grounded. Typically, only one end is grounded, and the drain wire at the other end is isolated by folding it back and taping over it. This prevents the electrical noise collected by the shield from recirculating through the system and interfering with the control signals.

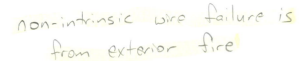
non-intrinsic wire failure is from exterior fire

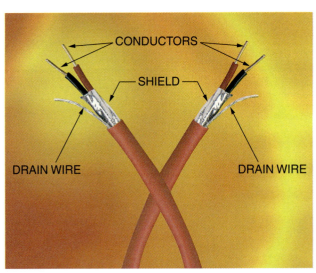
(A) FIRE ALARM CABLE

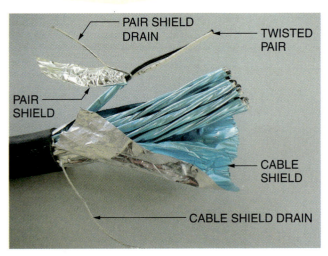
(B) INSTRUMENTATION CABLE

Figure 13 ◆ Fire alarm and instrumentation cable.

5.0.0 ♦ LOW-VOLTAGE CONNECTORS AND TERMINALS

Low-voltage control circuits typically use compression-type crimp connectors. These connectors are often color coded by wire size.

5.1.0 Crimp Connectors for Screw Terminals

Compression-type connectors for connecting conductors to screw terminals for low-voltage circuits include those in which hand tools indent or crimp tube-like sleeves that hold a conductor. Proper crimping action changes the size and shape of the connector and deforms the conductor strands enough to provide good electrical conductivity and mechanical strength.

Figure 14 shows the basic structure of a crimp connector. The crimp barrel receives the wire and is crimped to secure it in place. The V's or dimples inside the barrel improve the wire-to-terminal conductivity and also increase the termination tensile strength. Most crimp connectors have nylon or vinyl insulation covering the barrel to reduce the possibility of shorting to adjacent terminals. The insulation is color coded according to the connector's wire range to reduce the problem of wire-to-connector mismatch. An inspection hole is provided at the end of the barrel to allow visual inspection of the wire position. For the smaller wire sizes, a sleeve is crimped over the conductor insulation in the process of crimping the barrel, providing strain relief for the conductor.

The barrel is connected to the terminal tongue, which physically connects the wire to the termination point, such as a terminal screw. Information about the connector size and conductor range is usually stamped on the tongue by the manufacturer. Tongue styles vary depending on termination requirements. *Figure 15* shows standard tongue styles. The styles most frequently used are the ring tongue and flanged or locking fork. These types are preferred because the terminals will not slip off the terminal screw as the screw is tightened. They are also compatible with most vendor-supplied termination points.

5.2.0 Color Codes

Most manufacturers color code the barrel insulation to provide quick identification for installation and as an aid to inspection. Different colors, or a combination of colors, have special meanings. Although manufacturers vary, common or standard colors have been accepted. *Table 3* lists typical color codes.

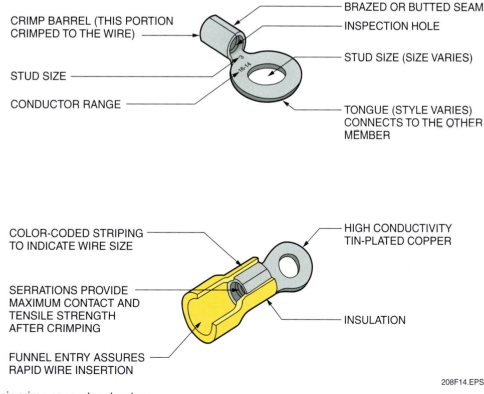

Figure 14 ♦ Basic crimp connector structure.

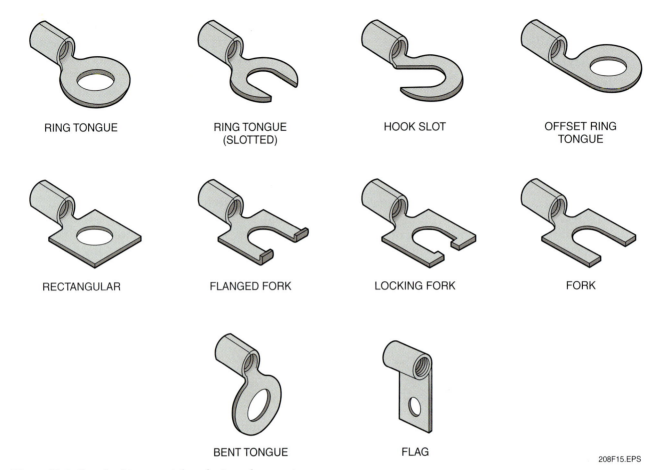

Figure 15 ◆ Standard tongue styles of crimped connectors.

Table 3	Typical Color Codes
AWG Wire Size	Color Code
22–16	Red
16–14	Blue
12–10	Yellow

Color combinations are sometimes varied to indicate the class or grade rating of an individual lug or splice. Some manufacturers use a clear plastic or other suitable insulation on the crimp barrel with a colored line to indicate wire size range.

6.0.0 ◆ GUIDELINES FOR INSTALLING CONNECTORS

Before beginning work on incoming line connections, refer to all drawings and specifications dealing with the project at hand. Details of terminations are usually furnished on larger installations. Guidelines for installing the different types of connectors covered in this module are given in the following sections.

6.1.0 Installing Compression Connectors

The task of fastening a compression connector to a wire requires the use of the proper connector, crimping tool, and installation procedure. Always use the correct tool for the connector and follow the manufacturer's instructions.

6.1.1 Crimping Tools

With a compression connector, an electrical connection between a wire and a terminal can be made by tightly compressing the crimp barrel with an ordinary pair of pliers. However, such a connection would not necessarily be made to the required pressure or in the correct location to ensure a good connection. A crimping tool is required to produce consistently good connections.

Figure 16 shows the relationship between the amount of crimping force and the mechanical and electrical performance. The maximum mechanical strength (A) occurs at a lower crimping force than the maximum electrical performance (B). The point of intersection (C) represents the ideal

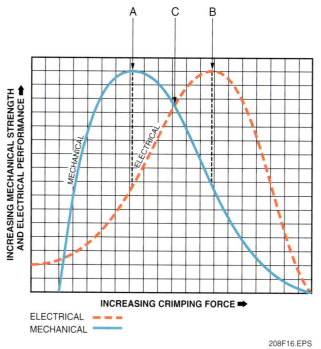

Figure 16 ♦ Mechanical strength versus electrical performance of a crimped connector.

crimping force. Using a crimping die that is too large results in poor electrical performance, and using a die that is too small produces a weak mechanical connection.

A simple pliers-type crimping tool was the earliest crimping tool developed and continues to be used for repair operations or where only a few installations are to be made. These tools are similar in construction to ordinary mechanics' pliers except the jaws are specially shaped, as shown in *Figure 17*.

It takes a tremendous amount of force at the lug to make a good crimp. Ratchet tools provide a means of increasing the human output force. The typical force capability of a hand for repetitive operations is 75 pounds for adult men and 50 pounds for adult women. The amount the tool multiplies the hand force is termed the **mechanical advantage (MA)** of the tool.

Simple pliers are basically constant-MA tools; the MA is the same whether the crimp is being started or finished. By adjusting linkage or cam mechanisms connecting the handles to the crimp dies, the MA can be varied so that it is low at the start and high at the finish of the crimp stroke. Thus, when the handles start to close from the open position and little or no crimping is being done, the MA is low. As the handles are closed farther, the crimp dies begin to compress the terminal, and the MA increases. In this manner, the MA is patterned to the crimp force requirements and distributed so that a high MA is achieved over the portion of the cycle where required.

Figure 18 shows a high MA type of tool equipped with a ratchet control. The ratchet mechanism prevents opening the tool and removing the crimped terminal before the handle has been closed all the way and the crimp completed. These provide a consistent, reliable crimp that meets the terminal manufacturer's requirements.

The crimp dies of this tool are interchangeable and may contain two or three positions for crimping different size terminals. These dies may be color coded to be used with a color-coded terminal lug for easy identification and to ensure proper crimping force. The crimp die of the tool determines the completed crimp configuration. There are a variety of configurations in use such as the simple nest and indenter type of die or the more complicated four-indent die.

Figure 19 shows hand-operated and hydraulic crimping tools typical of those used to crimp connectors for stranded wires ranging from No. 8 AWG to 750 kcmil. These tools normally develop about 12 tons of compression force at 10,000 pounds per square inch (psi). Guidelines for the use of these tools are described here. Many other tools operate in a similar manner.

Figure 17 ♦ Hand crimpers.

Figure 18 ♦ Leveraged crimping tool.

MODULE 26208-08 ♦ CONDUCTOR TERMINATIONS AND SPLICES 8.15

DIE SET

HAND-OPERATED

HYDRAULIC

208F19.EPS

Figure 19 ♦ Crimping tools used to crimp large connectors.

 NOTE
Some manufacturers design a crimping tool that must be used with their connectors. Use the crimping tool recommended by the manufacturer.

Several different configurations may work equally well for some applications, while for others, a certain shape is superior. Many considerations affect the determination of crimp die configuration, including the type of terminal (size, shape, material, and function), as well as the type and size of wires to be accommodated.

To use a hand-operated crimping tool, proceed as follows:

Step 1 Select the proper die for use with the connector to be crimped. Do not operate the tool without the die.

Step 2 Push the die release button on the C-head and slide one of the die halves into position until the retainer snaps. Insert the other die half in the piston body by pushing the die release button and sliding the die in until the retainer snaps.

Step 3 Place the tool C-head in position over the connector to be crimped. Pump the handle until compression is complete, as indicated by the dies touching at their flat surfaces nearest the throat of the C-head.

Step 4 Retract the ram and remove the connector after completion of the crimp. This is done by raising the pump handle slightly, rotating it clockwise until it stops, and then pushing the handle down in a pumping motion until the pressure release snaps.

 WARNING!
Always read and follow the manufacturer's instructions when using power tools.

To operate a hydraulic crimping tool, proceed as follows:

Step 1 Connect the hydraulic pump to the crimping tool using a suitable hydraulic hose.

Step 2 Select the proper dies for use with the connector to be crimped. Do not operate the tool without the dies.

Step 3 Push the die release button on the C-head and slide one of the die halves into position until the retainer pin snaps. In a similar manner, install the other die half in the piston body.

Step 4 Place the tool C-head in position over the connector to be crimped. Operate the remote pump until compression is complete, as indicated by the dies touching on the frame side.

Step 5 Release the pressure at the hydraulic pump to retract the lower die half, and then remove the connector from the tool.

In addition to hand-operated and hydraulic crimpers, battery-operated and corded crimping tools are also available (*Figures 20* and *21*). Battery-operated tools offer the advantage of freedom of movement, while corded tools allow extensive use without worrying about battery changes. Universal crimping tools are also available (*Figure 22*). They operate in the same way as other crimpers but do not require separate dies.

Figure 20 ◆ Battery-operated crimping tool.

Figure 21 ◆ Corded crimping tool.

Figure 22 ◆ Universal crimping tool.

6.1.2 General Compression (Crimp) Connector Installation Procedure

To make a crimp connection, proceed as follows:

Step 1 Select a crimp connector of proper size and appropriate material for the wire size you are using. Copper connectors should be used with copper wires and aluminum connectors with aluminum wires. Dual-rated connectors may be used with both copper and aluminum wires.

Step 2 Use a suitable wire stripper to remove the insulation from the end of the wire. Be careful not to nick the wire. Strip the insulation back far enough so that the bare conductor will go fully into the connector. Make sure not to strip off too much insulation; it should fit close to the connector when the wire is fully inserted into the connector.

Step 3 Clean the stripped portion of the wire. Use a wire brush for large wire sizes. Also clean the related unplated terminal pad and the surface to which the connector will be attached.

Step 4 Obtain the crimping tool and dies made for the type and size of connector to be crimped.

Step 5 Insert the stripped end of the wire completely into the connector. Position the crimping tool in place over the connector, and then operate the tool to fully crimp the connector, as directed in the tool manufacturer's instructions. Make sure that the crimping tool jaws are fully closed, indicating that a full compression crimp has been made. If multiple crimps are required (*Figure 23*), crimp from the lug back to the barrel base, rotating the crimper as necessary to avoid deforming the connector.

Step 6 Using a bolt or screw and washers (if required), secure the crimped connector and attached wire to the correct terminal in the equipment. Tighten the terminal bolt and torque to the level specified by the equipment manufacturer. Too little or too much torque can adversely affect the performance of the connection.

Table 4 lists some torque values typical of those used for tightening common sizes of steel and aluminum terminal bolts.

Figure 23 ◆ Multiple crimps.

Table 4 Recommended Tightening Torques for Various Bolt Sizes

Steel Hardware		Aluminum Hardware	
Bolt Size	Recommended Torque (Inch-Pounds)	Bolt Size	Recommended Torque (Inch-Pounds)
¼–20	80	½–13	300
⁵⁄₁₆–18	180	⅝–11	480
⅜–16	240	¾–10	650
½–13	480	—	—
⅝–11	660	—	—
¾–10	1,900	—	—

6.2.0 Installing Mechanical Terminals and Connectors

The procedure for installing mechanical connectors is basically the same as that described above for compression connectors, with the following exceptions. Before installing the mechanical connector on the wire, apply an oxide-inhibiting joint compound to the conductor to prevent the formation of surface oxides once the connection is made (also apply the compound to any terminal pad). Following this, install the connector on the wire, and then tighten the connector bolt or screw to the torque level specified by the connector manufacturer. Proper torque is important. Too much torque may sever the wires or break the connector, while too little torque can cause overheating and failure.

6.3.0 Installing Specialized Cable Connectors

There are a wide variety of cables that require specially designed connectors, commonly called terminators, to secure them to equipment enclosures. Normally, these specialized connectors are supplied with installation instructions. This section will introduce you to one type of specialized connector designed for use with metal-clad (Type MC) cable. Its construction and installation are typical of many specialized connectors.

Type MC cable is a factory assembly of one or more insulated circuit conductors with or without optical fiber members. It is enclosed in a metallic sheath of interlocking tape or a smooth or corrugated tube. Throughout the industry, there is increasing use of Type MC cable installed in trays and on racks instead of non-armored cable in conduit. At the appropriate locations, the cables are routed from the trays or racks, then along the structure to the various items of equipment. *NEC Article 330* governs the installation of Type MC cable. There are several types of connectors that can be used with Type MC cable. The specific connector used is determined by the size and type of cable and the application.

For the purpose of an example, the procedure for installing one manufacturer's weatherproof

INSIDE TRACK — Tightening Torques

Many terminations and types of equipment are marked with tightening torque information. For items of equipment, torque information is often marked on the equipment and/or listed in the manufacturer's installation instructions. When specific torque requirements are given, always follow the manufacturer's recommendations.

In cases where no tightening requirements are given, guidelines for tightening screw and bolt-type mechanical compression connectors can be found in the *National Electrical Code® Handbook* comments pertaining to *NEC Section 110.14*. Other good sources of tightening information are manufacturers' catalogs and product literature.

connector designed for use with Type MC cable is given here:

Step 1 Select the correct connector size. This is normally done by comparing the physical dimensions of the cable to a cross-reference table given in the manufacturer's product literature and/or installation instructions.

Step 2 Strip back the jacket and armor of the cable as needed to meet equipment requirements. Expose the cable armoring further by stripping the cable jacket for a specified distance (L), as shown in *Figure 24*. This distance will be found in the manufacturer's installation instructions.

Step 3 Make sure that the jacket seal and retaining spring are in their uncompressed state. If necessary, loosen the connector body and the compression nut. It is not necessary to separate the connector parts.

Step 4 Screw the connector body into the equipment if it has a threaded entry, or secure it with a locknut if it has an unthreaded entry.

Step 5 Pass the cable through the connector until the armor makes contact with the end stop. If it is not possible for the insulated wires to pass through the end stop, remove the end stop so that the wires can move past it and the armor can make contact with the integral end stop within the entry component.

Step 6 Tighten the connector body to compress the retaining spring and secure the armor. Normally, this is hand-tight plus one and a half full turns.

Step 7 Tighten the outer compression nut to form a seal on the cable jacket. Normally, this is hand-tight plus one full turn.

Step 8 If appropriate, terminate the individual wires contained in the cable using compression or mechanical connectors.

6.4.0 Installing Control and Signal Cables/Conductors

Cable/conductor termination first involves organizing the cables/conductors by destination and mounting or installing the appropriately sized termination panels or devices. Then the cables/conductors must be formed, supported, and dressed to length including appropriate slack. Once this has been accomplished, the cables/conductors are properly labeled and then terminated to an appropriate connection device. See *Figure 25*.

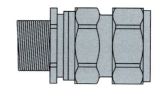
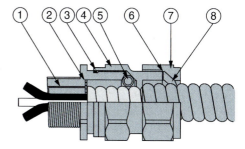

1. ENTRY COMPONENT
2. END STOP
3. O-RING
4. CONNECTOR BODY
5. RETAINING SPRING
6. WASHER
7. JACKET SEAL
8. COMPRESSION NUT

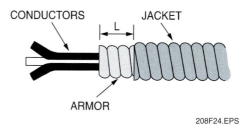

Figure 24 ◆ Weatherproof connector used with Type MC cable.

6.4.1 Crimping Procedure

Prior to making a crimped connection, verify the following items:

- The size and type of wire are correct.
- The connector and wire materials are compatible (compare cable and lug designations for material type).

NOTE

Conductors of dissimilar metals should not be intermixed in a terminal lug or splice connector unless the device is listed and labeled for this purpose.

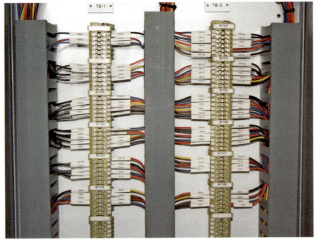

Figure 25 ♦ Labeled and terminated control cable.

- The correct crimp tool and die for the selected terminal and conductor are being used.
- The tool is in proper working order.

The installation of a compression terminal should be as follows:

Step 1 Select the proper terminal for the wire size being terminated, as specified by the terminal manufacturer. This ensures the proper fit of the lug and that the amperage capacity of the lug equals that of the conductor.

Step 2 Verify that the terminal stud size matches the terminal screw size.

Step 3 Select the proper crimping tool and die for the terminal being used. This minimizes the possibility of overcrimping or undercrimping the terminal.

Step 4 Train the wire strands if the strands are fanned. Grooming the conductor provides a proper fit in the crimp barrel.

Step 5 Insert the terminal in the proper die nest. Some insulated terminals have increased wire ranges that overlap with other size terminals, such as a terminal covering wire sizes No. 14 through 18 AWG. When crimping this type of terminal, use the middle wire the terminal will accept to determine the correct nest to use. For example, when crimping a No. 18 AWG wire in a 14–18 terminal, use the nest marked 16–14 on the tool. Overcrimping will occur if the terminal is crimped in the nest marked 18–22. When crimping an uninsulated terminal, the top of the crimp barrel should be positioned facing the indenter. This will produce the best crimp and facilitate visual inspection of the crimp after connection to the terminal point. When crimping an insulated terminal, insert the terminal in the color-coded die corresponding to the insulation color.

Step 6 Close the crimp tool handles slightly to secure the terminal while inserting the conductor.

Step 7 Insert the stripped wire into the barrel of the terminal until the insulation butts firmly against the forward stop. Make sure that the insulation is not actually in the barrel. The strands of the conductor should be clearly visible through the inspection opening of the terminal. Make sure all strands of the conductor are inserted into the barrel of the terminal.

Step 8 Complete the crimp by closing the handles until the mechanical cycle has been completed and the ratchet releases.

Step 9 Remove the termination from the tool and examine it for proper crimping.

6.4.2 Termination Inspection

The final step of the crimping procedure is inspecting the termination to ensure that it is electrically and mechanically sound. It must be tight with no loose strands or uninsulated conductor showing. Inspect uninsulated terminals for correct positioning, centering, and size of the crimp indent. Ensure that the terminal wire size stamped on each terminal matches the conductor that was crimped.

Acceptable and unacceptable positioning of a crimp indent is illustrated in *Figure 26*. The crimp barrel is designed to provide the best mechanical

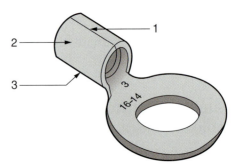

1 – ACCEPTABLE – INDENT ON SEAM (TOP)
2 – UNACCEPTABLE – INDENT ON SIDE
3 – UNACCEPTABLE – INDENT ON BOTTOM

Figure 26 ♦ Indent position.

strength when the indent is properly placed on the top (seam) of the barrel. An indent on the side can split the terminal seam, thereby reducing both the electrical and mechanical qualities of the termination.

The centering of a crimp indent is very important. A poor connection will result if a crimp is placed over either the belled mouth or the inspection hole of the terminal, as shown in *Figure 27*. A crimp over the belled mouth will compress the conductor insulation and result in poor or no electrical continuity. Crimping over the inspection hole reduces both electrical continuity and the holding capability of the terminal lug.

Crimping with an incorrect die changes the electrical and mechanical qualities of the termination. The crimp barrel should not be excessively distorted or show any cracks, breaks, or other damage to the base metal.

The position of the conductor is also very important. Tongue terminals should have the end of the conductor flush with and extending beyond the crimp barrel no more than $\frac{1}{32}$" to ensure proper crimp-to-wire contact area, as shown in *Figure 28*. Conductors extending more than $\frac{1}{32}$" past the inspection hole may interfere with the terminal screw. If a conductor is too short and does not reach the inspection hole, the connection will not provide enough contact surface area, which increases current density and may cause the lug to slip off the conductor.

Terminal lugs with a belled mouth must have the conductor insulation butted against the tapered edge of the crimp barrel, as shown in *Figure 28*. Terminal lugs without a belled mouth should not have any exposed conductor. The conductor insulation is butted to the terminal. Any exposed conductor reduces the overall strength of the connection.

Insulated terminals should be inspected in the same manner as uninsulated terminals: check for proper positioning, centering, and type of crimp. In addition, inspect the terminal insulation for breaks, cracks, holes, or any other damage. Any damage to the insulation is unacceptable. As with uninsulated terminals, inspect insulated terminals to ensure the proper terminal size for each individual conductor. Terminal conductor and stud sizes are usually stamped on the terminal tongue. They can also be checked against the manufacturer's color code.

6.4.3 Terminal Block Connections

A great variety of terminal blocks are available for use, including clamp-type, spring-loaded, and screw-type terminal blocks (*Figure 29*).

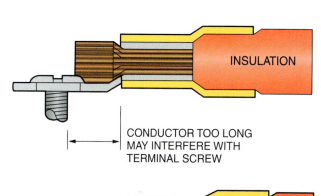

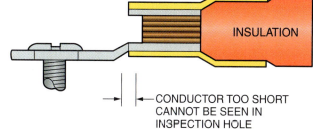

UNACCEPTABLE

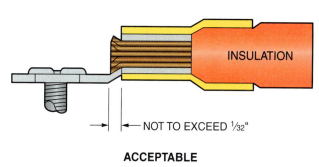

NOT TO EXCEED $\frac{1}{32}$"

ACCEPTABLE

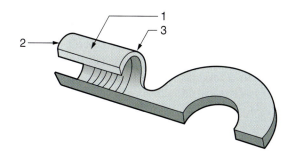

1 – ACCEPTABLE – CENTERED OVER SERRATIONS
2 – UNACCEPTABLE – OVER BELLED MOUTH
3 – UNACCEPTABLE – OVER INSPECTION HOLE

Figure 27 ♦ Crimp centering.

Figure 28 ♦ Conductor positioning.

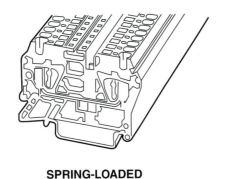
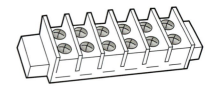
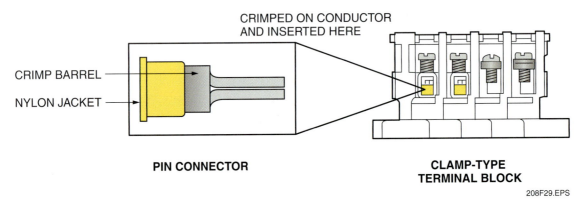

Figure 29 ♦ Terminal blocks.

- *Spring-type terminal block* – To install stripped wires in a spring-loaded terminal block, release the spring contact using a flat-blade screwdriver and insert the wire into the hole on top of the block. After the wire is inserted, remove the screwdriver. The wire is clamped in place by the spring contact.
- *Clamp-type terminal block* – To install stripped wires in a clamp-type terminal block, insert the wires into the boxed terminal and tighten the screw to clamp the wire in place.
- *Screw-type terminal block* – To install stripped wires in a common screw-type terminal block, curve the wires around the screw or fit them with a terminal lug that is inserted under the screw. Once the wire or terminal lug is under the head of the screw, tighten the screw to secure the wire.

NOTE

When installing wires in a common screw-type terminal block, only two terminals are permitted at any one terminal point. Also, only one flanged fork terminal may be used. When two types of terminals are located at one point, the bottom terminal should be installed upside down. This will provide easier installation and a neater appearance.

Care should be taken to strip the cable jacket to a point as close as possible to the first termination of the cable, but not to interfere with other terminations originating from that cable, as shown in *Figure 30*. Place the cable identification tag at that point on the cable jacket and make sure it can be read easily.

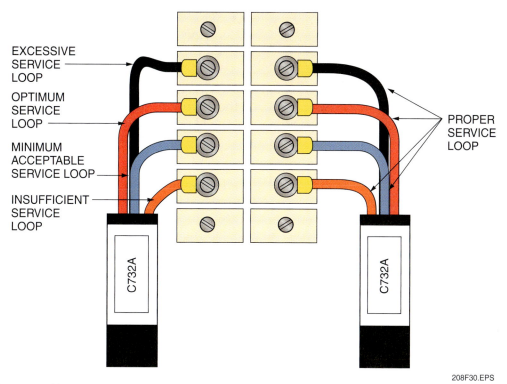

Figure 30 ♦ Routing cabling.

When multiple cables are installed, tie them neatly to a support without blocking access to the lower terminal blocks or interfering with the connection or disconnection of other wires.

Individual or multi-conductor cables should be routed parallel or at right angles to the frame or wireways. Take care to ensure that wires do not come in contact with sharp edges. Position shielded and coaxial cables on the outer perimeter of a cable bundle whenever possible and keep wire crossover to a minimum.

6.4.4 Cable/Conductor Routing and Inspection Considerations

Cable/conductor routing and inspection considerations include the following:

- *Wire bends* – For all wire and cable routing, use a minimum bend radius of three times the outside diameter of the wire or cable unless otherwise specified by the cable manufacturer.
- *Neatness of routing* – Provide adequate clearance for movement of mechanical parts. Allow enough slack in the cable for servicing plugs and terminal boards. Support all cables to prevent strain on the conductors or terminals. Strain can cause breaks in the terminations and render the system inoperable. Ensure that cables do not cross maintenance openings or otherwise interfere with normal operation and replacement of components.
- *Protection of wires or cable* – Provide clearance between the conductors and any heat-radiating components to prevent heat-related deterioration of the insulation. Protect all wires against abrasion. Do not route wire over sharp screws, lugs, terminals, or openings. Use grommets, chase nipples, or similar means to protect the insulation when passing wire through a metal partition.
- *Service loops* – Allow sufficient slack at the termination of each wire to permit one repair (such as cutting off a terminal and re-terminating it).
- *Jumper wires* – Bare jumper wires are not permitted. Use only insulated wire for jumpers.
- *Terminal bending limits* – Do not bend terminals more than 30 degrees above or below the termination point, as shown in *Figure 31*.

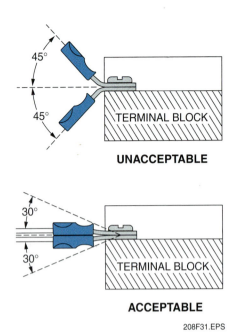

Figure 31 ◆ Terminal bend radius.

7.0.0 ◆ BENDING CABLE AND TRAINING CONDUCTORS

Training is the positioning of cable so that it is not under tension. Bending is the positioning of cable that is under tension. When installing cable or any large conductors, the object is to limit the tension so that the cable's physical and electrical characteristics are maintained for the expected service life. Training conductors, rather than bending them, also reduces the tension on lugs and connectors, extending their service life considerably.

All bends made in cable must comply with the NEC®. The minimum bending radius is determined by the cable diameter and, in some instances, by the construction of the cable. For example, NEC Section 330.24 states that bends in Type MC cable must be made so that the cable is not damaged, and the radius of the curve of the inner edge of any bend shall not be less than the following:

- *Smooth sheath* – Ten times the external diameter of the metallic sheath for cable not more than ¾" in external diameter, twelve times the external diameter of the metallic sheath for cable more than ¾" but not more than 1½" in external diameter, and fifteen times the external diameter of the metallic sheath for cable more than 1½" in external diameter.
- *Interlocked-type armor or corrugated sheath* – Seven times the external diameter of the metal sheath.
- *Shielded conductors* – Twelve times the overall diameter of the individual conductors or seven times the overall diameter of the multi-conductor cable, whichever is greater.

Two common types of cable bending tools are the ratchet bender and the hydraulic bender. The ratchet cable bender in *Figure 32* bends 600V copper or aluminum conductors up to 500 kcmil, and the hydraulic bender in *Figure 33* is designed for cables from 350 kcmil through 1,000 kcmil. In addition, the hydraulic bender is capable of one-shot bends up to 90° and automatically unloading the cable when the bend is finished. Either type simplifies and speeds cable installation.

Conductors at terminals or conductors entering or leaving cabinets or cutout boxes and the like must comply with certain NEC® requirements, many of which are covered in NEC Article 312. The bending radii for various sizes of conductors that do not enter or leave an enclosure through the wall opposite its terminal are shown in *Table 5*. When using this table, the bending space at terminals must be measured in a straight line from the end of the lug or wire connector (in the direction that the wire leaves the terminal) to the wall, barrier, or obstruction, as shown in *Figure 34*.

An unshielded cable can tolerate a sharper bend than a shielded cable. This is especially true of cables having helical metal tapes, which, when bent too sharply, can separate or buckle and cut into the insulation. This causes increased electrical

Figure 32 ◆ Ratchet bender.

Figure 33 ♦ Hydraulic bender.

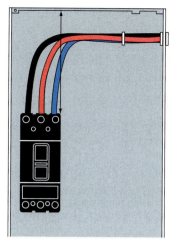

When using *NEC Table 312.6(A)*, bending space at terminals must be measured in a straight line from the end of the lug or wire connector (in the direction that the wire leaves the terminals) to the wall, barrier, or obstruction.

Figure 34 ♦ Bending space at terminals is measured in a straight line.

Table 5 Minimum Wire Bending Space for Conductors Not Entering or Leaving Opposite Wall [Data from *NEC Table 312.6(A)*]

AWG or Circular-Mil Size of Wire	Wires per Terminal				
	1	2	3	4	5
14–10	Not Specified	—	—	—	—
8–6	1½	—	—	—	—
4–3	2	—	—	—	—
2	2½	—	—	—	—
1	3	—	—	—	—
1/0–2/0	3½	5	7	—	—
3/0–4/0	4	6	8	—	—
250 kcmil	4½	6	8	10	—
300–350 kcmil	5	8	10	12	—
400–500 kcmil	6	8	10	12	14
600–700 kcmil	8	10	12	14	16
750–900 kcmil	8	12	14	16	18
1,000–1,250 kcmil	10	—	—	—	—
1,500–2,000 kcmil	12	—	—	—	—

Reprinted with permission from *NFPA 70*, the *National Electrical Code*®. Copyright © 2007, National Fire Protection Association, Quincy, MA 02269. This reprinted material is not the complete and official position of the National Fire Protection Association on the referenced subject, which is represented only by the standard in its entirety.

stress. The problem is compounded by the fact that most tapes are under jackets that conceal such damage. Damaged cable may initially pass acceptance testing, but often fails prematurely at the shield/insulation interface.

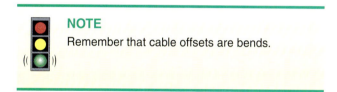

NOTE

Remember that cable offsets are bends.

When conductors enter or leave an enclosure through the wall opposite its terminals (*Figure 35*), *NEC Table 312.6(B)* applies. See *Table 6*. When using this table, the bending space at terminals must be measured in a straight line from the end of the lug or wire connector in a direction perpendicular to the enclosure wall. For removable and lay-in wire terminals intended for only one wire, the bending space in the table may be reduced by the number of inches shown in parentheses.

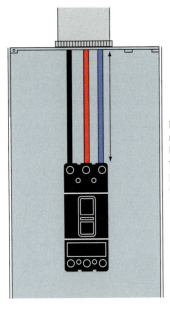

Figure 35 ♦ Conductors entering an enclosure opposite the conductor terminals.

Bending space at terminals must be measured in a straight line from the end of the lug or wire connector in a direction perpendicular to the enclosure wall. Use the values in *NEC Table 312.6(B)*.

8.0.0 ♦ *NEC®* TERMINATION REQUIREMENTS

There are many *NEC®* requirements governing the termination of conductors as well as the installation of enclosures containing conductors. *NEC Sections 110.14 and 312.6* cover most installations and terminations. However, other sections, such as *NEC Sections 300.15 and 430.10*, also apply for specific applications.

8.1.0 Overcurrent Protection

In general, all ungrounded conductors from a transformer secondary require overcurrent protection to comply with *NEC Section 240.21*. The conductors from the transformer secondary are the feeder to the service. The secondary conductors are protected by the transformer primary protection (assuming tap rules and other requirements are met) until the first overcurrent device (point of service) in the secondary circuit.

8.1.1 Overcurrent Protection from Upstream Devices

Motor control centers that are fed from protective devices in a switchboard or other switchgear are not required to have a main breaker or disconnect switch in the motor control center (MCC). *Figure 36* shows feeders from a 480V switchboard terminating in the main lugs only (MLO) incoming line compartment of an MCC. This connection is made directly to the horizontal bus system distributing power to the vertical bus in each vertical section of the MCC.

8.1.2 Overcurrent Protection within Equipment

Figure 37 shows a circuit interrupter and fuses serving as the main disconnect and overcurrent protection for an MCC fed directly from a transformer secondary. It also illustrates the *NEC®* requirements for conductors entering enclosures and the use of enclosures for routing or tapping conductors. The load side of the overcurrent protective device is connected to the MCC busbar system. Where the main disconnect is rated at 400A or less, the load connection may be made with stab connections to vertical bus sections connecting to the main horizontal bus.

Table 6 Minimum Wire Bending Space for Conductors Entering or Leaving Opposite Wall [Data from *NEC Table 312.6(B)*]

AWG or Circular-Mil Size of Wire	Wires per Terminal			
	1	2	3	4 or More
14–10	Not Specified	—	—	—
8	1½	—	—	—
6	2	—	—	—
4	3	—	—	—
3	3	—	—	—
2	3½	—	—	—
1	4½	—	—	—
1/0	5½	5½	7	—
2/0	6	6	7½	—
3/0	6½ (½)	6½ (½)	8	—
4/0	7 (1)	7½ (1½)	8½ (½)	—
250	8½ (2)	8½ (2)	9 (1)	10
300	10 (3)	10 (2)	11 (1)	12
350	12 (3)	12 (3)	13 (3)	14 (2)
400	13 (3)	13 (3)	14 (3)	15 (3)
500	14 (3)	14 (3)	15 (3)	16 (3)
600	15 (3)	16 (3)	18 (3)	19 (3)
700	16 (3)	18 (3)	20 (3)	22 (3)
750	17 (3)	19 (3)	22 (3)	24 (3)
800	18	20	22	24
900	19	22	24	24
1,000	20	—	—	—
1,250	22	—	—	—
1,500–2,000	24	—	—	—

Reprinted with permission from *NFPA 70*, the *National Electrical Code®*. Copyright © 2007, National Fire Protection Association, Quincy, MA 02269. This reprinted material is not the complete and official position of the National Fire Protection Association on the referenced subject, which is represented only by the standard in its entirety.

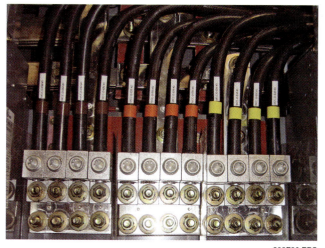

Figure 36 ♦ Incoming feeders connected to horizontal bus.

8.1.3 Short Circuit Bracing

All incoming lines to either incoming line lugs or main disconnects must be braced to withstand the mechanical force created by a high fault current. If the cables are not anchored sufficiently or the lugs are not tightened correctly, the connections become the weakest part of a panelboard or motor control center when a fault develops. In most cases, each incoming line compartment is equipped with a two-piece spreader bar located at a certain distance from the conduit entry. This spreader bar should be used along with appropriate lacing material to tie cables together where they can be bundled and to hold them apart where they must be separated. In other words, the incoming line cables should first be positioned and then anchored in place.

Manufacturers of electrical panelboards and motor control centers normally furnish detailed information on recommended methods of short circuit bracing; follow this information exactly.

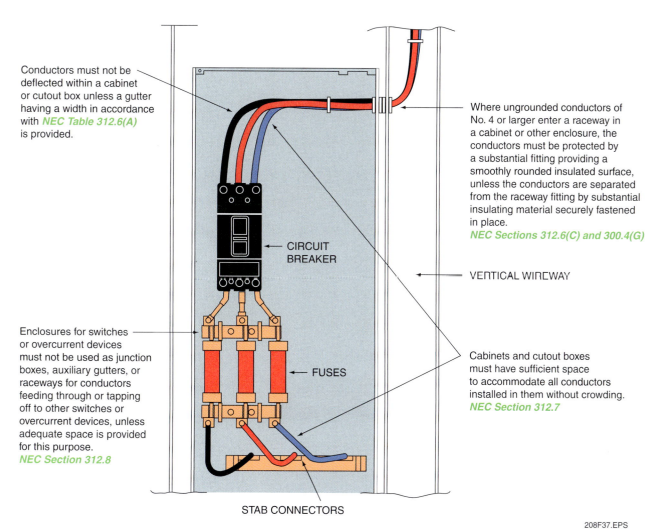

Figure 37 ♦ MCC fed directly from a transformer secondary.

MODULE 26208-08 ♦ CONDUCTOR TERMINATIONS AND SPLICES 8.27

9.0.0 ◆ TAPING ELECTRICAL JOINTS

When it is not practical to protect a spliced joint by some other means, electrical tape may be used to insulate the joint. Joints must be taped carefully to provide the same quality of insulation over the splice as over the rest of the conductors.

Various types of electrical insulating tapes are available for use in specific applications. Some common types of electrical tape include vinyl plastic tape, linerless rubber tape, high-temperature silicone rubber tape, and glass cloth tape. Electrical tapes made of vinyl plastic are widely used as primary insulation on joints made with thermoplastic-insulated wires. They are used for splices up to 600V and for fixture and wire splices up to 1,000V. Depending on the product, they are made for indoor use, outdoor use, or both.

Linerless rubber splicing tape provides for a tight, void-free, moisture-resistant insulation without loss of electrical characteristics. It is typically used as primary insulation with all solid dielectric cables through 70kV. Other applications include jacketing on high-voltage splices and terminals, moisture-sealing electrical connections, busbar insulations, and end sealing high-voltage cables.

High-temperature silicone rubber tapes are used as a protective overwrap for terminating high-voltage cables. Glass cloth electrical tapes provide a heat-stable insulation for hot-spot applications such as furnace and oven controls, motor leads, and switches. They are also used to reinforce insulation where heavy loads cause high heat and breakdown of insulation, such as in motor control exciter feeds, etc. All-metal braid tapes are also available. These are used to continue electrostatic shielding across a splice. When taping a splice, begin by selecting the correct tape for the job. Always follow the tape manufacturer's recommendations.

A general procedure describing one method of taping a splice or joint, such as encountered when connecting motor lugs, is shown in *Figure 38*. A method for taping a split-bolt connector is shown in *Figure 39*. Prior to taping, make sure the joined lugs are securely fastened together with the appropriate hardware. Use pieces of a suitable filler tape or putty to fill any voids around the lugs and hardware and eliminate any sharp edges. This also helps to provide a smooth, even surface that makes taping easier.

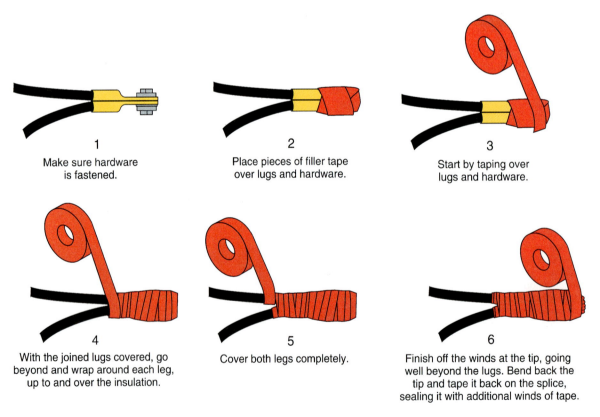

1 Make sure hardware is fastened.

2 Place pieces of filler tape over lugs and hardware.

3 Start by taping over lugs and hardware.

4 With the joined lugs covered, go beyond and wrap around each leg, up to and over the insulation.

5 Cover both legs completely.

6 Finish off the winds at the tip, going well beyond the lugs. Bend back the tip and tape it back on the splice, sealing it with additional winds of tape.

Figure 38 ◆ Typical method for taping motor lug connections.

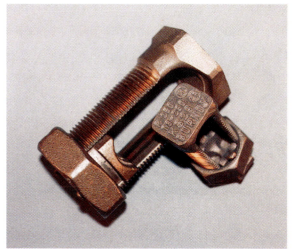

(A) SPLIT-BOLT CONNECTOR

Once the split-bolt connector has been installed and tightened securely on the conductors, cut pieces of filler tape and place over each side of the splice.

Wrap both pieces around the connector, using moderate finger pressure to shape the filler tape.

Wrap the covered connector with plastic tape.

(B) TAPING PROCEDURE

Figure 39 ◆ Typical method of taping a split-bolt connector.

Electrical Tape

Most non-electricians think of electrical tape as only the simple black vinyl variety found in nearly every home toolbox. Electrical tape actually comes in a wide range of colors to be used for labeling various conductors when making terminations.

 NOTE

For all splices and joints where it is likely that the tape will have to be removed at some future date to perform work on the joint, apply an upside down (that is, adhesive side up) wrap of tape to the joint before applying the final layers of insulating tape to the joint in the usual manner. This will keep the area free of tape residue and facilitate the removal of the tape later on, if necessary.

10.0.0 ◆ MOTOR CONNECTION KITS

Motor connection kits are available to insulate bolted splice connections, such as those in motor terminal boxes. These kits eliminate the need for taping and the use of filler tape or putty. To aid in joint reentry during rework, the insulator strips off easily, leaving a clean bolt area and thus eliminating the need to remove old tape and putty. Motor connection kits are available for use with stub (butt splice) connections (*Figure 40*) where there is insufficient room to make in-line connections. They are also made to insulate in-line splice connections where space permits. These insulating kits use a high-voltage mastic that seals the splice against moisture, dirt, and other contaminants.

One type of motor connection kit insulator is heat-shrinkable. It installs easily using heat from a propane torch to shrink the insulator in a manner similar to that of heat-shrink tubing (*Figure 41*). Another type of kit used for insulating stub connections comes in the form of an elastomeric insulating cap that is cold-applied by rolling it over the stub splice. Always follow the kit manufacturer's recommendations when selecting a kit to use for a particular application. Typically, the kit is selected based on the size of the motor feeder cable.

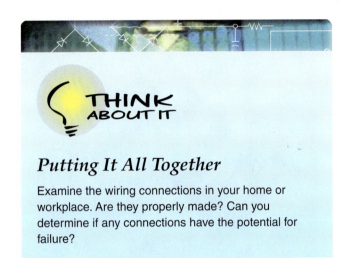

Putting It All Together

Examine the wiring connections in your home or workplace. Are they properly made? Can you determine if any connections have the potential for failure?

STUB CONNECTION

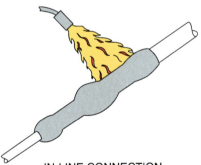

IN-LINE CONNECTION

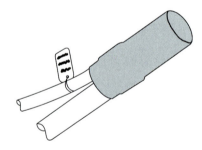

STUB ROLL-ON INSULATING
CAP CONNECTION

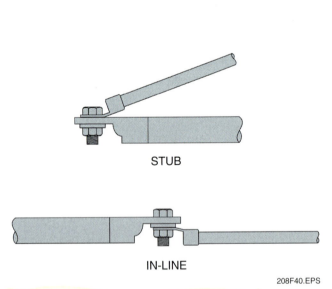

Figure 40 ◆ Stub and in-line splice connections.

Figure 41 ◆ Motor connection kits installed on splices.

Review Questions

1. A poorly made splice is acceptable as long as it provides a path for current flow.
   a. True
   b. False

2. Which of the following best describes the term stripping as it applies to conductor splicing?
   a. Removal of the insulation from conductors
   b. Removal of the packing material from the carton
   c. Removal of any excess strands of wire from a splice
   d. Removal of the pulling ring from lead sheath conductors

3. What is the diameter, in inches, of 2,000 kcmil wire?
   a. 0.89
   b. 0.99
   c. 1.29
   d. 1.63

4. All of the following are types of stripping except _____.
   a. end terminations
   b. window cuts
   c. spiral cuts
   d. indent cuts

5. Which of the following best describes the purpose of a reducing connector?
   a. To temporarily connect a circuit
   b. To join two different size conductors
   c. To join conductors of the same size
   d. To make a 90° bend in parallel conductors

6. Information about connector size and conductor range is _____.
   a. stamped on the connector tongue
   b. color coded
   c. printed on the insulation
   d. stamped on the connector barrel

7. The most frequently used tongue styles are _____.
   a. ringed tongue and locking fork
   b. slotted and flanged fork
   c. hook shot and rectangular
   d. fork and flag

8. When using a crimping tool, use _____.
   a. the maximum crimping force possible
   b. a lower mechanical force to achieve maximum electrical performance
   c. the highest force needed for the maximum mechanical strength
   d. an amount of force between optimum mechanical and electrical performance

9. Using a crimp die that is too large will _____.
   a. destroy the crimp barrel
   b. weaken the conductor
   c. provide a weak mechanical connection
   d. provide poor electrical performance

10. The crimp indent is properly placed on the _____ of the barrel.
    a. right side
    b. bottom
    c. seam
    d. left side

11. Which of the following tools is best for positioning cable?
    a. Split-bolt connectors
    b. A heat gun
    c. A power connector indenter
    d. Wire-bending tools (either ratchet or hydraulic)

Review Questions

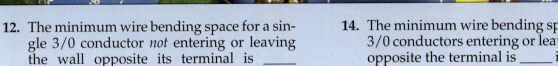

12. The minimum wire bending space for a single 3/0 conductor *not* entering or leaving the wall opposite its terminal is ____ inches.
    a. 2
    b. 3
    c. 4
    d. 5

13. The minimum wire bending space for two 1/0 conductors *not* entering or leaving the wall opposite the terminal is ____ inches.
    a. 2
    b. 3
    c. 4
    d. 5

14. The minimum wire bending space for three 3/0 conductors entering or leaving the wall opposite the terminal is ____ inches.
    a. 7
    b. 8
    c. 9
    d. 10

15. Which of the following types of electrical tape is best for taping most joints for voltages up to 600V?
    a. High-temperature silicone rubber
    b. Rubber
    c. Glass cloth
    d. Vinyl plastic

Summary

A system is only as good as the weakest link. Poor quality terminations and splices weaken a system and can result in failure or fire. Good terminations and splices are mechanically and electrically secure, well insulated, and should last as long as the conductor is in service.

Before a connection is made, the conductor must be stripped using the appropriate tool. Even minor damage to the conductor or insulation can affect the strength of the system.

Crimp connectors or terminals are used for many types of wire and cable. They are convenient for terminating conductors at terminal boards, control wiring terminals, and the like.

Lugs are provided for the larger wire sizes on panelboards and motor control centers. It is very important to tighten these lugs properly to provide a sound electrical connection as well as to provide short circuit bracing.

There are a variety of insulating tapes available to seal spliced joints. Always match the tape with the application.

Notes

Trade Terms Introduced in This Module

AL-CU: An abbreviation for aluminum and copper, commonly marked on terminals, lugs, and other electrical connectors to indicate that the device is suitable for use with either aluminum conductors or copper conductors.

Amperage capacity: The maximum amount of current that a lug can safely handle at its rated voltage.

Connection: That part of a circuit that has negligible impedance and joins components or devices.

Connector: A device used to physically and electrically connect two or more conductors.

Drain wire: A wire that is attached to a coaxial connector to allow a path to ground from the outer shield.

Grooming: The act of separating the braid in a coaxial conductor.

Insulating tape: Adhesive tape that has been manufactured from a nonconductive material and is used for covering wire joints and exposed parts.

Lug: A device for terminating a conductor to facilitate the mechanical connection.

Mechanical advantage (MA): The force factor of a crimping tool that is multiplied by the hand force to give the total force.

Pressure connector: A connector applied using pressure to form a cold weld between the conductor and the connector.

Reducing connector: A connector used to join two different sized conductors.

Shielding: The metal covering of a cable that reduces the effects of electromagnetic noise.

Splice: The electrical and mechanical connection between two pieces of cable.

Strand: A group of wires, usually stranded or braided.

Terminal: A device used for connecting cables.

Termination: The connection of a cable.

Additional Resources

This module is intended to be a thorough resource for task training. The following reference work is suggested for further study. This is optional material for continued education rather than for task training.

National Electrical Code® Handbook, Latest Edition. Quincy, MA: National Fire Protection Association.

CONTREN® LEARNING SERIES – USER UPDATE

NCCER makes every effort to keep these textbooks up-to-date and free of technical errors. We appreciate your help in this process. If you have an idea for improving this textbook, or if you find an error, a typographical mistake, or an inaccuracy in NCCER's Contren® textbooks, please write us, using this form or a photocopy. Be sure to include the exact module number, page number, a detailed description, and the correction, if applicable. Your input will be brought to the attention of the Technical Review Committee. Thank you for your assistance.

Instructors – If you found that additional materials were necessary in order to teach this module effectively, please let us know so that we may include them in the Equipment/Materials list in the Annotated Instructor's Guide.

Write: Product Development and Revision
National Center for Construction Education and Research
3600 NW 43rd St., Bldg. G, Gainesville, FL 32606

Fax: 352-334-0932

E-mail: curriculum@nccer.org

Craft _____ Module Name _____

Copyright Date _____ Module Number _____ Page Number(s) _____

Description _____

_____

_____

_____

_____

(Optional) Correction _____

_____

_____

_____

(Optional) Your Name and Address _____

_____

_____

Grounding and Bonding

1600 Glenarm Place, Denver Colorado

Greiner Electric won an ABC Pyramid Award for electrical contract work on this high-end downtown apartment building in Denver, Colorado.

26209-08

26209-08
Grounding and Bonding

Topics to be presented in this module include:

1.0.0	Introduction	9.2
2.0.0	Purpose of Grounding and Bonding	9.3
3.0.0	NEC® Requirements	9.3
4.0.0	Short Circuit Versus Ground Fault	9.4
5.0.0	Types of Grounding Systems	9.5
6.0.0	NEC® Requirements	9.9
7.0.0	Equipment Grounding	9.16
8.0.0	Bonding Service Equipment	9.22
9.0.0	Effective Grounding Path	9.24
10.0.0	Grounded Conductor	9.24
11.0.0	Separately Derived Systems	9.25
12.0.0	Grounding at More Than One Building	9.27
13.0.0	Systems Over 1,000 Volts	9.28
14.0.0	Testing for Effective Grounds	9.28
15.0.0	Measuring Earth Resistance	9.28
16.0.0	Three-Point Test for Single Electrode/Triad	9.33

Overview

One of the most comprehensive articles of the *National Electrical Code®* is *Article 250*, also known as the Grounding and Bonding Article. It is so comprehensive because of the important role that grounding plays in the safe operation of electrical systems.

NEC Article 250 describes two forms of grounding: system grounding and equipment grounding. System grounding is the intentional connection of one of the current-carrying conductors to a grounding electrode, while equipment grounding is the physical, conducting connection of any noncurrent-carrying metal parts to a grounding electrode. The interconnection of all noncurrent-carrying metal parts to each other is called bonding. Once the bonding system is connected to the grounding electrode, the equipment grounding system is effectively grounded.

Note: NFPF 70®, *National Electrical Code®*, and *NEC®* are registered trademarks of the National Fire Protection Association, Inc., Quincy, MA 02269. All *National Electrical Code®* and *NEC®* references in this module refer to the 2008 edition of the *National Electrical Code®*.

Objectives

When you have completed this module, you will be able to do the following:

1. Explain the purpose of grounding and bonding and the scope of *NEC Article 250*.
2. Distinguish between a short circuit and a ground fault.
3. Define the *National Electrical Code® (NEC®)* requirements relating to bonding and grounding.
4. Distinguish between grounded systems and equipment grounding.
5. Use *NEC Table 250.66* to size the grounding electrode conductor for various AC systems.
6. Explain the function of the grounding electrode system and determine the grounding electrodes to be used.
7. Define electrodes and explain the resistance requirements for electrodes using *NEC Section 250.56*.
8. Use *NEC Table 250.122* to size the equipment grounding conductor for raceways and equipment.
9. Explain the function of the main and system bonding jumpers in the grounding system and size the main and system bonding jumpers for various applications.
10. Size the main bonding jumper for a service utilizing multiple service disconnecting means.
11. Explain the importance of bonding equipment in clearing ground faults in a system.
12. Explain the purposes of the grounded conductor (neutral) in the operation of overcurrent devices.

Trade Terms

Auxiliary electrodes
Bonding
Effective ground fault path
Equipment bonding jumper
Equipment grounding conductor
Ground
Ground current
Ground grids
Ground mats
Ground resistance
Ground rod
Grounded
Grounded conductor
Grounding clip
Grounding conductor
Grounding connections
Grounding electrode
Grounding electrode conductor
Main bonding jumper
Neutral
Resistivity
Separately derived system
Short circuit
Step voltage
System grounding
Touch voltage
Ungrounded conductors

Required Trainee Materials

1. Pencil and paper
2. Appropriate personal protective equipment
3. Copy of the latest edition of the *National Electrical Code®*

209CMAP.EPS

MODULE 26209-08 ◆ GROUNDING AND BONDING 9.1

Prerequisites

Before you begin this module, it is recommended that you successfully complete *Core Curriculum; Electrical Level One;* and *Electrical Level Two*, Modules 26201-08 through 26208-08.

This course map shows all of the modules in *Electrical Level Two*. The suggested training order begins at the bottom and proceeds up. Skill levels increase as you advance on the course map. The local Training Program Sponsor may adjust the training order.

1.0.0 ◆ INTRODUCTION

The grounding system is a major part of the electrical system. Its purpose is to protect life and equipment against the various electrical faults that can occur. It is sometimes possible for higher-than-normal voltages to appear at certain points in an electrical system or in the electrical equipment connected to the system. Proper grounding ensures that the electrical charges that cause these high voltages are channeled to earth or ground before damaging equipment or causing danger to human life. Therefore, a circuit is grounded to limit the voltage on the circuit and improve overall operation of the electrical system and continuity of service.

When we refer to ground, we are talking about being connected to earth. If a conductor is connected to the earth or to some conducting body that extends the ground connection, such as a driven ground rod (electrode) or cold-water pipe, the conductor is said to be grounded. The neutral in a three- or four-wire service, for example, is intentionally grounded and therefore becomes a grounded conductor. However, a wire used to connect this neutral conductor to a grounding electrode or electrodes is referred to as a grounding electrode conductor. Note the difference in the two meanings; one is grounded, while the other is grounding.

The equipment grounding conductor is the conductive path used to connect the noncurrent-carrying metal parts of equipment, raceways, and other enclosures to the system grounded conductor, the grounding electrode conductor, or both, at the service equipment or at the source of a separately derived system. See *Figure 1*.

This module is designed to explain what a ground is and to teach proper grounding and bonding techniques. *NEC*® regulations and the testing of grounding systems will be thoroughly

Neutral Conductor Size

The grounded neutral conductor is permitted to be smaller than the hot ungrounded phase conductors only when the neutral conductor is properly and adequately sized to carry the maximum load imbalance as computed according to *NEC Sections 215.2, 220.61, and 230.42*.

It must be pointed out that per *NEC Section 250.24(C)* the grounded conductor must not be smaller than the size of the grounding electrode conductor. The purpose of the grounded conductor is not only to carry the unbalanced load, but also to provide a low-impedance path back to the source (utility) to complete a circuit and allow enough current to flow to open the overcurrent devices under ground fault conditions. This is important when you have three-phase services with little or no neutral loads to use as a basis for sizing the grounded conductor.

NEC Section 215.2 (feeders) and *NEC Section 230.42* (service-entrance conductors) reference *NEC Article 220* for minimum rating and size computation requirements. *NEC Article 220* states that the feeder neutral load must be the maximum imbalance of the load, which is the maximum net computed load between the neutral and any one ungrounded conductor.

In summary, both the neutral current, as determined in *NEC Articles 215, 220, and 230*, and the minimum grounding electrode size, as determined in *NEC Article 250*, must be used to determine the actual neutral wire size.

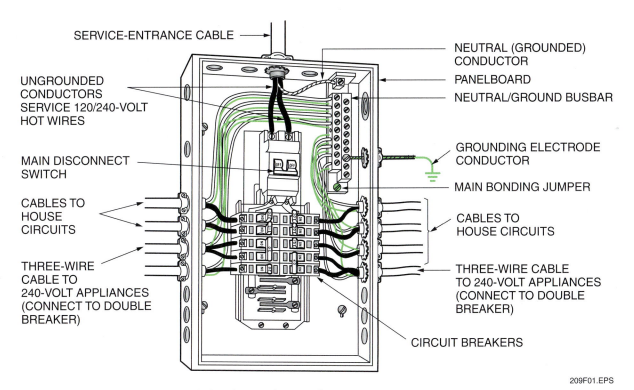

Figure 1 ♦ Panelboard showing grounded and grounding conductors.

covered. Upon completion of this module, you should have a solid foundation in the principles of grounding and bonding, one of the most important aspects of an electrical system for the protection of life and property.

2.0.0 ♦ PURPOSE OF GROUNDING AND BONDING

Systems are solidly grounded to limit the voltage to ground during normal operation and to prevent excessive voltages due to lightning and line surges. They are also grounded to stabilize the voltage to ground during normal operation. Conductive materials enclosing electrical conductors or equipment, or that form a part of the equipment, are grounded to limit the voltage to ground on these materials. The conductive materials are bonded (connected together) to establish electrical continuity and conductivity. This facilitates the operation of overcurrent devices under ground fault conditions.

3.0.0 ♦ NEC® REQUIREMENTS

NEC Article 250 is the primary governing article for the proper use and installation of grounding and bonding. This article covers general requirements for grounding and *bonding* of electrical

What's wrong with this picture?

installations. It presents the following specific requirements:

- Systems, circuits, and equipment that are required, permitted, or not permitted to be grounded
- Circuit conductor to be grounded on grounded systems
- Location of *grounding connections*
- Types and sizes of grounding and bonding conductors and electrodes
- Methods of grounding and bonding
- Conditions under which guards, isolation, or insulation may be substituted for grounding

NEC Section 250.4(A)(5) requires that the path to ground from circuits, equipment, and metal enclosures shall:

- Be permanent and continuous
- Have the capacity to safely conduct any fault current likely to be imposed on it
- Have sufficiently low impedance to facilitate the operation of the protective devices in the circuit or ground detector for high-impedance grounded systems
- Be capable of safely carrying the maximum ground fault likely to be imposed
- Not rely on the earth as an effective ground fault current path

Permanent, reliable, and continuous grounding systems are vital to the safety of electrical systems. Intermittent connections are likely to be unpredictable and may result in hazardous situations.

The minimum size grounding electrode conductor, **equipment bonding jumpers**, and equipment grounding conductor is necessary to ensure that the proper capacity to protect the system is in place. Methods to determine the proper size of these conductors will be discussed later in this module.

A low-impedance conductor is necessary because the higher the impedance, the more resistance there is to current flow. The current should flow with the least amount of resistance to ensure that personnel and equipment are protected when a fault does occur.

4.0.0 ◆ SHORT CIRCUIT VERSUS GROUND FAULT

Short circuits and ground faults are commonly misunderstood. Both faults stem from a failure of insulation resistance. It is important to have a common language for better understanding of these conditions.

4.1.0 Short Circuit

A short circuit is a conducting connection, whether intentional or accidental, between any of the conductors of an electrical system, whether it is from line-to-line or from line-to-ground. See *Figure 2*.

The failure might occur from one phase conductor to another phase conductor or from one phase conductor to the grounded conductor or neutral. The maximum value of fault current is dependent on the available capacity the system can deliver to the point of fault. The maximum value of short circuit current from line-to-ground will vary depending upon the distance from the source to the fault. The available short circuit current is further limited by the impedance of the arc where one is established, plus the impedance of the conductors to the point of short circuit.

4.2.0 Ground Fault

A ground fault, as defined in *NEC Section 250.2*, is the unintentional electrically conducting connection between the normally current-carrying conductor of an electrical circuit, and the normally noncurrent-carrying conductors, metallic enclosures, metallic raceways, metallic equipment, or earth. While not specified in the *NEC®*, an unintentional electrically conducting connection between a grounded conductor of an electrical circuit and the normally noncurrent-carrying conductors, metallic enclosures, metallic raceways, metallic equipment, or earth would also be considered a ground fault. While the first instance may result in a large amount of fault current to flow in a properly installed grounded system, the latter may result in parallel paths being formed between the grounded conductor and the grounding system. These parallel paths would result in an unwanted current flow on the grounding system. See *Figure 3*.

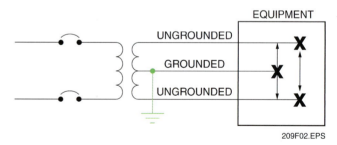

Figure 2 ◆ Short circuit.

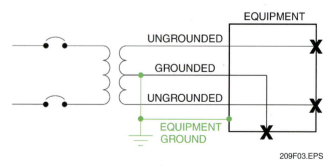

Figure 3 ◆ Ground fault.

5.0.0 ♦ TYPES OF GROUNDING SYSTEMS

The two general classifications of protective grounding are system and equipment grounding.

System grounding and bonding relates to the service-entrance equipment and its interrelated and bonded components. That is, one system conductor is grounded to limit voltages due to lightning, line surges, or unintentional contact with higher voltage lines, as well as to stabilize the voltage to ground during normal operation.

Equipment grounding conductors are used to connect the noncurrent-carrying metal parts of equipment, conduit, outlet boxes, and other enclosures to the system grounded conductor, the grounding electrode conductor, or both, at the service equipment or at the source of a separately derived system. Equipment grounding conductors are bonded to the system grounded conductor to provide a low impedance path for fault current that will facilitate the operation of overcurrent devices under ground fault conditions. *NEC Article 250* covers general requirements for grounding and bonding.

5.1.0 Single-Phase Systems

To better understand a complete grounding system, we will examine a conventional residential or small commercial system beginning at the power company's high-voltage lines and transformer, as shown in *Figure 4*. The pole-mounted transformer is fed with a two-wire 7,200V system that is transformed and stepped down to a three-wire, 120/240V, single-phase electric service suitable for residential use. A wiring diagram of the transformer connections is shown in *Figure 5*. Note that the voltage between Leg A and Leg B is 240V.

However, by connecting a third wire (neutral) on the secondary winding of the transformer between the other two, the 240V splits in half, giving 120V between either Leg A or Leg B and the neutral conductor. Consequently, 240V is available for household appliances such as ranges, hot water heaters, clothes dryers, and the like, while 120V is available for lights, small appliances, televisions, and similar appliances.

Referring again to the diagram in *Figure 5*, conductors A and B are **ungrounded conductors**, while the neutral is a grounded conductor. If only 240V loads were connected, the neutral or grounded conductor would carry no current. However, since 120V loads are present, the neutral will carry the unbalanced load and become a current-carrying conductor. For example, if Leg A carries 60A and Leg B carries 50A, the neutral conductor would carry only 10A (60 − 50 = 10). This is why the *NEC*® sometimes allows the neutral conductor in an electric service to be smaller than the ungrounded conductors.

The typical pole-mounted service entrance is normally routed by a grounded (neutral) messenger cable from a point on the pole to a point on the building being served, terminating at the service drop. Service-entrance conductors are routed

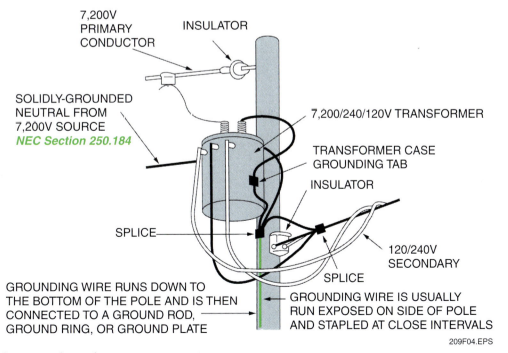

Figure 4 ♦ Pole-mounted transformer.

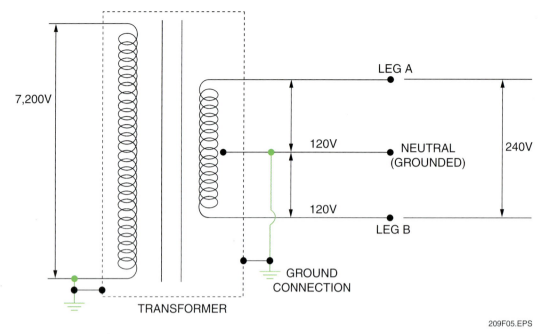

Figure 5 ♦ Wiring diagram of a 7,200V to 120/240V single-phase transformer connection.

between the meter housing and the main service switch or panelboard. This is the point where most systems are grounded—the neutral bus in the main panelboard. See *Figure 6*.

WARNING!
Exercise extreme caution when disconnecting a ground. Never grab a disconnected ground wire with one hand and the grounding electrode with the other. Your body will act as a conductor for any fault current; the results could be fatal.

5.2.0 Grounding Requirements

NEC Sections 250.20 and 250.21 provide the grounding requirements for AC systems. The *NEC®* should always be referenced when working with grounded and ungrounded systems. There are various exceptions to the requirements that will not be covered in detail in this module.

5.2.1 Systems Less Than 50 Volts

Grounding is required:

- Where supplied by transformers if the supply voltage to the transformer exceeds 150V to ground
- Where supplied by transformers if the transformer supply system is ungrounded
- Where installed as overhead conductors outside of the building
- In other circuits and systems provided they comply with the provisions of *NEC Article 250*

Figure 7 shows examples of these requirements.

5.2.2 50-Volt to 1,000-Volt Systems

AC systems of 50V to 1,000V supplying premises wiring and premises wiring systems shall be grounded under any of the following conditions:

- Where the system can be grounded so that the maximum voltage to ground on the ungrounded conductors does not exceed 150 volts
- Where the system is three-phase, four-wire, wye-connected and the neutral is used as a circuit conductor
- Where the system is three-phase, four-wire, delta-connected and the midpoint of one phase winding is used as a circuit conductor
- In other circuits and systems provided they comply with the provisions of *NEC Article 250*

Figure 8 shows examples of 50V to 1,000V grounded applications.

5.2.3 AC Systems 1kV and Over

AC systems over 1,000V must be grounded if they supply mobile or portable equipment as covered in *NEC Section 250.188*. Other AC systems over 1,000V are permitted (but not required) to be grounded.

Other circuits and systems shall be permitted to be grounded. If such systems are grounded, they shall comply with the provisions of *NEC Article 250*.

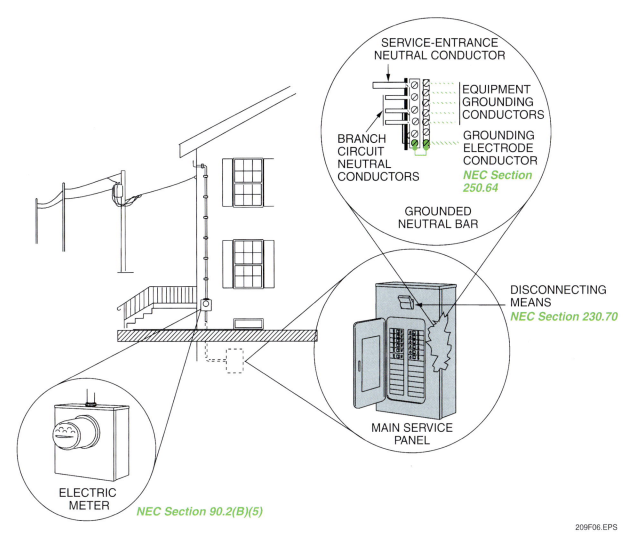

Figure 6 ◆ Typical service entrance and related service equipment.

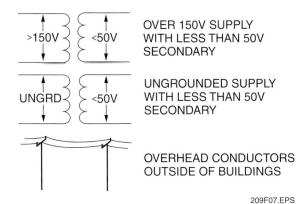

Figure 7 ◆ Systems less than 50V that must be grounded.

5.2.4 Separately Derived Systems

A separately derived system is a premises wiring system whose power is derived from a source of electrical energy other than a service. These electrical systems must be grounded if the system that is derived complies with the conditions previously described. The separately derived system must then be grounded as specified in *NEC Section 250.30*.

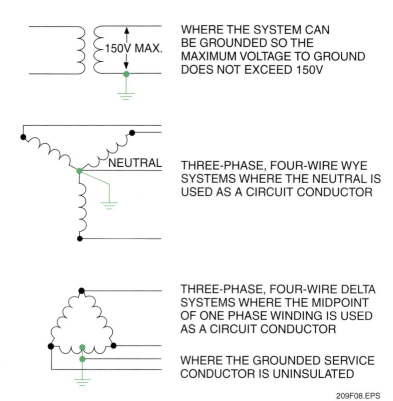

Figure 8 ♦ Systems that must be grounded.

Examples of separately derived systems include, but are not limited to:

- Transformers with no direct electrical connection between the primary and secondary
- Generator systems that supply power where the neutral is not connected to the utility system, such as for carnivals, rock crushers, or batch plants
- Generator systems used for emergency, required standby, or optional standby power that have all conductors, including a neutral, isolated from the neutral or grounded conductor of another system (usually via a transfer switch)
- AC or DC systems derived from inverters or rectifiers

Systems Permitted to Be Grounded but Not Required to Be Grounded

NEC Section 250.21 states that the following AC systems of 50V to 1,000V shall be permitted to be grounded but shall not be required to be grounded:

- Electric systems used exclusively to supply industrial electric furnaces for melting, refining, tempering, and the like
- Separately derived systems used exclusively for rectifiers that supply only adjustable-speed drives
- Separately derived systems supplied by transformers that have a primary voltage rating of less than 1,000V, provided that all of the conditions listed in *NEC Section 250.21* are met
- High-impedance grounded neutral systems as specified in *NEC Section 250.36*

5.2.5 Circuits That Cannot Be Grounded

According to *NEC Section 250.22*, only four types of circuits are not permitted to be grounded:

- Circuits for cranes that operate over combustible fibers in Class III locations, as provided in *NEC Section 503.155*
- For healthcare facilities, those isolated power circuits in hazardous (classified) locations as provided in *NEC Sections 517.61 and 517.160*
- Circuits for electrolytic cells as provided in *NEC Article 668*
- Secondary circuits on lighting systems as provided in *NEC Article 411.5(A) and 680.23(A)(2)*

> **NOTE**
> Equipment located or used within the electrolytic cell line working zone or associated with the cell line DC power circuits are not required to comply with *NEC Article 250*.

Figure 9 shows some examples of these circuits.

CRANES THAT OPERATE OVER COMBUSTIBLE FIBERS IN CLASS III LOCATIONS

ISOLATED POWER SYSTEMS IN HEALTHCARE FACILITIES FOR HAZARDOUS INHALATION ANESTHETIZING AND WET LOCATIONS

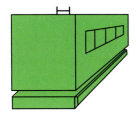
CIRCUITS FOR ELECTROLYTIC CELLS

SECONDARY CIRCUITS ON LIGHTING SYSTEMS

Figure 9 ◆ Circuits that are not permitted to be grounded.

6.0.0 ◆ NEC® REQUIREMENTS

The grounding equipment requirements established by Underwriters Laboratories, Inc. (UL) have served as the basis for approval for grounding in the *NEC®*. The *NEC®*, in turn, provides the grounding standard for the Occupational Safety and Health Administration (OSHA).

All electrical systems must be grounded and bonded in a manner prescribed by the *NEC®* to protect personnel and equipment. A grounded system must limit the voltage on the electrical system and protect it from:

- Lightning
- Voltage surges higher than that for which the circuit is designed
- An increase in maximum potential to ground due to abnormal voltages

6.1.0 Grounding Methods

The requirements for grounding of services are found in *NEC Sections 250.24(A), (B), (C), and (D)*. Methods of grounding an electric service are covered in *NEC Article 250, Part III*. In general, all of the following electrodes that are present must be bonded together to form the grounding electrode system:

- An underground metal water pipe in direct contact with the earth for no less than 10'
- The metal frame of a building as described in *NEC Section 250.52(A)(2)*
- An electrode encased by at least 2" of concrete, located within and near the bottom of a concrete foundation or footing that is in direct contact with the earth

> **NOTE**
> This electrode must be at least 20' long and must be made of electrically conductive coated steel reinforcing bars or rods of not less than ½" in diameter, or consisting of at least 20' of bare copper conductor not smaller than No. 4 AWG wire size.

- A ground ring encircling the building or structure, in direct contact with the earth at a depth not less than 2.5' below grade
- Rod, pipe, or plate electrodes
- Other local metal underground systems or structures

NOTE

This ring must consist of at least 20' of bare copper conductor not smaller than No. 2 AWG wire size. See *Figure 10*.

Grounding electrode systems used in industrial buildings will frequently use all of the methods shown in *Figure 10*, and the methods used will often surpass the *NEC®*, depending upon the manufacturing process and the calculated requirements made by plant engineers.

Figure 11 shows a floor plan of a typical industrial grounding system.

The building in *Figure 12* has a metal underground water pipe that is in direct contact with the earth for more than 10', so this is one valid grounding source. The building also has a metal underground gas piping system, but this may not be used as a grounding electrode [*NEC Section 250.52(B)*].

NEC Section 250.53(D)(2) further states that the underground water pipe must be supplemented by an additional electrode of a type specified in *NEC Sections 250.52(A)(2) through 250.52(A)(8)*. Since a grounded metal building frame, concrete-encased electrode, or ground ring is not available in this application, *NEC Section 250.50* must be used in determining the supplemental electrode.

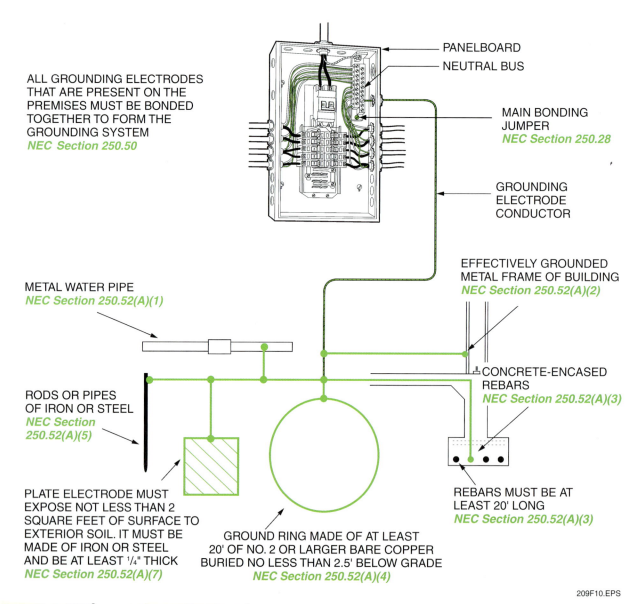

Figure 10 ◆ *NEC®*-approved grounding electrodes.

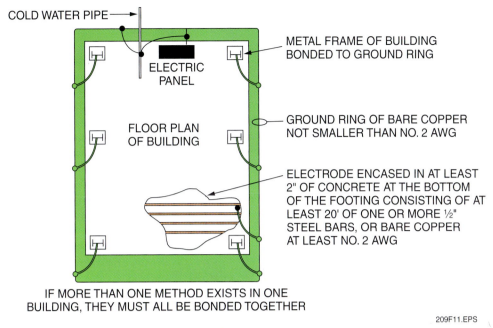

Figure 11 ♦ Floor plan of the grounding system for an industrial building.

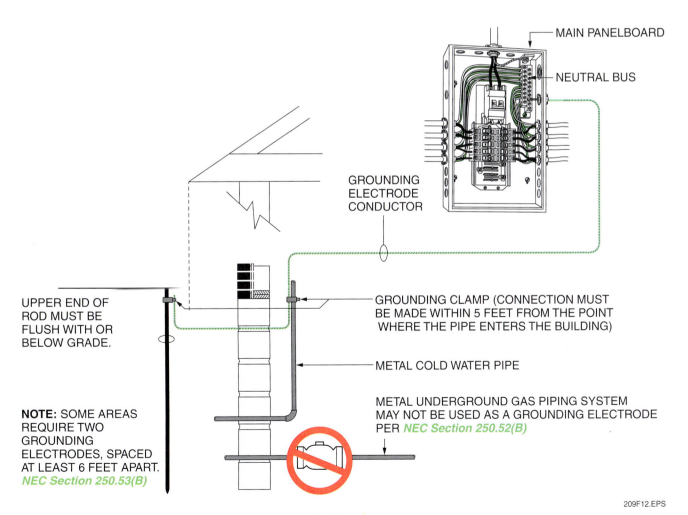

Figure 12 ♦ Grounding requirements for non-industrial buildings.

In most cases, this supplemental electrode will consist of either a driven rod (*Figure 13*) or pipe electrode, the specifications for which are as follows:

- Withstand and dissipate repeated surge circuits
- Provide corrosion resistance to various soil chemistries to ensure continuous performance for the life of the equipment being protected
- Provide rugged mechanical properties for easy driving with minimum effort and rod damage

An alternate method to the pipe or rod method is a plate electrode. Each plate electrode must expose not less than two square feet of surface to the surrounding earth. Plates made of iron or steel must be at least ¼" thick, while plates of nonferrous metal like copper need only be .06" thick.

Rod, pipe, and plate electrodes must have a resistance to ground of 25 ohms (Ω) or less. If not, they must be augmented by an additional electrode spaced not less than 6' from the first. Many locations require two electrodes regardless of the resistance to ground. This is not an *NEC®* requirement, but is required by some power companies and local ordinances in some cities and counties. Always check with the local inspection authority for such rules that may go beyond the requirements of the *NEC®*.

6.2.0 Grounding Electrode Conductor

The grounding electrode conductor is the sole connection between the grounding electrode and the grounded system conductor for a grounded system, or the sole connection between the grounding electrode and the service equipment enclosure for an ungrounded system. See *NEC Section 250.64*, which describes grounding electrode conductor installation.

A common grounding electrode conductor is required to ground both the circuit grounded conductor and the equipment grounding conductor. *Figure 14* shows the common connection for this system.

6.2.1 Sizing

The grounding electrode conductor must be sized in accordance with *NEC Table 250.66*. The size of this conductor is based on the size of the service-entrance conductor.

The size of the service-entrance conductor is determined from the ampacity of that system.

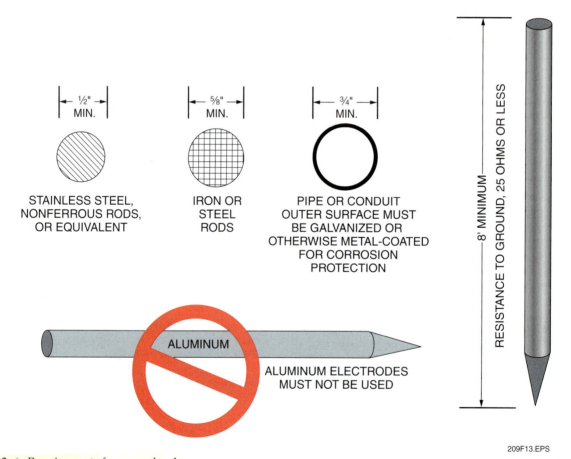

Figure 13 ◆ Requirements for ground rods.

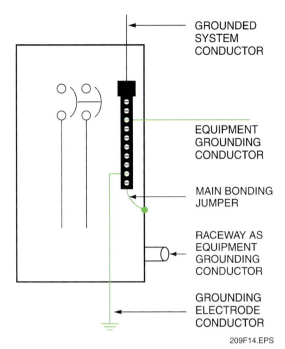

Figure 14 ◆ Grounding electrode connection for grounded system.

NEC Table 310.16 is used to determine the size of the service conductor from the ampacity.

The majority of applications use 75°C copper conductors. This information is necessary when using *NEC Table 310.16*. For a 100A service application, the size of the service conductor is No. 3 AWG copper.

Now that the size of the service conductors is known, the size of the grounding electrode conductor can be determined from *NEC Table 250.66*.

Using the No. 3 AWG from the 100A service example, it can be determined from the table that a No. 8 AWG copper grounding electrode conductor is required.

Another example that is fairly common would be to determine the size of the grounding electrode conductor for a 200A, 208V, three-phase system.

Step 1 Determine the size of the service conductor from *NEC Table 310.16*. It is 3/0 AWG.

Step 2 Go down the left column of *NEC Table 250.66* and find 3/0.

Step 3 Go across to find the size of the grounding electrode conductor. It is a No. 4 AWG conductor.

NOTE
Remember, it is the total size of the service conductor that determines the size of the grounding electrode conductor.

6.2.2 Installation and Protection

NEC Section 250.64(B) requires the grounding electrode conductor or its enclosure to be securely fastened to the surface on which it is carried.

No. 4 AWG or larger conductors require protection where exposed to physical damage. If free from exposure to physical damage, No. 6 AWG conductors may be run along the surface of the building and be securely fastened. Otherwise, the conductor must be protected by installation in rigid or intermediate metal conduit, rigid nonmetallic conduit, EMT, or cable armor. Smaller conductors shall be protected in conduit or armor See *NEC Section 250.64(E)* when using a ferrous enclosure or raceway to enclose a grounding electrode conductor.

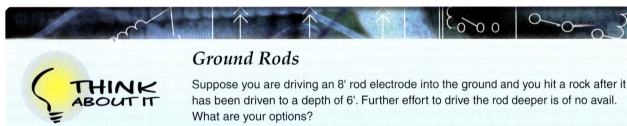

Ground Rods
Suppose you are driving an 8' rod electrode into the ground and you hit a rock after it has been driven to a depth of 6'. Further effort to drive the rod deeper is of no avail. What are your options?

Sizing Electrode Conductors
NEC Table 310.16 shows that the service conductor size (copper wire at 75°C) for a 150A service is 1/0 AWG. Which of the three AWG copper electrode conductor sizes listed below would be used with this service?

No. 4
No. 6
No. 8

6.3.0 Grounding Electrode Conductors

The grounding electrode conductor, which connects the grounded conductor and the panelboard neutral bus to ground, must be copper, aluminum, or copper-clad aluminum. Furthermore, the material selected must be resistant to any corrosive condition existing at the installation, or it must be suitably protected against corrosion. The conductor may be either solid or stranded, and covered or bare, but it must be in one continuous length without a splice or joint, except for the following conditions:

- Splices in busbars are permitted.
- Where a service consists of more than one single enclosure, it is permissible to connect taps to the grounding electrode conductor.
- Bonding jumper(s) from grounding electrode(s) and GECs shall be permitted to be connected to an aluminum or copper busbar not less than ¼" × 2". See *NEC Section 250.64(F)(3)*.

WARNING!
Exothermic welding is a hazardous process that should be performed only by qualified personnel. Refer to your company's safety procedures and the manufacturer's recommendations before using exothermic welding equipment.

- Per *NEC Section 250.64*, grounding electrode conductors may be spliced at any location by means of irreversible compression-type connectors listed for the purpose or using the exothermic welding process (*Figure 15*).

The size of grounding electrode conductors depends on service-entrance size; that is, the size of the largest service-entrance conductor or equivalent for parallel conductors. *NEC Table 250.66* gives the proper sizes of grounding electrode conductors for various sizes of electric services. *Figure 16* shows the various protection guidelines.

6.4.0 Other Electrodes

When an electrode that meets the requirements of *NEC Section 250.50* is not present, other electrodes may be used. These may be the rod and pipe electrodes discussed earlier, other listed electrodes, plate electrodes, or other local metal underground systems such as piping and underground tanks. *NEC Sections 250.52(A)(5) through 250.52(A)(8)* provide information and requirements for these electrodes. *Figure 17* shows other electrodes.

The specific requirements for other electrodes are:

- *Local systems* – Local metallic underground systems such as piping, tanks, metal well casings, etc.

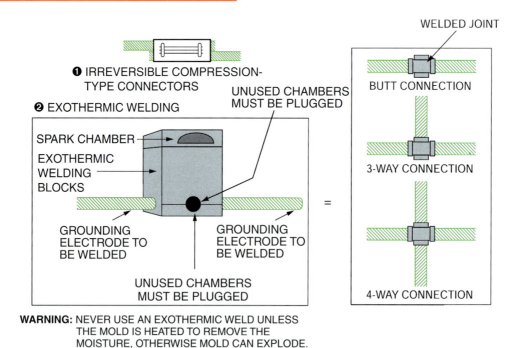

Figure 15 ◆ Methods of splicing grounding conductors.

PROTECTION OF GROUNDING ELECTRODE CONDUCTORS

Grounding electrode conductor or enclosure must be securely fastened to the surface.

No. 4 or larger – If exposed to severe physical damage.

No. 6 – Run along surface, securely fastened or protected.

Smaller than No. 6 – Must be protected from damage (must be enclosed in rigid metal conduit).

Bare aluminum – Not allowed in contact with masonry, where subject to corrosive conditions, or within 18 inches of the earth.

209F16.EPS

Figure 16 ◆ Protection of grounding electrode conductors.

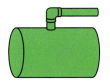
LOCAL METALLIC UNDERGROUND SYSTEMS SUCH AS PIPING, METAL WELL CASINGS, AND TANKS

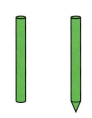
PIPE OR CONDUIT ELECTRODES NOT LESS THAN 8' LONG

PLATE ELECTRODES MINIMUM ¼" THICK IF IRON OR STEEL; 0.06" IF NONFERROUS METAL

209F17.EPS

Figure 17 ◆ Other electrodes.

- *Pipe electrodes* – Pipe or conduit electrodes shall be not less than 8' in length nor smaller than ¾" trade size, and if of iron or steel, shall be galvanized or metal-coated for corrosion protection.
- *Rod and pipe electrodes* – Electrodes of steel or iron shall be at least ⅝" in diameter. Rods of nonferrous metal or stainless steel that are less than ⅝" in diameter shall be listed and must be at least ½" in diameter.
- *Plate electrodes* – Electrodes shall have at least two square feet of surface in contact with exterior soil. If of iron or steel, the plate shall be at least ¼" thick. If of nonferrous metal, it shall be at least 0.06" thick.

NOTE
Underground metal gas piping systems are not permitted to be used as grounding electrodes. However, this does not eliminate the requirement that metal gas piping systems be bonded.

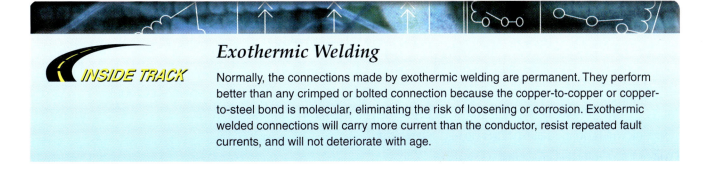

Exothermic Welding

Normally, the connections made by exothermic welding are permanent. They perform better than any crimped or bolted connection because the copper-to-copper or copper-to-steel bond is molecular, eliminating the risk of loosening or corrosion. Exothermic welded connections will carry more current than the conductor, resist repeated fault currents, and will not deteriorate with age.

Plate Electrodes

Why does the *NEC®* require that 2' square electrode plates made of iron or steel be at least ¼" (0.25") thick, while 2' square electrode plates made of copper (a nonferrous metal) need only be 0.06" thick?

6.4.1 Installation of Rod, Pipe, and Plate Electrodes

Rod and pipe electrodes must be installed so at least 8' are in contact with the soil. Rods must be driven vertically unless rock obstruction is encountered. If an obstruction is encountered, the rod may be driven at not more than a 45° angle to clear the rock. The other possibility is for the rod to be buried in a trench that is at least 2.5' deep.

The upper end of the rod must be at or below surface level. The rod may be above ground, but at least 8' of rod must be beneath the surface.

7.0.0 ◆ EQUIPMENT GROUNDING

Figure 18 summarizes the equipment grounding rules for most types of equipment required to be grounded. These general *NEC®* regulations apply to all installations except for specific equipment (special applications) as indicated in the code. The *NEC®* also lists specific equipment that is to be grounded regardless of voltage.

In all occupancies, major appliances and many hand-held appliances and tools are required to be grounded. The appliances include refrigerators, freezers, air conditioners, clothes dryers, washing machines, dishwashing machines, sump pumps, and electrical aquarium equipment. Other tools likely to be used outdoors and in wet or damp locations must be grounded or have a system of double insulation.

7.1.0 Equipment Grounding Conductor

The equipment grounding conductor (EGC) is the conductor used to connect the noncurrent-carrying metal parts of equipment, raceways, and

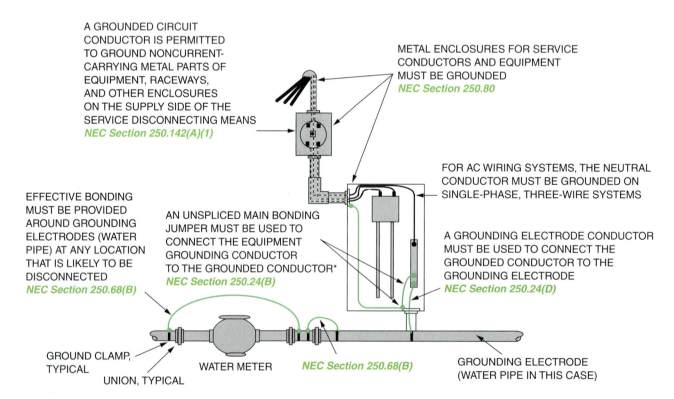

Figure 18 ◆ Equipment grounding summary.

other enclosures to the system grounded conductor and/or the grounding electrode conductor at the service equipment or at the source of a separately derived system.

7.1.1 Grounded and Ungrounded Systems

The EGC or path must extend from the farthest point on the circuit to the service equipment where it is connected to the grounded conductor. *Figure 19* shows an EGC and its path to ground when a ground fault occurs. It is the path for ground faults within the system back to the source that completes the circuit, allowing enough current to flow to ensure that overcurrent devices will trip.

7.1.2 Sizing an EGC

EGCs carry fault current from the load to the grounded terminal bar of the service equipment. The size of the EGC is determined from the size of the overcurrent device that is protecting that particular system. *NEC Table 250.122* is used to determine the size of the EGC.

For example, if the ampacity is 400A, the overcurrent device is set at 400A. Use *NEC Table 250.122* and go down the left column until the desired ampacity or the next highest ampacity is found. For the 400A system, the size of the EGC is No. 3 copper wire or No. 1 aluminum or copper-clad wire.

NOTE
The size of the EGC is determined by the overcurrent protective device.

Determine the size of the EGC for a flexible metal conduit connection to a motor supplied with a 30A circuit.

Step 1 Using *NEC Section 348.60*, verify that an EGC is needed.

Step 2 Since this is an installation that requires flexibility, an EGC is required.

Step 3 The 30A circuit will have a 30A overcurrent protective device feeding the circuit.

Step 4 Find 30A in *NEC Table 250.122*.

Step 5 A No. 10 copper or No. 8 aluminum or copper-clad EGC is required for 30A.

Figure 20 shows the requirements for using flexible metal conduit as an equipment ground.

Another example to analyze would be a liquidtight flexible metal conduit connection to a machine supplied by a 100A circuit.

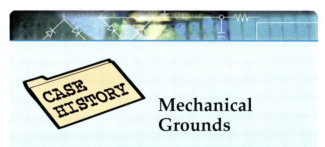

Mechanical Grounds

A technician was electrocuted while servicing a refrigeration unit for a walk-in cooler. The insulation on one of the power conductors inside the flexible metal conduit was damaged, resulting in electrical arcing to a conduit connector on the unit starter box. The conduit connection from the unit to the starter box was loose, thereby effectively disconnecting the mechanical ground from the unit. As the technician was servicing the unit, the thermostat caused the starter to close, energizing the surfaces of the unit and fatally shocking the technician.

The Bottom Line: This accident might have been prevented if one or more of the following procedures had been implemented:

- Proper preventive maintenance checks of all wiring and connections
- Proper lockout/tagout procedures
- Equipment grounding conductor provided along with the power feed conductors

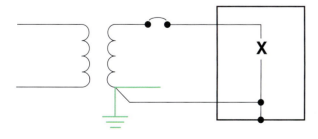

GROUND FAULT IN GROUNDED SYSTEM

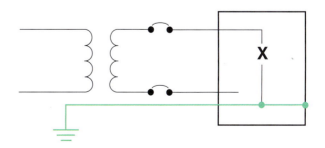

GROUND FAULT IN UNGROUNDED SYSTEM

Figure 19 ◆ Ground faults.

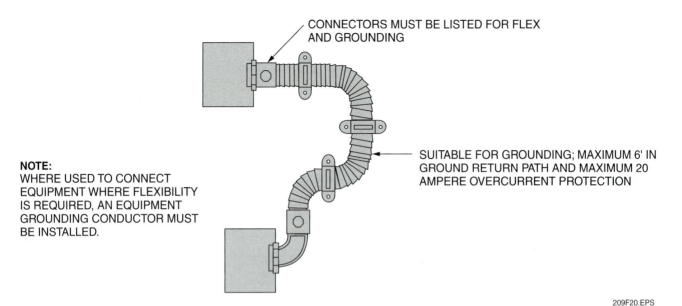

Figure 20 ♦ Installation requirements for flexible metal conduit.

Figure 21 details the requirements for using the conduit as the **grounding conductor**. After determining the requirements and the ampacity of this application, a No. 8 copper EGC would be required.

7.2.0 Grounding Enclosures

As previously mentioned, equipment grounding covers the metallic noncurrent-carrying parts of an electrical system. Such parts include metallic conduit, outlet boxes, enclosures, and frames on motors, and other electrically operated equipment. These items are bonded together to ensure operation of overcurrent devices in case of ground faults.

A bare or green-insulated grounding wire is usually attached to the metal frame or cabinet of the equipment. When connected to the circuit, this grounding wire is attached to the equipment grounding system that, in turn, was originally bonded to the system neutral busbar at the service-entrance equipment. When properly connected, if a live ungrounded wire should make contact with the frame or cabinet of a motor, appliance, or other metallic object in the system, a ground fault will occur and open the overcurrent

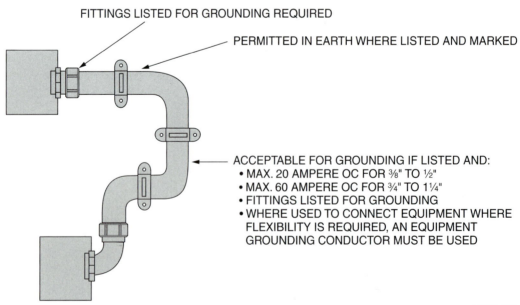

Figure 21 ♦ Installation requirements for liquidtight flexible metal conduit.

9.18 ELECTRICAL LEVEL TWO ♦ TRAINEE GUIDE

protective device that is protecting the circuit. Consequently, equipment grounding is one of the best methods of protecting life and property should a ground fault occur.

7.2.1 Grounding Outlet Boxes and Devices

All receptacles used in residential, commercial, or industrial applications must be of the grounding type, which means that one of the receptacle openings connects the equipment ground to the appliance, tool, or other apparatus that is connected to the receptacles.

Ensuring continuity of the equipment grounding system to each receptacle can be handled in different ways, depending upon the wiring method used. For example, in most residential wiring systems, Type NM cable will be used. If metallic device boxes are used, the bare or green-insulated grounding wire must be attached to the box. This is accomplished by using either a grounding clip, as shown in *Figure 22*, or a screw designed for this purpose that is secured to the outlet box. NEC Section 250.8 lists acceptable methods for connecting grounding conductors to enclosures.

Figure 23 shows methods of providing a grounding conductor to duplex receptacles using several wiring methods. For example, if nonmetallic boxes are used with Type NM cable, no connection to the box itself is required. However, the equipment grounding conductor must be connected to the grounding terminal on the receptacle. If two or more cables enter the nonmetallic box, each of the equipment grounding wires must be connected with an approved connector, and one wire must then be attached to the grounding terminal on the receptacle. The terminal itself may not contain more than one grounding wire. Therefore, if more than one grounding wire enters the box, they must be spliced independently from the device, and then only one conductor may be attached to the device.

When metallic boxes are used, these equipment grounding wires must be attached to the box, along with the wiring device (receptacle). As mentioned previously, this may be accomplished with either a grounding clip or a grounding screw (NEC Section 250.8). See also NEC Section 250.146.

Any of the following are recognized by NEC Section 250.118 as being adequate for use as an equipment grounding conductor:

- A copper or other corrosion-resistant conductor (solid or stranded; insulated, covered, or bare; and in the form of a wire or a busbar of any shape)
- Rigid metal conduit
- Intermediate metal conduit
- Liquidtight flexible metal conduit
- Electrical metallic tubing
- Flexible metal conduit and tubing as permitted by NEC Sections 250.118(5), 250.118(6), and 250.118(7)
- Armor of Type AC cable
- The copper sheath of mineral-insulated, metal-sheathed cable
- The metallic sheath or the combined metallic sheath and grounding conductors of Type MC cable
- Cable trays as permitted in NEC Sections 392.3(C) and 392.7
- Cable bus framework as permitted in NEC Section 370.3
- Other electrically continuous metal raceways listed for grounding

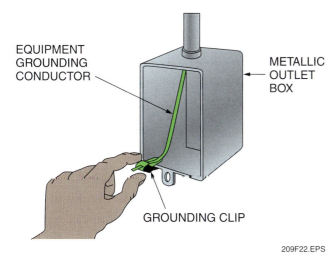

Figure 22 ◆ Grounding clip.

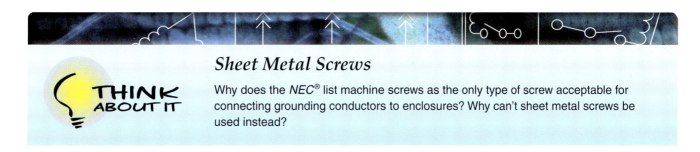

Sheet Metal Screws

Why does the *NEC®* list machine screws as the only type of screw acceptable for connecting grounding conductors to enclosures? Why can't sheet metal screws be used instead?

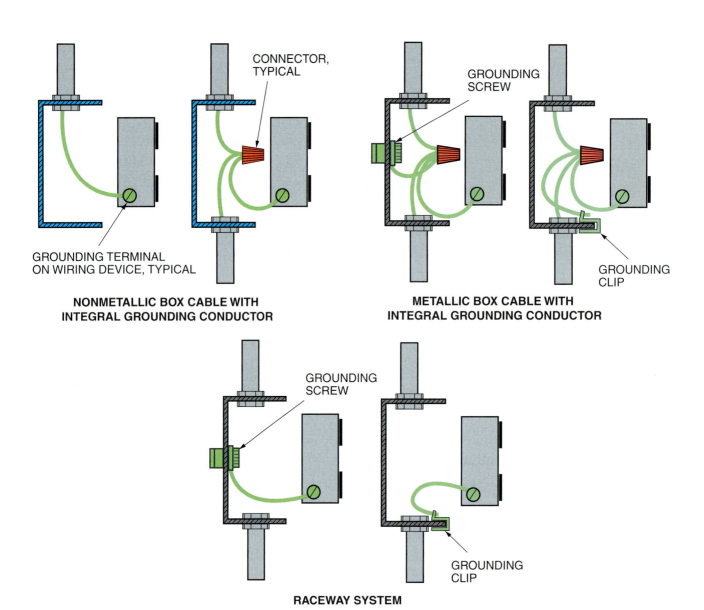

Figure 23 ◆ Grounding receptacles with different wiring methods.

7.3.0 Main Bonding Jumper

When a listed switchboard or panelboard is installed, the main bonding jumper that is provided with the equipment is rated for the size of conductors used for the service. The proper size of the main bonding jumper is verified by the listing agency and can be used without calculating the size.

Since the main bonding jumper must carry the full ground fault current of the system back to the grounded service conductor, its size must relate to the rating of the service conductors that supply the service. *NEC Table 250.66* is used to determine the size of the main bonding jumper. *Figure 24* shows how to calculate the size of the main bonding jumper when parallel runs are used.

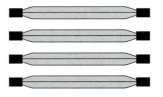

4 – 250 KCMIL ALUMINUM CONDUCTORS
4 × 250 = 1,000 KCMIL
NEC Table 250.66
2/0 COPPER OR 4/0 ALUMINUM

Figure 24 ◆ Main bonding jumper for parallel runs.

Grounding and Bonding

When measuring voltage at a grounded duplex receptacle installed in a metal outlet box, you measure 120VAC between the hot and neutral terminals, 120VAC between the hot and ground terminals, and 120VAC between the hot terminal and the outlet box enclosure. You also measure 0VAC between the neutral and ground terminals. Does this represent a short circuit or ground fault?

Where the multiple conductors are larger than the maximum given in *NEC Table 250.66*, the jumper cannot be less than 12.5% of the area at the largest phase conductor [*NEC Section 250.28(D)*].

An example of three 500 kcmil copper conductors gives a total area of 1,500 kcmil. Because this is larger than that of *NEC Table 250.66*, we must take 12.5% of the total area. This is 187.5 kcmil. Since this is not a standard size conductor, we must refer to *NEC Chapter 9, Table 8*.

The table shows that the next larger conductor size in circular mils is 211,600. This gives a conductor of 4/0 AWG.

Bonding Jumpers

Use a bonding jumper to provide continuity at cable tray joints.

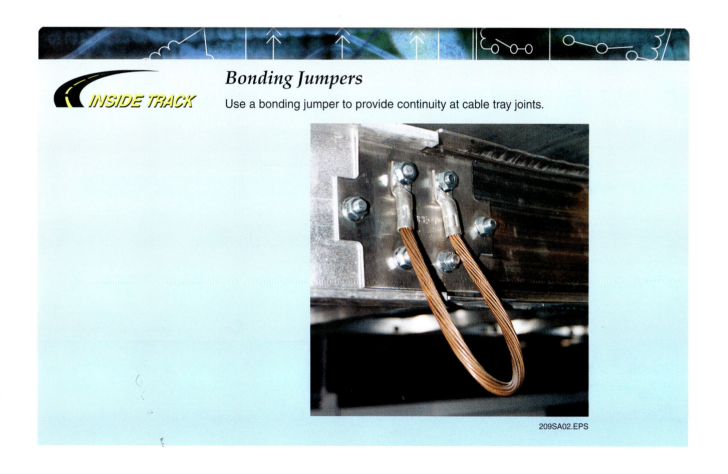

209SA02.EPS

8.0.0 ♦ BONDING SERVICE EQUIPMENT

Electrical continuity is the key to successful clearing of ground fault currents. This continuity between service equipment and enclosures must be maintained by bonding. The items that are required to be bonded together are listed in *NEC Section 250.92(A)*. They include:

- The service raceways, cable trays, cable bus framework, or service cable armor or sheath
- All service equipment enclosures containing service conductors, including meter fittings, boxes, etc., interposed in the service raceway or armor
- Any metallic raceway or armor that encloses the grounding electrode conductor

NEC Section 250.92(B) defines the methods permitted to be used to ensure continuity at service equipment. These methods are:

- Bonding equipment to the grounded service conductor in a manner provided by *NEC Section 250.8*
- Connections utilizing threaded couplings or threaded bosses on enclosures when made up wrenchtight
- Threadless couplings and connectors when made up wrenchtight
- Other approved devices, such as bonding-type locknuts

The exception to this requirement is found in *NEC Section 250.84(A)*. This refers to an underground service cable that is metallically connected to the underground service conduit. The code states that if a service cable contains a metal armor, and if the service cable also contains an uninsulated grounded service conductor that is in continuous electrical contact with its metallic armor, then the metal covering of the cable is considered to be adequately grounded.

8.1.0 Bonding Multiple Service Disconnecting Means

Multiple disconnecting means can take several forms. *NEC Article 230* discusses the various forms.

Because this bonding jumper is located on the supply side of the service disconnect, the basic rule for sizing the equipment bonding jumper for bonding these various configurations is found in *NEC Section 250.102(C)*. This section requires the use of *NEC Table 250.66* for selecting the size. However, if the size of the service conductor is larger than 1,100 kcmil copper or 1,750 kcmil aluminum, the bonding jumper must have a cross-sectional area not less than 12.5% of the largest phase conductor. Where the service conductors are paralleled in two or more raceways, the equipment bonding jumper shall be run in parallel. The size of each bonding jumper shall be based on the largest phase conductor in each raceway or cable.

Table 1 shows various sizes of bonding jumpers based on the size of the service-entrance conductor.

Table 1 Bonding of Multiple Service Disconnecting Means

Service-Entrance Conductor	Bonding Jumper (Copper)
500 kcmil in service mast	1/0
1,000 kcmil in wireway	2/0
300 kcmil to 300A service	No. 2
3/0 to 200A service	No. 4
No. 2 to 125A service	No. 8

Multiple Load Centers

For installations in which the main service panel is located at a distance from areas that have many circuits and/or a heavy load concentration, subpanels may be installed near these loads. Doing so allows the branch circuit wiring runs to be shorter, resulting in lower line voltage losses than would occur if the branch circuits had been run all the way back to the main panel.

8.2.0 Bonding of Enclosures and Equipment

The bonding requirements for an installation depend on the system voltage, number of raceways, and type of equipment in use.

8.2.1 Bonding of Services Over 250 Volts

NEC Section 250.97 defines the methods permitted to be used on circuits over 250V. The methods permitted for these circuits are the same as for services, with the exception of using the service grounded conductors beyond the service.

Figure 25 shows examples of these bonding methods.

8.2.2 Bonding Multiple Raceway Systems

NEC Section 250.102(D) allows the use of a single conductor to bond two or more raceways or cables where the bonding jumper is sized in accordance with *NEC Table 250.122* for the largest overcurrent device supplying circuits therein. *Figure 26* shows bonding of multiple raceways.

It is acceptable to install one equipment bonding jumper individually from each raceway to the equipment grounding terminals of the equipment. This is shown in *Figure 27*. Each bonding conductor is sized per *NEC Table 250.122*.

8.2.3 Installation of Equipment Bonding Jumper

The equipment bonding jumper is permitted to be installed either inside or outside of a raceway or enclosure. When the jumper is installed outside, the length is limited to not more than 6'. The bonding jumper must also follow the raceway routing. This is vital to keep the impedance of the equipment bonding jumper as low as possible.

Figure 28 shows a bonding jumper installed inside a nonmetallic enclosure.

NOTE

The equipment bonding jumper must be sized for the overcurrent device.

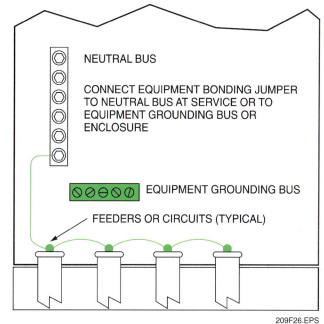

Figure 26 ♦ Bonding multiple raceways.

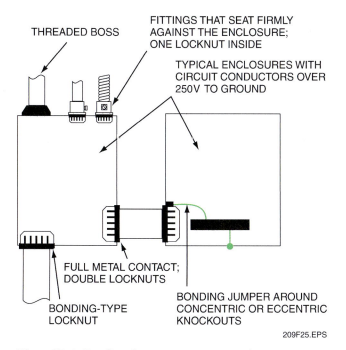

Figure 25 ♦ Bonding for circuits over 250 volts.

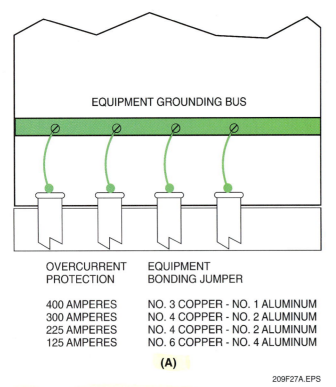
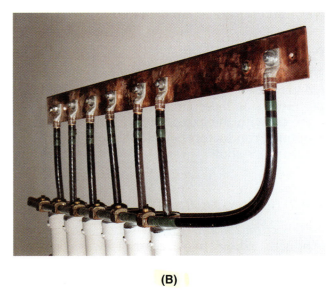

Figure 27 ♦ Individual bonding jumpers.

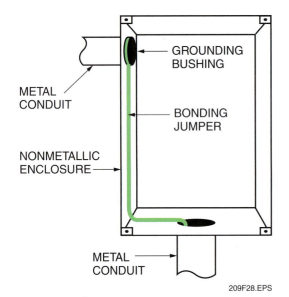

Figure 28 ♦ Bonding jumper.

9.0.0 ♦ EFFECTIVE GROUNDING PATH

As previously stated, to have effective grounding for both a grounded and an ungrounded system, the grounding system shall have an **effective ground fault path** to provide a permanent and continuous path, sized to safely conduct any fault current that is likely to occur, and it shall have impedance that is low enough to limit the voltage to ground and to facilitate the operation of the circuit protective devices in the circuit.

Low impedance means that the grounding path and the circuit conductors must always be within the same raceway, or the raceway may provide the grounding path.

9.1.0 Clearing Ground Faults

A ground fault is a different circuit that becomes dependent on the impedance of the return path back to the source. The amount of current flow depends on the impedance. The higher the current, the harder it becomes to maintain voltage in the system. It is at this voltage that the short circuit capacity plays its role by attempting to maintain voltage. As the system tries to maintain voltage, its current will increase, and when the current reaches the overcurrent setting of the device, the system will be protected by the opening of the overcurrent device.

10.0.0 ♦ GROUNDED CONDUCTOR

The neutral in a grounded system serves two main purposes:

- It permits utilization of power at line-to-neutral voltage. This will allow the current-carrying conductor to carry any unbalanced current.
- It provides a low-impedance return path for the flow of fault current to the source to facilitate the operation of the overcurrent devices in the circuit.

Separately Derived Systems

The requirements of *NEC Section 250.30* pertaining to the grounding of separately derived AC systems are most commonly applied to 480Y/277V transformers that transform a 480V supply to a 208Y/120V system for lighting and appliance loads. The requirements provide for a low-impedance path to ground, so that line-to-ground faults on circuits supplied by the transformer will result in sufficient current flow to operate the overcurrent devices.

Figure 29 shows the path to ground for this neutral. If the neutral is not needed for voltage requirements, it still must be run to the service and connected to the grounded electrode conductor and equipment grounding conductor at the service.

11.0.0 ◆ SEPARATELY DERIVED SYSTEMS

Figure 30 shows a transformer-type separately derived system. This system has no direct electrical connection with the premises' source of electrical power. *NEC Section 250.20(D)* discusses whether or not a separately derived system is required to be grounded. Each voltage level that is required to be grounded must meet the requirements of *NEC Section 250.30(A)(1)* and have a system bonding jumper that is used to connect the equipment grounding conductors of the separately derived system to the grounded conductor. The grounding electrode conductor is connected to the grounded conductor at the same point where the system bonding jumper is connected.

In *Figure 30*, the supply to the transformer will be at one voltage level while the secondary is at either a higher or a lower voltage level.

11.1.0 Grounding of Separately Derived Systems

For separately derived systems that are required to be grounded, an equipment grounding conductor must be supplied with the primary circuit. This will provide a low-impedance fault-current path from the transformer case to the main service.

Figure 29 ◆ Grounded service conductor run to service.

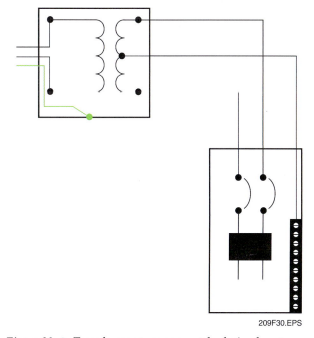

Figure 30 ◆ Transformer-type separately derived system.

Along with the equipment grounding conductor, *NEC Section 250.30(A)(1)* requires a bonding jumper be installed for the derived system required to be grounded. This bonding jumper may be placed in either the transformer or the panel, but not in both. Installing it in both locations would create a parallel path with the grounded conductor, resulting in an unwanted current flow on the grounding conductor. Next, a grounding electrode conductor must be installed and connected at the same point. This is illustrated in *Figure 31*.

The grounding electrode for separately derived systems must be:

- As near as practicable to the system
- The nearest available grounded structural metal member or the nearest available grounded water pipe
- Other electrodes where the above are not available (*NEC Sections 250.50 and 250.52*)

11.2.0 Generator-Type Separately Derived System

A simple generator-type separately derived system is shown in *Figure 32*.

The way to determine that a generator is a separately derived system is to examine the transfer switch. If the neutral and all phase conductors are switched, then the system is separately derived. If the neutral is not switched, but solidly connected, then the system is not a separately derived system.

The neutral bonding jumper must be installed either at the generator or any point in between. A grounding electrode conductor must be installed between the neutral and a grounding electrode.

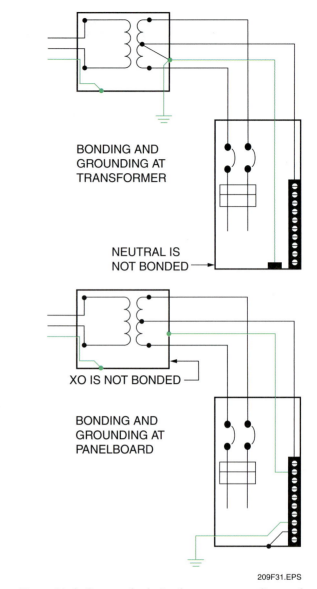

Figure 31 ◆ Separately derived system grounding and bonding locations.

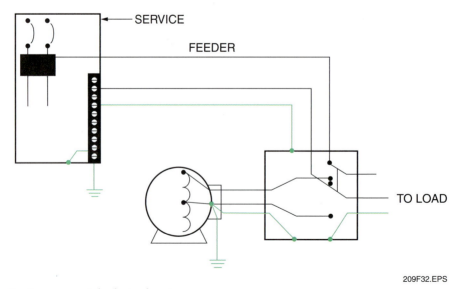

Figure 32 ◆ Generator-type separately derived system.

When the generator is not a separately derived system, the neutral bonding jumper must be removed from the generator and the neutral must not be grounded. Grounding is accomplished through direct connection to the neutral of the premises wiring.

12.0.0 ◆ GROUNDING AT MORE THAN ONE BUILDING

NEC Section 250.32 provides the method for grounding electrical systems at additional buildings on the premises.

The grounded circuit conductor and the equipment grounding conductor must both be extended to the second building. The grounded conductor is terminated on an insulated bus isolated from the metal cabinet, while the equipment grounding conductor from the main building is terminated on an equipment grounding terminal bus, directly connected to the cabinet. The equipment grounding conductor would be sized using *NEC Table 250.122* based on the overcurrent device supplying the second structure. All grounding electrodes present at the second building would connect to the grounding bus. The GEC in the second structure is sized using *NEC Table 250.66* based on the size of the feeder conductors. *Table 2* shows the number and type of conductors that must be taken from the first structure to where service is located in the second structure.

When no grounding electrodes are at the additional structures, a grounding electrode must be installed. These electrodes include underground metal pipes, the grounded metal frame of the structure, or concrete-encased electrodes. Rod, pipe, or plate electrodes may also be permitted, in accordance with *NEC Section 250.52*.

By exception and in existing installations only, the *NEC®* allows a method whereby no equipment ground is extended to the second building (*Figure 33*). In this case, the grounded conductor is extended to and terminated on a terminal bus that is bonded to the equipment, similar to the connection made at the grounded conductor at the service equipment. The grounded system will be connected to all grounding electrodes at the second building. The grounded conductor is sized using *NEC Sections 220.61 and 250.122*. This method may only be used when an equipment grounding conductor is not run with the supply to the second building, there are no continuous metallic paths bonded to the grounding system in both buildings, and ground fault protection has not been installed on the common electrical service in the first building. These additional provisions prevent the possibility of parallel current paths between grounded and grounding conductors, as well as nuisance tripping of a ground fault protective device.

Table 2 Equipment Grounding Conductor Not Installed

System	Ungrounded	Grounded	Equipment Ground
120V	1	1	1
120/240V	2	1	1
208/120V	3	1	1
480/277V	3	1	1

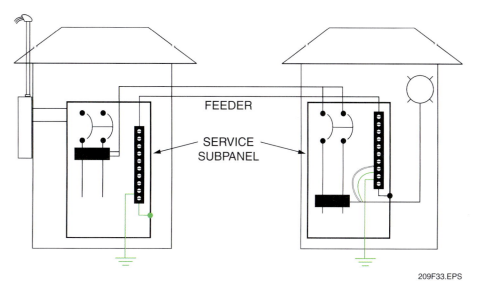

Figure 33 ◆ Grounding neutral at second building.

13.0.0 ♦ SYSTEMS OVER 1,000 VOLTS

NEC Article 250, Part X discusses the grounding requirements for high-voltage systems. It also lists a few additional requirements for derived neutral systems, solidly grounded neutral systems, and impedance-grounded neutral systems.

14.0.0 ♦ TESTING FOR EFFECTIVE GROUNDS

An earth ground resistance tester (*Figure 34*) may be used to make soil resistivity measurements or to measure the resistance to earth of the installed grounding electrode system.

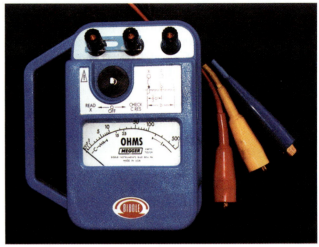

Figure 34 ♦ Earth ground resistance tester.

NOTE
An ordinary ohmmeter is not used to measure the resistance of a grounding electrode to earth because it does not provide for adequate levels of current and voltage.

WARNING!
Ground testers can be hazardous to both equipment and personnel if improperly used. Always check with your supervisor before using a ground tester.

One use of the ground tester is for testing electrical systems after they are installed and before normal voltage is applied. This test is made after all the conductors, fuses or circuit breakers, panelboards, outlets, etc., are in place and connected. The current used for testing is produced by a small generator within the ground tester that generates DC power, either by turning a crank handle (also a part of the ground tester), or by using a small electric DC motor within the ground tester.

The test is made by connecting the terminals to the two points between which the test is to be made and then rapidly turning the handle on the ground tester. The resistance in ohms can then be read from the meter dial. Satisfactory insulation resistance values will vary under different conditions, and the charts supplied with the ground tester should be consulted for the proper value for a particular installation.

15.0.0 ♦ MEASURING EARTH RESISTANCE

An earth ground is commonly used as an electrical conductor for system returns. Although the resistivity of the earth is high compared to a metal conductor, its overall resistance can be quite low because of the large cross-sectional area of the electrical path.

Connections to the earth are made with grounding electrodes, ground grids, and ground mats. The resistance of these devices varies proportionately with the earth's resistivity, which in turn depends on the composition, compactness, temperature, and moisture content of the soil.

A good grounding system limits system-to-ground resistance to an acceptably low value. This protects personnel from a dangerously high voltage during a fault in the equipment. Furthermore, equipment damage can be limited by using this ground current to operate protective devices.

Ground testers measure the ground resistance of a grounding electrode or ground grid system. Some of the major purposes of ground testing are to verify the adequacy of a new grounding system, detect changes in an existing system, and determine the presence of hazardous step voltage and touch voltage.

In addition to personnel safety considerations, ground testing also provides information for equipment insulation ratings. Equipment can be damaged by an overvoltage that exceeds the rating of the insulation system. *Figure 35* depicts a poorly grounded system where the ground resistance (R_1) is 10Ω.

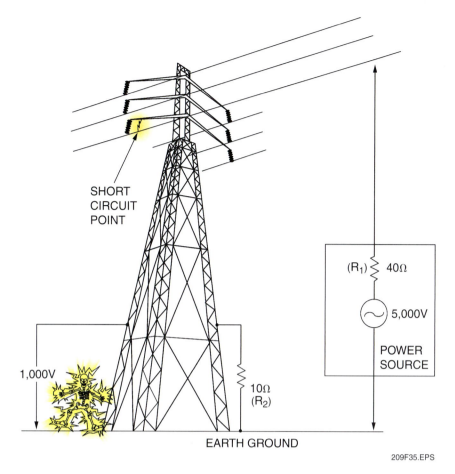

Figure 35 ♦ Poorly grounded system.

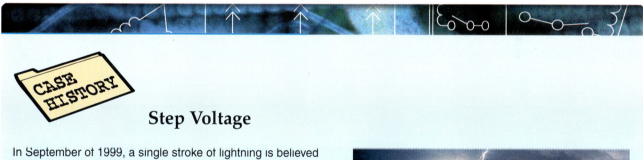

Step Voltage

In September of 1999, a single stroke of lightning is believed to have killed a herd of 56 elk on Mount Evans in Colorado. This happened because when lightning strikes the earth, it disperses in concentric rings of equal potential (equipotential zones). Because the earth has resistance, the voltage drops as it travels away from the strike point. Step voltage is the difference in potential that exists when you step between these equipotential zones. In this case, the hind legs of the elk were at a different potential than their front legs, resulting in current flow.

Bottom Line: During a lightning strike or any other condition that creates zones of different potential, your best bet is to remain in one spot, crouched low with your feet together and your head bent forward. If you must move, hop with your feet together, moving much like a bird on a wire.

Assuming a power source resistance (R_2) of 40Ω, a short circuit between the 5,000V power line and the steel tower would produce 100A of short circuit current (I).

$$I = \frac{E}{R_1 + R_2} = \frac{5,000}{40 + 10} = 100A \text{ of short circuit current}$$

A person touching the tower would be subjected to the voltage (E') developed across the ground resistance:

$$E' = IR = 100 \times 10 = 1,000 \text{ volts}$$
between power and ground

Statistics vary widely concerning what may be considered a dangerous voltage. This depends largely on body resistance and other conditions. However, to limit the touch voltage for this situation to 100V, the ground resistance for the tower would have to be less than 1Ω.

15.1.0 How the Ground Tester Works

Several methods are used for measuring the resistance to earth of a grounding electrode; one of the most common is the fall-of-potential method. See *Figure 36*.

In this method, **auxiliary electrodes** 1 and 2 are placed at sufficient distances from grounding electrode X. A current (I) is passed through the earth between the grounding electrode X and auxiliary current electrode 2 and is measured by the ammeter. The voltage drop (E) between the grounding electrode X and the auxiliary potential electrode 1 is indicated on the voltmeter. Resistance (R) can therefore be calculated as follows:

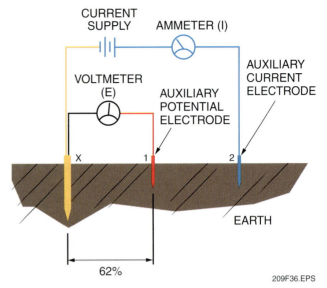

Figure 36 ◆ Fall-of-potential method of testing.

$$R = \frac{E}{I}$$

Certain problems may arise in measuring with the simple system shown in *Figure 36*:

- Natural currents in the soil caused by electrolytic action can cause the voltmeter to read either high or low, depending on polarity.
- Induced currents in the soil, instrument, or electrical leads can cause vibration of the meter pointer, interfering with readability.
- Resistance in the auxiliary electrode and electrical leads can introduce error into the voltmeter reading.

Touch Voltage

If a system is ungrounded and the metal parts become energized, a potential difference will exist between the system and ground. This is known as touch voltage. For example, a 10-year-old boy was swimming at a community pool and exited the pool to buy a snack from a vending machine. The grounding prong had been removed from the plug and the cord was pinched under the machine's metal leg, exposing the conductor and causing the metal frame of the unit to become energized. As soon as he touched the metal, he received a fatal shock.

Bottom Line: Make sure the entire grounding system is intact for cord and plug connected equipment. In case of damage to the cord, the equipment grounding conductor could be damaged and deficient. When servicing any motor, check to ensure that the cord is not pinched. Never remove the grounding prong from a three-prong plug. Because of this and similar incidents, *NEC Section 422.51* now requires GFCI protection for cord and plug connected vending machines.

INSIDE TRACK

Clamp-On Ground Testers

This clamp-on ground tester allows for checking ground grid connections from 0.10Ω to 1,200Ω without the need to disconnect the conductors.

Most ground testers use a null balance metering system. Unlike the separate voltmeter and ammeter method, this instrument provides a readout directly in ohms, thus eliminating calculation. Although the integrated systems of the ground testers are sophisticated, they still perform the basic functions for fall-of-potential testing.

15.2.0 Current Supply

As in the simple circuit, the ground tester also has a current supply circuit (see *Figure 37*). This may be traced from grounding electrode X through terminal X, potentiometer R1, the secondary of power transformer T1, terminal 2, and auxiliary current electrode 2. This produces a current in the earth between electrodes X and 2.

When switch S1 is closed, battery B energizes the coil of vibrator V. Vibrator reed V1 begins oscillating, thereby producing an alternating current in the primary and secondary windings of T1. The negative battery terminal is connected alternately across first one and then the other half of the primary winding.

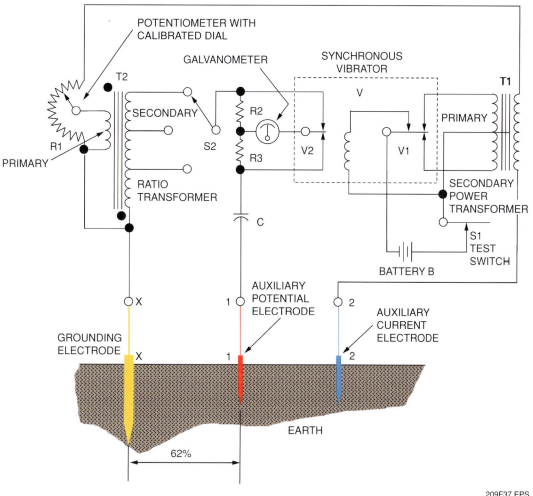

Figure 37 ◆ Three-point testing using a ground tester.

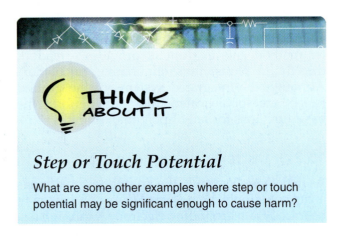

Think About It

Step or Touch Potential

What are some other examples where step or touch potential may be significant enough to cause harm?

15.3.0 Voltmeter Circuit

The voltmeter circuit can be traced from grounding electrode X through terminal X, the T2 secondary, switch S2, resistors R2 and R3 (paralleled by the meter and V2 contacts), capacitor C, terminal 1, and auxiliary potential electrode 1. The current in the earth between grounding electrode X and auxiliary current electrode 2 creates a voltage drop due to the earth's resistance. With auxiliary potential electrode 1 placed at any distance between grounding electrode X and auxiliary current electrode 2, the voltage drop causes a current in the voltmeter circuit through balanced resistors R2 and R3. The voltage drop across these resistors causes galvanometer M to deflect from zero center scale.

Vibrator reed V2 operates at the same frequency as V1, thereby functioning as a mechanical rectifier for galvanometer M. The vibrator is tuned to operate at 97.5 hertz (Hz), a frequency unrelated to commercial power line frequencies and their harmonics. Thus, currents induced in the earth by power lines are rejected by most ground testers and have virtually no effect on their accuracy. Stray direct current in the earth is blocked out of the voltmeter circuit by capacitor C.

The current in the primary of T2 can be adjusted with potentiometer R1. Primary current in T2 induces a voltage in the secondary of T2, which is opposite in polarity to the voltage drop caused by current in the voltmeter circuit.

With R1 adjusted so the primary and opposing secondary voltages of T2 are equal, current in the voltmeter circuit is zero, and the galvanometer reads zero. The resistance of grounding electrode X can then be read on the calibrated dial of the potentiometer.

With no current in the voltmeter circuit, the lead resistance of the auxiliary potential electrode 1 has no bearing on accuracy. (With no current, there is no voltage drop in the leads.) Resistance to earth of the current electrode results only in a reduction of current, and consequently, a loss of sensitivity. Therefore, the auxiliary electrodes need only be inserted into the earth 6" to 8" to make sufficient contact. In some locations, where the soil is very dry, it may be necessary to pour water around the current electrode to lower the resistance to a practical value.

15.4.0 Nature of Earth Electrode Resistance

Current in a grounding system is primarily determined by the voltage and impedance of the electrical equipment. However, the resistance of the grounding system is very important in determining the voltage rise between the ground electrode and the earth as well as the voltage gradients that will occur in the vicinity when current is present.

Three components constitute the resistance of a grounding system:

- Resistance of the conductor connecting the ground electrode
- Contact resistance between the ground electrode and the soil
- Resistance of the body of earth immediately surrounding the electrode

Resistance of the connecting conductor can be dealt with separately since this is a function of the conductor cross-sectional area and length. Contact resistance is usually negligibly small if the electrode is free from paint or grease. Therefore, the main resistance is that of the body of earth immediately surrounding the electrode. Current from a grounding electrode flows in all directions via the surrounding earth. It is as though the current flows through a series of concentric spherical shells, all of equal thickness.

The shell immediately surrounding the electrode has the smallest cross-sectional area, and therefore its resistance is highest. As the distance from the electrode is increased, each shell becomes correspondingly larger; thus, the resistance becomes smaller. Finally, a distance from the electrode is reached where additional shells do not add significantly to the total resistance. From a theoretical viewpoint, total resistance is included only when the distance is infinite. For practical purposes, only the volume that contributes the major part of the resistance need be considered. This is known as the effective resistance area and depends on electrode diameter and driven depth.

16.0.0 ♦ THREE-POINT TEST FOR SINGLE ELECTRODE/TRIAD

To measure the resistance of a grounding electrode with most ground testers, an auxiliary current electrode and an auxiliary potential electrode are required (see *Figures 38* and *39*). The current electrode is placed a suitable distance from the grounding electrode under test, and the potential electrode is then placed at 62% of the current electrode distance.

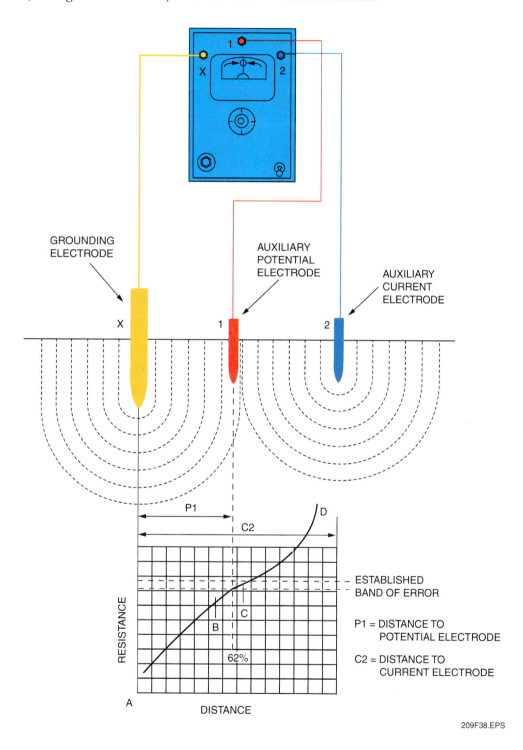

Figure 38 ♦ Plotted curve showing insufficient electrode spacing.

NOTE

The 62% figure has been arrived at by empirical data gathered by many authorities on ground resistance measurement, and in some cases, has been computed based on analysis of an equivalent hemisphere. Testing on large systems requires engineering assistance to determine proper testing procedures and electrode spacing.

If the current electrode is too near the grounding electrode, their effective resistance areas will overlap, as shown in *Figure 38*. If a series of measurements are made with the potential electrode driven at various distances in a straight line between the current electrode and grounding electrode, the readings will yield a curve as illustrated in *Figure 38*.

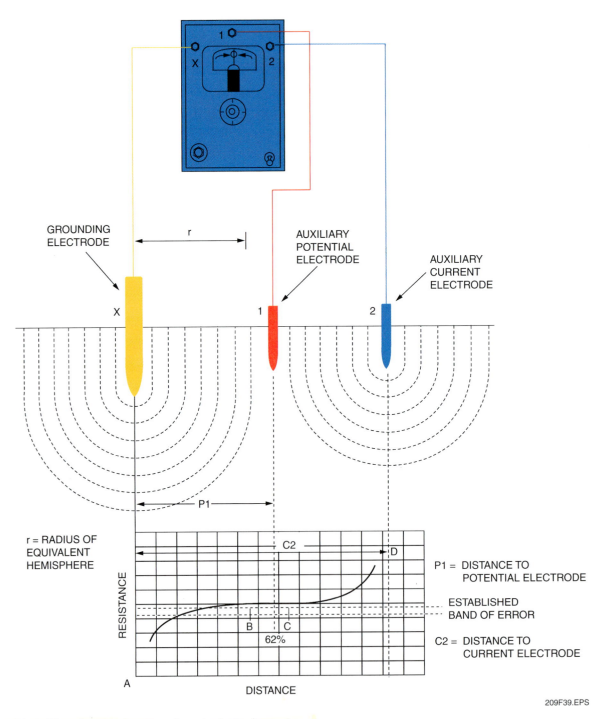

Figure 39 ◆ Plotted curve showing adequate electrode spacing.

9.34 ELECTRICAL LEVEL TWO ◆ TRAINEE GUIDE

Size and Depth of Ground Electrode

What effect do the diameter and driven depth of a ground electrode have on its resistance?

In *Figure 39*, a curve is plotted with the current electrode at a sufficient distance from the grounding electrode. Note that the curve is relatively flat between points B and C, which are usually considered to be at ±10' with respect to the 62% point. Usually, a tolerance is established for the maximum allowable deviation for the second and third readings with relation to the initial reading at 62%. This tolerance is a certain ±% of the initial reading, such as ±1%, ±2%, etc.

No definite distance from the current electrode can be forecast since the optimum distance is based on the homogeneity of the earth, depth of the grounding electrode, diameter, etc. However, for a starting point for a single driven grounding electrode, the effective radius of the equivalent hemisphere can be computed. The curve of *Figure 40* can then be used for initial placement of the auxiliary electrodes. Using the practical method of moving the auxiliary potential electrode 10' to either side of the 62% point, a curve can be plotted to determine if the current electrode spacing is adequate, as depicted by the curve flattening out between points B and C in *Figure 39*.

For example, assume that a 1" (2.54 cm) diameter grounding electrode is buried 10' (3.0 m) deep. The equivalent hemisphere radius is 1.7' (51.8 cm). Using *Figure 40*, the current electrode would be established at approximately 90' (27 m), and the potential electrode at approximately 55' (17 m) for the initial reading. The results obtained when the potential electrode is moved (to 45' [14 m] and then to 65' [20 m]) will determine if the current electrode, and consequently the potential electrode, must be spaced at a greater distance.

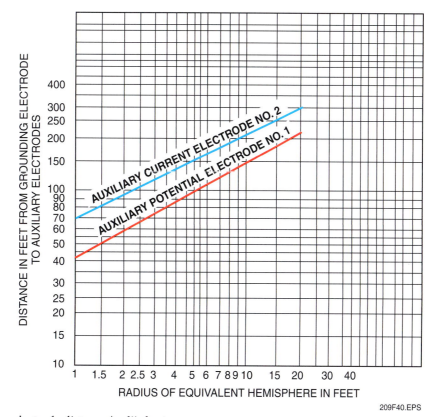

Figure 40 ◆ Auxiliary electrode distance/radii chart.

Poor Soil Conductivity

What can be done to lower the ground electrode (rod) resistance when it is driven into soil with poor conductivity?

It is usually advisable to plot a complete curve for each season of the year. See *Figure 41*. These curves should be retained for comparison purposes. Measurements at established intervals in the future need only be made at the 62% point and, if desired, 10' on each side, providing there is no erratic deviation from the original curve. Serious deviation, other than seasonal, could mean corrosion has eaten away some of the electrode.

NOTE

Safety grounds should be applied in such a way that a zone of equal potential is formed in the work area. This equipotential zone is formed when fault current is bypassed around the work area by metallic conductors. In this situation, the worker is bypassed by the low-resistance metallic conductors of the safety ground.

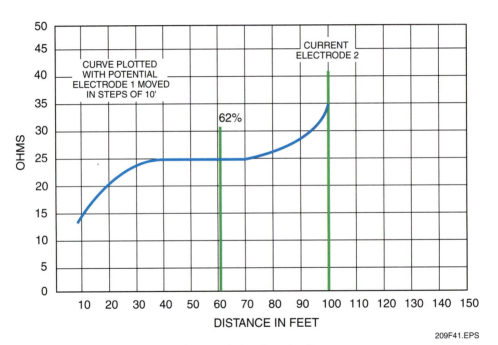

Figure 41 ◆ Typical grounding resistance curve to be recorded and retained.

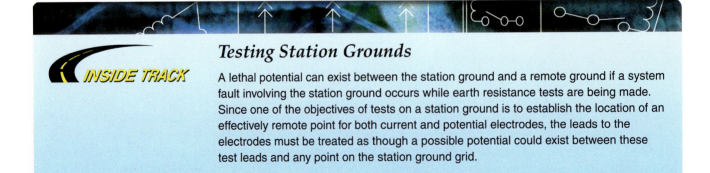

Testing Station Grounds

A lethal potential can exist between the station ground and a remote ground if a system fault involving the station ground occurs while earth resistance tests are being made. Since one of the objectives of tests on a station ground is to establish the location of an effectively remote point for both current and potential electrodes, the leads to the electrodes must be treated as though a possible potential could exist between these test leads and any point on the station ground grid.

9.36 ELECTRICAL LEVEL TWO ◆ TRAINEE GUIDE

Putting It All Together

A low-impedance ground is essential to the performance of any electrical protection system. The ground must dissipate electrical transients and surges in order to minimize the chance of damage or injury. Proper grounding, which includes bonding and connections, protects personnel from the danger of shock and protects equipment and buildings from hazardous voltages. Proper grounding also contributes to the reduction of electrical noise and provides a reference for circuit conductors to stabilize their voltage to ground during normal operation. Examine the grounding system of the electrical service at your home or workplace. Is it adequate? If not, how can it be corrected to meet current *NEC®* requirements?

Review Questions

Refer to the following figure when answering Questions 1 through 4.

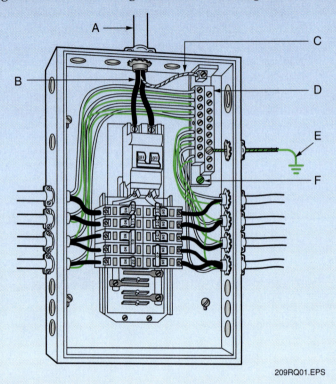

1. *Item B* in the figure represents the _____.
   a. grounded conductor
   b. ungrounded conductors
   c. grounding electrode conductor
   d. main bonding jumper

2. *Item E* in the figure represents the _____.
   a. grounded conductor
   b. ungrounded conductors
   c. grounding electrode conductor
   d. main bonding jumper

3. *Item C* in the figure represents the _____.
   a. grounded conductor
   b. ungrounded conductors
   c. grounding electrode conductor
   d. main bonding jumper

4. *Item F* in the figure represents the _____.
   a. grounded conductor
   b. ungrounded conductors
   c. grounding electrode conductor
   d. main bonding jumper

5. While grounding prevents excessive voltage due to lightning or line surges, it does not stabilize voltage during normal operation.
   a. True
   b. False

6. The _____ conductor(s) are grounded on a single-phase, three-wire, 120/240V system.
   a. hot
   b. current-carrying
   c. neutral
   d. black

7. The neutral conductor is always larger than the ungrounded conductors.
   a. True
   b. False

Review Questions

8. Residential electrical services are grounded by connecting the grounded conductor to a grounding electrode at the _____.
   a. pole-mounted transformer
   b. service drop connection
   c. neutral bus in the main panelboard
   d. meter enclosure

9. The use of _____ as a grounding electrode is in violation of the *NEC*®.
   a. copper cold-water pipe
   b. galvanized pipe
   c. an underground gas line
   d. a grounding ring

10. The maximum distance allowed by the *NEC*® to connect a grounding conductor to a water pipe after the pipe enters the building is _____.
    a. 5'
    b. 10'
    c. 15'
    d. 20'

11. A _____ is used to connect a ground rod with a grounding conductor.
    a. grounding clip
    b. butt-and-slide connector
    c. wire nut
    d. grounding clamp

12. A _____ is the conducting connection, whether intentional or accidental, between any of the conductors of an electrical system.
    a. ground fault
    b. bonding
    c. grounded conductor
    d. short circuit

13. All of the following are suitable materials for ground rods *except* _____.
    a. ½" solid stainless steel rod
    b. ¾" galvanized pipe
    c. ⅝" copperweld solid rod
    d. ½" solid aluminum rod

14. Which of the following copper AWG sizes should be used for the grounding conductor for a 200A service using 3/0 copper conductors?
    a. No. 2
    b. No. 4
    c. No. 6
    d. No. 8

15. When a metal water pipe is used as a grounding electrode, _____ must be provided at the water meter.
    a. bonding jumpers
    b. a floor drain
    c. a grounding locknut
    d. a grounding bushing

Summary

There is no subject in the electrical industry more important than grounding. It is the chief means of protecting life and property from electrical hazards. It also ensures proper operation of the system and helps other protective devices to function properly.

Before a system is grounded, it must first be bonded together. Bonding is the interconnection of all noncurrent-carrying metal components in a system. A bonded system is then connected to a grounding electrode.

The term grounded means connected to earth (ground) by a conductor or to some conducting body that extends the ground connection, such as a ground rod. The earth as a whole is properly classed as a conductor. For convenience, its electric potential is assumed to be zero. When a metal object is grounded, it too is thereby forced to take the same zero potential as the earth. Therefore, the main purpose of grounding is to ensure that the grounded object cannot take on a potential differing sufficiently from earth potential to be hazardous.

Notes

Trade Terms Introduced in This Module

Auxiliary electrodes: Metallic electrodes pushed or driven into the earth to provide electrical contact for the purpose of performing measurements on grounding electrodes or ground grid systems.

Bonding: Connected to establish electrical continuity and conductivity.

Effective ground fault path: An intentionally constructed, permanent, low-impedance electrically conductive path designed and intended to carry current under ground-fault conditions from the point of a ground fault on a wiring system to the electrical supply source and that facilitates the operation of the overcurrent protective device or ground-fault detectors on a high-impedance grounded system.

Equipment bonding jumper: The connection between two or more portions of the equipment grounding conductor.

Equipment grounding conductor: The conductive path installed to connect the normally noncurrent-carrying metal parts of equipment together and to the system grounded conductor, or to the grounding electrode conductor, or both.

Ground: A conducting connection, whether intentional or accidental, between an electrical circuit or equipment and the earth.

Ground current: Current in the earth or grounding connection.

Ground grids: System of grounding electrodes interconnected by bare cables buried in the earth to provide lower resistance than a single grounding electrode.

Ground mats: System of bare conductors, on or below the surface of the earth, connected to a ground or ground grid to provide protection from dangerous touch voltage.

Ground resistance: The ohmic resistance between a grounding electrode and a remote or reference grounding electrode that are spaced such that their mutual resistance is essentially zero.

Ground rod: A metal rod or pipe used as a grounding electrode.

Grounded: Connected to ground or to a conductive body that extends the ground connection.

Grounded conductor: A system or circuit conductor that is intentionally grounded.

Grounding clip: A spring clip used to secure a bonding conductor to an outlet box.

Grounding conductor: A conductor used to connect equipment or the grounded circuit of a wiring system to a grounding electrode or electrodes.

Grounding connections: Connections used to establish a ground; they consist of a grounding conductor, a grounding electrode, and the earth surrounding the electrode.

Grounding electrode: A conducting object through which a direct connection to earth is established.

Grounding electrode conductor: A conductor used to connect the system grounded conductor or the equipment to a grounding electrode or to a point on the grounding electrode system.

Main bonding jumper: The connection between the grounded circuit conductor and the equipment grounding conductor at the service.

Neutral: The conductor connected to the neutral point of a system that is intended to carry current under normal conditions.

Resistivity: Resistance between opposite faces of a unit cube. Expressed in ohm-centimeters or ohms per cubic centimeter.

Separately derived system: A premises wiring system whose power is derived from a source of electric energy or equipment other than a service. Such systems have no direct electrical connection, including a solidly connected ground circuit conductor, to supply conductors originating in another system.

Short circuit: An often unintended low-resistance path through which current flows around, rather than through, a component or circuit.

Step voltage: The potential difference between two points on the earth's surface separated by a distance of one pace, or about three feet.

System grounding: Intentional connection of one of the circuit conductors of an electrical system to ground potential.

Touch voltage: The potential difference between a grounded metallic structure and a point on the earth's surface equal to the normal maximum horizontal reach—approximately three feet.

Ungrounded conductors: Conductors in an electrical system that are not intentionally grounded.

Additional Resources

This module is intended to present thorough resources for task training. The following reference works are suggested for further study. These are optional materials for continued education rather than for task training.

American Electrician's Handbook, Latest Edition. New York: Croft and Summers, McGraw-Hill.

National Electrical Code® Handbook, Latest Edition. Quincy, MA: National Fire Protection Association.

CONTREN® LEARNING SERIES – USER UPDATE

NCCER makes every effort to keep these textbooks up-to-date and free of technical errors. We appreciate your help in this process. If you have an idea for improving this textbook, or if you find an error, a typographical mistake, or an inaccuracy in NCCER's Contren® textbooks, please write us, using this form or a photocopy. Be sure to include the exact module number, page number, a detailed description, and the correction, if applicable. Your input will be brought to the attention of the Technical Review Committee. Thank you for your assistance.

Instructors – If you found that additional materials were necessary in order to teach this module effectively, please let us know so that we may include them in the Equipment/Materials list in the Annotated Instructor's Guide.

Write: Product Development and Revision
National Center for Construction Education and Research
3600 NW 43rd St., Bldg. G, Gainesville, FL 32606

Fax: 352-334-0932

E-mail: curriculum@nccer.org

Craft _____ Module Name _____

Copyright Date _____ Module Number _____ Page Number(s) _____

Description
_____
_____
_____
_____

(Optional) Correction
_____
_____
_____

(Optional) Your Name and Address
_____
_____
_____

Circuit Breakers and Fuses

Inn at the Ballpark, Houston, TX

HOAR Construction won an ABC Excellence in Construction award in the Renovation division for its work in completely gutting a 150,000-square-foot condemned downtown office space and rebuilding it into a 201-room, baseball-theme boutique hotel.

26210-08

26210-08
Circuit Breakers and Fuses

Topics to be presented in this module include:

1.0.0	Introduction	10.2
2.0.0	Circuit Breaker Ratings	10.3
3.0.0	Ground Fault Current Circuit Protection	10.11
4.0.0	Fuses	10.14
5.0.0	Overcurrents	10.17
6.0.0	Guide for Sizing Fuses	10.19
7.0.0	Safety	10.21
8.0.0	Coordination	10.21

Overview

The primary function of fuses and circuit breakers is to protect people and equipment from excessive current by an unintentional load increase or fault condition. A fault condition can be defined as current finding an unintended path back to its point of origin. A fault may occur from one current-carrying conductor to another (short circuit) or from a current-carrying conductor to ground (ground fault).

There are several ratings associated with circuit breakers and fuses. These protective devices have a maximum voltage rating, current trip rating, interrupting capacity rating, and some have a current-limiting rating. Current-limiting devices react much faster to clear the short circuit or fault current. This added characteristic greatly reduces the amount of short circuit or fault current that is allowed to pass through before the device opens, thus reducing the levels of available fault current at the next device downstream.

Ground fault circuit interrupters (GFCI) do not provide overcurrent protection. They are devices that recognize a ground fault condition and open the circuit in which they are connected.

Note: *NFPF 70*®, *National Electrical Code*®, and *NEC*® are registered trademarks of the National Fire Protection Association, Inc., Quincy, MA 02269. All *National Electrical Code*® and *NEC*® references in this module refer to the 2008 edition of the *National Electrical Code*®.

Objectives

When you have completed this module, you will be able to do the following:

1. Explain the necessity of overcurrent protection devices in electrical circuits.
2. Define the terms associated with fuses and circuit breakers.
3. Describe the operation of a circuit breaker.
4. Apply the *National Electrical Code®* (*NEC®*) requirements for overcurrent devices.
5. Describe the operation of single-element and time-delay fuses.

Trade Terms

Cartridge fuse
Circuit breaker
Dual-element fuse
Edison-base
Fault current
Frame size
Fuse
Fuse link
Molded-case circuit breaker
Nonrenewable fuse
Overcurrent protection
Plug fuse
Pole
Short circuit

Required Trainee Materials

1. Pencil and paper
2. Appropriate personal protective equipment
3. Copy of the latest edition of the *National Electrical Code®*

Prerequisites

Before you begin this module, it is recommended that you successfully complete the following: *Core Curriculum; Electrical Level One; Electrical Level Two*, Modules 26201-08 through 26209-08.

This course map shows all of the modules in *Electrical Level Two*. The suggested training order begins at the bottom and proceeds up. Skill levels increase as you advance on the course map. The local Training Program Sponsor may adjust the training order.

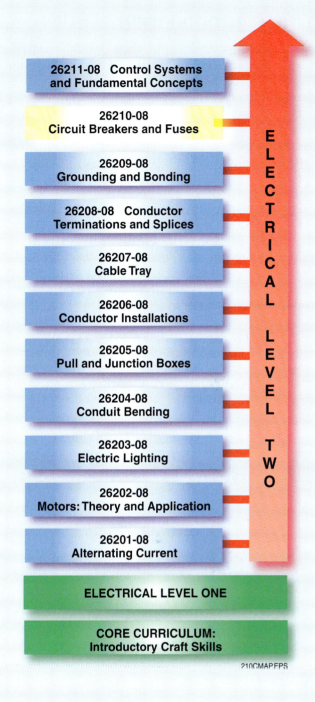

1.0.0 ♦ INTRODUCTION

All electrical circuits and their related components are subject to destructive overcurrents. Harsh environments, general deterioration, accidental damage, damage from natural causes, excessive expansion, and overloading of the electrical system are all factors that contribute to the occurrence of such overcurrents. Reliable protective devices prevent or minimize costly damage to transformers, conductors, motors, equipment, and the many other components and loads that make up the complete electrical system. Therefore, reliable circuit protection is essential to avoid the severe monetary losses that can result from power blackouts and prolonged downtime. To protect electrical conductors and equipment against abnormal operating conditions and their consequences, protective devices are used in the circuits.

Fuses and circuit breakers are two types of automatic overload devices that are normally used in electrical circuits to prevent fires and the destruction of the circuit and its associated equipment.

Basically, a circuit breaker is a device used for closing and interrupting a circuit between separable contacts under both normal and abnormal conditions. This is done manually by switching the handle to the ON or OFF positions. However, the circuit breaker is also designed to open a circuit automatically on a predetermined overload or fault current without damage to itself or its associated equipment. As long as a circuit breaker is applied within its rating, it will automatically interrupt any fault and, therefore, must be classified as an inherently safe overcurrent protective device.

The internal arrangement of a circuit breaker is shown in *Figure 1*, while its external operating characteristics are shown in *Figure 2*. Note that the handle on a circuit breaker resembles an ordinary toggle switch. On an overload, the circuit breaker opens itself, or trips. Once tripped, the handle jumps to the middle position (*Figure 2*). To reset, turn the handle to the OFF position and then turn it as far as it will go beyond this position (RESET); finally, turn it to the ON position.

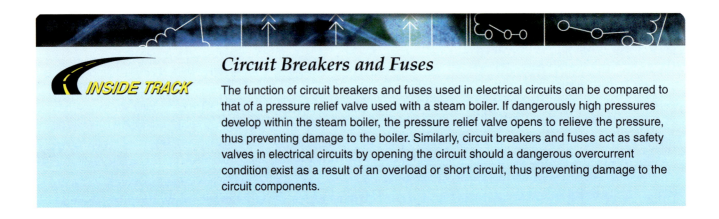

Circuit Breakers and Fuses

INSIDE TRACK

The function of circuit breakers and fuses used in electrical circuits can be compared to that of a pressure relief valve used with a steam boiler. If dangerously high pressures develop within the steam boiler, the pressure relief valve opens to relieve the pressure, thus preventing damage to the boiler. Similarly, circuit breakers and fuses act as safety valves in electrical circuits by opening the circuit should a dangerous overcurrent condition exist as a result of an overload or short circuit, thus preventing damage to the circuit components.

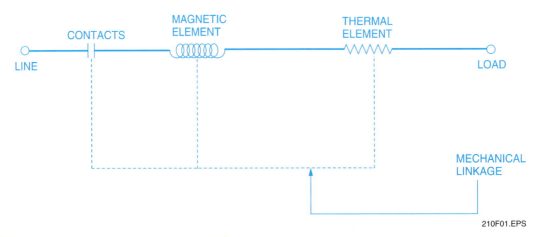

Figure 1 ♦ Internal arrangement of a circuit breaker.

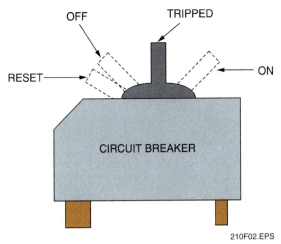

Figure 2 ♦ Operating characteristics of a circuit breaker.

A standard molded-case circuit breaker usually contains:

- A set of contacts
- A magnetic trip element
- A thermal trip element
- Line and load terminals
- Bussing used to connect these individual parts
- An enclosing housing of insulating material

The circuit breaker handle manually opens and closes the contacts and resets the automatic trip units after an interruption. Some circuit breakers also contain a manually operated push-to-trip testing mechanism.

Circuit breakers are classified according to given current ranges. Each group is classified by the largest ampere rating of its range. These groups are:

- 15A–100A
- 125A–225A
- 250A–400A
- 500A–1,000A
- 1,200A–2,000A

Therefore, they are classified as 100A, 225A, 400A, 1,000A, and 2,000A frames. These numbers are commonly referred to as frame classifications or frame sizes and are terms applied to groups of molded-case circuit breakers that are physically interchangeable with each other.

2.0.0 ♦ CIRCUIT BREAKER RATINGS

The established voltage rating of a circuit breaker is based on its clearances or space, both through air and over surfaces between components of the electrical circuit and between the electrical components and ground. Circuit breaker voltage ratings indicate the maximum electrical system voltage on which they can be applied. Underwriters Laboratories, Inc. (UL) recognizes only the ratings listed in *Table 1* for molded-case circuit breakers.

A circuit breaker can be rated for either alternating current (AC) or direct current (DC) system applications or for both. Single-pole circuit breakers, rated at 120/240VAC or 125/250VDC, can be used singly and in pairs on three-wire circuits having a neutral connected to the mid-point of the load. Single-pole circuit breakers rated at

Table 1 Molded-Case Circuit Breaker Ratings Recognized by Underwriters Laboratories

For Alternating Current	For Direct Current
120V	125V
120/240V	—
240V	250V
277V	600V
277/480V	—
480V	—
600V	—

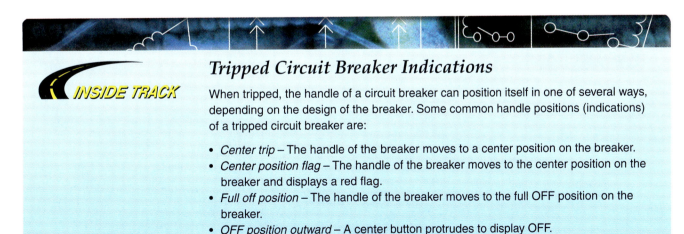

Tripped Circuit Breaker Indications

When tripped, the handle of a circuit breaker can position itself in one of several ways, depending on the design of the breaker. Some common handle positions (indications) of a tripped circuit breaker are:

- *Center trip* – The handle of the breaker moves to a center position on the breaker.
- *Center position flag* – The handle of the breaker moves to the center position on the breaker and displays a red flag.
- *Full off position* – The handle of the breaker moves to the full OFF position on the breaker.
- *OFF position outward* – A center button protrudes to display OFF.

Circuit Breakers

Circuit breakers are available in a wide variety of sizes and types to suit various applications.

120/240VAC or 125/250VDC can also be used in pairs on a two-wire circuit connected to the ungrounded conductors of a three-wire system. Two-pole circuit breakers rated at 120/240VAC or 125/250VDC can be used only on a three-wire, direct current, or single-phase, alternating current system having a grounded neutral. Circuit breaker voltage ratings must be equal to or greater than the voltage of the electrical system on which they are used.

Circuit breakers have two types of current ratings. The first—and the one that is used most often—is the continuous current rating. The second is the fault current interrupting capacity.

2.1.0 Current Rating

The rated continuous current of a device is the maximum current (in amperes) that it will carry continuously without exceeding the specified limits of observable temperature rise. The continuous current ratings of circuit breakers are established based on standard UL ampere ratings. As listed in *NEC Section 240.6 (A)*, these range from 15A to 6,000A. The ampere rating of a circuit breaker is located on the handle of the device (see *Figure 3*).

Generally, the circuit breaker current rating must be equal to or less than the load circuit conductor current-carrying capacity (ampacity).

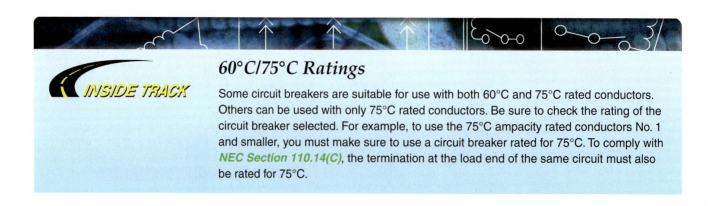

60°C/75°C Ratings

Some circuit breakers are suitable for use with both 60°C and 75°C rated conductors. Others can be used with only 75°C rated conductors. Be sure to check the rating of the circuit breaker selected. For example, to use the 75°C ampacity rated conductors No. 1 and smaller, you must make sure to use a circuit breaker rated for 75°C. To comply with *NEC Section 110.14(C)*, the termination at the load end of the same circuit must also be rated for 75°C.

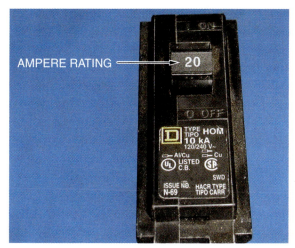

Figure 3 ♦ Circuit breaker current rating.

Most overcurrent protective devices are labeled with the following current ratings:

- Normal current rating
- Interrupting rating

2.2.0 Interrupting Capacity Rating

The amperage interrupting capacity (AIC) of a circuit breaker is the maximum short circuit current at which the breaker will safely interrupt the circuit. The AIC is at the rated voltage and frequency.

NEC Section 110.9 states that equipment intended to interrupt current at fault levels (fuses and circuit breakers) must have an interrupting rating sufficient for the nominal circuit voltage and the current that is available at the line terminals of the equipment.

Equipment intended to break current at other than fault levels must have an interrupting rating at nominal circuit voltage sufficient for the current that must be interrupted.

These *NEC®* statements mean that fuses and circuit breakers (and their related components) that are designed to break fault or operating currents (open the circuit) must have a rating sufficient to withstand such currents. This *NEC®* section emphasizes the difference between clearing fault level currents and clearing operating currents. Protective devices such as fuses and circuit breakers are designed to clear fault currents and, therefore, must have short circuit interrupting ratings that are sufficient for fault levels.

Equipment such as contactors and safety switches have interrupting ratings for currents at levels other than fault levels. Therefore, the interrupting rating of electrical equipment is now divided into two parts:

- Current at fault (short circuit) levels
- Current at operating levels

Most people are familiar with the normal current-carrying ampere rating of fuses and circuit breakers. If an overcurrent protective device is designed to open a circuit when the circuit load exceeds 20A for a given time period, as the current approaches 20A, the overcurrent protective device begins to overheat. If the current barely exceeds 20A, the circuit breaker will open normally or a fuse link will melt after a given period of time with little, if any, arcing. If 40A of current are instantaneously applied to the circuit, the overcurrent protective device will open faster, but again with very little arcing. However, if a fault current occurs that runs the amperage up to 5,000A, an explosion effect would occur within the protective device. One simple indication of this explosion effect is demonstrated by the blackened windows of blown plug fuses.

If this fault current exceeds the interrupting rating of a fuse or circuit breaker, the protective device can be damaged or destroyed; such current can also cause severe damage to equipment and injure personnel. Therefore, selecting overcurrent protective devices with the proper interrupting capacity is extremely important in all electrical systems.

Consider the following analogy using a dammed stream (*Figure 4*) as an example of interrupting

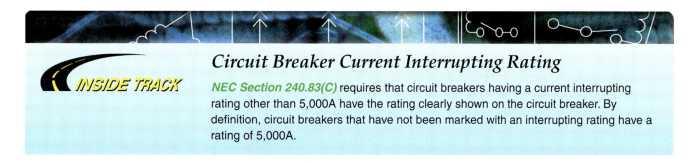

Circuit Breaker Current Interrupting Rating

INSIDE TRACK

NEC Section 240.83(C) requires that circuit breakers having a current interrupting rating other than 5,000A have the rating clearly shown on the circuit breaker. By definition, circuit breakers that have not been marked with an interrupting rating have a rating of 5,000A.

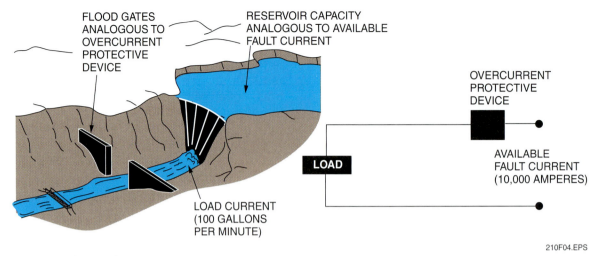

Figure 4 ◆ Normal current operation.

rating. Let the reservoir capacity be the available fault current in an electrical circuit; the flood gates (located downstream from the dam) be the overcurrent protective device in the circuit rated at 10,000 gallons per minute (10,000A); and the stream of water coming through the discharge pipes in the dam be the normal load current. Our drawing shows a normal flow of 100 gallons per minute (100A). Also, note the bridge downstream from the flood gates. This bridge will represent downstream circuit components or equipment connected to the circuit.

Figure 5 shows this same diagram with a fault in the dam—creating a water short circuit that allows 50,000 gallons per minute to flow (50,000 fault circuit amperes). Such a situation destroys the flood gates because of inadequate interrupting rating. The overcurrent protective device in the circuit will also be destroyed. With the flood gates damaged, this surge of water continues downstream, wrecking the bridge. Similarly, the downstream components may not be able to withstand the let-through current in an electrical circuit.

Figure 6 shows the same situation, but with adequate interrupting capacity. Note that the flood gates have adequately contained the surge of water and restricted the let-through current to an amount that can be withstood by the bridge or the components downstream.

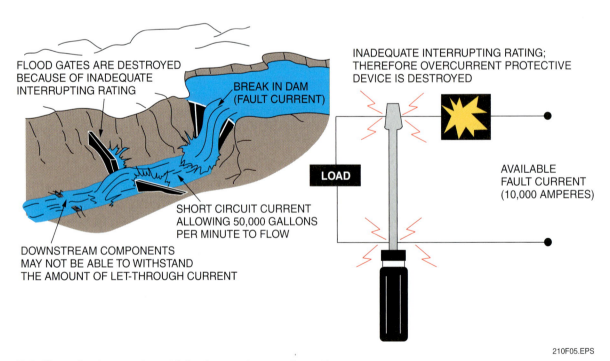

Figure 5 ◆ Short circuit operation with inadequate interrupting rating.

Current Interrupting Rating

THINK ABOUT IT — When replacing a circuit breaker, is it necessary to take into account the current interrupting rating of the replacement breaker? If so, why?

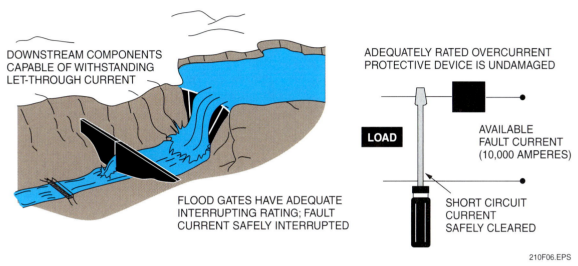

Figure 6 ◆ Short circuit operation with adequate interrupting rating.

There are several factors that must be considered when calculating the required interrupting capacity of an **overcurrent protection** device. *NEC Section 110.10* states that the overcurrent protective devices, the total impedance, the component short circuit current ratings, and other characteristics of the circuit to be protected shall be so selected and coordinated to permit the circuit protective devices used to clear a fault to do so without extensive damage to the electrical components of the circuit. This fault shall be assumed to be either between two or more of the circuit conductors, or between any circuit conductor and the grounding conductor or enclosing metal raceway.

The component short circuit rating is a current rating given to conductors, switches, circuit breakers, and other electrical components, which, if exceeded by fault currents, will result in extensive damage to the component. Short circuit damage can be the result of heat generated or the electromechanical force of a high-intensity magnetic field. The rating is expressed in terms of time intervals and/or current values.

The intent of the *NEC*® is that the design of a system must be such that short circuit currents cannot exceed the short circuit current ratings of the components selected as part of the system. Given specific system components and the level of available short circuit currents that could occur, overcurrent protection devices (mainly fuses and/or circuit breakers) must be used that will limit the let-through current to levels within the withstand ratings of the system components.

In most large commercial and industrial installations, it is necessary to calculate the available short circuit current at various points in a system to determine if the equipment meets the requirements of *NEC Sections 110.9 and 110.10*. There are a number of methods used to determine the short circuit requirements in an electrical system. Some give approximate values; others require extensive computations and are quite exacting. A simple, yet accurate method is the point-by-point method, which will be discussed in more detail later in this module.

An overcurrent protection device must be selected with an interrupting capacity that is equal to or greater than the available short circuit current at the point where the circuit breaker or fuse is applied in the system. The breaker interrupting capacity is based on tests to which the breaker is subjected. There are two such tests; one is set up by Underwriters Laboratories, Inc. (UL) and the other by the National Electrical Manufacturers Association (NEMA). The NEMA tests are self-certification tests, while the UL tests are certified by unbiased witnesses. UL tests have been limited to a maximum of 10,000A in the past, so the emphasis was placed on the NEMA tests with higher ratings. The UL tests now include the NEMA tests plus other ratings. Consequently, emphasis is now placed on the UL tests. It is important to remember that NEMA does not list devices. Two organizations that both test and list devices are UL and the Canadian Standards Association (CSA).

UL requires that molded-case circuit breakers open within a certain period of time during overcurrent conditions. For example, UL requires that a 240V molded-case circuit breaker trip at 300% of its rated continuous current within 50 seconds. If the breaker were rated at 600V, it would be required to trip within 70 seconds at 300% rated current. Currents between 7 and 15 times the rated current (depending on voltage rating and frame size) are handled as overcurrents by the thermal magnetic trip element. Currents above this level activate the magnetic trip element.

The interrupting capacity of a circuit breaker is based on its rated voltage. Where the circuit breaker can be used on more than one voltage, the interrupting capacity will be shown for each voltage level. For example, the LA-type circuit breaker has 42,000A symmetrical interrupting capacity at 240V, 30,000A symmetrical at 480V, and 22,000A symmetrical at 600V.

2.2.1 Standard Interrupting Capacity

Standard interrupting capacity circuit breakers can be identified by their black operating handles and black printed interrupting rating labels. The interrupting rating of a circuit breaker is as important in application as the voltage and current ratings and should be considered each time a breaker is applied. In residential applications, the available fault current is seldom higher than 10,000A or even near this value. The QO and A1 breakers (15A–150A) with the interrupting rating of 10,000A are used in these applications. Higher interrupting ratings are available when required.

2.2.2 High Interrupting Capacity

Where still higher interrupting capacity than the standard ratings discussed previously is required in the 15A–100A FA-type circuit breaker, the FH (H for high interrupting capacity) is available (*Figure 7*). The FH-type circuit breaker has an interrupting rating of 65,000A symmetrical at 240VAC. The continuous current ratings are duplicated in these breakers (15A–100A), but the interrupting capacity has been increased to satisfy the need for greater interrupting capacity. This type of breaker is applied in installations where higher fault currents are available (such as large

Figure 7 ♦ High interrupting capacity circuit breaker.

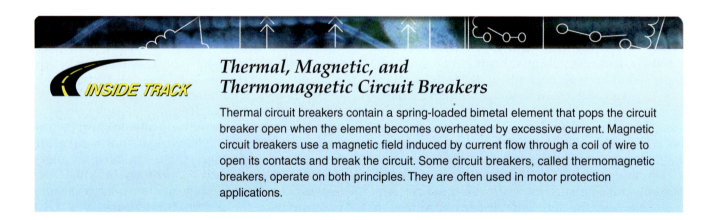

Thermal, Magnetic, and Thermomagnetic Circuit Breakers

Thermal circuit breakers contain a spring-loaded bimetal element that pops the circuit breaker open when the element becomes overheated by excessive current. Magnetic circuit breakers use a magnetic field induced by current flow through a coil of wire to open its contacts and break the circuit. Some circuit breakers, called thermomagnetic breakers, operate on both principles. They are often used in motor protection applications.

industrial plants). These breakers are built with a case material that will withstand the higher shocks from heat and interrupting forces. Circuit breakers with high interrupting capacity are available in the FH, KH, LH, MH, NH, and PH types.

Three ampere ratings (15A, 20A, and 30A) are also available in the QH high interrupting capacity breakers. These breakers satisfy conditions of lighting circuits supplied from these high available fault current circuit breakers.

2.2.3 Current-Limiting Circuit Breakers

The need for current limitation is the result of the increasingly higher available fault currents associated with the growth and interconnection of modern power systems. To meet this need, current-limiting circuit breakers have been developed. This type of breaker operates extremely fast to provide downstream protection for other types of fuses with as little as 10,000 AIC on systems with 100,000A available fault current.

There are circuit breakers available for almost every need in any electrical system. For example, in circuits where the breaker must be highly sensitive to amperage changes (in an environment where the temperature changes are great), an ambient-compensating circuit breaker may be used. Its design permits it to compensate for temperature variation.

A current-limiting circuit breaker is basically a conventional, common-trip, thermomagnetic circuit breaker with an independent limiter section in series with each pole. For purposes of explanation, each limiter can be represented by a set of contacts shunted by a transformable resistor.

At currents below a threshold of about 1,000A, the current-limiting circuit breaker performs in a manner similar to conventional thermomagnetic circuit breakers. In the event of an overload or minor fault current, only the breaker section contacts open to interrupt the circuit. The limiter contacts remain closed.

For 100A frame current-limiting circuit breakers, the threshold of current limitation occurs in the range of 1,000A to 2,000A. Above this point, the limiter contacts open first—starting the limiting action within ¼ of a millisecond of fault initiation. A substantial arc voltage is generated across the limiter contacts. Rapid generation of this arc voltage is essential to drastically limit the peak let-through current in the faulted circuit.

The rapid rising equivalent resistance of the arc across the limiter contacts forces the current to transfer to the alternate path provided by the limiter resistor. This special resistor has a high positive temperature coefficient of resistance. As current is transferred to it, the temperature of the resistor increases rapidly with a consequent rapid rise in resistance. The resistor dissipates and limits let-through energy and increases the short circuit power factor of the circuit nearly to unity, resulting in an easier, more reliable, transient-free interruption.

During the extremely short period of operation of the limiter section, the breaker section tripping mechanism is activated and its contacts begin to open. After the arc between the limiter contacts has been extinguished and all the current is flowing through the limiter resistor, the breaker contacts interrupt the already drastically limited fault current. Because the fault current and system voltage are nearly in phase at this point, complete interruption of the circuit takes place near the first voltage zero of the AC wave.

When a current-limiting circuit breaker trips to clear a faulted circuit, it can be reset in the same manner as a conventional molded-case circuit breaker. This is true whether tripping was due to low-level overload currents or major faults involving the breaker's current-limiting mode of operation. The limiter contacts are automatically closed after circuit interruption. Breaker contacts are reclosed by moving the handle to the extreme OFF position and then to ON. There are no fusible elements in this type of circuit breaker.

Series-Rated Circuit Breakers

The *NEC®* recognizes series-rated circuit breakers and requires that the end use equipment be marked with the series combination rating. Such a series rating of devices is specific to individual manufacturers. For this reason, it is especially important to consult listing information or product literature for details of how such a combination rating was derived, especially for retrofit installations.

Regular maintenance of circuit breakers and their enclosures is necessary to obtain the best service and performance.

Since every circuit breaker failure represents a potential hazard to other equipment in the system, it is difficult to calculate the risks involved in prolonging maintenance inspections. One of the chief causes of circuit breaker failure is high heat caused by loose connections at the load side of the breaker. Another cause of circuit breaker failure is defective loads, that is, electrical equipment that cycles too frequently, which in turn causes the circuit breaker to overheat. Therefore, in the majority of circuit breaker failures, heat is the biggest culprit.

2.3.0 Labeling Considerations

NEC Sections 110.22(A) and 240.86(A) require special marking for UL-recognized, series-rated systems (see *Figure 8*). For example, *NEC Section 110.22(B)* states that where circuit breakers or fuses are applied in compliance with the series combination ratings marked on the equipment by the manufacturer, the equipment enclosure(s) shall be legibly marked in the field to indicate the equipment has been applied with a series combination rating. The marking shall be readily visible and state, "CAUTION—SERIES COMBINATION SYSTEM RATED _____ AMPERES. IDENTIFIED REPLACEMENT COMPONENTS REQUIRED."

Complete coverage of engineered electrical distribution systems giving complete details of these *NEC*® requirements are presented in Electrical Level Three and will not be covered in depth in this module.

2.4.0 *NEC*® Regulations

The following are critical *NEC*® sections with which you should become familiar. These are not the only *NEC*® requirements of importance when dealing with overcurrent protection devices, but they are considered by many to be the most important.

- *NEC Section 110.3(B), Installation and Use* – Listed or labeled equipment shall be used and installed in accordance with any instructions included in the listing or labeling.
- *NEC Section 110.9, Interrupting Rating* – Equipment intended to interrupt current at fault levels shall have an interrupting rating sufficient for the nominal circuit voltage and the current that is available at the terminals of the equipment. Equipment intended to interrupt current at other than fault levels shall have an interrupting rating at nominal circuit voltage sufficient for the current that must be interrupted.
- *NEC Section 110.10, Circuit Impedance and Other Characteristics* – The overcurrent protection devices, total impedance, component short circuit current ratings, and other characteristics of the circuit to be protected shall be so selected and coordinated as to permit the circuit protective devices that are used to clear a fault to do so without extensive damage to the electrical components of the circuit. The fault shall be assumed to be either between two or more of the circuit conductors, or between any circuit conductor and the grounding conductor or enclosing metal raceway.

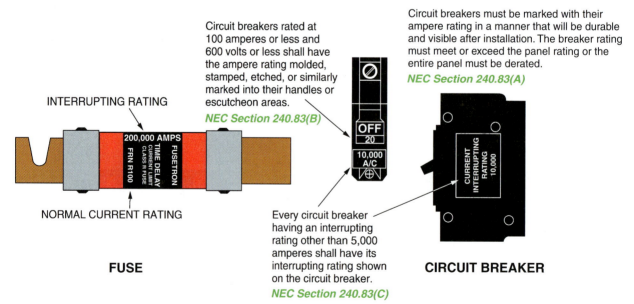

Figure 8 ♦ Most overcurrent protective devices are labeled with two current ratings.

- *NEC Section 240.1, Scope (FPN)* – Overcurrent protection for conductors and equipment is provided to open the circuit if the current reaches a value that will cause an excessively dangerous temperature in conductors or conductor insulation. See also *NEC Sections 110.9 and 110.10* for requirements for interrupting ratings and protection against fault currents.
- *NEC Section 240.2, Definition of Current-Limiting Overcurrent Protective Devices* – A current-limiting overcurrent protection device is a device which, when interrupting currents in its current-limiting range, will reduce the current flowing in the faulted circuit to a magnitude that is substantially less than that obtainable in the same circuit if the device were replaced with a solid conductor having comparable impedance.
- *NEC Section 240.83(C), Interrupting Rating* – Every circuit breaker having an interrupting rating other than 5,000A shall have its interrupting rating shown on the circuit breaker.
- *NEC Section 240.83(E), Voltage Marking* – Circuit breakers shall be marked with a voltage rating that is not less than the nominal system voltage that is indicative of their capability to interrupt fault currents between phases or from phase to ground.
- *NEC Section 240.85, Applications* – A circuit breaker with a straight voltage rating, such as 240V or 480V, may be applied in a circuit in which the nominal voltage between any two conductors does not exceed the circuit breaker's voltage rating; except that a two-pole circuit breaker cannot be used for protecting a three-phase corner-grounded delta circuit unless it is marked single-phase/three-phase to indicate such suitability. A circuit breaker with a slash rating such as 120V/240V or 480Y/277V shall be permitted to be applied on a solidly grounded circuit where the nominal voltage of any conductor to ground does not exceed the lower of the two values of the circuit breaker's voltage rating, and the nominal voltage between any two conductors does not exceed the higher value of the circuit breaker's voltage rating.
- *NEC Section 250.4(A)(1), General Requirements* – Systems and circuit conductors are grounded to limit voltages due to lightning, line surges, or unintentional contact with higher voltage lines, and to stabilize the voltage to ground during normal operation. System equipment and circuit conductors are solidly grounded to the system grounded conductor to facilitate overcurrent device operation in case of fault currents.

Installing Circuit Breakers

In addition to using the correct size circuit breaker (trip current and interrupt current), what are some other factors that should be taken into consideration when selecting and installing a circuit breaker in a panel?

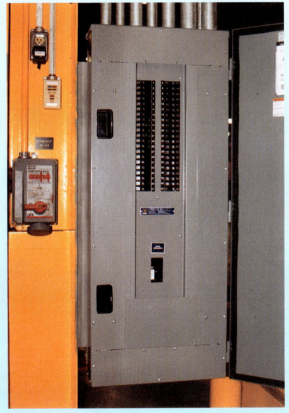

3.0.0 ◆ GROUND FAULT CURRENT CIRCUIT PROTECTION

Fault current protection is extremely important to anyone who works with and uses electrically operated equipment. A fault current exists when an unintended path is established between an ungrounded conductor and ground. This situation can occur not only from worn or defective electrical equipment, but also from accidental misuse of equipment that is in good working order.

Listed HACR Circuit Breakers

HACR circuit breakers are UL listed for use in group motor applications such as heating, air conditioning, and refrigeration (HACR) applications. HACR breakers have a built-in time delay that allows a higher current than its rating to momentarily flow in the circuit. This compensates for the large starting current drawn by such loads. The equipment being protected must also be marked by the manufacturer as being suitable for protection by this type of breaker. (Note: The *NEC*® now permits the use of any inverse-time circuit breaker in heating, air conditioning, and refrigeration equipment comprising group motor installations.)

Will a conventional overcurrent device (fuse or circuit breaker) detect a fault current and open the circuit before irreparable harm is done? Before answering this question, a study of the effects of current on the human body is in order.

The hand-to-hand body resistance of an adult lies between 1,000Ω and 4,000Ω, depending on moisture, muscular structure, and voltage. The average value is 2,100Ω at 240VAC and 2,000Ω at 120VAC.

Using Ohm's law, the current resulting from the above-listed average hand-to-hand resistance values is 120 milliamperes (mA) at 240VAC and 60mA at 120VAC. The effects of 60Hz alternating current on a normal healthy adult are as follows (note that the current is in milliamperes):

- *More than 5mA* – Generally painful shock.
- *More than 15mA* – Sufficient to cause freezing to the circuit for 50% of the population.
- *More than 30mA* – Breathing difficulty (possible suffocation).
- *50mA to 100mA* – Possible ventricular fibrillation or very rapid uncoordinated contractions of the ventricles of the heart resulting in loss of synchronization between the heartbeat and pulse beat. Once ventricular fibrillation occurs, it usually continues and death will happen within a few minutes unless special defibrillation equipment is used.
- *100mA to 200mA* – Certain ventricular fibrillation.
- *Over 200mA* – Severe burns and muscle contractions. The heart is more likely to stop than fibrillate.

GFCI Circuit Protection

Electricity to operate portable power tools at a construction site was supplied by a temporary service. A panel box on the pole had two duplex receptacles in waterproof boxes. One was a regular receptacle and the other a GFCI receptacle. The pole had not been inspected by the city and was not in compliance with code requirements. (It was not grounded.)

A worker used extension cords plugged into the regular receptacle to supply a power tool. The site was wet, humidity was high, and the worker was sweating. Reportedly, he was getting shocks whenever he used the tool. Unknown to the worker, the tool had a ground fault. As he was climbing down to the ground using a piece of metal floor truss as a makeshift ladder, he received a shock, causing him to fall from the floor truss into a puddle of water while still holding onto the tool. He was electrocuted.

The Bottom Line: This accident could have been prevented if the temporary electrical service at the construction site had complied with all local regulations and OSHA standards, including proper grounding. Also, OSHA requires that all 120V, single-phase 15A and 20A receptacles at construction sites that are not part of the permanent wiring be GFCI protected.

Another factor that contributed to this accident was the continued use of a power tool that should have been taken out of service immediately at the occurrence of the first shock. Also, the worker should have used an approved ladder instead of the floor truss to climb down to the ground.

A conventional overcurrent device will not open a circuit without causing harm. Here is why.

The current that would flow from a defective electric drill through the metal housing and through the human body to ground would be 60mA, calculated using the above value for 120VAC. The 60mA current is less than one-half of 1% of the rating of 15A circuit breaker or fuse, and yet it approaches the current level, which may produce ventricular fibrillation. Obviously, the standard circuit breaker or fuse will not open the circuit under such low levels of current flow.

Ground fault circuit interrupters (GFCIs) are Class A protective devices built in accordance with *UL Standard No. 943* for ground fault circuit interrupters. UL defines a Class A device as one that will trip when a fault current to ground is 6mA or more. Also, Class A devices must not trip below 4mA.

Class A GFCIs provide a self-contained means of testing the fault current circuitry, as required by UL. To test, simply push the test button, and the device will respond with a trip indication. UL requires that the current generated by the test circuit shall not exceed 9mA. Also, UL requires the device to be functional at 85% of the rated voltage.

Knowing these facts, OSHA and other codes require the use of fault current circuit interrupters to be used on certain circuits, namely, all construction sites where temporary service is in use for electric power tools. Other required uses include:

- Circuits for electrically operated pool covers
- Power or lighting circuits for swimming pools, fountains, and similar locations
- Receptacles in both commercial and residential garages
- Receptacles in residential bathrooms
- Receptacles installed outdoors in public spaces
- Receptacles installed in residential crawl spaces or unfinished basements
- All kitchen countertop receptacles or any receptacles mounted within 6' of a wet bar sink
- Receptacles installed in bathhouses
- Receptacles installed in bathrooms of commercial or industrial buildings
- Receptacles installed on the roof of any building
- Dockside receptacles
- Branch circuits derived from autotransformers

There are two types of GFCI receptacles. The feed-through type will provide GFCI protection to the remainder of the outlets in a circuit, similar to a GFCI circuit breaker. The second, a nonfeed-through type, offers protection only at the point of installation. Both types are available in various styles.

When installed properly, GFCIs continuously monitor the current in the two conductor wires of the circuit—both the grounded and ungrounded conductors. The current in these two conductors should always be equal. If the GFCI senses a difference between them of more than 6mA, it assumes that the difference is due to fault current and automatically trips the circuit. Power is interrupted within 1/40 of a second or less, which is fast enough to prevent injury to anyone in normal health. Note that the ungrounded and grounded conductors are constantly being monitored. If at any time the sensor detects a difference in current flow between these conductors, this difference will be amplified. See *Figure 9*. The amplified signal will then cause a circuit breaker to open to de-energize the circuit.

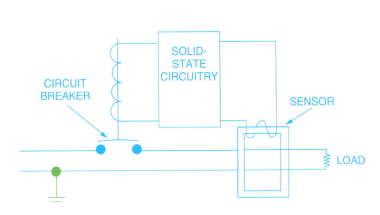
(A) DIAGRAM OF A GFCI CIRCUIT MONITORING A TWO-WIRE CIRCUIT

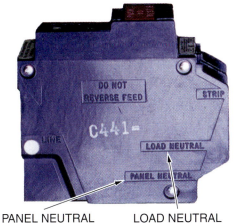
(B) GFCI CIRCUIT BREAKER

Figure 9 ◆ GFCI protection.

4.0.0 ◆ FUSES

A fuse is the simplest device for opening an electric circuit when excessive current flows due to an overload or such fault conditions as grounds and short circuits. A fusible link or links encapsulated in a tube and connected to contact terminals comprise the fundamental elements of the basic fuse. The electrical resistance of the link is so low that it simply acts as a conductor, and every fuse is intended to be connected in series with each phase conductor so that current flowing through the conductor or to any load must also pass through the fuse. The continuous current rating of the fuse in amps establishes the maximum amount of current the fuse will carry without opening. When the circuit current flow exceeds this value, an internal element (link) in the fuse melts due to the heat of the current flow and opens the circuit. Fuses are manufactured in a wide variety of types and sizes with different current ratings, different abilities to interrupt fault currents, various speeds of operation (either quick-opening or time-delay types), different internal and external constructions, and voltage ratings for both low-voltage (600V and below) and medium-voltage (over 600V) circuits.

- *Voltage rating* – Most low-voltage power distribution fuses have 250V or 600V ratings (other ratings are 125V and 300V). The voltage rating of a fuse must be at least equal to the circuit voltage. It can be higher but never lower. For instance, a 600V fuse can be used in a 208V circuit. The voltage rating of a fuse is a function of or depends upon its ability to open a circuit under an overcurrent condition. Specifically, the voltage rating determines the ability of the fuse to suppress the internal arcing that occurs after a fuse link melts and an arc is produced. If a fuse is used with a voltage rating lower than the circuit voltage, arc suppression will be impaired and, under some fault current conditions, the fuse may not safely clear the overcurrent.
- *Ampere rating* – Every fuse has a specific ampere rating. In selecting the ampacity of a fuse, consideration must be given to the type of load and code requirements. The ampere rating of a fuse should normally not exceed the current-carrying capacity of the circuit. For instance, if a conductor is rated to carry 20A, a 20A fuse is the largest that should be used in the conductor circuit. However, there are some specific circumstances under which the ampere rating is permitted to be greater than the current-carrying capacity of the circuit. A typical example is the motor circuit; **dual-element fuses** are generally permitted to be sized up to 175% and nontime-delay fuses up to 300% of the motor full-load amperes. Generally, the ampere rating of a fuse and switch combination should be selected at 125% of the load current. There are exceptions, such as when the fuse-switch combination is approved for continuous operation at 100% of its rating. A protective device must be able to withstand the destructive energy of a short circuit. If a fault current exceeds a level beyond the capability of the protective device, the device may actually rupture and cause severe damage. Thus, it is important when applying a fuse or circuit breaker to use one that can sustain the largest potential short circuit currents. The rating that defines the capacity of a protective device to maintain its integrity when reacting to fault currents is termed its interrupting rating. The interrupting rating of most branch circuit, molded-case circuit breakers typically used in residential service entrance boxes is 10,000A. The rating is usually expressed as 10,000 AIC. Larger, more expensive circuit breakers may have ratings of 14,000 AIC or higher. In contrast, most modern, current-limiting fuses have an interrupting capacity of 200,000A and are commonly used to protect the lower-rated circuit breakers. **NEC Section 110.9** requires equipment intended to interrupt current at fault levels to have an interrupting rating sufficient for the current that must be interrupted.

INSIDE TRACK

Hurricane and Flood Damage Victims

The U.S. Consumer Product Safety Commission has issued a warning to hurricane and flood damage victims that circuit breakers, GFCIs, and fuses that have been submerged under water must be replaced to avoid electrocutions, explosions, and fires. This also applies to panels and electrical devices in general. This is because any flood water and silt trapped inside the devices may prevent them from performing properly. In addition, silt, sand, and salt can also allow tracking to occur between conductors, buses, and connection points.

- *Time-delay ratings* – The time-delay rating of a fuse is established by standard UL tests. All fuses have an inverse time current characteristic. That is, the fuse will open quickly on high currents and after a period of time delay on low overcurrents. Specific types of fuses are made to have specific time delays. The basic UL requirement for Class RK-1, RK-15, and J fuses that are marked time delay is that the fuse must carry a current equal to five times its continuous rating for a period of not less than ten seconds. UL has not developed time-delay tests for all fuse classes. Fuses are available for use where a time delay is needed along with current limitation on high-level short circuits. In all cases, the manufacturer's literature should be consulted to determine the degree of time delay in relation to the operating characteristics of the circuit being protected.

4.1.0 Plug Fuses

Plug fuses have a screw-shell base and are commonly used in dwellings for circuits that supply lighting, heating, and appliances. Plug fuses are supplied with standard screw bases (Edison-base) or Type S bases. An Edison-base fuse consists of a strip of fusible (capable of being melted) metal in a small porcelain or glass case, with the fuse strip or link visible through a window in the top of the fuse. The screw base corresponds to the base of a standard medium-base incandescent lamp. Edison-base fuses are permitted only as replacements in existing installations; all new work must use Type S fuses in accordance with NEC Section 240.51(B), which states that plug fuses of the Edison-base type may only be used for replacements in existing installations where there is no evidence of overfusing or tampering.

The chief disadvantage of the Edison-base plug fuse is that it is made in several ratings (from 0 to 30A), all with the same size base—permitting unsafe replacement of one rating by a higher rating. Type S fuses were developed to reduce the possibility of over-fusing a circuit (inserting a fuse with a rating greater than that required by the circuit). There are 15 classifications of Type S fuses, from 0 to 30A. Each Type S fuse has a base of a different size and a matching adapter. Once an adapter is screwed into a standard Edison-base fuseholder, it locks into place and is not readily removed without destroying the fuseholder. As a result, only a Type S fuse with a size the same as that of the adapter may be inserted. Two types of plug fuses are shown in *Figure 10*; a Type S adapter is also shown.

EDISON-BASE FUSE
(MAXIMUM RATING 30 AMPERES)

TYPE S
FUSE ADAPTER

TYPE S FUSE
(MAXIMUM RATING 30 AMPERES)

Figure 10 ◆ Standard Edison-base fuse on top; Type S with adapter shown on bottom.

Plug fuses are also made in time-delay types that permit a longer period of overload flow before operation, such as on motor inrush current and other higher-than-normal currents. They are available in ratings up to 30A, both in Edison-base and Type S.

Their principal use is in motor circuits, where the starting inrush current to the motor is much higher than the running or continuous current. The time-delay fuse will not open on the inrush of high starting current. If, however, the high current persists, the fuse will open the circuit just as if a short circuit or heavy overload current had developed. All Type S fuses are time-delay fuses.

Plug fuses are normally permitted to be used in circuits of no more than 125V between phases, but they may also be used where the voltage between any ungrounded conductor and ground is not more than 150V. The screwshell of the fuseholder for plug fuses must be connected to the load side

Plug Fuses

Plug fuses are used in the following applications:

- *W series* – Residential loads and nonmotor industrial circuits
- *TL series* – Normal-duty motor loads
- *T series* – When extra time delay or sensitivity to temperature is needed
- *S series* – Time-delay fuses with tamper-resistant dimensions to prevent substitution of incorrect fuses

The condition of plug fuses can be checked visually by looking into the fuse window to observe the condition of the fusible link. If the element is broken and/or the window is black or discolored, the fuse is blown and must be replaced. Note that the condition of the window gives an indication of why the fuse blew. If the element is broken but the window is clear, this indicates an overload occurred; if the window is black or discolored, this indicates a short circuit occurred.

circuit conductor; the base contact is connected to the line side or conductor supply. A disconnecting means (switch) is not required on the supply side of a plug fuse.

The plug fuse is a nonrenewable fuse, or one-time fuse; that is, once it has opened the circuit because of a fault or overload, it cannot be used again or renewed. For safe and effective restoration of circuit operation, it must be replaced by a new fuse of the same rating and characteristics.

4.2.0 Cartridge Fuses

In most industrial and commercial applications, cartridge fuses are used because they have a wider range of types, sizes, and ratings than plug fuses. Older cartridge fuses were provided with a means to renew the fuse by unscrewing the end caps and replacing the links. These fuses presented safety hazards in that it was easy to tamper with them to override the safety feature. Their production has been discontinued. Two types of cartridge fuses are shown in *Figure 11*.

Single-element cartridge fuses – The basic component of this fuse is the link. Depending upon the ampere rating of the fuse, the single-element fuse may have one or more links. They are electrically connected to the end blades (or ferrules) and enclosed in a tube or cartridge surrounded by an arc-quenching filler material.

Under normal operation, when the fuse is operating at or near its ampere rating, it simply functions as a conductor. However, if an overload current occurs and persists for more than a short interval of time, the temperature of the link eventually reaches a level that causes a restricted

Figure 11 ◆ Cartridge fuses.

segment of the link to melt; as a result, a gap is formed and an electric arc established. As the arc causes the link metal to burn back, the gap becomes progressively larger. The electrical resistance of the arc eventually reaches such a high level that the arc cannot be sustained and is extinguished; the fuse will have then completely cut off all current flow in the circuit. Suppression or quenching of the arc is accelerated by the filler material.

Single-element fuses have a very high speed of response to overcurrents. They provide excellent short circuit component protection. However, temporary harmless overloads or surge currents may cause nuisance openings unless these fuses are oversized. They are best used, therefore, in circuits that are not subject to heavy transient surge currents and the temporary overload of circuits with inductive loads such as motors, transformers, and solenoids. Because single-element fuses have a high speed of response to short circuit currents, they are particularly suited for the protection of circuit breakers with low interrupting ratings.

Dual-element cartridge fuses – Unlike single-element fuses, the dual-element fuse can be applied in circuits subject to temporary motor overload and surge currents to provide both high performance short circuit and overload protection. Oversizing in order to prevent nuisance openings is not necessary. The dual-element fuse contains two distinct types of elements. Electrically, the two elements are series connected. The fuse links, similar to those used in the single-element fuse, perform the short circuit protection function; the overload element provides protection against low-level overcurrents or overloads and will hold an overload that is five times greater than the ampere rating of the fuse for a minimum time of ten seconds.

The overload section consists of a copper heat absorber and a spring-operated trigger assembly. The heat-absorber strip is permanently connected to the first short circuit link and to the second short circuit link on the opposite end of the fuse by the S-shaped connector of the trigger assembly. The connector electronically joins the one short circuit link to the heat absorber in the overload section of the fuse. These elements are joined by a calibrated fusing alloy. An overload current causes heating of the short circuit link connected to the trigger assembly. Transfer of heat from the short circuit link to the heat absorber in the mid-section of the fuse begins to raise the temperature of the heat absorber. If the overload is sustained, the temperature of the heat absorber eventually reaches a level that permits the trigger spring to fracture the calibrated fusing alloy and pull the connector free. The short circuit link is electrically disconnected from the heat absorber, the conducting path through the fuse is opened, and the overload current is interrupted. A critical aspect of the fusing alloy is that it retains its original characteristic after repeated temporary overloads without degradation.

Dual-element fuses may also be used in circuits other than motor branch circuits and feeders, such as lighting circuits and those feeding mixed lighting and power loads. The low-resistance construction of the fuses offers cooler operation of the equipment, which permits higher loading of fuses in switches and panel enclosures without heat damage and without nuisance openings from accumulated ambient heat.

4.3.0 Fuse Markings

It is a requirement of the *NEC*® that any cartridge fuses used for branch circuit or feeder protection must be plainly marked, either by printing on the fuse barrel or by a label attached to the barrel that shows the following:

- Ampere rating
- Voltage rating
- Interrupting rating (if other than 10,000A)
- Current limiting, where applicable
- The name or trademark of the manufacturer

4.4.0 UL Fuse Classes

Fuses are tested and listed by Underwriters Laboratories, Inc. in accordance with established standards of construction and performance. There are many varieties of miscellaneous fuses used for special purposes or for supplementary protection of individual types of electrical equipment. However, here we will chiefly be concerned with those fuses used for the protection of branch circuits and feeders on systems operating at 600V or below. Cartridge fuses in this category include UL Classes H, K, G, J, R, T, CC, and L.

5.0.0 ♦ OVERCURRENTS

An overcurrent is either an overload current or a short circuit current. The overload current is an excessive current relative to normal operating current, but one that is confined to the normal conductive paths provided by the conductors and other components and loads of the electrical system.

A short circuit (*Figure 12*) is probably the most common cause of electrical problems. It is an undesired current path that allows the electrical current to bypass the load on the circuit. Sometimes the short is between two wires due to faulty insulation, or it can occur between a wire and a grounded object, such as the metal frame of a motor.

5.1.0 Overloads

Overloads are most often between one and six times the normal current level. Usually, they are caused by harmless temporary surge currents that occur when motors are started up or transformers are energized. Since they are of brief duration, any temperature rise is trivial and has no harmful effect on the circuit components. Therefore, it is important that protective devices (fuses and circuit breakers) do not react to them.

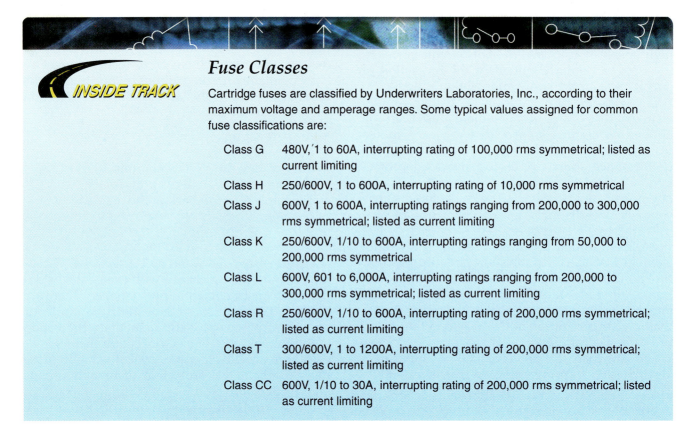

> ### Fuse Classes
>
> Cartridge fuses are classified by Underwriters Laboratories, Inc., according to their maximum voltage and amperage ranges. Some typical values assigned for common fuse classifications are:
>
> Class G   480V, 1 to 60A, interrupting rating of 100,000 rms symmetrical; listed as current limiting
>
> Class H   250/600V, 1 to 600A, interrupting rating of 10,000 rms symmetrical
>
> Class J   600V, 1 to 600A, interrupting ratings ranging from 200,000 to 300,000 rms symmetrical; listed as current limiting
>
> Class K   250/600V, 1/10 to 600A, interrupting ratings ranging from 50,000 to 200,000 rms symmetrical
>
> Class L   600V, 601 to 6,000A, interrupting ratings ranging from 200,000 to 300,000 rms symmetrical; listed as current limiting
>
> Class R   250/600V, 1/10 to 600A, interrupting rating of 200,000 rms symmetrical; listed as current limiting
>
> Class T   300/600V, 1 to 1200A, interrupting rating of 200,000 rms symmetrical; listed as current limiting
>
> Class CC 600V, 1/10 to 30A, interrupting rating of 200,000 rms symmetrical; listed as current limiting

Figure 12 ◆ Short circuit.

Continuous overloads can result from motor defects (such as worn motor bearings), overloaded equipment, or too many loads on one circuit. Such sustained overloads are destructive and must be cut off by protective devices before they damage the electrical distribution system or affect the system loads. However, since they are of relatively low magnitude compared to short circuit currents, removal of the overload current within a few seconds will generally prevent equipment damage. A sustained overload current results in overheating of conductors and other components and will cause deterioration of insulation, which may eventually result in severe damage and short circuits if not interrupted.

Current-Limiting Fuses

Current-limiting fuses have a much faster response to short circuit currents and are typically used to provide protection to mains, feeders, circuit breakers, and other components that lack an adequate interrupting rating.

6.0.0 ♦ GUIDE FOR SIZING FUSES

General guidelines for sizing fuses are given here for most circuits that will be encountered on conventional systems. Some specific applications may warrant other fuse sizing; in these cases, the load characteristics and appropriate *NEC®* sections should be considered. The selections shown here are not, in all cases, the maximum or minimum ampere ratings permitted by the *NEC®*. Demand factors as permitted by the *NEC®* are not included here.

6.1.0 Dual-Element Time-Delay Fuses

Application considerations for dual-element time-delay fuses include the following:

- *Main service* – Each ungrounded service-entrance conductor shall have a fuse in series with a rating not higher than the ampacity of the conductor, except as permitted in *NEC Section 230.90(A), Exceptions*. As specified in *NEC Section 230.91*, the service fuses must either be part of the service disconnecting means or located immediately adjacent to it.
- *Feeder circuit with no motor load* – The fuse size must be at least 125% of the continuous load plus 100% of the non-continuous load.
- *Feeder circuit with all motor loads* – Select the largest branch circuit overcurrent device and add 100% of the full-load current of all other motors connected on the same feeder. The *NEC®* will not permit you to size up; you must size down.
- *Feeder circuit with mixed loads* – Select the largest motor branch circuit fuse and add 100% of the full-load current of all other motors plus 125% of the continuous, non-motor load plus 100% of the non-continuous, non-motor load.
- *Branch circuit with no motor load* – The fuse size must be at least 125% of the continuous load plus 100% of the non-continuous load.

6.2.0 Nontime-Delay Fuses

Application considerations for nontime-delay fuses include the following:

- *Main service* – Service-entrance conductors shall have a short circuit protective device in each ungrounded conductor, either on the load side of or as an integral part of the service-entrance switch. The protective device shall be capable of detecting and interrupting all values of current in excess of its trip setting or melting point, which can occur at its location. A fuse rated in continuous amperes not to exceed three times the ampacity of the conductor, or a circuit breaker with a trip setting of not more than six times the ampacity of the conductors, shall be considered as providing the required short circuit protection.
- *Feeder circuit with no motor load* – The fuse size must be at least 125% of the continuous load plus 100% of the non-continuous load.
- *Feeder circuit with all motor loads* – Select the largest branch circuit overcurrent device and add 100% of the full-load current of all other motors connected on the same feeder. The *NEC®* will not permit you to size up; you must size down.
- *Feeder circuit with mixed loads* – Select the largest motor branch circuit fuse and add 100% of the full-load current of all other motors plus 125% of the continuous, non-motor load plus 100% of the non-continuous, non-motor load.
- *Branch circuit with no motor load* – The fuse size must be at least 125% of the continuous load plus 100% of the non-continuous load.

When sizing fuses for a given application, a schematic drawing of the system will help tremendously. Such drawings do not have to be detailed; a single-line schematic will suffice, such as the one shown in *Figure 13*.

For example, determine the feeder ampacity, motor overload protection, branch circuit short circuit and fault current protection, and feeder protection for one 25hp squirrel-cage induction motor (full starting voltage, nameplate current 31.6A, service factor 1.15, Code letter F), and two 30hp wound rotor induction motors (nameplate primary current 36.4A, nameplate secondary current 65A, 40°C rise), on a 460V, three-phase, 60Hz electrical supply.

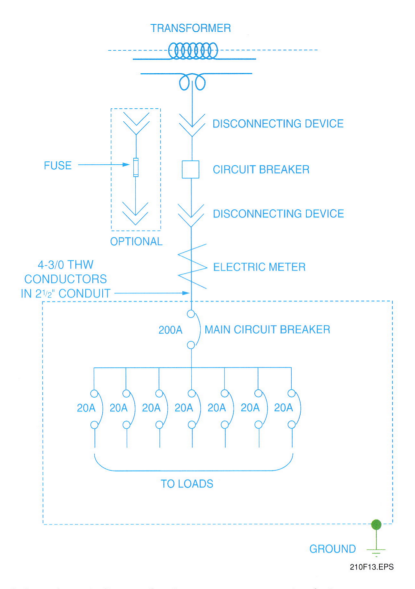

Figure 13 ♦ Typical single-line schematic diagram showing overcurrent protective devices.

- *Feeder ampacity* – The full-load current value used to determine the ampacity of conductors for the 25hp motor is 34A [*NEC Section 430.6(A) and Table 430.250*]. A full-load current of 34A × 1.25 = 42.5A (*NEC Section 430.22*). The full-load current value used to determine the ampacity of primary conductors for each 30hp motor is 40A [*NEC Section 430.6(A) and Table 430.250*]. A full-load primary current of 40A × 1.25 = 50A (*NEC Section 430.22*). A full-load secondary current of 65A × 1.25 = 81.25A [*NEC Section 430.23(A)*]. The feeder ampacity will be (40A × 1.25) + 40A + 34A = 124A (*NEC Section 430.24*).
- *Overload* – Where protected by a separate overload device, the 25hp motor with a nameplate current of 31.6A must have overload protection of not over 39.5A [*NEC Sections 430.6(A) and 430.32(A)(1)*]. Where protected by a separate overload device, the 30hp motor with a nameplate current of 36.4A must have overload protection of not over 45.5A [*NEC Section 430.6(A)*]. If the overload protection is not sufficient to start the motor or carry the load, it may be increased according to *NEC Section 430.32(C)*. For a motor marked thermally protected, overload protection is provided by the thermal protector [*NEC Sections 430.7(A)(13) and 430.32(A)(2)*].
- *Branch circuit short circuit and fault current* – The branch circuit of the 25hp motor must have short circuit and fault current protection of not over 300% for a nontime-delay fuse (*NEC Table 430.52*) or 3.00 × 34A = 102A. The next smallest standard size fuse is 100A. The fuse size may be increased to 110A [*NEC Section 430.52(C)(1), Exception No. 1 and NEC Section 240.6*]. Where the maximum value of branch circuit short

circuit and fault current protection is not sufficient to start the motor, the value for a nontime-delay fuse may be increased to 400% [*NEC Section 430.52(C)(1), Exception No. 2(a)*]. If a time-delay fuse is to be used, see *NEC Section 430.52(C)(1), Exception No. 2(b)*.

- *Feeder circuit* – The maximum rating of the feeder short circuit and fault current protection device is based on the sum of the largest branch circuit protective device (110A fuse) plus the sum of the full-load currents of the other motors, or 110A + 40A + 40A = 190A. The nearest standard fuse that does not exceed this value is 175A [*NEC Section 430.62(A)*].

7.0.0 ◆ SAFETY

Overcurrent protection offers one of the greatest means of protection against electrical faults to both equipment and personnel. However, there are certain precautionary measures that must be observed at all times:

- Make certain the switch or circuit breaker is open before making any inspections or changes to either the overcurrent device or the circuit it is protecting.
- Make certain that all components are compatible with each other and with the system on which they are used; that is, manufacturer, type, normal current ratings, interrupting capacity, voltage (especially voltage), etc.
- Use the proper tools for the job.
- When an overcurrent protection device opens a circuit, determine the reason before changing fuses or resetting the circuit breaker. A severe fault may not damage the circuit or equipment the first time, but repeated hits with high inrush current may do so the second or third time.
- Make sure that adjustable circuit breakers are calibrated correctly (refer to the manufacturer's instructions for calibration details). An improperly calibrated circuit breaker is considered by many to be the most dangerous of circuit breaker faults.
- Use the test breaker to trip devices one phase at a time.
- Never substitute a fuse or circuit breaker that is rated higher than the circuit conductors.
- When working on circuits protected by overcurrent protective devices, lock out the device so that it cannot be inadvertently closed. If locking out the device or switch is not possible, tagging is the next best means to help prevent someone from closing the device while the circuit is being worked on.

8.0.0 ◆ COORDINATION

Coordination is the name given to the time-current relationship among a number of overcurrent devices connected in series, such as fuses in a main feeder, subfeeder, and branch circuits. Safety is the prime consideration in the operation of fuses; however, coordination of the characteristics of fuses is also an important factor in large and complex electrical systems. Every fuse must be properly rated for continuous current and overloads and for the maximum short circuit current the electrical system could feed into a fault on the load side of the fuse, but this is not enough. It might still be possible for a fault on a feeder to open the main service fuse before the feeder fuse opens. Alternately, a branch circuit fault might open the feeder fuse before the branch circuit device opens. Such applications are said to be uncoordinated or nonselective—the fuse closest to the fault is not faster than the one farthest from the fault. When a fault on a feeder opens the main service fuse instead of the feeder fuse, the entire electrical system is taken out of service instead of just the one faulted feeder. Effective coordination minimizes the extent of electrical outage when a fault occurs. It therefore minimizes loss of production, interruption of critical continuous processes, and loss of vital facilities.

Selective coordination is the selection of overcurrent devices with time/current characteristics that ensure the clearing of a fault or short circuit by the device nearest the fault on the line side of the fault. A fault on a branch circuit is cleared by the branch circuit device. The subfeeder, feeder, and main service overcurrent devices will not operate. Alternately, a fault on a feeder is opened

Cartridge Fuse Removal

After the ON-OFF handle of a general-purpose disconnect switch is set to OFF and the door has been opened, is there any danger involved in the removal or installation of the cartridge fuses?

by the feeder fuse without opening any other fuse on the supply side of the feeder. With selective coordination, only the faulted part of the system is taken out of service, which represents the condition of minimum outage.

With proper selective coordination, every device is rated for the maximum fault current it might be called upon to open. Coordination is achieved by studying the curve of current versus the time required for the operation of each device. The selection is then made so that the device nearest any load is faster than all devices closer to the supply, and each device going back to the service entrance is faster than all devices closer to the supply. The main service fuses must have the longest opening time for any branch or feeder fault.

Other Types of Circuit Breakers

In addition to circuit breakers that protect against overloads and ground fault short circuit protection (overcurrents) some circuit breakers are designed with additional capabilities or other functions. The following are some examples of these breakers:

- *Shunt trip circuit breaker* – In addition to the normal operating devices that protect against overcurrents, this breaker has a built-in electric coil that causes it to open the breaker contacts when the coil is energized by an outside source. Typical sources may include fire suppression circuits, pushbuttons, or alarm circuits.
- *Arc fault circuit breaker* – In addition to the normal operating devices that protect against overcurrents, this breaker includes electronic circuits to monitor current flow within the breaker. If it detects a pattern of small continuous surges or spikes in that current flow, the breaker will operate and open the circuit. This pattern of surges or spikes is typical of currents in a short, high-resistance arcing circuit, such as a frayed electrical cord or a loose connection. *NEC Section 210.12(B)* requires these types of breakers to be used to protect all 120V, 15A and 20A branch circuits in residential family rooms, dining rooms, living rooms, parlors, libraries, dens, bedrooms, sunrooms, recreation rooms, closets, hallways, or similar rooms or areas.
- *GFCI breaker* – In addition to the normal operating devices that protect against overcurrents, this breaker includes circuitry to monitor currents on both conductors in the circuit. If there is an imbalance of those currents of 6mA or more, the breaker will operate and open the circuit. This imbalance would indicate that a current path other than that intended has formed, possibly through equipment or persons, and is flowing to ground. GFCI breakers have different ranges to enable protection of people, equipment, or both.
- *Switched neutral breaker* – In addition to the normal operating devices that protect against overcurrents, this breaker disconnects the neutral or grounded conductor simultaneously with all ungrounded conductors. The use of breakers that switch grounded conductors is limited to installations such as fuel dispensing equipment as identified in *NEC Section 514.11*.
- *Non-automatic breaker* – This breaker has no devices to protect the circuit against overcurrent. It is used as a means of manually disconnecting circuits by operating the handle. Its use would be similar to a non-fused disconnect switch. It is also known as a molded case switch.

Putting It All Together

Examine the electrical service in your home. Does it use fuses or circuit breakers? What are the listed amperages? Are there any double breakers? What devices do they serve? Does your furnace use a separate fused disconnect? If the fuses/breakers aren't labeled, take some time to label them now. Can you visualize the internal circuits of your home based on the areas that are linked to the same breaker?

Review Questions

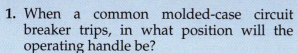

1. When a common molded-case circuit breaker trips, in what position will the operating handle be?
   a. Halfway between the ON and OFF positions
   b. In the ON position
   c. In the OFF position
   d. In the RESET position

2. What is the highest voltage rating for circuit breakers used on DC systems that UL recognizes?
   a. 240V
   b. 277V
   c. 480V
   d. 600V

3. Where is the ampere rating normally marked on circuit breakers?
   a. On the side of the molded case
   b. On the handle
   c. Beneath the terminal lug
   d. On the bottom of the molded case

4. Which of the following is the most important consideration when working with overcurrent protection devices?
   a. The brand name
   b. The type of project
   c. The correct interrupting capacity
   d. The age of the device

5. What is normally the highest short circuit current in residential applications?
   a. 10,000A
   b. 30,000A
   c. 50,000A
   d. 100,000A

6. Every circuit breaker having an interrupting rating other than _____ must have its interrupting rating shown on the circuit breaker.
   a. 1,000A
   b. 5,000A
   c. 10,000A
   d. 65,000A

7. Which of the following best describes where the *NEC®* allows the use of Edison-base fuses?
   a. On new installations only
   b. On circuits rated at over 40A
   c. On circuits rated under 40A
   d. For replacements in existing installations

8. What must be used in conjunction with Type S plug fuses when they are installed in a new fuse holder?
   a. A Type S adapter
   b. A copper penny
   c. A circuit breaker connected in series
   d. A setscrew holding device

9. On renewable cartridge fuses, which part of the fuse is normally renewed?
   a. The barrel
   b. The link
   c. The ferrule
   d. The ferrule ring

10. Which of the following best describes a dual-element fuse?
    a. A fuse having an overcurrent limit and a time delay before activation
    b. A fuse with two attachment points
    c. A fuse having two ferrules at each end
    d. A fuse with two barrels

Summary

Personal safety is the most important issue when completing electrical installations, and the protection of buildings and equipment comes second. Much of this protection is provided by overcurrent protection. This is why the *NEC*®, UL, NEMA, and other organizations put forth so much effort in regulating the type and methods of overcurrent protection.

To fully understand the purpose of overcurrent protection devices, you must first understand the meaning and importance of common electrical terms relating to overcurrent protection. Furthermore, you must understand the key *NEC*® requirements regarding overcurrent protection. When making electrical installations, always check the working drawings to make certain that all parts of the electrical system—that is, short circuit currents, fault currents, interrupting capacities, grounding electrode conductors, electrical grounding conductors, and the like—comply with the latest edition of the *NEC*®.

Also verify that all conductors (circuit, feeder, service, grounding electrode conductors, equipment grounding conductors, and bonding conductors) have adequate capacity to safely conduct any fault current that is likely to be imposed on them.

Determine let-through current values (peak and rms) when current-limiting overcurrent devices are used. Furthermore, apply current-limiting data to protect downstream electrical components that have interrupting ratings and/or withstand ratings less than the available fault current at any given point in the system.

Notes

Trade Terms Introduced in This Module

Cartridge fuse: A fuse enclosed in an insulating tube to confine the arc when the fuse blows.

Circuit breaker: A device that is designed to open a circuit automatically at a certain overcurrent. It can also be used to manually open and close the circuit.

Dual-element fuse: A fuse having two fuse characteristics; the usual combination is an overcurrent limit and a time delay before activation.

Edison-base: The standard screw base used for ordinary lamps and Edison-base plug fuses.

Fault current: The current that exists when an unintended path is established between an ungrounded conductor and ground.

Frame size: A method used to classify circuit breakers according to given current ranges.

Fuse: A protective device that opens a circuit when the fusible element is severed by heating due to a fault current or overcurrent passing through it.

Fuse link: The fusible part of a cartridge fuse.

Molded-case circuit breaker: A circuit breaker enclosed in an insulating housing.

Nonrenewable fuse: A fuse that must be replaced after it interrupts a circuit.

Overcurrent protection: De-energizing a circuit whenever the current exceeds a predetermined value; the usual devices are fuses, circuit breakers, or magnetic relays.

Plug fuse: A type of fuse that is held in position by a screw thread contact instead of spring clips, as is the case with a cartridge fuse.

Pole: That portion of a device associated exclusively with one electrically separated conducting path of the main circuit or device.

Short circuit: The current that exists when an unintended path is created between any two components in a circuit.

Additional Resources

This module is intended to present thorough resources for task training. The following reference work is suggested for further study. This is optional material for continued education rather than for task training.

National Electrical Code® Handbook, Latest Edition. Quincy, MA: National Fire Protection Association.

CONTREN® LEARNING SERIES – USER UPDATE

NCCER makes every effort to keep these textbooks up-to-date and free of technical errors. We appreciate your help in this process. If you have an idea for improving this textbook, or if you find an error, a typographical mistake, or an inaccuracy in NCCER's Contren® textbooks, please write us, using this form or a photocopy. Be sure to include the exact module number, page number, a detailed description, and the correction, if applicable. Your input will be brought to the attention of the Technical Review Committee. Thank you for your assistance.

Instructors – If you found that additional materials were necessary in order to teach this module effectively, please let us know so that we may include them in the Equipment/Materials list in the Annotated Instructor's Guide.

Write: Product Development and Revision
National Center for Construction Education and Research
3600 NW 43rd St., Bldg. G, Gainesville, FL 32606

Fax: 352-334-0932

E-mail: curriculum@nccer.org

Craft _____ Module Name _____

Copyright Date _____ Module Number _____ Page Number(s) _____

Description _____

_____

_____

_____

_____

(Optional) Correction _____

_____

_____

_____

(Optional) Your Name and Address _____

_____

_____

_____

Control Systems and Fundamental Concepts

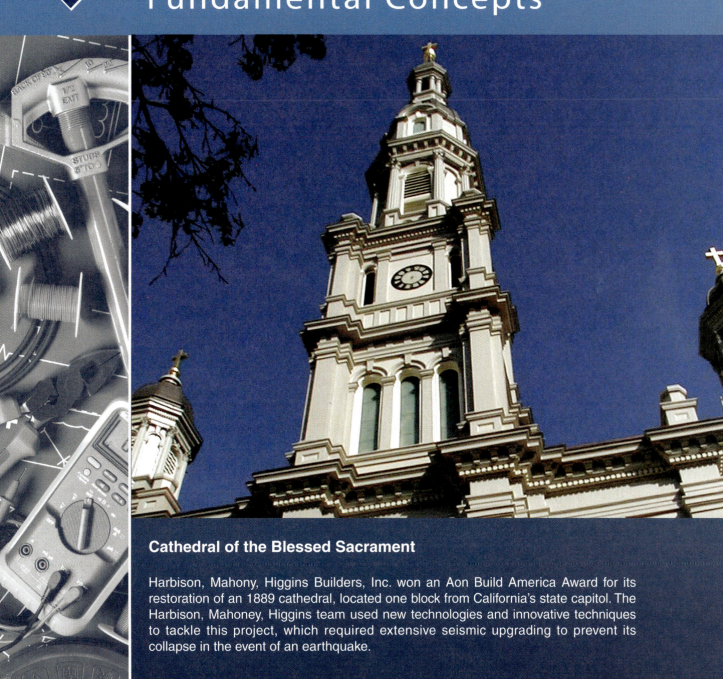

Cathedral of the Blessed Sacrament

Harbison, Mahony, Higgins Builders, Inc. won an Aon Build America Award for its restoration of an 1889 cathedral, located one block from California's state capitol. The Harbison, Mahoney, Higgins team used new technologies and innovative techniques to tackle this project, which required extensive seismic upgrading to prevent its collapse in the event of an earthquake.

26211-08

26211-08
Control Systems and Fundamental Concepts

Topics to be presented in this module include:

1.0.0	Introduction	11.2
2.0.0	Magnetic Contactors	11.2
3.0.0	Relays	11.9
4.0.0	Solid-State Relays	11.13
5.0.0	Overload Relays	11.15
6.0.0	Protective Enclosures	11.18
7.0.0	Low-Voltage Remote Control Switching	11.18
8.0.0	Relay Troubleshooting	11.30

Overview

Contactors and relays are switching devices typically equipped with more than one set of contacts that are opened or closed by electromagnetic energy such as a coil. The coil usually gets its control current from an outside source such as a control circuit. Whenever current flows through the coil, the magnetic fields created by the current flow cause a mechanical action to occur that changes the state of the contacts. Since the circuits that are attached to the contacts are electrically isolated from the coil circuit, the contacts can control circuits with a much higher voltage and current flow than that of the control circuit.

Contactors can control more than one circuit at a time and are limited only by the number of contacts operated by the coil. A common multi-circuit application for contactors is the control of lighting. Contactors are regulated by their ratings. Depending on the type of contactor, typical ratings may include maximum horsepower, continuous current, types of lighting loads, maximum resistance heat loads, maximum transformer kVA, and ratings for switching capacitors.

Relays are contactors that are used to control circuits of lower voltage and amperage ratings. These devices are commonly applied in control circuits and are referred to as control relays. Control relays are available in a variety of configurations and functions, including timers and timing relays, solid-state relays, and overload relays.

Note: *NFPF 70*®, *National Electrical Code*®, and *NEC*® are registered trademarks of the National Fire Protection Association, Inc., Quincy, MA 02269. All *National Electrical Code*® and *NEC*® references in this module refer to the 2008 edition of the *National Electrical Code*®.

Objectives

When you have completed this module, you will be able to do the following:

1. Describe the operating principles of contactors and relays.
2. Select contactors and relays for use in specific electrical systems.
3. Explain how mechanical contactors operate.
4. Explain how solid-state contactors operate.
5. Install contactors and relays according to the *NEC*® requirements.
6. Select and install contactors and relays for lighting control.
7. Read wiring diagrams involving contactors and relays.
8. Describe how overload relays operate.
9. Connect a simple control circuit.
10. Test control circuits.

Trade Terms

Ambient temperature
Break
Inductive load
Instantaneous
Latch coil
Make
Mechanically held relay
Normally closed (N.C.) contacts
Normally open (N.O.) contacts
Shading coil
Solid-state relay
Unlatch coil

Required Trainee Materials

1. Pencil and paper
2. Appropriate personal protective equipment
3. Copy of the latest edition of the *National Electrical Code*®

Prerequisites

Before you begin this module, it is recommended that you successfully complete the following: *Core Curriculum*; *Electrical Level One*; and *Electrical Level Two*, Modules 26201-08 through 26210-08.

This course map shows all of the modules in *Electrical Level Two*. The suggested training order begins at the bottom and proceeds up. Skill levels increase as you advance on the course map. The local Training Program Sponsor may adjust the training order.

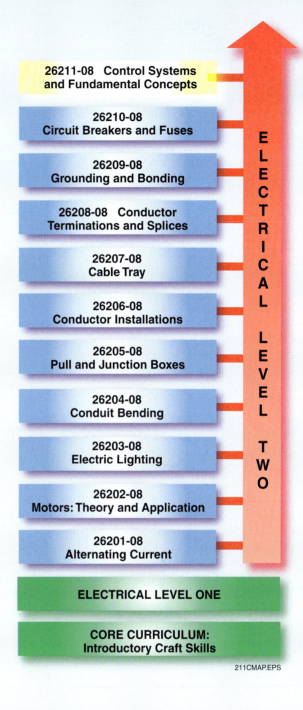

MODULE 26211-08 ♦ CONTROL SYSTEMS AND FUNDAMENTAL CONCEPTS 11.1

1.0.0 ◆ INTRODUCTION

The general classification of contactor covers a type of electromagnetic apparatus that is designed to handle relatively high currents. The conventional contactor is identical in appearance, construction, and current-carrying ability to the equivalent National Electrical Manufacturers Association (NEMA) size magnetic motor starter. The magnet assembly and coil, contacts, holding circuit interlock, and other structural features are the same.

The significant difference is that the contactor does not provide overload protection. Contactors, therefore, are used to *make* or *break* the circuit to high-current, non-motor loads, or are used in motor circuits if overload protection is separately provided. A typical application of the latter is in a reversing motor starter.

A relay is an electromagnetic device whose contacts are used in the control circuits of magnetic starters, contactors, solenoids, timers, and other relays. Relays are generally used to amplify the contact capability or multiply the switching functions of a pilot device.

A tremendous variety of relays are used for lighting, motor, and HVAC control circuits. The magnetic relay is still common, although solid-state control devices are rapidly finding their way into all types of applications.

Besides controlling electrical circuits, some types of relays are also used for circuit and equipment protection. For example, thermal overload relays are intended to protect motors, controllers, and branch circuit conductors against excessive heating due to prolonged motor overcurrents up to and including locked rotor currents. These relays sense motor current by converting this current to heat in a resistance element. The heat generated is used to open a set of contacts in series with a starter coil, causing the motor to be disconnected from the line.

There are also a great many electrical protective relays installed in modern power systems, including:

- Overcurrent (time delay and *instantaneous*)
- Over-/under-voltage
- Directional overcurrent
- Percentage differential

The operating characteristics of these protective relays are all similar and are covered in this module. However, the sensing, adjustment, and actuating functions are quite different and are beyond the scope of this module. For specific instructions, always consult the manufacturer's literature.

2.0.0 ◆ MAGNETIC CONTACTORS

The typical magnetic contactor (*Figure 1*) contains a coil of wire wound around an iron core or a laminated iron core (as shown) and an armature. The coil acts as an electromagnet, and when the proper

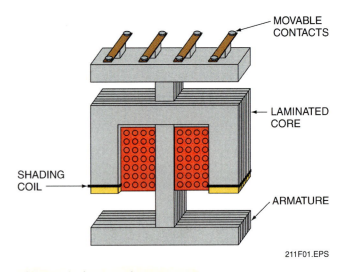

Figure 1 ◆ A magnetic contactor.

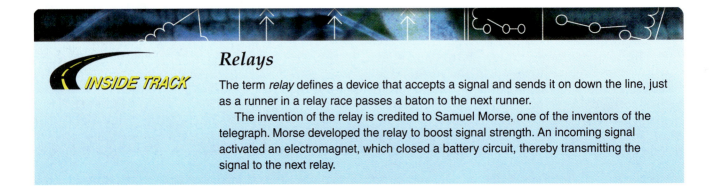

Relays

The term *relay* defines a device that accepts a signal and sends it on down the line, just as a runner in a relay race passes a baton to the next runner.

The invention of the relay is credited to Samuel Morse, one of the inventors of the telegraph. Morse developed the relay to boost signal strength. An incoming signal activated an electromagnet, which closed a battery circuit, thereby transmitting the signal to the next relay.

electric current is applied, it attracts or picks up the armature. Since the movable contacts are attached via an iron bar to the armature, when the armature moves upwards, the movable contacts make contact with a set of stationary contacts to close and complete the circuit.

When power is being applied to the coil, the device is said to be energized. The motion of the armature moves the contact points together or apart, according to their arrangement. When the power is disconnected from the coil, a spring returns the armature to its original position. When no power is being applied, the device is said to be in its de-energized or normal condition. Contacts that are together when the relay is de-energized are said to be **normally closed (N.C.) contacts.** Contacts that are apart when the relay is de-energized are said to be **normally open (N.O.) contacts.**

Notice that a **shading coil** has been added to the iron core in *Figure 1*. Shading coils are used with AC relays to prevent contact chatter and hum. They operate in the same way as in the shaded-pole motor. In general, the shading coil is used to provide a continuous magnetic flow to the armature when the voltage of the AC waveform is zero.

The relay operates on the solenoid principle; that is, it converts electrical energy to linear motion, as shown in *Figure 2*. In this example, a coil of wire is wound around an iron core. When

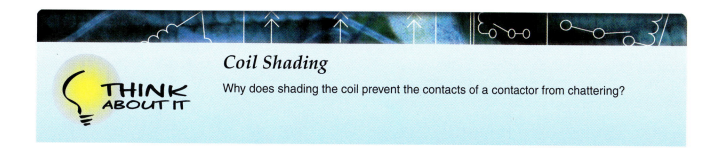

Coil Shading
THINK ABOUT IT — Why does shading the coil prevent the contacts of a contactor from chattering?

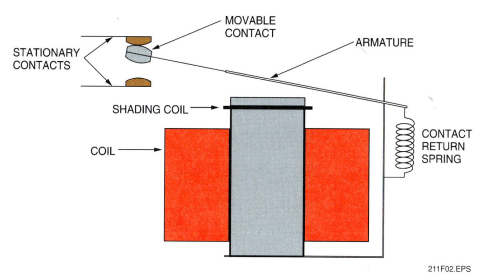

Figure 2 ◆ Contactor operating on the solenoid principle.

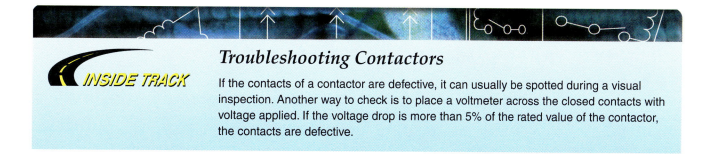

Troubleshooting Contactors
INSIDE TRACK — If the contacts of a contactor are defective, it can usually be spotted during a visual inspection. Another way to check is to place a voltmeter across the closed contacts with voltage applied. If the voltage drop is more than 5% of the rated value of the contactor, the contacts are defective.

current flows through the coil, a magnetic field develops in the iron core. The magnetic field of the iron core attracts the movable arm, known as the armature. In its present position, the movable contact makes connection with a stationary contact. This contact is normally closed. When the armature is attracted to the iron core, the movable contact breaks the connection with the normally closed stationary contact and makes the connection with another stationary contact. This contact is normally open. In a wiring diagram, the movable contact would be the common, and the stationary contacts would be labeled normally open and normally closed.

2.1.0 Lighting Contactors

Filament-type lamps (e.g., tungsten, infrared, and quartz) have inrush currents of approximately 15 to 17 times their normal operating currents. To prevent welding of their contacts on the high initial current, standard motor control contactors must be derated if used to control this type of load.

Table 1 lists electrical ratings for AC magnetic contactors and starters. Referring to this table, a NEMA Size 1 contactor has a continuous current rating of 27A, but if it is used to switch certain lighting loads, it must be derated to 15A. The standard contactor, however, need not be derated for resistance heating or fluorescent lamp loads that do not impose as high an inrush current.

Lighting contactors differ from standard contactors in that the contacts are rated for high inrush currents. A holding circuit interlock is not normally provided because this type of contactor is frequently controlled by a two-wire pilot device, such as a time clock or photoelectric relay.

Unlike standard contactors, lighting contactors are not horsepower-rated or categorized by NEMA size, but are designated by ampere ratings (20A, 30A, 60A, 100A, 200A, and 300A). It should be noted that lighting contactors are specialized in their application and should not be used on motor loads.

A basic circuit using a lighting contactor and serving as a control for a high-amperage lamp is shown in *Figure 3*. Its purpose is to control the lamp from two different locations. The relay coil is indicated by the standard circular symbol and the relay contacts are indicated in the diagram by two parallel lines. The contacts are drawn separated from the coil to show their function. In reality, they are next to the coil in the relay itself.

Closing the N.O. pushbutton energizes the coil of relay R1, and both sets of normally open R1 contacts close. The R1-2 contacts provide a closed circuit to light the lamp. The R1-1 contacts act as holding or seal-in contacts for the relay coil, so that the coil remains energized when the N.O. pushbutton is released.

The lamp will remain lighted until either of the N.C. pushbuttons is momentarily pressed. Pressing either of these buttons opens the circuit to the coil. The coil will drop out or become de-energized and will open the contacts of the relay. The circuit through the lamp is now open and the lamp is turned off. The lamp remains off until one of the N.O. pushbuttons is pressed again. One set of N.O. and N.C. pushbuttons (on and off) is placed at one

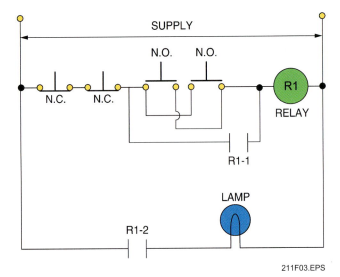

Figure 3 ♦ Pushbutton control circuit with a magnetic relay.

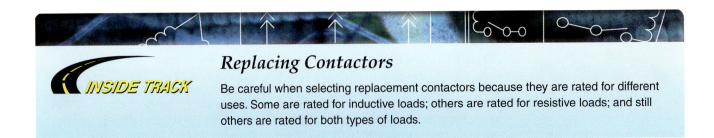

location where remote control of the lamp is desired, and another set is placed in another convenient location. The lamp can be turned on and off from either of these locations.

Another application of a magnetic relay is its use in an emergency lighting circuit. The lamps in such a circuit should be lighted instantly after the failure of the main power supply. A simple emergency lighting system controlling one lamp is shown in *Figure 4*. The main power is supplied through the power supply terminals to the relay coil. The relay has a set of normally open contacts and a set of normally closed contacts. As long as the main power is available, it energizes the relay coil, which in turn keeps the N.O. contacts closed. The power is therefore supplied to the transformer, whose primary is connected in series with the N.O. contacts. The secondary of the transformer keeps the storage battery charged through the rectifier. Since the coil is energized, the N.C. contacts are open and the emergency lamp is not lighted.

In the event of main power failure or low voltage in the line, the relay will drop out, the N.O. contacts will open, and the N.C. contacts will close. The lamp will be connected to the battery and give light as long as the relay coil remains de-energized. Such relay and circuit arrangements are frequently used in the emergency battery packs found in many commercial, industrial, and public buildings of all types.

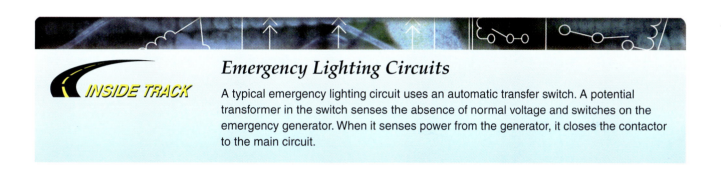

Emergency Lighting Circuits

A typical emergency lighting circuit uses an automatic transfer switch. A potential transformer in the switch senses the absence of normal voltage and switches on the emergency generator. When it senses power from the generator, it closes the contactor to the main circuit.

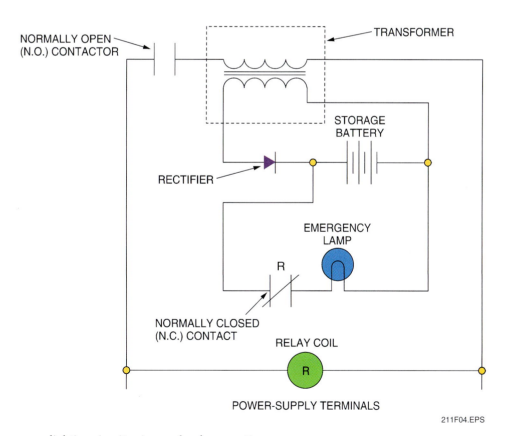

Figure 4 ◆ Emergency lighting circuit using a relay for operation.

Table 1 Electrical Ratings for AC Magnetic Contactors and Starters *(Reprinted with the permission of the National Electrical Manufacturers Association)*

NEMA Size	Volts	Maximum Horsepower Rating—Nonplugging and Nonjogging Duty		Maximum Horsepower Rating—Plugging and Jogging Duty		Continuous Current Rating, Amperes—600 Volt Max.	Service-Limit Current Rating, Amperes	Tungsten and Infrared Lamp Load, Amperes—250 Volts Max.	Resistance Heating Loads, kW—other than Infrared Lamp Loads		KVA Rating for Switching Transformer Primaries at 50 to 60 Cycles		3-Phase Rating for Switching Capacitors
		Single Phase	Poly-Phase	Single Phase	Poly-Phase				Single Phase	Poly-Phase	Single Phase	Poly-Phase	Kvar
00	115	⅓	—	—	—	9	11	5	—	—	—	—	—
	200	—	1½	—	—	9	11	5	—	—	—	—	—
	230	1	1½	—	—	9	11	5	—	—	—	—	—
	380	—	1½	—	—	9	11	—	—	—	—	—	—
	460	—	2	—	—	9	11	—	—	—	—	—	—
	575	—	2	—	—	9	11	—	—	—	—	—	—
0	115	1	—	½	—	18	21	10	—	—	0.9	1.2	—
	200	—	3	—	1½	18	21	10	—	—	—	1.4	—
	230	2	3	1	1½	18	21	10	—	—	1.4	1.7	—
	380	—	5	—	1½	18	21	—	—	—	—	2	—
	460	—	5	—	2	18	21	—	—	—	1.9	2.5	—
	575	—	5	—	2	18	21	—	—	—	1.9	2.5	—
1	115	2	—	1	—	27	32	15	3	5	1.4	1.7	—
	200	—	7½	—	3	27	32	15	—	9.1	—	3.5	—
	230	3	7½	2	3	27	32	15	6	10	1.9	4.1	—
	380	—	10	—	5	27	32	—	—	16.5	—	4.3	—
	460	—	10	—	5	27	32	—	12	20	3	5.3	—
	575	—	10	—	5	27	32	—	15	25	3	5.3	—
1P	115	3	—	1½	—	36	42	24	—	—	—	—	—
	230	5	—	3	—	36	42	24	—	—	—	—	—
	115	3	—	2	—	45	52	30	5	8.5	1	4.1	—
	200	—	10	—	7½	45	52	30	—	15.4	—	6.6	11.3
	230	7½	15	5	10	45	52	30	10	17	4.6	7.6	13

Group	Voltage												
2	380	—	25	—	15	45	52	—	—	28	—	9.9	21
	460	—	25	—	15	45	52	—	20	34	5.7	12	26
	575	—	25	—	15	45	52	—	25	43	5.7	12	33
3	115	7½	—	—	—	—	—	60	10	17	4.6	7.6	—
	200	—	25	—	15	90	104	60	—	31	—	13	23.4
	230	15	30	—	20	90	104	60	20	34	8.6	15	27
	380	—	50	—	30	90	104	—	—	56	—	19	43.7
	460	—	50	—	30	90	104	—	40	68	14	23	53
	575	—	50	—	30	90	104	—	50	86	14	23	67
4	200	—	40	—	25	135	156	120	—	45	—	20	34
	230	—	50	—	30	135	156	120	30	52	11	23	40
	380	—	75	—	50	135	156	—	—	86.7	—	38	66
	460	—	100	—	60	135	156	—	60	105	22	46	80
	575	—	100	—	60	135	156	—	75	130	22	46	100
5	200	—	75	—	60	270	311	240	—	91	—	40	69
	230	—	100	—	75	270	311	240	60	105	28	46	80
	380	—	150	—	125	270	311	—	—	173	—	75	132
	460	—	200	—	150	270	311	—	120	210	40	91	160
	575	—	200	—	150	270	311	—	150	260	40	91	200
6	200	—	150	—	125	540	621	480	—	182	—	79	139
	230	—	200	—	150	540	621	480	120	210	57	91	160
	380	—	300	—	250	540	621	—	—	342	—	148	264
	460	—	400	—	300	540	621	—	240	415	86	180	320
	575	—	400	—	300	540	621	—	300	515	86	180	400
7	230	—	300	—	—	810	932	720	180	315	—	—	240
	460	—	600	—	—	810	932	—	360	625	—	—	480
	575	—	600	—	—	810	932	—	450	775	—	—	600
8	230	—	450	—	—	1215	1400	1080	—	—	—	—	360
	460	—	900	—	—	1215	1400	—	—	—	—	—	720
	575	—	900	—	—	1215	1400	—	—	—	—	—	900

The following example demonstrates how a relay and one single-pole, 15A toggle switch can be used to control 60A of lighting.

Figure 5 shows a diagram of a lighting contactor controlled by a single-pole switch. The contactor coil operates six sets of contacts, each of which is connected to a lighting circuit with lamps totaling 1,200W (10A). When the single-pole switch is turned to the ON position (contacts closed), the coil is energized and moves the armature, which in turn closes all of the contacts simultaneously. When these contacts are closed, this completes the circuit to all of the lamps. Thus, all of the lamps will come on at the same time.

Six single-pole switches could have been used in place of the contactor, but simultaneous switching of the lamps would not be possible.

2.2.0 Reversing Motor Starter

Reversing the direction of motor shaft rotation is often required. Three-phase squirrel cage motors can be reversed by reconnecting any two of the three line connections to the motor. By interwiring two contactors, an electromagnetic method of making the reconnection can be obtained.

As seen in the power circuit in *Figure 6*, the contacts (F) of the forward contactor, when closed, connect lines L1, L2, and L3 to the motor terminals T1, T2, and T3, respectively. As long as the forward contacts are closed, mechanical and electrical interlocks prevent the reverse contactor from being energized.

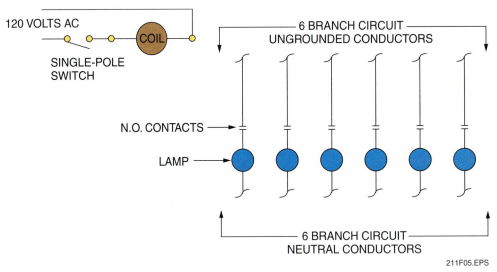

Figure 5 ◆ Several lighting circuits controlled by one contactor.

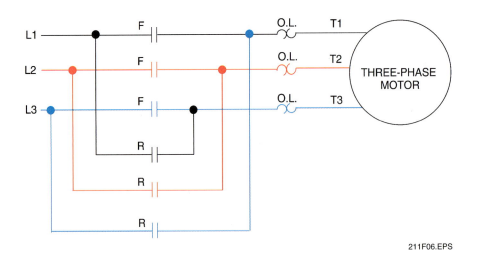

Figure 6 ◆ Three-pole reversing starting diagram.

Relays with Many Contacts

A common example of a relay that controls many circuits is the defrost control on a large refrigeration unit, such as a walk-in cooler. When a timing relay closes the defrost circuit, one set of N.C. contacts opens to shut the fans off, another set of N.C. contacts opens to turn the compressor off, and a set of N.O. contacts closes to turn on the heaters. A set of N.C. contacts then opens to activate a solenoid that mechanically impedes the refrigerant.

When the forward contactor is de-energized, the second contactor can be picked up, closing its contacts (R), which reconnect the lines to the motor. Note that by running through the reverse contacts, L1 is connected to motor terminal T3, and L3 is connected to motor terminal T1. The motor will now run in the opposite direction.

Whether operating through either the forward or reverse contactor, the power connections are run through an overload relay assembly, which provides motor overload protection. A magnetic reversing starter, therefore, consists of a starter and a set of contactors, suitably interwired, with electrical and mechanical interlocking to prevent the coil of both units from being energized at the same time.

Manual reversing starters (employing two manual starters) are also sometimes used. As in the magnetic version, the forward and reverse switching mechanisms are mechanically interlocked, but since coils are not used in manually operated equipment, electrical interlocks are not furnished.

2.3.0 Mechanically Held Contacts

In a conventional contactor, current flow through the coil creates a magnetic pull to seal in the armature and maintain the contacts in a switched position; that is, the N.O. contacts will be held closed while the N.C. contacts will be held open. Because the contactor action is dependent on the current flow through the coil, the contactor is described as electrically held. As soon as the coil is de-energized, the contacts will revert to their initial position.

Versions of mechanically held relays and contactors are also used in some applications. The action is accomplished through the use of two coils and a latching mechanism. Energizing one coil (called the latch coil) through a momentary signal causes the contacts to switch, and a mechanical latch holds the contacts in this position even though the initiating signal is removed. The coil is then de-energized. To restore the contacts to their initial position, a second coil (called the unlatch coil) is momentarily energized.

Mechanically held relays and contactors are used where the hum of an electrically held device would be objectionable, such as in auditoriums, hospitals, churches, etc.

3.0.0 ◆ RELAYS

A control relay is an electromagnetic device that is similar in operating characteristics to a contactor. The contactor, however, is generally employed to switch power circuits or relatively high-current loads.

Relays, with few exceptions, are used in control circuits, and consequently, their lower ratings (15A maximum at 600V) reflect the reduced current levels at which they operate.

Contactors generally have from one to five poles. Although both normally open and normally closed contacts are frequently encountered, the great majority of applications use normally open contacts. There is little, if any, conversion of contact operation on the job.

As compared to contactors, it is not uncommon to find relays used in applications requiring 10 or 12 poles per device, with various combinations of normally open and normally closed contacts. In addition, some relays have convertible contacts, permitting changes to be made in the field from N.O. to N.C. operation, or vice versa, without requiring modification kits or additional components.

3.1.0 Control Relays

A control relay is an electromagnetic device whose contacts are used in the control circuits of magnetic starters, solenoids, timers, and other relays. Relays are generally used to amplify the contact capability or multiply the switching functions of a pilot device.

The wiring diagrams in *Figures 7* and *8* demonstrate how a relay amplifies contact capacity. *Figure 7* represents a current amplification. The relay and starter coil voltages are the same, but the ampere rating of the temperature switch is too low to handle the current drawn by the starter coil (M). A relay is interposed between the temperature switch and the starter coil. The current drawn by the relay coil (CR) is within the rating of the

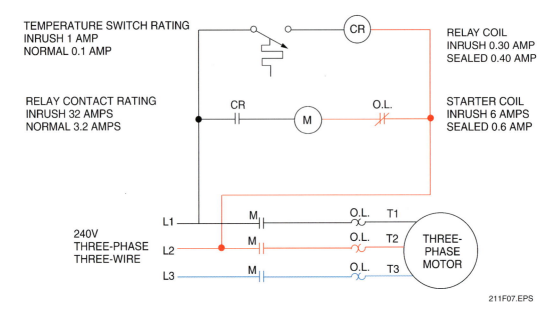

Figure 7 ♦ Relay used to amplify contact capacity.

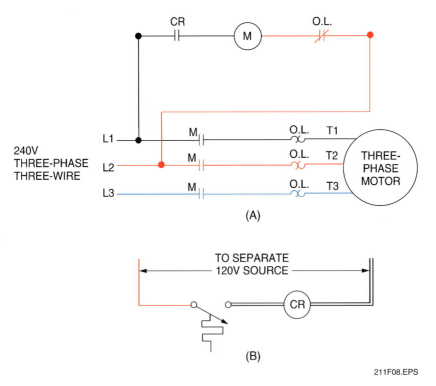

Figure 8 ♦ Relay used for voltage amplification.

11.10 ELECTRICAL LEVEL TWO ♦ TRAINEE GUIDE

temperature switch, and the relay contact (CR) has a rating that is adequate for the current drawn by the starter coil.

Figure 8 represents a voltage amplification. A condition may exist in which the voltage rating of the temperature switch is too low to permit its direct use in a starter control circuit operating at a higher voltage. In this application, the coil of the interposing relay and the pilot device are both wired to a low-voltage source of power that is compatible with the rating of the pilot device. The relay contact, with its higher voltage rating, is then used to control the operation of the starter.

Figure 9 represents another use of relays, which is to multiply the switching functions of a pilot device with a single or limited number of contacts. In this circuit, a single-pole pushbutton contact can, through the use of an interposing six-pole relay, control the operation of a number of different loads such as a pilot light, starter, contactor, solenoid, and timing relay.

Relays are commonly used in complex controllers to provide the logic required to set up and initiate the proper sequencing and control of a number of interrelated operations.

Relays differ in voltage ratings (up to 600V), number of contacts, contact convertibility, physical size, and attachments for accessory functions such as mechanical latching and timing.

In selecting a relay for a particular application, one of the first steps should be a determination of the control voltage at which the relay will operate. Once the voltage is known, the relays that have the necessary contact rating can be further reviewed and a selection made on the basis of the number of contacts and other characteristics required.

Figures 10 and *11* show an 8-pin plug-in relay with its base and wiring diagram.

3.2.0 Timers and Timing Relays

A pneumatic timer or timing relay is similar to a control relay, except that it contains contacts that are designed to operate at a specific time interval after the coil is energized or de-energized. A delay on energization is also referred to as an on delay. A delay on de-energization is called an off delay.

A timed function is useful in applications such as the lubricating system of a large industrial machine, in which a small oil pump must deliver lubricant to the bearings of the main motor for a set period of time before the main motor starts.

In pneumatic timers, the timing is accomplished by the transfer of air through a restricted orifice. The amount of restriction is controlled by an adjustable needle valve, permitting changes to be made in the timing period.

Timing relays are often motor-driven by a clock-type mechanism, but solid-state timing circuits will also be encountered.

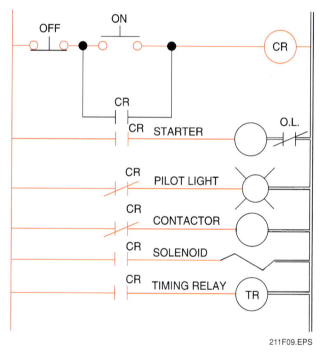

Figure 9 ◆ Relay used to multiply the switching functions of a pilot device.

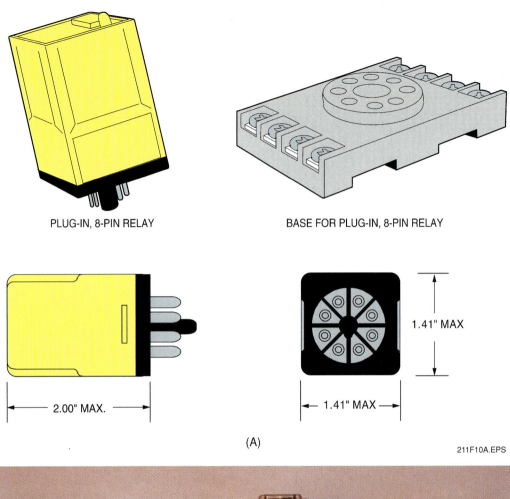

(A)

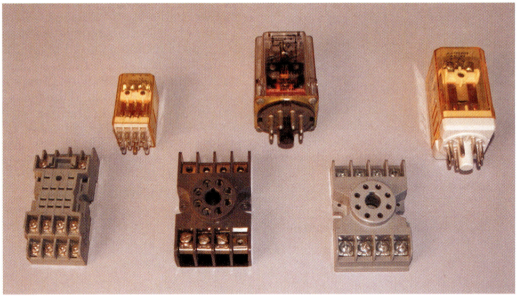

(B)

Figure 10 ◆ Plug-in relays.

11.12 ELECTRICAL LEVEL TWO ◆ TRAINEE GUIDE

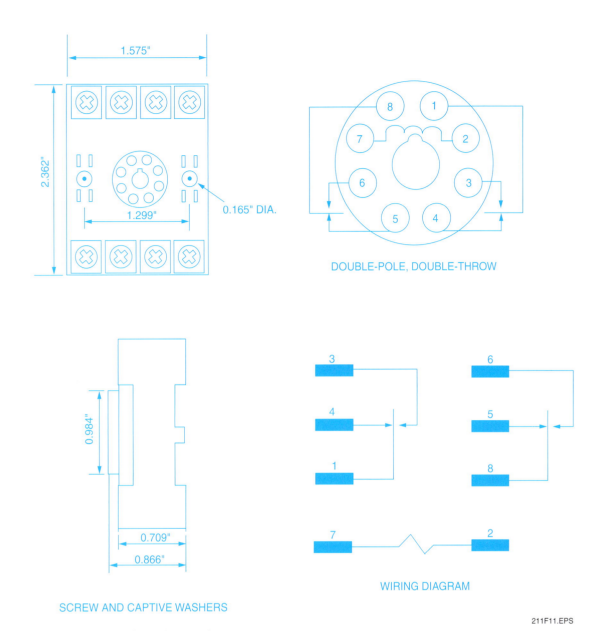

Figure 11 ♦ Wiring diagram for a plug-in relay.

4.0.0 ♦ SOLID-STATE RELAYS

Like magnetic relays, **solid-state relays** are also used in switching applications (*Figure 12*). They are available in different styles and ratings, including time-delay relays. One common use for a solid-state relay is the I/O track of a programmable controller, which is a controller that can be changed by reprogramming rather than requiring an extensive rewiring of the control system. This type of relay can be used to switch a 120V signal to a 5VDC signal.

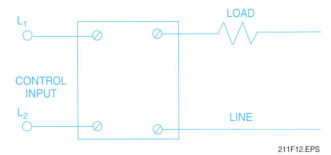

Figure 12 ♦ Basic solid-state relay.

Some of the advantages of solid-state relays include the following:

- No moving parts
- Resistant to shock and vibration
- Sealed against dirt and moisture
- Control input voltage isolated from relay-controlled circuit

NOTE
Solid-state relays are much more sensitive to loads than equivalent mechanical devices. Always follow the manufacturer's application instructions.

Figure 13 shows typical solid-state relay control circuits. These include:

- *Power transistors for DC loads* – In this DC circuit, a power transistor is used to connect the load to the line, as shown in *Figure 13(A)*. This is an opto-isolated relay with a light-emitting diode (LED) connected to the input or control voltage. This is a common circuit arrangement in solid-state relays, and relays that use this coupling method are referred to as opto-isolated because the load side of the relay is optically isolated from the control side of the relay. Since a light beam (rather than electricity) is used to control the relay, no voltage spikes or electrical noise produced on the load side of the relay can be transmitted to the control side of the relay. When the LED is activated by the input voltage, a photo-detector connected to the base of the transistor turns the transistor on

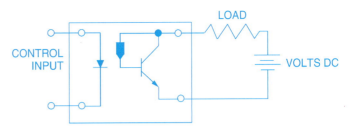

(A) POWER TRANSISTOR USED TO CONTROL A DC LOAD

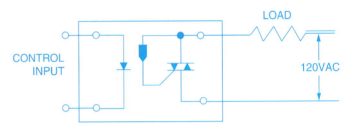

(B) TRIAC USED TO CONTROL AN AC LOAD

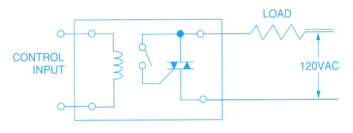

(C) REED RELAY CONTROLS THE OUTPUT

Figure 13 ◆ Solid-state relay circuits.

Plug-In Relays

Plug-in relays make troubleshooting and replacement of relays much easier. Instead of having to connect and disconnect wiring, a new relay can simply be snapped into place.

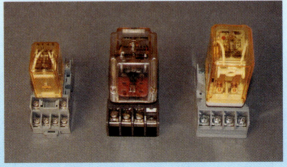

11.14 ELECTRICAL LEVEL TWO ◆ TRAINEE GUIDE

Timing Relays

Some of the simplest examples of timing relays are the plug-in devices that turn a household lamp on and off or the clocks that control a night-rate water heater. These relays are usually mechanical switches. A more sophisticated example of a timing relay is the relay that energizes half the windings of a split-wound, three-phase motor a microsecond after initially energizing the motor.

Solid-State Relays

Several advantages of solid-state relays are longer life, higher reliability, higher speed switching, and high resistance to shock and vibration. The absence of mechanical contacts eliminates contact bounce, arcing when contacts open, and hazards from explosives and flammable gases.

---

and connects the load to the line via this optical coupling.

- *Triac-controlled AC relays* – In this AC circuit, a triac is connected to the load in the place of a power transistor. Refer to *Figure 13(B)*. This circuit also uses an LED as the control device. When the photo-detector detects the LED, it triggers the gate of the triac and connects the load to the line.
- *Reed-controlled AC relays* – Another method of controlling solid-state relays in AC circuits is to use a small reed relay to control the output, as shown in *Figure 13(C)*. In this circuit, small sets of reed contacts are connected to the gate of the triac and the control circuit is connected to the coil of the reed relay. When the control voltage sends a current through the coil, the magnetic field produced around the coil of the relay closes the reed contacts and activates the triac. Unlike opto-isolated relays, this type of solid-state relay uses a magnetic field to isolate the control circuit from the load circuit.

Solid-state relays normally have control voltages from 3V to 32V. Triac-controlled relays commonly have load voltage ratings of 120VAC to 240VAC, with current ratings between 5A and 25A.

Many solid-state relays incorporate a zero switching feature, which means that if the relay is programmed to turn off when the AC voltage is in the middle of a cycle, it will continue to conduct until the AC voltage returns to zero and turns off. Zero switching applications include some inductive loads such as transformers. In normal operation, the core material of a transformer can be left saturated on one end of the flux swing if power is removed from the primary winding when the AC voltage is at its positive or negative peak. This can cause inrush currents of up to 600% of the normal operating current when power is restored to the primary. If the circuit is equipped with a zero switching feature, the power will return to zero before the relay is de-energized.

5.0.0 ♦ OVERLOAD RELAYS

To protect a motor and its related circuits from accidental or prolonged overloads, either the starter or the motor should be equipped with automatic devices that will open the circuit in the event of an overload. This protection may be provided by fuses, circuit breakers, or overload relays.

Overcurrent protection must be provided in the line of every motor circuit, but additional protection should be provided in the form of magnetic overload relays. These are used in both manual and automatic starters.

Figure 14 shows a simplified bimetallic overload relay. The load current passes through a heater element in the overload relay. The heater element is replaceable and contains a resistance. When the current exceeds a certain value, the heat radiated from the element causes a bimetallic strip to spring away, interrupting the control circuit or other mechanical action. The heater element is sized to the full load current of the motor. A small overload may have a long delay, while a larger overload will have a shorter delay.

Most overload relays used on modern motor starters are thermally operated and usually consist of two strips of metal that are welded together. Each has a different degree of thermal expansion. If this bimetallic strip is heated, it will deflect sufficiently to trip two normally closed contacts,

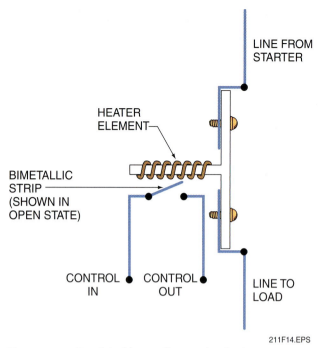

Figure 14 ♦ Simplified bimetallic overload relay.

The ideal overload protection for a motor is an element with current-sensing properties that are very similar to the heating curve of the motor it protects and will act to open the motor circuit when the full-load current is exceeded. The operation of the protective device should be such that the motor is allowed to carry harmless overloads but is quickly removed from the line when an overload has persisted for too long.

Fuses are not the best choice to provide overload protection. Their basic function is to protect against short circuits (overcurrent protection). Motors draw a high inrush current when starting, and conventional single-element fuses have no way of distinguishing between this temporary and harmless inrush current and a damaging overload. Such fuses, if chosen on the basis of motor full-load current, would blow every time the motor started. On the other hand, if a fuse were chosen so that it was large enough to pass the starting or inrush current, it would not protect the motor against small, sustained overloads that might occur later.

Dual-element or time-delay fuses can provide motor overload protection but have the disadvantage of being nonrenewable and must be replaced once tripped.

The overload relay is the heart of motor protection. It has inverse trip time characteristics, permitting it to hold in during the accelerating period (when inrush current is drawn) yet providing protection on small overloads above the full-load current when the motor is running. Unlike dual-element fuses, overload relays are renewable and can withstand repeated trip and reset cycles without needing replacement. They cannot, however, take the place of overcurrent protective equipment.

An overload relay consists of a current-sensing unit connected in the line to the motor, plus a mechanism that is actuated by the sensing unit and serves to directly or indirectly break the circuit. In a manual starter, an overload trips a mechanical latch, causing the starter contacts to open and disconnect the motor from the line. In magnetic starters, an overload opens a set of

which in turn will open the circuit connected to the holding coil of the magnetic contactor, causing the main contacts to open. An advantage of this protective device is that it provides a time delay, which prevents the circuit from being opened by momentary high starting currents and short overloads. At the same time, it protects the motor from prolonged overloading.

Conditions of motor overload may be caused by an overload on driven machinery, by a low line voltage, or by an open line in a poly-phase system, which results in single-phase operation. Under any conditions of overload, a motor draws excessive current that causes overheating. Since the motor winding insulation will deteriorate when subjected to overheating, there are established limits on motor operating temperatures. To protect a motor from overheating, overload relays are employed on a motor control to limit the amount of current drawn. This is known as overload protection or running protection.

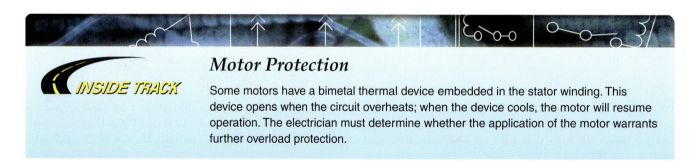

Motor Protection

Some motors have a bimetal thermal device embedded in the stator winding. This device opens when the circuit overheats; when the device cools, the motor will resume operation. The electrician must determine whether the application of the motor warrants further overload protection.

contacts within the overload relay itself. These contacts are wired in series with the starter coil in the control circuit of the magnetic starter. Breaking the coil circuit causes the starter contacts to open, disconnecting the motor from the line.

Overload relays can be classified as being either thermal or magnetic. Magnetic overload relays react only to current excesses and are not affected by temperature. As the name implies, thermal overload relays rely on the rising temperatures caused by the overload current to trip the overload mechanism. Thermal overload relays can be further subdivided into two types: melting alloy (eutectic) and bimetallic.

The melting alloy assembly of a heater element and solder pot are shown in *Figure 15*. The motor load current passes through the heater element. When it exceeds the designated capacity of the heater, it melts an alloy solder pot. The ratchet wheel will then be allowed to turn in the molten pool, and a tripping action of the starter control circuit results, stopping the motor. A cooling-off period is required to allow the solder pot to harden before the overload relay assembly may be reset and the motor service restored.

Melting alloy thermal units are interchangeable and of a one-piece construction, which ensures a constant relationship between the heater element and solder pot and allows for factory calibration, making them virtually tamperproof in the field. These important features are not possible with any other type of overload relay construction. Many types of interchangeable thermal units are available to provide precise overload protection from any full-load current.

Bimetallic overload relays have two major advantages: first, the automatic reset feature is desirable when devices are mounted in locations that are not easily accessible for manual operation; and, second, these relays can easily be adjusted to trip within a range of 85% to 115% of the nominal trip rating of the heater unit. This feature is useful when the recommended heater size might result in unnecessary tripping, while the next larger size would not give adequate protection. The ambient temperature affects overload relays operating on the principle of heat.

Ambient-compensated bimetallic overload relays were designed for one particular situation, that is, when the motor is at a constant temperature and the controller is located separately and has a varying temperature. In this case, if a standard thermal overload relay were used, it would not trip consistently at the same level of motor current if the controller temperature changed. A thermal overload relay is always affected by the surrounding temperature. To compensate for temperature variations, an ambient-compensated overload relay is applied. Its trip point is not affected by temperature and it performs consistently at the same value of current.

Melting alloy and bimetallic overload relays are designed to approximate the heat actually generated in the motor. As the motor temperature increases, so does the temperature of the thermal unit. Motor and relay heating curves show this relationship. From such graphs (obtainable from the manufacturers of overload relays), you can see that, no matter how high the current drawn, the overload relay will provide protection, yet the relay will not trip out unnecessarily.

HEAT WINDING
(HEAT-PRODUCING ELEMENT)

FRONT

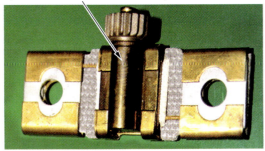

SOLDER POT
(HEAT-SENSITIVE ELEMENT)

BACK

Figure 15 ◆ Melting alloy thermal unit.

When selecting thermal overload relays, the following must be considered:

- Motor full-load current
- Type of motor
- Difference in ambient temperature between the motor and controller

Motors of the same horsepower and speed do not all have the same full-load current, and the motor nameplate must always be referred to and checked to obtain the full-load amperes for a particular motor. Thermal unit selection tables are published on the basis of continuous-duty motors operating under normal conditions with a 1.15 service factor. These tables are shown in manufacturers' catalogs and also appear on the inside of the door or cover of the motor controller. These values are selected to properly protect the motor and allow it to develop its full horsepower, allowing for the service factor if the ambient temperature is the same at the motor as at the controller. If the temperatures are not the same, or if the motor service factor is less than 1.15, a special procedure is required to select the proper thermal unit.

Standard overload relay contacts are closed under normal conditions and open when the relay trips. An alarm signal is sometimes required to indicate when a motor has stopped due to an overload trip. Also, with some machines, particularly those associated with continuous processing, it may be required to signal an overload condition, rather than have the motor and process stop automatically. This is done by fitting the overload relay with a set of contacts that close when the relay trips, thus completing the alarm circuit. These contacts are appropriately called alarm contacts.

A magnetic overload relay has a movable magnetic core inside a coil that carries the motor current. The flux set up inside the coil pulls the core upward. When the core rises far enough, it trips a set of contacts on the top of the relay. The movement of the core is slowed by a piston working in an oil-filled dashpot mounted below the coil. This produces an inverse-time characteristic. The effective tripping current is adjusted by moving the core on a threaded rod. The tripping time is varied by uncovering oil bypass holes in the piston. Because of the time and current adjustments, the magnetic overload relay is sometimes used to protect motors having long accelerating times or unusual duty cycles.

6.0.0 ◆ PROTECTIVE ENCLOSURES

The correct selection and installation of an enclosure for a particular application can contribute considerably to extended service life and trouble-free operation. To shield live parts from accidental contact, some form of enclosure is always necessary. This function is usually filled by a general-purpose sheet steel cabinet. Frequently, however, dust, moisture, or explosive gases make it necessary to employ a special enclosure to protect the motor controller from corrosion or the surrounding equipment from explosion. In selecting and installing control apparatus, it is always necessary to carefully consider the conditions under which the apparatus must operate; there are many applications where a general-purpose enclosure does not afford protection.

Underwriters Laboratories, Inc., has defined the requirements for protective enclosures according to hazardous conditions, and the National Electrical Manufacturers Association (NEMA) has standardized enclosures from these requirements. These enclosures were discussed in detail in an earlier module.

7.0.0 ◆ LOW-VOLTAGE REMOTE CONTROL SWITCHING

In applications where lighting must be controlled from several points, where there is a complexity of lighting or power circuits, or where flexibility is desirable in certain systems, low-voltage remote control relay systems have been applied. Basically, these systems use special low-voltage components operated from a transformer to switch relays, which in turn control the standard line voltage circuits. Because the control wiring does not carry the line load directly, small, lightweight cable can be used. It can be installed wherever and however convenient—placed behind moldings, stapled to woodwork, buried in shallow plaster channels, or installed in holes bored in wall studs. A basic circuit of a remote control switching system is shown in *Figure 16*. In both circuits, the relay permits positive control for on and off. It can be located near the load or installed in centrally located distribution panel boxes, depending on the application.

Step-Down Transformers

Step-down transformers like the one shown here can provide low voltage for remote control switching circuits.

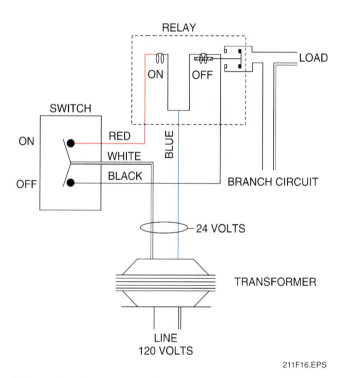

Figure 16 ◆ Basic circuit of a remote control switching system.

Because no line voltage flows through the control circuits and low voltage is used for all switch and relay wiring, it is possible to place the controls at a great distance from the source or load, thus offering many advantages.

7.1.0 Advantages of Remote Control Switching

Since branch circuits go directly to the loads in this type of system, no line voltage switch legs are required. This saves costly runs of larger conductor through all switches and saves installation time and costs if multipoint switches are used.

Under certain conditions, the *NEC®* allows low-voltage conductors to be run in open spaces above ceilings and through wall partitions without further protection. An outlet box is not required at the switch locations. When the electricians rough-in the wiring, they merely secure a plaster ring of the correct depth at each switch location and usually wrap the low-voltage cable around a nail driven behind the plate. The switch and its cover are then installed after the wall is finished.

Low-voltage switching is especially useful in rewiring existing buildings, because the small cables are as easy to run as telephone wires. They are easy to hide behind baseboards or even behind quarter-round molding. The cable can be run exposed without being very noticeable because of its small size. The small, flexible wires are easily fished in partitions.

Figure 17 shows several types of remote control circuits. In the circuits shown, any number of on-off switches can be connected to provide control from many remote points. In a typical installation, outside lights could be controlled from any area within the building without the need to run full-size cable for three- or four-way switch legs through the building. By running a low-voltage line from each relay to a central point, such as an exit door, all relays can be operated from one spot using a selector switch; this allows quick, convenient control of all exterior and interior lighting and power circuits.

A motor-driven rotary switch is available that allows control of 25 (or more) separate circuits with the push of one button. Dimming can also be accomplished by using the motorized control unit together with a modular incandescent dimming system.

7.2.0 System Components

The relay is the heart of a remote control switching system. It employs a split low-voltage coil to move the line voltage contact armature to the ON (latched) position. As illustrated in *Figure 18*, the ON coil moves the armature to the right when a 24VAC control signal is impressed across its leads.

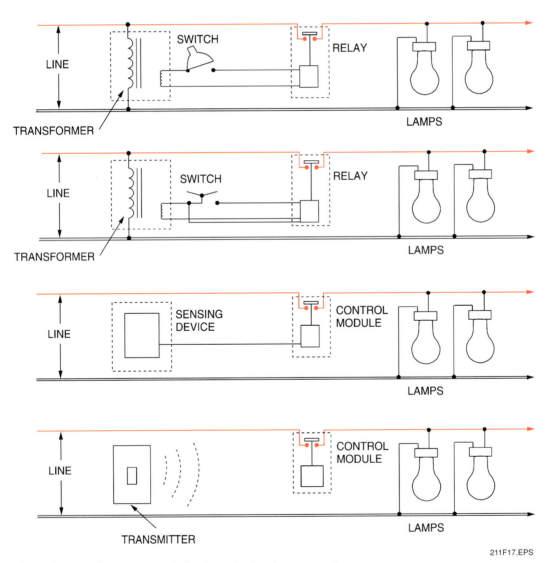

Figure 17 ♦ Several types of remote control circuits and related components.

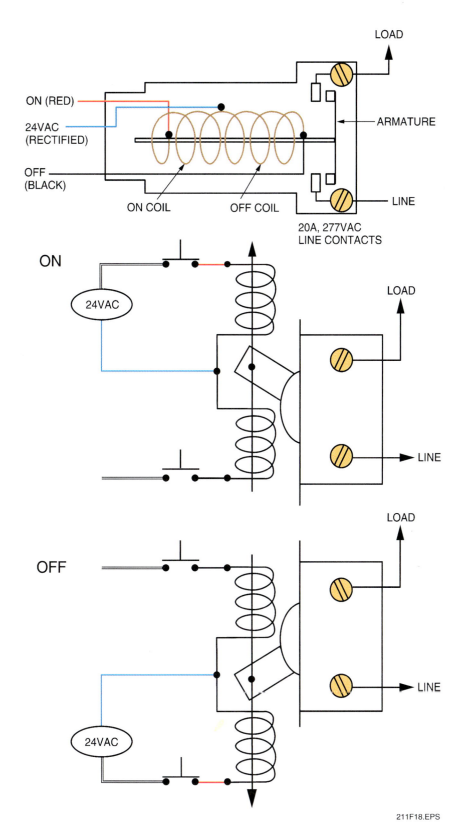

Figure 18 ♦ Relay operation.

This is similar to a magnet attracting the handle of a standard single-pole switch to the ON position when energized. The armature (handle) latches in the ON position and will remain there until the OFF coil is energized, drawing the armature into the OFF position. *Figure 19* shows how a low-voltage relay is installed in an outlet box.

7.2.1 Power Supplies

Transformers designed for use on remote control switching systems supply 24VAC power. This is rectified to 24VDC to operate the relays and pilot lights. The AC input voltage to the power supply can vary by manufacturer, with the more common voltages being 120V, 208V, 240V, and 277V. The type of voltage used depends on the system. *Figure 20* shows a control transformer along with connection details.

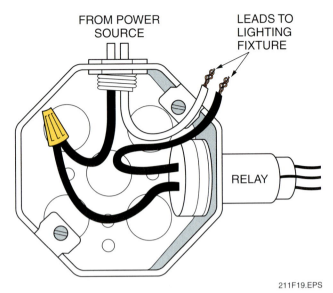

Figure 19 ◆ Relay installed in an outlet box.

7.2.2 Switches

The standard low-voltage switch uses a rocker or two-button configuration to provide a momentary single-pole, double-throw action. Pushing the ON button completes the circuit to the ON coil of the relay, shifting the contact armature to the corresponding position. When the button is released, the relay remains in that position.

The pulse operation allows any number of switches to be wired in parallel, as shown in *Figure 21*. A group of relays could also be wired for common switch control by paralleling their control leads. Those relays would operate as a group.

Pilot switches include a lamp wired between the switch common (white) and a pilot terminal. The auxiliary contact in the relay provides power to drive this lamp when the relay is energized.

7.2.3 Master Sequencer

The master sequencer allows relays to be controlled as a group while still allowing individual switch control for each. When the master switch is turned ON, the sequencer pulses each of its ON relay outputs sequentially. A local switch can control an individual relay without affecting any other.

A second input channel allows time clocks, building automation system outputs, photocells, or other maintained contact devices to also control the sequencer. This provides simple automation coupled with local override of individual loads. See *Figure 22*.

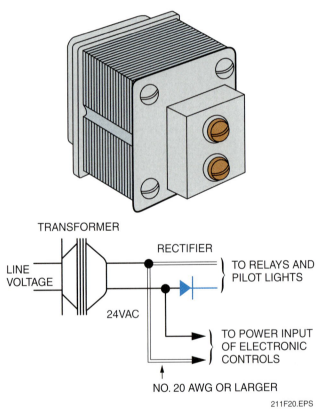

Figure 20 ◆ Control transformer with power-supply wiring diagram.

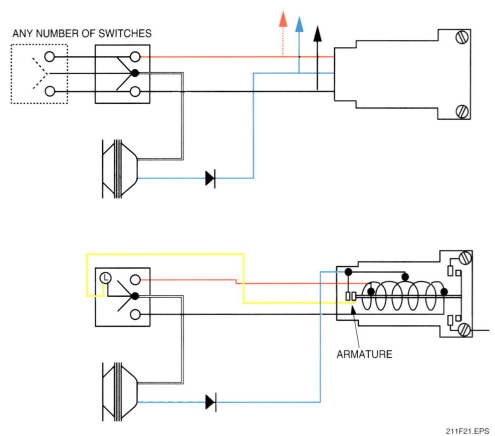

Figure 21 ◆ Switch operation.

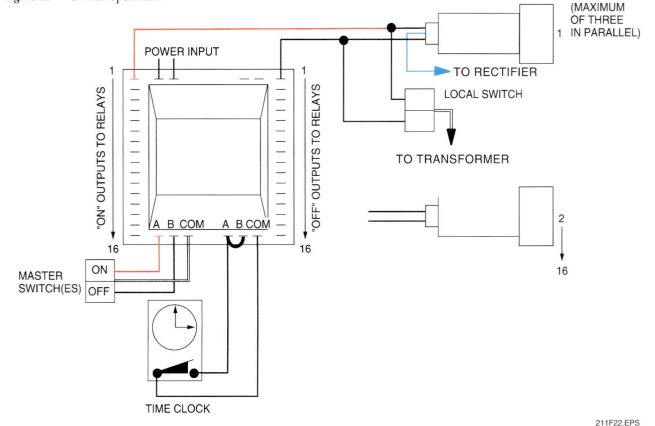

Figure 22 ◆ Master sequencer connections.

MODULE 26211-08 ◆ CONTROL SYSTEMS AND FUNDAMENTAL CONCEPTS 11.23

7.3.0 Basic Operation of Remote Control Switching

Figure 23 shows a single-switch control of a light or a group of lights in one area. This is the basic circuit of relay control and is similar to single-pole conventional switch circuits. The schematic wiring diagram of this circuit is shown in *Figure 24*.

The addition of a time clock and/or photoelectric cell allows applications in which any number of control circuits are turned on or off at desired intervals (e.g., the dusk-to-dawn control of outdoor lighting circuits).

Multipoint switching, or the switching of a single circuit from two or more switches, is shown in *Figure 25*, along with its related wiring diagram. In conventional wiring, multipoint switching requires costly three-way and four-way switches, plus the extension of switch legs and traveler wires through all switches. One of the greatest advantages of relay switching is the low cost of adding additional switch points.

Figure 26 shows many lighting circuits being controlled from one location. In conventional line voltage switches, this is accomplished by ganging the individual switches together. The same procedure can be employed with remote control switches. When more than six switches are required, it is usually desirable to install a master-selector switch that permits the control of individual circuits.

Master control of many individual and isolated circuits is possible. Where more than 12 circuits are to be controlled from a single location and the selection of individual circuits is not required, a motor master control unit automatically sweeps 25 circuits from either ON or OFF at the touch of a single master switch, which can be placed at one or more locations.

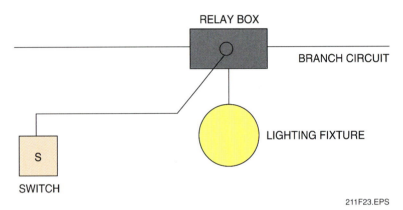

Figure 23 ◆ Single-switch control.

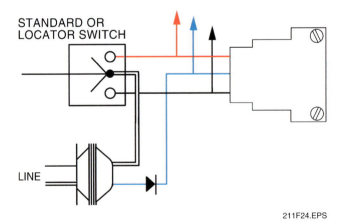

Figure 24 ◆ Schematic wiring diagram of single switches controlling each circuit.

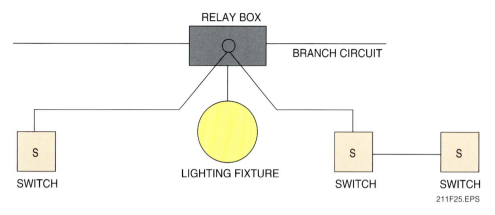

Figure 25 ♦ Diagram of multipoint switching of a single circuit.

7.4.0 Planning a Remote Control Switching System

Low-voltage lighting control systems can be installed in both residential and commercial installations. The systems installed in each type of location may differ. The manufacturer's instructions and recommendations should always be consulted and adhered to, as should the blueprint specifications. The first step in the installation procedure, whether the building is residential or commercial, is to develop a plan. When laying out a remote control switching system for a building, drawing symbols such as the ones shown in *Figure 27* are normally used to identify the various components. The drawing may be simple, requiring only the following:

- The location of each low-voltage control with a dashed line drawn from the switch to the outlet to be controlled
- A schematic wiring diagram showing all components
- The following note:
  Furnish and install complete remote control wiring system for control of lighting and other equipment as indicated on the drawings, diagrams, and schedules. System shall be complete with transformers, rectifiers, relays, switches, master-selector switches, wallplates, and wiring. All remote control wiring components shall be of the same type and installed in accordance with the recommendations of the manufacturer. Except where otherwise indicated, all remote control wiring shall be installed in accordance with NEC Article 725, Class 2.

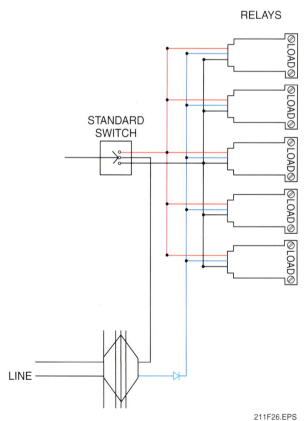

Figure 26 ♦ Switching circuits being controlled from one location.

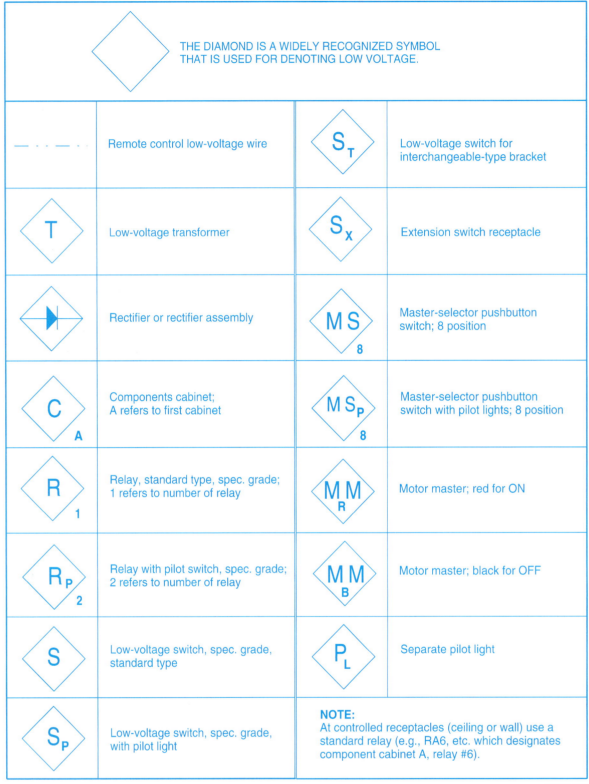

Figure 27 ♦ Recommended symbols for low-voltage drawings.

7.5.0 Design of Low-Voltage Switching

Many times the low-voltage lighting control system will be designed by architects, especially in new installations. However, in existing buildings, the electrician may be expected to design and install the system. If this situation occurs, work from the set of job blueprints, and use the following general procedure:

Step 1 Assign each light or receptacle a specific relay. This can be accomplished by using the letters R1 for relay 1, R2 for relay 2, and so on.

Step 2 Group together all the lights and outlets that are to be switched together.

Step 3 Determine the number of relays that can be controlled by each control module.

Step 4 Divide the number of relays by the number computed in Step 3. This is the number of control modules needed to control the relays.

Step 5 Determine the location of the switch points on the building blueprints.

Step 6 Indicate what lights and receptacles are being controlled by what relay. This step may be accomplished by indicating the group of lights on the blueprint. However, it is easier if a switch schedule is used.

Step 7 Review the blueprints or schedule to determine the number and size of the switching stations.

Figure 28 shows the floor plan of a building. Symbols are used to show the location of all lighting fixtures just as they are used to show the location of conventional switches. Line voltage circuits feeding the lighting outlets are indicated in the same way (by solid lines) that they are in conventional switching. All of the remote control switches, however, are indicated by the symbol S_L, instead of S, S_3, S_4, etc. Lines from these remote control switches to the lighting fixtures that they control are shown by dashed lines instead of the conventional solid line.

The schematic wiring diagram in *Figure 29* shows further details of all components and the related wiring connections. This wiring diagram aids the contractor or electrician in the installation of the system and leaves little doubt as to exactly what is required.

A brief note or specification, such as described earlier, completes the design of the low-voltage remote control system.

On the floor plan in *Figure 28*, notice that the door switches controlling the closet lights use conventional switches since low-voltage switches of this type are uncommon.

7.6.0 Installations

Before starting, consult both the *NEC*® and local codes for installation requirements. Once the layout plan has been developed and materials are acquired, the installation of the system can begin. The rough-in instructions presented here are only an example. As with the previous sections of this task module, it is important to follow the manufacturer's recommendations and blueprint specifications.

- *Enclosure* – The enclosure should be placed in a location that is easily accessible but hidden from sight. If at all possible, install a unit that is prewired. This will decrease installation time and also limit the possibility of mistakes. Note that even though the panel may be prewired, the supply cables will still have to be installed. This should be done according to the manufacturer's recommendations.

  Since heat will be produced by the enclosed control modules, vents are placed around the enclosure to facilitate cooling. The location chosen for the enclosure should not hinder the cooling capabilities for which the panels are designed. The manufacturer may require that the high-voltage portion of the enclosure panel be mounted in the upper right-hand corner. This is due to the *NEC*® requirement that is designed to keep low-voltage conductors separated from high-voltage conductors. The panel itself should be mounted at eye level between two studs.

  The power supply should be mounted on the enclosure and the power connections made. The number of power supplies needed per system will depend on the size of the system and the capabilities of the power supply. Some power supplies are capable of supplying up to 10 modules if the average number of LEDs per channel does not exceed 2 or 2.5. Be sure to check the manufacturer's specifications to ensure that the power supply will not be overloaded.

- *Switch points* – The first step when installing the switch points is to mark their location. As with ordinary switches, the switch points should be located as specified by the plans or specifications. Once the location has been determined, mount either a plaster ring or a nonmetallic box. A box is not needed in most frame construction installations.

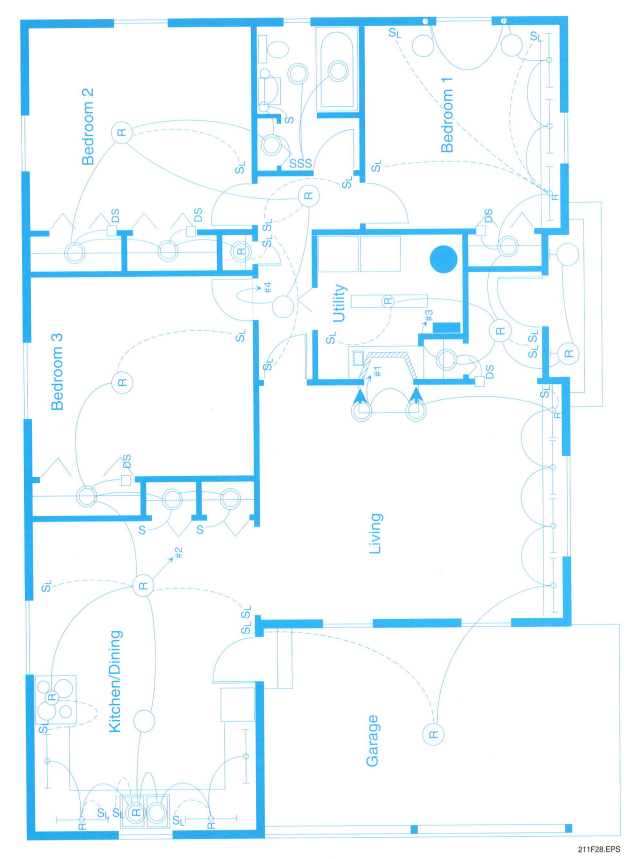

Figure 28 ♦ Floor plan of a residence using low-voltage switching.

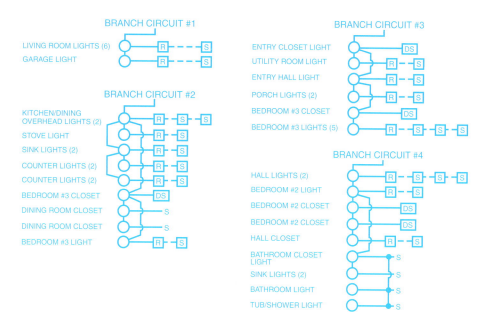

Figure 29 ◆ Schematic wiring diagram of lighting circuits.

A plaster ring or Romex® staple should then be positioned partway in the stud, behind the plaster ring or nonmetallic box. The preparation for running the low-voltage wiring is the same as for running 120V wiring. If the edge of the hole is closer than 1¼" from the nearest edge of the stud, a 1/16" steel nail plate or bushing must be used to protect the cable from future damage by nails or screws.

The wires are then pulled from the switch station to the panel enclosure. They should come from the left side if the manufacturer recommends that the enclosure be installed with the high-voltage section in the upper right-hand corner. There should be approximately 10" to 12" of extra wire left at each switch location and 12" to 18" left at the panel enclosure.

Multiconductor cable is frequently used due to its simpler installation. Run one wire (low-voltage lead) from the enclosure to the farthest switching point and anchor the wire to the plaster ring. This wire will supply the low-voltage power to each of the switches. Working back toward the enclosure, loop the wire down and anchor it to each switch location. Take care to leave a sufficient amount of wire for connecting the switch. The low-voltage lead is then connected to the appropriate power supply lead at the enclosure. The process is repeated for the DC common lead. When this wiring method is used, the DC positive and common lead will have the same color throughout the system and future connection mistakes will be minimized.

The next step is to select a multiple-switch circuit. Pull a different colored wire from the enclosure to the farthest switching location in this circuit. Working back toward the panel enclosure as before, loop the wire to the other switch locations in the multiple-switch circuit. Thus, there are two leads going to each switch with one of the leads common to all the switches in the system. However, if the switches contain a pilot light, two additional No. 18 wires must be run to the LED. The wires are connected to the appropriate pilot light transformer located in the panel enclosure.

If possible, use different colored wires for each run. This will limit confusion in the future when connecting the individual circuits to power. Wire schedules should be used to limit hookup mistakes and provide a clearer picture of the system.

Remember that the colors for the control wiring should not be the same as the colors for the DC power and common conductors. Additional marking methods should also be used for future reference. Note that if the wires are marked using adhesive tabs or punch cards, they should be placed in a plastic bag and shoved behind the ring or securely into the plastic box. This will prevent the markings from being covered by paint or plaster when the walls are finished.

The low-voltage wiring should also be kept apart from the high-voltage wiring whenever possible. If the cables must be placed in parallel runs, they should be separated by at least 6", otherwise the noise generated from the high-voltage conductors could seriously interfere with the operation of the low-voltage system.

- *Power and finish trim* – Connecting the circuits to power is now a fairly simple task. Each wire from the switch is connected to a relay with the other side of the relay connected to the common of the DC supply.

When the walls of the building are finished, the switches are connected to the conductors located at each switch point location. Wall plates are then added. A common type of wall plate used with low-voltage lighting control systems requires no screws since it can be snapped into place.

8.0.0 ◆ RELAY TROUBLESHOOTING

Electromagnetic contactors and relays of various types are used to make and break circuits that control and protect the operation of such devices as electric motors, combustion and process controls, alarms, and annunciators. Successful operation of a relay is dependent on maintaining the proper interaction between its solenoid and its mechanical elements—springs, hinges, contacts, dashpots, and the like.

The operating coil is usually designed to operate the relay from 80% to 110% of rated voltage. At low voltage, the magnet may not be strong enough to pull in the armature against the armature spring and gravity, while at high voltages excessive coil temperatures may develop. Increasing the armature spring force or the magnet gap will require higher pull-in voltage or current values and will result in higher drop-out values.

When sufficient voltage is applied to the operating coil, the magnetic field builds up, and the armature is attracted and begins to close. The air gap is shortened, increasing the magnetic attraction and accelerating the closing action, so that the armature closes with a snap, closing normally open contacts and opening normally closed contacts. Binding at the hinges or excessive armature spring force may cause a contactor or relay that normally has snap action to make and break sluggishly. This condition is often encountered in relays that are rarely operated. Contactors and relays should be operated and observed from time to time to make sure their parts are working freely with proper clearances and spring actions.

Troubleshooting charts are provided in *Table 2*. Knowing these symptoms and their causes should help you to quickly spot and repair contactor and relay problems.

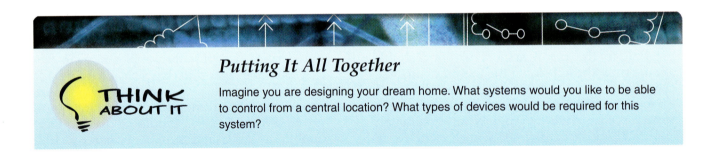

Putting It All Together

Imagine you are designing your dream home. What systems would you like to be able to control from a central location? What types of devices would be required for this system?

Table 2 Contactor and Relay Troubleshooting Chart (1 of 2)

Symptoms	Probable Cause	Action or Items to Check
Failure to pull in	Either no voltage or low voltage at coil terminals	Blown fuse, open line switch, break in wiring Line voltage below normal Overload relay open or set too low Tripping toggle (non-automatic breakers) fouled Control level or start button in off position Pull-in circuit open, shorted, or grounded Contacts in protective or controlling circuit open or one of their pigtail connections broken
	Operating coil open or grounded	Inspect and test coil
	Loose or disconnected coil lead wire	Inspect and correct
	Excessive magnet gap, improper alignment	Inspect and correct
	Armature obstructed or gumming deposits between armature and pole face	Inspect and correct
	Binding caused by deformed or gummy hinge	Replace if bent; degrease if gummed up
	Excessive armature spring force	Lessen spring force
	Normally closed contacts welded together	Replace contacts
Failure of equipment to start with contactor closed	One contact not closing Contacts burned Contact pigtail connection broken	Replace set of contacts Replace Replace
Failure to drop out	Operating coil is energized	Contacts in controlling or protective tripping circuits closed, shorted, or shunted Tripping devices defective, such as overload tripping toggles do not strike release, undervoltage relay plunger stuck or out of adjustment, defective stop button, or defective time-delay escape mechanism (closed air vent) Current supplied over an unintentional path due to grounds, defective insulation, pencil markings, moisture, or lacquer chipped off relay base
	Residual magnetism excessive due to armature closed tightly against pole face	Adjust or replace
	Armature obstructed or gumming deposits between armature and pole face	Clear and clean
	Binding caused by deformed or gummy hinge	Replace or degrease
	Contact pressure spring or armature spring too weak or improperly adjusted	Replace or adjust
	Improper mounting position (upside down)	Mount correctly
	Normally open contacts welded together	Replace contacts
Time-delay relays operate too fast	Air escapes too freely due to holes in bellows or open air vent	Escape mechanism faulty Worn dashpot plunger, dashpot oil too thin
	Non-magnetic shim in air gap too large or armature spring too strong	Adjust or replace
	Magnets out of adjustment	Adjust

Table 2 Contactor and Relay Troubleshooting Chart (2 of 2)

Symptoms	Probable Cause	Action or Items to Check
Contacts pitted or discolored	Contacts overheated from overload	Reduce load and replace contacts
	Contacts not fitted properly	Refit
	Barriers broken from rough usage or breakers closing with too much force	Replace barriers; check for high voltage
	Wiping action of contacts on closing is insufficient	Adjust
	Excessive chatter or hum	Check causes listed below
	Exposure to weather, dripping water, salt air, or vibration	Use suitable NEMA enclosure
Excessive chatter or hum	Vibration of surrounding devices communicated to relay	Arrange differently
	Relay receiving contradictory signals	Correct
	Bouncing of controlling or protective contacts	Adjust
	Line voltage fluctuating or insufficient	Consider using buck-and-boost transformers
	Excessive coil circuit resistance	Lessen resistance
	Armature spring or contact pressure spring too strong	Lessen
	Excessive coil voltage drop on closing	Adjust
	Missing shading coil	Replace the contactor
	Improper antifreeze pin location	Locate in correct position
	Free movement of armature hindered due to deformed parts or dirt	Clean and replace appropriate parts
Excessive coil temperature	Excessive current or voltage	Reduce load or adjust taps on transformer
	Short circuit in coil	Replace coil

Review Questions

1. The function of a shading coil in a contactor is to _____.
   a. keep ultraviolet light off the contacts
   b. prevent contact chatter and hum
   c. act as a holding circuit to keep the contactor energized when power is removed
   d. help the contacts return to their normal position when the contactor becomes de-energized

2. What is the major difference between a contactor and a motor starter?
   a. Contactors are always larger than motor starters for the same current-carrying capacity.
   b. All contactors are suitable for full-load lighting loads; starters are not.
   c. A contactor always has a disconnect while a starter does not.
   d. The starter provides overload protection while a contactor does not.

3. Which of the following best describes a control relay?
   a. A device designed to handle relatively high currents
   b. A fuse with two attachment points
   c. An electromagnetic device used in control circuits
   d. A fuse with two barrels

4. Which of the following best describes a relay that is used for motor protection?
   a. Amplifying relay
   b. Thermal overload relay
   c. Solenoid
   d. Contactor

5. The highest voltage rating for magnetic relays is _____.
   a. 240V
   b. 277V
   c. 480V
   d. 600V

6. The approximate inrush current of filament-type lamps is _____ times the normal operating current.
   a. 15 to 17
   b. 20 to 30
   c. 30 to 40
   d. 50 to 60

7. What is the normal position of a stop pushbutton that is used in conjunction with contactors?
   a. Normally open
   b. Normally closed
   c. Normally closed 50% of the time and open 50% of the time
   d. Normally open 70% of the time and closed 30% of the time

8. The normal maximum amperage rating of control relays at 600V is _____.
   a. 15A
   b. 20A
   c. 0A
   d. 65A

9. When selecting overload relays for motors, which of the following must be considered?
   a. The motor manufacturer
   b. The cost of the motor
   c. The length of time the motor has been in use
   d. The motor full-load current

10. Special enclosures are required to protect motor controllers from corrosion.
    a. True
    b. False

Summary

The general classification of contactor covers a type of electromagnetic apparatus that is designed to handle relatively high currents and is identical in appearance, construction, and current-carrying ability to the equivalent NEMA size magnetic starter. The magnet assembly and coil, contacts, holding circuit interlock, and other structural features are the same. The significant difference is that the contactor does not provide overload protection.

A relay is an electromagnetic device whose contacts are used in the control circuits of magnetic starters, contactors, solenoids, timers, and other relays. Relays are also used extensively in remote control switching, especially where an outlet or a group of outlets is to be controlled from several locations.

Notes

Trade Terms Introduced in This Module

Ambient temperature: The temperature of the medium (usually air) surrounding a device and into which the heat losses in the device are dissipated.

Break: The opening of closed contacts to interrupt an electrical circuit.

Inductive load: A load in which the voltage leads the current. Electromagnet coils such as those used in relays and starters are inductive loads.

Instantaneous: Contacts that make or break a circuit immediately upon coil energization or de-energization.

Latch coil: The coil of a latching relay that, when energized, causes the relay to assume the switched or latched position.

Make: The closing of open contacts to establish an electrical circuit.

Mechanically held relay: A latching relay in which the switched or latched position is maintained by the interference of a mechanical part that moves to a non-interfering position when the unlatch coil is energized, allowing the relay to return to the normal or unlatched position.

Normally closed (N.C.) contacts: A combination of a stationary contact and a movable contact, which are engaged when the coil is de-energized. In the case of a latching relay, the contacts that are engaged after the unlatch coil is energized.

Normally open (N.O.) contacts: A combination of a stationary contact and a movable contact, which are disengaged when the coil is de-energized. In the case of a latching relay, the contacts that are disengaged after the unlatch coil is energized.

Shading coil: A shorted turn surrounding a portion of the pole of an AC electromagnet, producing by mutual inductance a delay in the change of the magnetic field in that part, thereby tending to prevent chatter and reduce hum.

Solid-state relay: A semiconductor switch having no moving parts. These devices control power to the load and are analogous to relay contacts.

Unlatch coil: The coil of a latching relay that, when energized, causes the relay to assume the normal or unlatched position.

Additional Resources

This module is intended to present thorough resources for task training. The following reference work is suggested for further study. This is optional material for continued education rather than for task training.

National Electrical Code® Handbook, Latest Edition. Quincy, MA: National Fire Protection Association.

CONTREN® LEARNING SERIES – USER UPDATE

NCCER makes every effort to keep these textbooks up-to-date and free of technical errors. We appreciate your help in this process. If you have an idea for improving this textbook, or if you find an error, a typographical mistake, or an inaccuracy in NCCER's Contren® textbooks, please write us, using this form or a photocopy. Be sure to include the exact module number, page number, a detailed description, and the correction, if applicable. Your input will be brought to the attention of the Technical Review Committee. Thank you for your assistance.

Instructors – If you found that additional materials were necessary in order to teach this module effectively, please let us know so that we may include them in the Equipment/Materials list in the Annotated Instructor's Guide.

Write: Product Development and Revision
National Center for Construction Education and Research
3600 NW 43rd St., Bldg. G, Gainesville, FL 32606

Fax: 352-334-0932

E-mail: curriculum@nccer.org

Craft _____ Module Name _____

Copyright Date _____ Module Number _____ Page Number(s) _____

Description _____

_____

_____

_____

(Optional) Correction _____

_____

_____

(Optional) Your Name and Address _____

_____

_____

_____

Glossary

AL-CU: An abbreviation for aluminum and copper, commonly marked on terminals, lugs, and other electrical connectors to indicate that the device is suitable for use with either aluminum conductors or copper conductors.

Ambient temperature: The temperature of the medium (usually air) surrounding a device and into which the heat losses in the device are dissipated.

Amperage capacity: The maximum amount of current that a lug can safely handle at its rated voltage.

Approximate ram travel: The distance the ram of a hydraulic bender travels to make a bend. To simplify and speed bending operations, many benders are equipped with a scale that shows ram travel. Using a simple table (supplied with many benders), the degree of bend can easily be converted to inches of ram travel. This measurement, however, can only be approximated because of the variation in springback of the conduit being bent.

Armature: The rotating windings of a DC motor.

Auxiliary electrodes: Metallic electrodes pushed or driven into the earth to provide electrical contact for the purpose of performing measurements on grounding electrodes or ground grid systems.

Back-to-back bend: Any bend formed by two 90° bends with a straight section of conduit between the bends.

Ballast: A circuit component in fluorescent and HID lighting fixtures that provides the required voltage surge at startup and then controls the subsequent flow of current through the lamp during operation.

Barrier strip: A metal strip constructed to divide a section of cable tray so that certain kinds of cable may be separated from each other.

Basket grip: A flexible steel mesh grip that is used on the ends of cable and conductors for attaching the pulling rope. The more force exerted on the pull, the tighter the grip wraps around the cable.

Bending protractor: Made for use with benders mounted on a bending table and used to measure degrees; also has a scale for 18, 20, 21, and 22 shots when using it to make a large sweep bend.

Bending shot: The number of shots needed to produce a specific bend.

Bonding: Connected to establish electrical continuity and conductivity.

Branch circuit: The circuit conductors between the final overcurrent device protecting the circuit and the outlet(s).

Break: The opening of closed contacts to interrupt an electrical circuit.

Brush: A conductor between the stationary and rotating parts of a machine. It is usually made of carbon.

Cable grip: A device used to secure ends of cables to a pulling rope during cable pulls.

Cable pulley: A device used to facilitate pulling conductor in cable tray where the tray changes direction. Several types are available (single, triple, etc.) to accommodate almost all pulling situations.

Capacitance: The storage of electricity in a capacitor; capacitance produces an opposition to voltage change. The unit of measurement for capacitance is the farad (F) or microfarad (µF).

Capstan: The turning drum of a cable puller on which the rope is wrapped and pulled. An increase in the number of wraps increases the pulling force of the cable puller.

Cartridge fuse: A fuse enclosed in an insulating tube to confine the arc when the fuse blows.

Circuit breaker: A device designed to open and close a circuit by nonautomatic means and to open the circuit automatically on a predetermined overcurrent without injury to itself when properly applied within its rating.

Circuit breaker: A device that is designed to open a circuit automatically at a certain overcurrent. It can also be used to manually open and close the circuit.

Clevis: A device used in cable pulls to facilitate connecting the pulling rope to the cable grip.

Color rendering index (CRI): A measurement of the way a light source reproduces color. The higher the index number (0–100), the closer colors are to how an object appears in full sunlight or incandescent light.

Commutator: A device used on electric motors or generators to maintain a unidirectional current.

Concentric bending: The process of making 90° bends in parallel runs of conduit. This requires increasing the radius of each conduit from the inside of the bend toward the outside.

Conductor support: The act of providing support in vertical conduit runs to support the cables or conductors. The *NEC*® gives several methods in which cables may be supported, including wedges in the tops of conduits, supports to change the direction of cable in pull boxes, etc.

Conduit: Piping designed especially for pulling electrical conductors. Types include rigid, IMC, EMT, PVC, aluminum, and other materials.

Conduit body: A separate portion of a conduit or tubing system that provides access through a removable cover (or covers) to the interior of the system at a junction of two or more sections of the system or at a terminal point of the system. Boxes such as FS and FD or larger cast or sheet metal boxes are not classified as conduit bodies.

Conduit piston: A cylinder of foam rubber that fits inside the conduit and is then propelled by compressed air or vacuumed through the conduit run to pull a line, rope, or measuring tape. Also called a mouse.

Connection: That part of a circuit that has negligible impedance and joins components or devices.

Connector: A device used to physically and electrically connect two or more conductors.

Continuous duty: Operation at a substantially constant load for an indefinitely long time.

Controller: A device that serves to govern, in some predetermined manner, the electric power delivered to the apparatus to which it is connected.

Cross: A four-way section of cable tray used when the tray assembly must branch off in four different directions.

Degree indicator: An instrument designed to indicate the exact degree of bend while it is being made.

Developed length: The amount of straight pipe needed to bend a given radius. Also, the actual length of the conduit that will be bent.

Dip tolerance: The ability of an HID lamp or lighting fixture circuit to ride through voltage variations without the lamp extinguishing and cooling down.

Direct rod suspension: A method used to support cable tray by means of threaded rods and hanger clamps. One end of the threaded rod is secured to an overhead structure, while the other end is connected to hanger clamps that are attached to the cable tray side rails.

Drain wire: A wire that is attached to a coaxial connector to allow a path to ground from the outer shield.

Dropout: Cable leaving the tray assembly and traveling directly downward; that is, the cable is not routed into a conduit or channel.

Dropout plate: A metal plate used at the end of a cable tray section to ensure a greater cable bending radius as the cable leaves the tray assembly.

Dual-element fuse: A fuse having two fuse characteristics; the usual combination is an overcurrent limit and a time delay before activation.

Duty: Describes the length of operation. There are four designations for circuit duty: continuous, periodic, intermittent, and varying.

Edison-base: The standard screw base used for ordinary lamps and Edison-base plug fuses.

Effective ground fault path: An intentionally constructed, permanent, low-impedance, electrically conductive path designed and intended to carry current under ground-fault conditions from the point of a ground fault on a wiring system to the electrical supply source. It facilitates the operation of the overcurrent protective device or ground-fault detectors on a high-impedance grounded system.

Efficacy: The light output of a light source divided by the total power input to that source. It is expressed in lumens per watt (LPW).

Elbow: A 90° bend or fitting in a section of conduit or cable tray.

Equipment: A general term including material, fittings, devices, appliances, fixtures, apparatus, and the like used as a part of, or in connection with, an electrical installation.

Equipment bonding jumper: The connection between two or more portions of the equipment grounding conductor.

Equipment grounding conductor: The conductive path installed to connect the normally noncurrent-carrying metal parts of equipment together and to the system grounded conductor, or to the grounding electrode conductor, or both.

Expansion joints: Plates used at intervals along a straight run of cable tray to allow space for thermal expansion or contraction of the tray.

Explosion-proof: Designed and constructed to withstand an internal explosion without creating an external explosion or fire.

Fault current: The current that exists when an unintended path is established between an ungrounded conductor and ground.

Field poles: The stationary portion of a DC motor that produces the magnetic field.

Fish line: Light cord used in conjunction with vacuum/blower power fishing systems that attaches to the conduit piston to be pushed or pulled through the conduit. Once through, a pulling rope is attached to one end and pulled back through the conduit for use in pulling conductors.

Fish tape: A flat iron wire or fiber cord used to pull conductors or a pulling rope through conduit.

Fittings: Devices used to assemble and/or change the direction of cable tray systems.

Frame size: A method used to classify circuit breakers according to given current ranges.

Frequency: The number of cycles an alternating electric current, sound wave, or vibrating object undergoes per second.

Fuse: A protective device that opens a circuit when the fusible element is severed by heating due to a fault current or overcurrent passing through it.

Fuse link: The fusible part of a cartridge fuse.

Gain: The amount of pipe saved by bending on a radius as opposed to right angles. Because conduit bends in a radius and not at right angles, the length of conduit needed for a bend will not equal the total determined length. Gain is the difference between the right angle distances A and B and the shorter distance C—the length of conduit actually needed for the bend.

Grooming: The act of separating the braid in a coaxial conductor.

Ground: A conducting connection, whether intentional or accidental, between an electrical circuit or equipment and the earth.

Ground current: Current in the earth or grounding connection.

Grounded: Connected to ground or to a conductive body that extends the ground connection.

Grounded conductor: A system or circuit conductor that is intentionally grounded.

Ground grids: System of grounding electrodes interconnected by bare cables buried in the earth to provide lower resistance than a single grounding electrode.

Ground mats: System of bare conductors, on or below the surface of the earth, connected to a ground or ground grid to provide protection from dangerous touch voltage.

Ground resistance: The ohmic resistance between a grounding electrode and a remote or reference grounding electrode that are spaced such that their mutual resistance is essentially zero.

Ground rod: A metal rod or pipe used as a grounding electrode.

Grounding clip: A spring clip used to secure a bonding conductor to an outlet box.

Grounding conductor: A conductor used to connect equipment or the grounded circuit of a wiring system to a grounding electrode or electrodes.

Grounding connections: Connections used to establish a ground; they consist of a grounding conductor, a grounding electrode, and the earth surrounding the electrode.

Grounding electrode: A conducting object through which a direct connection to earth is established.

Grounding electrode conductor: A conductor used to connect the system grounded conductor or the equipment to a grounding electrode or to a point on the grounding electrode system.

Handhole: An enclosure used with underground systems to provide access for installation and maintenance.

Hertz (Hz): A unit of frequency; one hertz equals one cycle per second.

Horsepower: The rated output capacity of the motor. It is based on breakdown torque, which is the maximum torque a motor will develop without an abrupt drop in speed.

Hours: The duty cycle of a motor. Most fractional horsepower motors are marked continuous for around-the-clock operation at the nameplate rating in the rated ambient conditions. Motors marked one-half are for ½-hour ratings, and those marked one are for 1-hour ratings.

Impedance: The opposition to current flow in an AC circuit; impedance includes resistance (R), capacitive reactance (X_C), and inductive reactance (X_L). Impedance is measured in ohms.

Incandescence: The self-emission of radiant energy in the visible light spectrum resulting from thermal excitation of atoms or molecules such as occurs when an electric current is passed through the filament in an incandescent lamp.

Incident light: The light emitted from a self-luminous object such as the sun or an electric source.

Inductance: The creation of a voltage due to a time-varying current; also, the opposition to current change, causing current changes to lag behind voltage changes. The unit of measure for inductance is the henry (H).

Inductive load: A load in which the voltage leads the current. Electromagnet coils such as those used in relays and starters are inductive loads.

Inside diameter (ID): The inside diameter of conduit. All electrical conduit is measured in this manner. The outside dimensions, however, will vary with the type of conduit used.

Instantaneous: Contacts that make or break a circuit immediately upon coil energization or de-energization.

Insulating tape: Adhesive tape that has been manufactured from a nonconductive material and is used for covering wire joints and exposed parts.

Interlocked armor cable: Mechanically protected cable; usually a helical winding of metal tape formed so that each convolution locks mechanically upon the previous one (armor interlock).

Intermittent duty: Operation for alternate intervals of (1) load and no load; or (2) load and rest; or (3) load, no load, and rest.

Junction box: An enclosure where one or more raceways or cables enter, and in which electrical conductors can be, or are, spliced.

Kicks: Bends in a piece of conduit, usually less than 90°, made to change the direction of the conduit.

Ladder tray: A type of cable tray that consists of two parallel channels connected by rungs, similar in appearance to the common straight ladder.

Latch coil: The coil of a latching relay that, when energized, causes the relay to assume the switched or latched position.

Leg length: The distance from the end of the straight section of conduit to the bend, measured to the centerline or to the inside or outside of the bend or rise.

Lug: A device for terminating a conductor to facilitate the mechanical connection.

Lumen (lm): The basic measurement of light. One lumen is defined as the amount of light cast upon one square foot of the inner surface of a hollow sphere with a one-foot radius with a light source of one candela in its center.

Lumen maintenance: A measure of how a lamp maintains its light output over time. It may be expressed either numerically or as a graph of light output versus time.

Lumens per watt (LPW): A measure of the efficiency, or, more properly, the efficacy of a light source. The efficacy is calculated by taking the lumen output of a lamp and dividing by the lamp wattage. For example, a 100W lamp producing 1,750 lumens has an efficacy of 17.5 lumens per watt.

Luminaire (fixture): A complete lighting unit consisting of a lamp (or lamps) and ballasts (where applicable), together with the parts designed to distribute the light, position and protect lamps, and connect them to the power supply.

Luminance: The luminous intensity of any surface in a given direction per unit area of that surface as viewed from that direction. Measured in candela/m^2. The term luminance is commonly used to express brightness.

Main bonding jumper: The connection between the grounded circuit conductor and the equipment grounding conductor at the service.

Make: The closing of open contacts to establish an electrical circuit.

Mechanical advantage (MA): The force factor of a crimping tool that is multiplied by the hand force to give the total force.

Mechanically held relay: A latching relay in which the switched or latched position is maintained by the interference of a mechanical part that moves to a non-interfering position when the unlatch coil is energized, allowing the relay to return to the normal or unlatched position.

Micro: Prefix designating one-millionth of a unit. For example, one microfarad is one-millionth of a farad.

Mogul: A type of conduit body with a raised cover to provide additional space for large conductors.

Molded-case circuit breaker: A circuit breaker enclosed in an insulating housing.

Neutral: The conductor connected to the neutral point of a system that is intended to carry current under normal conditions.

Ninety-degree bend: A bend in a piece of conduit that changes its direction by 90°.

Nonrenewable fuse: A fuse that must be replaced after it interrupts a circuit.

Normally closed (N.C.) contacts: A combination of a stationary contact and a movable contact, which are engaged when the coil is de-energized. In the case of a latching relay, the contacts that are engaged after the unlatch coil is energized.

Normally open (N.O.) contacts: A combination of a stationary contact and a movable contact, which are disengaged when the coil is de-energized. In the case of a latching relay, the contacts that are disengaged after the unlatch coil is energized.

Offsets: Two equal bends made to avoid an obstruction blocking the run of the conduit.

One-shot shoe: A large bending shoe that is designed to make 90° bends in conduit.

Outside diameter (OD): The size of any piece of conduit measured on the outside diameter.

Overcurrent: Any current in excess of the rated current of equipment or the ampacity of a conductor. It may result from an overload, short circuit, or ground fault.

Overcurrent protection: De-energizing a circuit whenever the current exceeds a predetermined value; the usual devices are fuses, circuit breakers, or magnetic relays.

Overload: Operation of equipment in excess of the normal, full-load rating, or of a conductor in excess of rated ampacity, which, after a sufficient length of time, will cause damage or dangerous overheating. A fault, such as a short circuit or ground fault, is not an overload.

Peak voltage: The peak value of a sinusoidally varying (cyclical) voltage or current is equal to the root-mean-square (rms) value multiplied by the square root of two (1.414). AC voltages are usually expressed as rms values; that is, 120 volts, 208 volts, 240 volts, 277 volts, 480 volts, etc., are all rms values. The peak voltage, however, differs. For example, the peak value of 120 volts (rms) is actually $120 \times 1.414 = 169.71$ volts.

Periodic duty: Intermittent operation at a substantially constant load for a short and definitely specified time.

Pipe racks: Structural frames used to support the piping that interconnects equipment in outdoor industrial facilities.

Plug fuse: A type of fuse that is held in position by a screw thread contact instead of spring clips, as is the case with a cartridge fuse.

Pole: That portion of a device associated exclusively with one electrically separated conducting path of the main circuit or device.

Pressure connector: A connector applied using pressure to form a cold weld between the conductor and the connector.

Pull box: A sheet metal box-like enclosure used in conduit runs to facilitate the pulling of cables from point to point in long runs, or to provide for the installation of conduit support bushings needed to support the weight of long riser cables, or to provide for turns in multiple-conduit runs.

Radian: An angle at the center of a circle, subtending (opposite to) an arc of the circle that is equal in length to the radius.

Radius: The relative size of the bent portion of a pipe.

Raintight: Constructed or protected so that exposure to a beating rain will not result in the entrance of water under specified test conditions.

Reactance: The imaginary part of impedance. Also, the opposition to alternating current (AC) due to capacitance (X_C) and/or inductance (X_L).

Reducing connector: A connector used to join two different size conductors.

Reflected: The bouncing back of light waves or rays by a surface.

Reflected light: Light reflected from an object that does not have self-luminating properties but reflects light provided from another source.

Refracted/refraction: The bending of a light wave or ray as it passes obliquely from one medium to another of different density, or through layers of different density in the same medium.

Resistivity: Resistance between opposite faces of a unit cube. Expressed in ohm-centimeters or ohms per cubic centimeter.

Revolutions per minute (rpm): The approximate full-load speed at the rated power line frequency. The speed of a motor is determined by the number of poles in the winding. A four-pole, 60Hz motor runs at an approximate speed of 1,725 rpm. A six-pole, 60Hz motor runs at an approximate speed of 1,140 rpm.

Rise: The length of the bent section of conduit measured from the bottom, centerline, or top of the straight section to the end of the bent section.

Root-mean-square (rms): The square root of the average of the square of the function taken throughout the period. The rms value of a sinusoidally varying voltage or current is the effective value of the voltage or current.

Rotation: For single-phase motors, the standard rotation, unless otherwise noted, is counter-clockwise facing the lead or opposite shaft end. All motors can be reconnected at the terminal board for opposite rotation unless otherwise indicated.

Segment bend: Any bend formed by a series of bends of a few degrees each, rather than a single one-shot bend.

Segmented bending shoe: A smaller type of shoe designed for bending segmented bends only (always less than 15°).

Self-inductance: A magnetic field induced in the conductor carrying the current.

Separately derived system: A premises wiring system whose power is derived from a source of electric energy or equipment other than a service. Such systems have no direct electrical connection, including a solidly connected ground circuit conductor, to supply conductors originating in another system.

Setscrew grip: A cable grip, usually with built-in clevis, in which the cable ends are inserted in holes and secured with one or more setscrews.

Shading coil: A shorted turn surrounding a portion of the pole of an AC electromagnet, producing by mutual inductance a delay in the change of the magnetic field in that part, thereby tending to prevent chatter and reduce hum.

Sheave: A pulley-like device used in cable pulls in both conduit and cable tray systems.

Shielding: The metal covering of a cable that reduces the effects of electromagnetic noise.

Short circuit: The current that exists when an unintended path is created between any two components in a circuit.

Soap: Slang for wire-pulling lubricant.

Solid-state relay: A semiconductor switch having no moving parts. These devices control power to the load and are analogous to relay contacts.

Splice: The electrical and mechanical connection between two pieces of cable.

Springback: The amount, measured in degrees, that a bent conduit tends to straighten after pressure is released on the bender ram. For example, a 90° bend, after pressure is released, will pull back about 2° to 88°.

Starter: A circuit component of certain fluorescent lighting fixtures that is used to heat the lamp electrodes before the lamps are lighted.

Step voltage: The potential difference between two points on the earth's surface separated by a distance of one pace, or about three feet.

Strand: A group of wires, usually stranded or braided.

Stub-up: Another name for the rise in a section of conduit.

Sweep bend: A 90° bend with a radius larger than that produced by a standard one-shot shoe.

Swivel plates: Devices used to make vertical offsets in cable tray.

Synchronous speed: When the speed of the rotor is equal to the speed of the stator. The speed is determined by multiplying 120 times the frequency divided by the number of poles.

System grounding: Intentional connection of one of the circuit conductors of an electrical system to ground potential.

Take-up (comeback): The amount that must be subtracted from the desired stub length to make the bend come out correctly using a point of reference on the bender or bending shoe.

Tee: A section of cable tray that branches off the main section in two other directions.

Terminal: A device used for connecting cables.

Termination: The connection of a cable.

Thermal protector: A protective device for assembly as an integral part of a motor or motor compressor that, when properly applied, protects the motor against dangerous overheating due to overload or failure to start.

Touch voltage: The potential difference between a grounded metallic structure and a point on the earth's surface equal to the normal maximum horizontal reach—approximately three feet.

Trapeze mounting: A method of supporting cable tray using metal channel, such as Unistrut®, Kindorf®, etc., supported by two threaded rods, and giving the appearance of a swing or trapeze.

Tray cover: A flat piece of metal, fiberglass, or plastic designed to provide a solid covering that is needed in some locations where conductors in the tray system may be damaged.

Troffer: A recessed lighting fixture installed with the opening flush with the ceiling.

Trough: A type of cable tray consisting of two parallel channels (side rails) having a corrugated, ventilated bottom or a corrugated, solid bottom.

Ungrounded conductors: Conductors in an electrical system that are not intentionally grounded.

Unistrut®: A brand of metal channel used as the bottom bracket for hanging cable trays. Double Unistrut® adds strength and stability to the trays and also provides a means of securing future runs of conduit.

Unlatch coil: The coil of a latching relay that, when energized, causes the relay to assume the normal or unlatched position.

Varying duty: Operation at varying loads and/or intervals of time.

Wall mounting: A method of supporting cable tray systems using supports secured directly to the wall.

Waterproof: See *Weatherproof*.

Watertight: Constructed so that moisture will not enter the enclosure under specified test conditions.

Weatherproof: Constructed or protected so that exposure to the weather will not interfere with successful operation.

Wye: A section of cable tray that branches off the main section in one direction.

Figure Credits

Module 26201-08

Associated Builders and Contractors, Inc., module divider
Topaz Publications, Inc., module overview, 201F13C, 201SA04

Module 26202-08

Timothy A. Ely, module divider
Topaz Publications, Inc., module overview, 202F25, 202SA02, 202F40, 202SA03, 202SA04, 202F45B, 202SA05, 202SA06, 202F55B, 202SA07
Reprinted with the permission of the National Electrical Manufacturers Association, 202T02, 202T05
National Fire Protection Association, 202T06–202T09
Reprinted with permission from *NFPA 70*®-2008, the *National Electrical Code*®. Copyright © 2007, National Fire Protection Association, Quincy, MA 02269. This reprinted material is not the complete and official position of the National Fire Protection Association on the referenced subject, which is represented only by the standard in its entirety. *NFPA 70*®, *National Electrical Code*®, and *NEC*® are registered trademarks of the National Fire Protection Association, Inc., Quincy, MA 02269.

Module 26203-08

Topaz Publications, Inc., 203SA01, 203SA04–203SA06, 203F52
Courtesy of General Electric Company, 203F07–203F09
OSRAM SYLVANIA, module overview, 203F11–203F13, 203SA02, 203SA03, 203F14, 203F16–203F18
Leviton, 203F15
Hubbell Incorporated, 203F22, 203F33, 203F47, 203F51, 203A03–203A06
Intermatic, Inc., 203F53
Lighting Science Group Corporation, 203SA07
Tri-City Electrical Contractors, Inc., module divider

Module 26204-08

AGC of America, module divider
Topaz Publications, Inc., module overview, 204SA01, 204F01B (photo), 204F16, 204SA02, 204F26, 204F31, 204SA05, 204SA07, 204F52, 204F56
National Fire Protection Association, 204T01, 204T02
Reprinted with permission from *NFPA 70*®-2008, the *National Electrical Code*®. Copyright © 2007, National Fire Protection Association, Quincy, MA 02269. This reprinted material is not the complete and official position of the National Fire Protection Association on the referenced subject, which is represented only by the standard in its entirety. *NFPA 70*®, *National Electrical Code*®, and *NEC*® are registered trademarks of the National Fire Protection Association, Inc., Quincy, MA 02269.
John Traister, 204F02, 204F09, 204F10
Tim Ely, 204F03B (photo), 204F27
Greenlee Textron, Inc., a subsidiary of Textron, Inc., 204F15, 204F25, 204F28, 204F39–204F42, 204F44
Mike Powers, 204SA03
Tim Dean, 204SA04, 204SA06

Module 26205-08

AGC of America, module divider
Hubbell Wiegmann, 205F01
John Traister, 205F02, 205F18
Topaz Publications, Inc., 205F05–205F08, 205F12, 205SA02, 205SA03, 205F15, 205F20, 205F22
Tim Ely, 205SA01, 205F10, 205F11, 205F14, 205F21
Photo courtesy of Cooper Crouse-Hinds, module overview, 205SA04
Greenlee Textron, Inc., a subsidiary of Textron, Inc., 205SA05, 205F23, 205F24

Module 26206-08

AGC of America, module divider
Topaz Publications, Inc., 206F01, 206SA01, 206F12A, 206F12B, 206F14B, 206F25, 206SA08

Greenlee Textron, Inc., a subsidiary of Textron, Inc., module overview, 206F02, 206F03, 206SA02, 206F08, 206F10, 206SA03, 206F12C, 206F13, 206F14A, 206SA04, 206F16, 206SA05, 206F17–206F19, 206SA06, 206F20, 206F23, 206SA07, 206F29, 206F30

Tim Dean, 206F26

Module 26207-08

Associated Builders and Contractors, Inc., module divider

Topaz Publications, Inc., 207SA01, 207SA02, 207SA09, 207SA10

John Traister, 207F01–207F03, 207F18

Tim Dean, module overview, 207SA03, 207SA04, 207F09, 207F12, 207F15, 207SA07, 207SA08

Greenlee Textron, Inc., a subsidiary of Textron, Inc., 207SA05

Jim Mitchem, 207SA06

Tim Ely, 207SA11

Module 26208-08

Associated Builders and Contractors, Inc., module divider

Greenlee Textron, Inc., a subsidiary of Textron, Inc., module overview, 208F01, 208F02, 208SA01, 208F03, 208F06, 208SA02, 208F17–208F22, 208F32, 208F33

Topaz Publications, Inc., 208F08, 208F10, 208SA03, 208F13B, 208F39A, 208SA04

Belden Inc., 208F13A

Jim Mitchem, 208F23, 208F25, 208F36

National Fire Protection Association, 208T05, 208T06

Reprinted with permission from *NFPA 70*®-2008, the *National Electrical Code*®. Copyright © 2007, National Fire Protection Association, Quincy, MA 02269. This reprinted material is not the complete and official position of the National Fire Protection Association on the referenced subject, which is represented only by the standard in its entirety. *NFPA 70*®, *National Electrical Code*®, and *NEC*® are registered trademarks of the National Fire Protection Association, Inc., Quincy, MA 02269.

Module 26209-08

Associated Builders and Contractors, Inc., module divider

Mike Powers, 209SA01

John Traister, 209F04, 209F05, 209F10–209F13, 209F18, 209F22, 209F23

Topaz Publications, Inc., module overview, 209SA02, 209F27B, 209F34

National Oceanic and Atmospheric Administration/Department of Commerce, 209SA03

Greenlee Textron, Inc., a subsidiary of Textron, Inc., 209SA04

Module 26210-08

Associated Builders and Contractors, Inc., module divider

John Traister, 210F01, 210F02, 210F10

Topaz Publications, Inc., 210SA01, 210SA02, 210F09B, 210F11

Tim Dean, 210F03

Cooper Bussmann, 210F04–210F06

Tim Ely, module overview, 210F07

Module 26211-08

AGC of America, module divider

Reprinted with permission of the National Electrical Manufacturers Association, 211T01

Topaz Publications, Inc., module overview, 211F10B, 211SA01, 211SA02

Tim Dean, 211F15

Index

Figures are indicated with *f*.

A

Absorption of light, 3.3–3.4
AC (alternating current)
   capacitance and, 1.14, 1.15–1.21, 1.16*f*, 1.17*f*, 1.18*f*, 1.19*f*
   inductance in AC circuits, 1.12–1.15, 1.13*f*, 1.14*f*
   LC circuits, 1.27–1.29, 1.28*f*, 1.29*f*
   nonsinusoidal waveforms, 1.9–1.11, 1.10*f*
   phase relationships, 1.6–1.8, 1.8*f*, 1.21, 1.22*f*
   power in AC circuits, 1.32–1.34, 1.33*f*, 1.34*f*, 1.35*f*
   RC circuits, 1.25–1.27, 1.26*f*, 1.27*f*
   resistance in AC circuits, 1.11, 1.11*f*, 1.12*f*
   RLC circuits, 1.29–1.30, 1.30*f*, 1.31*f*
   RL circuits, 1.21, 1.23–1.25, 1.23*f*, 1.24*f*, 1.25*f*
   sine wave generation, 1.2–1.3, 1.4*f*
   sine wave terminology, 1.4–1.6
   symbol for AC voltage sources, 1.11, 1.11*f*
   transformers and, 1.35–1.43, 1.36*f*, 1.37*f*, 1.38*f*, 1.39*f*, 1.40*f*, 1.41*f*, 1.42*f*, 1.43*f*, 1.44
AC motors. *See also* Variable-speed drives
   capacitor-type induction motors, 2.33–2.34, 2.33*f*
   connections and terminal markings, 2.57–2.59, 2.57*f*, 2.58*f*, 2.59*f*
   multiple-speed induction motors, 2.37–2.39, 2.37*f*, 2.38*f*, 2.39*f*
   polyphase motor theory, 2.16–2.18, 2.17*f*
   shaded-pole induction motors, 2.31, 2.34, 2.35, 2.35*f*, 2.36*f*, 2.37
   single-phase AC motors, 2.16, 2.30–2.31, 2.31*f*, 2.32, 2.33–2.34, 2.33*f*, 2.34*f*, 2.35*f*, 2.36*f*, 2.37
   synchronous motors, 2.16, 2.24, 2.26–2.30, 2.27*f*, 2.28*f*, 2.29*f*
   three-phase induction motors, 2.18–2.24, 2.18*f*, 2.19*f*, 2.20*f*, 2.21*f*, 2.23*f*, 2.24*f*, 2.25*f*
Adjustable-speed drives, 1.32, 2.26
Adjustable speed loads, 2.40, 2.40*f*
AIC (amperage interrupting capacity), 10.5
Air cores, 1.26
Air gaps, troubleshooting, 11.30, 11.31
Airtight plugs, bending PVC conduit, 4.37, 4.38*f*
Alarm contacts, 11.18
AL-CU
   connectors marked with, 8.10
   defined, 8.34
All-in-one conduit bodies, 5.9
Allowable sidewall loads, 6.29, 6.31, 6.31*f*
Allowable tension on conductors, 6.29, 6.30
Allowable tension on pulling devices, 6.29, 6.30
Alternating current (AC). *See* AC (alternating current); AC motors
Alternations, 2.4

Aluminum
   ground rods and, 9.12*f*
   properties of, 8.9
Aluminum conduit fittings, 5.13–5.14, 5.14*f*
Aluminum connections, 8.9–8.10
Aluminum connectors, 8.9–8.10
Ambient-compensated overload relays, 11.17
Ambient temperatures, 11.17
   defined, 11.35
American National Standards Institute/Society of Cable Telecommunications Engineers. *See* ANSI/SCTE
Amortisseur windings, 2.27
Amount of travel method, segment bending techniques, 4.23, 4.24
Ampacity, of cable tray conductors, 7.22
Amperage capacity, 8.20
   defined, 8.34
Amperage interrupting capacity (AIC), 10.5
Ampere rating
   for circuit breakers, 10.4–10.5
   for fuses, 10.14
Amplifying contact capacity, relays, 11.10, 11.10*f*
Amplifying voltage, relays, 11.10*f*, 11.11
Amplitude
   heat and, 1.6
   values, 1.5–1.6, 1.5*f*
Anchoring cable pulling equipment, 6.20
ANSI/SCTE, requirements for handholes, 5.10, 5.11–5.12
Anti-dog devices, 4.17, 4.34
Apparent power, 1.32, 1.33*f*
Approximate ram travel, 4.22
   defined, 4.43
Arc fault circuit breakers, 10.22
Area, of cable trays, 7.2
Armature control, variable-speed drives, 2.43*f*, 2.44
Armature reaction, 2.8–2.9, 2.9*f*
Armatures, 2.3
   defined, 2.68
Auto-lag ballasts, 3.22, 3.23*f*
Automatic over-tension shutoffs, 6.23
Autotransformers, 1.41–1.42, 1.41*f*
Auxiliary electrodes
   defined, 9.41
   three-point testing and, 9.30, 9.30*f*, 9.31, 9.31*f*, 9.32, 9.33–9.36, 9.33*f*, 9.34*f*, 9.35*f*, 9.36*f*
Auxiliary winches, 7.25
Average values, 1.5–1.6, 1.5*f*

B

Back-to-back bends, 4.4, 4.12*f*
   defined, 4.43

Ballasts
   defined, 3.55
   fluorescent lamps and, 3.10
   fluorescent light fixtures and, 3.17–3.21, 3.17f, 3.18f, 3.20f
   HID lighting fixtures and, 3.21–3.22, 3.21f, 3.22f, 3.23f, 3.24
   testing procedure and checklist, 3.59
Barrier strip cable protectors, 7.12
Barrier strips
   carrier trays and, 7.2, 7.12
   defined, 7.28
Bases, for incandescent lamps, 3.7, 3.9f
Basket grips, 6.2, 6.2f, 6.11, 6.15f, 6.16
   defined, 6.37
Basket trays, 7.6
Battery-operated crimping tools, 8.17, 8.17f
Battery-powered knockout kits, 5.16, 5.16f
Bearings, motor nameplate information, 2.53, 2.54f
Bend degree protractors, 4.23
Bending cable, 8.24–8.25, 8.24f, 8.25f, 8.26f
Bending conduit
   electric benders, 4.17–4.18, 4.17f
   geometry and, 4.6–4.12, 4.7f, 4.8f, 4.9f, 4.10f, 4.12f
   hydraulic benders, 4.18–4.23, 4.19f, 4.20f, 4.21f, 4.22f
   mechanical benders, 4.12–4.15, 4.12f, 4.13f, 4.14f, 4.15f
   mechanical offsets, 4.16, 4.16f, 4.17
   mechanical stub-ups, 4.15–4.16
   NEC requirements for, 4.3–4.4, 4.3f
   PVC conduit bending, 4.36–4.38, 4.36f, 4.37f, 4.38f, 4.39
   PVC conduit installations, 4.35–4.36
   safety and, 4.2
   segment bending techniques, 4.23–4.33, 4.25f, 4.26f, 4.27f, 4.28f, 4.30f, 4.31f, 4.32f, 4.33f
   tricks of the trade, 4.34–4.35, 4.34f, 4.35f
   types of bends, 4.4–4.5, 4.4f, 4.5f, 4.6f
Bending gauges, 4.13
Bending protractors, 4.13, 4.13f
   defined, 4.43
Bending shoes, 4.13
Bending shots, 4.20
   defined, 4.43
Bending space, 8.24, 8.25f, 8.26, 8.26f
Bending tables, 4.25–4.26, 4.25f, 4.26f, 4.27f
Bends per run, *NEC* requirements for, 4.3–4.4
Bimetallic overload relays, 11.15–11.18, 11.16f
Bimetallic thermal devices, 11.16
Bi-pin lamps, 3.11
Blower/vacuum fish tape systems, 6.2, 6.11, 6.13f, 6.14f
Body resistance, 9.30, 10.12
Bonding
   defined, 9.41
   of enclosures and equipment, 9.23, 9.23f
   multiple raceway systems, 9.23, 9.23f
   multiple service disconnecting means, 9.22
   NEC requirements for, 9.3–9.4, 9.5, 9.22–9.23, 9.23f
   purpose of, 9.3
   service equipment, 9.22–9.23, 9.23f, 9.24f
   of services over 250 volts, 9.23, 9.23f
Bonding jumpers
   equipment bonding jumpers, 9.4, 9.23, 9.24f, 9.41
   main bonding jumpers, 9.20–9.21, 9.20f, 9.41
   materials for connecting, 9.14
Braking, dynamic braking, 2.41, 2.62–2.63
Braking systems, vertical cable pulls, 6.27
Branch circuits, 2.56
   defined, 2.68
Break
   circuits and, 11.2
   defined, 11.35

Brightness. *See* Luminance
Brushes, 2.5, 2.5f
   defined, 2.68
Brushless DC motors, 2.4, 2.16
Bull wheels, 6.29
Bushings, 5.14, 5.15f

C

Cable bending tools, 8.24, 8.24f, 8.25f
Cable blowing, 6.13
Cable connector installation, 8.18–8.19, 8.19f
Cable edge protection, 7.11
Cable exits, from cable trays, 7.9–7.10, 7.10f, 7.11f
Cable grips, 6.11, 6.15f
   defined, 6.37
Cable identification tags, 8.22
Cable inspections, 8.23
Cable installation, in cable trays, 7.19–7.22
Cable length meters, 6.32–6.33, 6.33f
Cable limitations, 6.28–6.32, 6.31f
Cable management systems, 7.3
Cable placement, cable trays, 7.9, 7.9f
Cable pullers
   hand-operated, 6.17, 6.19, 6.20f
   power-operated, 6.17, 6.18f, 6.19, 6.20, 6.20f, 6.23
   safety and, 6.19, 6.33
   setting up, 6.21, 6.21f, 6.22f
   types of, 6.19–6.20, 6.20f
Cable pulleys, 6.26, 7.9, 7.9f
   defined, 7.28
Cable pulling instruments, 6.32–6.33, 6.33f, 6.34
Cable pulling operations, 6.5, 6.6f
Cable reels, 6.7, 6.8f, 6.9f, 6.10f
Cable rollers, 6.27
Cable rollers with adjustable brackets, 6.29
Cable routing, 8.22–8.23, 8.23f
Cables, number allowed in cable trays, 7.20–7.22
Cable slitting and ringing tools, 8.5, 8.5f
Cable strippers, 8.3, 8.4, 8.4f, 8.5, 8.5f, 8.6f
Cable supports
   center hung support, 7.13, 7.14, 7.15, 7.15f
   direct rod suspension, 7.13, 7.15f
   load on, 7.7, 7.7f
   trapeze mounting, 7.13, 7.14f, 7.15, 7.15f
   for vertical cable trays, 7.11, 7.11f
   wall mounting, 7.13, 7.15f
Cable terminators, 6.16
Cable tray installation, 7.3
Cable trays
   basket trays, 7.6
   cable installation, 7.19–7.22
   center rail systems, 7.16, 7.16f
   channels for, 7.2, 7.5f, 7.14f
   cross section of, 7.4f
   drawings of, 7.22–7.23, 7.23f
   introduction to, 7.2, 7.3
   ladder trays, 7.2, 7.5
   loading of, 7.7–7.12, 7.7f, 7.8f, 7.9f, 7.10f, 7.11f, 7.12f, 7.13f
   longitudinal side rails, 7.2, 7.5f
   NEC requirements for, 7.2, 7.7, 7.10, 7.17, 7.17f, 7.18f, 7.19, 7.19f
   NEMA requirements for, 7.2
   pulling cable in, 6.27–6.28, 6.28f, 6.29, 7.24
   safety and, 7.25
   supports for, 7.7, 7.7f, 7.11, 7.11f, 7.13, 7.14, 7.14f, 7.15, 7.15f
   typical system, 7.4f

Capacitance
　alternating current and, 1.14, 1.15–1.21, 1.16f, 1.17f, 1.18f, 1.19f
　defined, 1.48
Capacitor-run motors, 2.31, 2.33–2.34, 2.33f
Capacitors
　calculating equivalent capacitance, 1.17–1.18, 1.17f, 1.18f
　capacitive reactance and, 1.20–1.21
　charging and discharging, 1.15–1.16, 1.16f
　determining capacitance of, 1.16–1.17
　specifications for, 1.18
　testing procedure and checklist, 3.59
　voltage/current relationship and, 1.18–1.20, 1.19f
Capacitor-start, capacitor-run motors, 2.31, 2.34, 2.35f, 2.36
Capacitor-start motors, 2.31, 2.33, 2.33f
Capacitor-type induction motors, 2.33–2.34, 2.33f
Capstans, 6.17, 6.18f
　defined, 6.37
Cartridge fuses, 10.16–10.17, 10.16f, 10.21
　defined, 10.25
Ceiling fans, installing, 3.34, 3.35f
Ceiling fixtures, 3.27, 3.34, 3.35f
CEMF (counter-electromotive force), 1.15, 2.9–2.10, 2.10f, 2.11f
Center hung support, for cable trays, 7.13, 7.14, 7.15, 7.15f
Center of developed length, segment bends, 4.20, 4.22, 4.22f
Center rail cable tray systems, 7.2, 7.16, 7.16f
Centers, of bends, 4.15, 4.22
Center-tapped transformers, 1.40, 1.40f
Chandeliers, installing, 3.33, 3.34f, 3.35
Channels, for cable trays, 7.2, 7.5f, 7.14f
Charting benders, 4.13–4.15, 4.14f, 4.15f
Chicago benders, 4.12, 4.12f, 4.16. *See also* Mechanical benders
Chisel edges, wire cutting, 8.5, 8.5f
Circles, bending conduit, 4.8–4.10, 4.8f, 4.9f, 4.10f
Circline fluorescent lamps, 3.13f
Circuit breaker ratings
　current rating, 10.4–10.5, 10.5f
　interrupting capacity rating, 10.5–10.10, 10.6f, 10.7f, 10.8f
　labeling, 10.10, 10.10f
　NEC requirements, 10.4, 10.10–10.11
　Underwriters Laboratories and, 10.3–10.4, 10.7–10.8
Circuit breakers
　as automatic overload devices, 10.2
　defined, 2.68, 10.25
Circuit breakers and fuses
　additional functions of, 10.22
　circuit breaker ratings, 10.3–10.11, 10.5f, 10.6f, 10.7f, 10.8f, 10.10f
　coordination of multiple devices, 10.21–10.22
　fuse sizing, 10.19–10.21, 10.20f
　fuse types, 10.14–10.17, 10.15f, 10.16f
　ground fault current circuit protection, 10.2, 10.11–10.13, 10.13f
　introduction to, 10.2–10.3, 10.2f, 10.3f
　labeling, 10.10
　NEC requirements for, 10.10–10.11, 10.15, 10.17, 10.19–10.21
　overcurrents, 10.17–10.18, 10.18f
　safety and, 10.21
　short circuit calculations, 9.30
Circuits that cannot be grounded, 9.9, 9.9f
Circuit tester/wire sorter, 6.33, 6.33f
Clamp-on ground testers, 9.31
Clamp-type terminal block connectors, 8.21, 8.22, 8.22f
Class I locations, 5.3
Class II locations, 5.3
Class III locations, 9.9
Cleaning conductors. *See* Stripping and cleaning conductors
Clevises, 6.11, 6.15f
　defined, 6.37
Coil diameters, inductance, 1.12, 1.13f
Coils. *See also* Windings
　latch coils, 11.9, 11.35
　shading coils, 11.2f, 11.3–11.4, 11.3f, 11.35
　three-coil ballasts, 3.22, 3.23f, 3.24
　unlatch coils, 11.9, 11.35
　winding coils, 1.12, 1.13f
Cold flow, 8.9
Color codes, for crimp connector insulation, 8.13–8.14
Color rendering indices (CRI), 3.15
　defined, 3.55
Color rendition, fluorescent lamps, 3.11
Colors, and light, 3.4–3.6, 3.5f
Color temperature, lamps, 3.11, 3.15
Combination cables, cable trays, 7.21–7.22
Combination couplings, 5.13–5.14, 5.14f
Comeback, 4.10
Communication circuits, 8.12
Communications, cable pulling, 6.19, 6.22
Commutators, 2.4–2.6, 2.4f
　defined, 2.68
Compact fluorescent lamps, 3.12, 3.13f, 3.37, 3.37f
Compound motors, 2.14–2.15, 2.14f, 2.16, 2.16f
Compression fittings, 5.13
Concentric bending, 4.16, 4.26–4.28, 4.27f
　defined, 4.43
Concentric bends, 4.8, 4.29
Concentric circles, 4.8, 4.8f
Concentric offsets, 4.31–4.33, 4.31f, 4.32f, 4.33f
Conductor installation
　cable-pulling equipment, 6.17–6.21, 6.18f, 6.20f, 6.21f, 6.22f
　cable pulling instruments, 6.32–6.33, 6.33f, 6.34
　complex installations, 6.34
　high-force cable pulling, 6.22–6.24, 6.24f, 6.25f, 6.26
　interrupting pulls and, 6.28, 6.32
　introduction to, 6.2
　physical limitations of cable, 6.28–6.32, 6.31f
　planning installations, 6.3–6.5, 6.4f, 6.6f
　pulling cable in cable trays, 6.27–6.28, 6.28f, 6.29, 7.24
　supporting conductors, 6.26–6.27, 6.26f
　wire pulling setup, 6.7–6.8, 6.7f, 6.8f, 6.9f, 6.10–6.11, 6.10f, 6.11f, 6.12, 6.13, 6.13f, 6.14f, 6.15–6.17, 6.15f
Conductor positioning, crimp connections, 8.21, 8.21f
Conductors
　allowable tension on, 6.29, 6.30
　cable trays and, 6.27–6.28, 6.28f, 7.19–7.22, 7.24
　entering or leaving enclosures, 8.24–8.25, 8.26f
　equipment grounding conductors, 9.2, 9.16–9.18, 9.16f, 9.18f, 9.41
　grounded conductors, 9.2, 9.3f, 9.24–9.25, 9.25f, 9.41
　grounded service conductors, 9.20, 9.22
　low-impedance conductors, 9.4, 9.24
　primary conductors, 10.20
　secondary conductors, 8.26
　service conductors, 9.13, 9.22, 9.25f
　service entrance conductors, 9.2, 9.12, 9.14, 9.22
　stripping control and signal cable/conductors, 8.5, 8.6f, 8.7, 8.7f
　stripping power cables and large conductors, 8.3, 8.4, 8.4f, 8.5, 8.5f
　stripping small conductors, 8.2–8.3, 8.3f
　ungrounded conductors, 9.5
Conductor supports, 6.26–6.27, 6.26f
　defined, 6.37

Conductor terminations and splices
   bending cable and training conductors, 8.24–8.25, 8.24f, 8.25f, 8.26f
   control and signal cable types, 8.12, 8.12f
   installing connectors, 8.14–8.23, 8.15f, 8.16f, 8.17f, 8.18f, 8.19f, 8.20f, 8.21f, 8.22f, 8.23f, 8.24f
   low-voltage connectors and terminals, 8.13–8.14, 8.13f, 8.14f
   motor connection kits, 8.30, 8.30f
   NEC requirements for, 8.8, 8.26–8.27, 8.27f
   stripping and cleaning conductors, 8.2–8.3, 8.3f, 8.4, 8.4f, 8.5, 8.5f, 8.6, 8.6f, 8.7, 8.7f
   taping electrical joints, 8.28, 8.28f, 8.29, 8.29f
   wire connections under 600 volts, 8.8–8.11, 8.8f, 8.9f, 8.10f
Conduit. *See also* Bending conduit
   defined, 4.43
Conduit bodies, 4.2, 5.7–5.10, 5.8f, 5.9f, 5.10f
   defined, 5.21
Conduit brushes, 6.8, 6.10f
Conduit pistons, 6.11, 6.14f
   defined, 6.37
Conduit swabs, 6.8, 6.10f
Condulets. *See* Conduit bodies
Connecting capacitors, 1.17–1.18, 1.17f, 1.18f
Connection diagrams, motor nameplate data, 2.49, 2.50f, 2.57, 2.57f, 2.62
Connections
   basic requirements for, 8.2
   defined, 8.34
Connector installation
   compression connectors, 8.14–8.17, 8.15f, 8.16f, 8.17f, 8.18f
   control and signal cables, 8.19–8.23, 8.20f, 8.21f, 8.22f, 8.23f, 8.24f
   mechanical terminals and connectors, 8.18
   specialized cable connectors, 8.18–8.19, 8.19f
Connectors, 8.2
   defined, 8.34
Consequent-pole motors, 2.37–2.39, 2.38f, 2.39f
Constant-wattage autotransformer (CWA) ballasts, 3.22, 3.23f
Constant-wattage isolated (CWI) ballasts, 3.22, 3.23f
Contactors and relays. *See also* Relays
   low-voltage remote control switching, 11.18–11.20, 11.21f, 11.22, 11.23f, 11.24–11.25, 11.26f, 11.27, 11.28f, 11.29–11.30, 11.29f
   magnetic contactors, 11.2–11.5, 11.2f, 11.3f, 11.4f, 11.5f, 11.6, 11.7, 11.8–11.9, 11.8f
   overload relays, 11.15–11.18, 11.16f, 11.17f
   protective enclosures for, 11.18
   relay types, 11.2, 11.9–11.11, 11.10f, 11.11f, 11.12f, 11.13f
   replacing, 11.4
   solid-state relays, 11.13–11.15, 11.13f, 11.14f
   troubleshooting, 11.3, 11.30, 11.31–11.32
Containment barriers, metal halide lamps, 3.14
Continuity, bonding service equipment, 9.22
Continuous duty, 2.49
   defined, 2.68
Contrast, 3.2
Control and signal cables
   installation, 8.19–8.23, 8.20f, 8.21f, 8.22f, 8.23f, 8.24f
   stripping, 8.5, 8.6f, 8.7, 8.7f
   types of, 8.12, 8.12f
Controllers, 2.14
   defined, 2.68
Control relays, 11.9–11.11, 11.10f, 11.11f, 11.12f, 11.13f
Controls for lighting
   occupancy sensors, 3.47, 3.47f
   photosensors, 3.47, 3.47f
   relay controlled lighting circuits, 3.49, 3.50f
   timers, 3.48, 3.48f
   zone lighting control circuit, 3.49, 3.49f
Conversions
   of decimals to fractions, 4.10–4.11
   metric conversions, 4.4, 7.30
Conveyor sheaves, 6.24, 6.25f
Cooling PVC conduit after bending, 4.37, 4.37f, 4.39
Coordination of multiple devices, circuit breakers and fuses, 10.21–10.22
Corded crimping tools, 8.17, 8.17f
Core characteristics, transformers, 1.35, 1.36, 1.36f, 1.37f
Core material, inductance, 1.12, 1.13f
Corneas, 3.2, 3.2f
Cosecants, 4.6–4.7
Cosines, 1.32, 4.6–4.7
Cotangents, 4.6–4.7
Counter-electromotive force (CEMF), 1.15, 2.9–2.10, 2.10f, 2.11f
Crimp centering, 8.21, 8.21f
Crimp connector installation, 8.17, 8.18, 8.18f, 8.19–8.21, 8.20f, 8.21f
Crimping force, 8.14, 8.15f
Crimping tool dies, 8.15–8.16, 8.16f, 8.21
Crimping tools, 8.8, 8.14–8.17, 8.15f, 8.16f, 8.17f, 8.18f
Crimp-on wire lugs, 8.8
Crimp-type terminals and connectors, 8.2, 8.13, 8.13f, 8.14f
Crosses
   cable trays and, 7.2, 7.6
   defined, 7.28
Current, in inductive AC circuits, 1.15, 1.21
Current-limiting circuit breakers, 10.9–10.10
Current-limiting fuses, 10.19
Current rating, for circuit breakers, 10.4–10.5, 10.5f, 10.10, 10.10f
Current supply circuits, in ground testers, 9.31, 9.31f
Current transformers, 1.42, 1.42f, 1.44
Curvature, feed-in setup, 6.24, 6.24f
CWA (constant-wattage autotransformer) ballasts, 3.22, 3.23f
CWI (constant-wattage isolated) ballasts, 3.22, 3.23f
Cycles, of AC voltage, 1.2–1.5, 1.4f

D

Damp and wet locations, pull and junction boxes, 5.2–5.3
Damper windings, 2.27
DC injection braking, 2.62
DC motors. *See also* Variable speed drives
   armature reaction, 2.8–2.9, 2.9f
   brushless DC motors, 2.4, 2.16
   components of, 2.2–2.6, 2.3f, 2.4f, 2.5f
   compound motors, 2.14–2.15, 2.14f, 2.16, 2.16f
   counter-electromotive force, 2.9–2.10, 2.10f, 2.11f
   as generators and brakes, 2.9, 2.43
   neutral plane, 2.6–2.7, 2.6f, 2.7f
   operating characteristics of, 2.15–2.16, 2.16f
   principles of, 2.2–2.12, 2.2f, 2.3f, 2.4f, 2.5f, 2.6f, 2.7f, 2.8f, 2.9f, 2.10f, 2.11f
   series motors, 2.12, 2.13–2.14, 2.13f, 2.15–2.16, 2.16f
   shunt motors, 2.11, 2.11f, 2.12–2.13, 2.12f, 2.15, 2.15f
   starting resistance, 2.10–2.12
   torque and, 2.3, 2.4, 2.4f, 2.13, 2.14
   two-loop DC motors, 2.8, 2.8f
   types of, 2.12–2.16, 2.12f, 2.13f, 2.14f, 2.15f, 2.16f
Decimals, converting to fractions, 4.10–4.11
Deflection under load, cable trays, 7.7
Degree indicator, 4.22
   defined, 4.43

Degrees, angular representation, 4.9
Delta-connected motors, 2.57, 2.57*f*, 2.58–2.59, 2.59*f*
Developed length, 4.9, 4.20
    defined, 4.43
Device boxes, as junction boxes, 5.2
Dielectric, 1.17, 2.16–2.17
Dielectric permittivity, capacitance, 1.17
Dimension sheets, 2.47, 2.47*f*
Dimensions of common wire sizes, 8.3, 8.4
Dimming ballasts, 3.20–3.21, 3.20*f*, 3.24
Dip tolerance, 3.22
    defined, 3.55
Direct current (DC). *See* DC motors
Direct rod suspension, 7.13, 7.15*f*
    defined, 7.28
Dissimilar materials, connectors, 8.19
Dog legs, eliminating, 4.16, 4.17, 4.34, 4.34*f*
Double-braided polyester composite pulling lines, 6.17
Double insulation, 9.16
Double squirrel cage rotors, 2.19
Double-throw switches, 11.22
Double wall heat-shrink tubing, 8.11
Double wall tubing, 8.11
Down cable pulls, 6.21, 6.21*f*
Drain wires, 8.12
    defined, 8.34
Drawings
    cable trays and, 7.22–7.23, 7.23*f*
    symbols for low-voltage drawings, 11.25, 11.26*f*
Drip-proof guarded motor enclosures, 2.45
Drip-proof motor enclosures, 2.45
Dropout plates, 7.10, 7.10*f*, 7.11*f*
    defined, 7.28
Dropouts, 7.10, 7.10*f*
    defined, 7.28
Drum lighting fixtures, installing, 3.32–3.33, 3.32*f*
Dual-element cartridge fuses, 10.17
Dual-element fuses, 10.14
    defined, 10.25
Dual-element time delay fuses, 10.19, 11.16
Dust- and ignition-proof motor enclosures, 2.46
Dust-tight, rain-tight (NEMA Type 3) enclosures, 5.3
Duty, 2.51
    defined, 2.68
Duty rating, motor nameplate data, 2.51
Dynamic braking, 2.41, 2.62–2.63
Dynamometers (force gauges), 6.33

E

Earth electrode resistance, 9.32
Earth ground resistance testers, 9.28, 9.28*f*
Eddy currents, 1.36
Edison-base, defined, 10.25
Edison-base fuses, 10.15, 10.15*f*
Effective ground fault path, 9.24
    defined, 9.41
Effective grounding path, 9.24
Effective grounds, testing for, 9.28
Effective resistance area, 9.32
Effective values, 1.6, 1.9
Efficacy, 3.6
    defined, 3.55
Efficiency. *See also* Going green
    ballasts and, 3.18–3.19, 3.20
    lamps and, 3.6, 3.7, 3.15
EGC (equipment grounding conductors), 9.2, 9.5, 9.16–9.18, 9.16*f*, 9.18*f*, 9.41
Elastic tape, layout of segment bends, 4.23

Elastomeric insulating caps, 8.30
Elbow rollers, 6.29
Elbows
    cable hangers, 7.11, 7.11*f*
    conduit and, 4.2, 4.4, 4.4*f*
    defined, 4.43, 7.28
    pulling elbows, 5.9, 5.10*f*
Electrical connection of lighting fixtures, 3.45–3.46, 3.46*f*
Electrical generation, 1.2–1.3, 1.2*f*
Electrical joints, taping, 8.28, 8.28*f*, 8.29, 8.29*f*
Electrical lighting. *See also* Lighting fixture installation; Lighting fixtures
    ballasts, 3.10, 3.17–3.22, 3.17*f*, 3.18*f*, 3.20*f*, 3.21*f*, 3.22*f*, 3.23*f*, 3.24, 3.55
    characteristics of light, 3.3–3.6, 3.4*f*, 3.5*f*
    controls for, 3.47–3.48, 3.47*f*, 3.48*f*
    energy management systems, 3.49, 3.49*f*, 3.50*f*, 3.51
    human vision and, 3.2, 3.3*f*
    lamps, 3.6–3.7, 3.6*f*, 3.8*f*, 3.9–3.17, 3.9*f*, 3.10*f*, 3.11*f*, 3.13*f*, 3.14*f*
    testing procedures and checklist, 3.59
Electrical tape, 8.28, 8.28*f*, 8.29, 8.29*f*
Electric conduit benders, 4.17–4.18, 4.17*f*
Electric shock
    ground fault circuit interrupters (GFCI) and, 9.30, 10.12–10.13
    isolation transformers and, 1.40–1.41, 1.41*f*
Electrocutions, 1.43, 10.12
Electrode location, three-point testing procedures, 9.32, 9.33–9.36, 9.33*f*, 9.34*f*, 9.35*f*, 9.36*f*
Electrolytic action, aluminum connections, 8.9
Electrolytic cell line working zones, grounding, 9.9
Electromagnetic induction, 1.35
Electromagnetism, 1.2
Electromotive force (EMF), 1.14
Electronic instant-start ballasts, 3.18, 3.19
Electronic rapid-start ballasts, 3.18, 3.19
Elevators, DC motors and, 2.13, 2.14
ELI the ICE man, 1.14, 1.15, 1.20, 1.21
Emergency lighting circuits
    ballasts and, 3.21
    relays and, 11.5, 11.5*f*
EMF (electromotive force), 1.14
EMT conduit fittings, 5.13, 5.13*f*
Encapsulated windings motor enclosures, 2.45
Enclosed motors, 2.46
Enclosures
    grounding enclosures, 9.18–9.19, 9.19*f*, 9.20*f*
    low-voltage switches and, 11.27
    motor enclosures, 2.45–2.46, 5.3
    NEMA requirements for, 5.3–5.4, 8.27*f*, 11.18
End of run, pull boxes, 5.6
End power feed tracks, 3.43, 3.43*f*, 3.44, 3.45*f*
End terminations, cable stripping, 8.2
Energy efficiency. *See* Going green
Energy management systems, 3.49, 3.49*f*, 3.50*f*, 3.51
Entrance ells (SLBs), 5.9, 5.10*f*
EPACT (National Energy Policy Act), 3.6
Equipment, defined, 2.68
Equipment bonding jumpers, 9.4, 9.23, 9.24*f*
    defined, 9.41
Equipment checklists, for wire pulling, 6.7, 6.7*f*
Equipment grounding
    equipment ground conductors and, 9.2, 9.5, 9.16–9.18, 9.16*f*, 9.18*f*
    grounding enclosures and, 9.18–9.19, 9.19*f*, 9.20*f*
    main bonding jumpers, 9.20–9.21, 9.20*f*

Equipment grounding conductors (EGC), 9.2, 9.5, 9.16–9.18, 9.16*f*, 9.18*f*
   defined, 9.41
Equipotential grounding, 9.29, 9.36
Equivalent hemisphere radii, 9.34–9.35
Eutectic thermal overload relays, 11.17, 11.17*f*
Exciting current, transformers, 1.37
Exothermic welding, 9.14, 9.14*f*, 9.15
Expansion and contraction, PVC conduit, 4.37–4.38
Expansion couplings, 4.37–4.38
Expansion joints, 7.11–7.12, 7.13*f*
   defined, 7.28
Expansion splice plates, 7.11–7.12, 7.13*f*
Explosion-proof, 5.2
   defined, 5.21
Explosion-proof motor enclosures, 2.46
Externally ventilated motor enclosures, 2.45
Eyes, structure of, 3.2, 3.2*f*

F

Failure under load, cable trays, 7.8, 7.8*f*
Fall-of-potential resistance testing, 9.30–9.31, 9.30*f*
Fan-cooled guarded motor enclosures, 2.45, 2.46
Fan-cooled motor enclosures, 2.46
Fan joist hangers, 3.34, 3.35*f*
Farads (F), 1.16
Fault current circuit protection, 10.2
Fault currents, 10.2
   defined, 10.25
FD boxes, 5.8–5.9
Feeders, coordination, 10.21–10.22
Feed-in setups, 6.24, 6.24*f*, 6.25*f*
Field control, DC motors, 2.43, 2.43*f*
Field of vision, 3.2
Field poles, 2.9–2.10
   defined, 2.68
Filament lamps. *See* Incandescent lamps
Filaments, 3.6–3.7, 3.8*f*
Fill, of cable trays, 7.7, 7.7*f*
Finish trim, low-voltage lighting controls, 11.30
Fire alarm cable, 8.12, 8.12*f*
Fires, poor electrical connections, 8.2
Fish lines, 6.11
   defined, 6.37
Fish poles, 6.12
Fish tape leaders, 6.2
Fish tapes, 6.2, 6.3, 6.4
   defined, 6.37
Fittings
   cable trays and, 7.2, 7.4*f*
   defined, 7.28
   for EMT conduit, 5.13, 5.13*f*
   locknuts and bushings, 5.14–5.17, 5.14*f*, 5.15*f*, 5.17*f*
   pull and junction boxes and, 5.12–5.16, 5.13*f*, 5.14*f*, 5.15*f*, 5.16*f*, 5.17*f*
   for rigid, aluminum, and IMC conduit, 5.13–5.14, 5.14*f*
Fixture whips, 3.39, 3.39*f*
FLA (full-load amps), 2.49–2.50, 2.51
Flange kits, recessed fixtures, 3.38, 3.38*f*
Fleming's first rule, 1.7
Flexible metal conduit, as equipment grounding conductor (EGC), 9.17–9.18, 9.18*f*
Floating canopy power feeds, 3.43*f*, 3.45*f*
Floating power feed tracks, 3.43, 3.43*f*
Floods, overcurrent and ground fault protective devices, 10.14

Floor boxes, 5.2
Floor plans, 3.28, 3.28*f*, 7.23, 7.23*f*, 9.11*f*, 11.27, 11.28*f*
Fluorescent lamp holders, 3.13, 3.13*f*
Fluorescent lamps, 3.10–3.13, 3.10*f*, 3.11*f*, 3.13*f*, 3.17
Fluorescent lighting fixtures
   ballasts, 3.17–3.21, 3.17*f*, 3.18*f*, 3.20*f*
   installing, 3.33, 3.33*f*
Fluorescent service checklist, 3.60
Fluorescent troffer installation, 3.38–3.39, 3.38*f*, 3.39*f*
Follow bars, 4.13
Foot-mounted motors, 2.47*f*
Formulas
   apparent power, 1.32, 1.33*f*
   arc of a quadrant, 4.9–4.10
   armature current, 2.10, 2.11, 2.12
   capacitive reactance, 1.20
   circumference, 4.8–4.9, 4.9*f*
   cross-section area of cables, 7.20
   gains, 4.10
   impedance of parallel LC circuit, 1.29
   impedance of parallel RC circuit, 1.27
   inductive reactance, 1.15
   magnitude of impedance, 1.21
   power dissipation, 1.11
   power factor, 1.33
   pulling tension, 6.30–6.31
   self-induced voltage, 1.11, 1.12
   short circuit calculations, 9.30
   slip, 2.21
   speed control of AC motors, 2.44
   synchronous speed of motor, 2.18, 2.22
   transformer turns ratio, 1.38
   trigonometric functions, 4.7, 4.25
   true power, 1.32
   voltage anywhere along sine waves, 1.3, 1.4*f*
Four-wire delta-connected secondaries, 9.6
Four-wire wye-connected secondaries, 9.6
Frame sizes
   circuit breakers and, 10.3
   defined, 10.25
Frequency
   capacitive reactance and, 1.20
   defined, 1.48
   impedance and, 1.21
   inductive reactance and, 1.15
   motor nameplate information and, 2.53
   sine waves and, 1.4–1.5, 1.4*f*, 1.6
FS boxes, 5.8–5.9
Full-load amps (FLA), motor nameplate data, 2.49–2.50, 2.51
Full-load torque, 2.22
Fuse classes, 10.17, 10.18
Fuse links, 10.5
   defined, 10.25
Fuses. *See also* Circuit breakers and fuses
   amperage ratings of, 10.14
   cartridge fuses, 10.16–10.17, 10.16*f*, 10.21, 10.25
   defined, 10.25
   fuse classes, 10.17, 10.18
   labeling, 10.10
   markings for, 10.17
   overload protection and, 2.54, 10.2, 11.16
   plug fuses, 10.5, 10.15–10.16, 10.15*f*, 10.25
   sizing, 10.19–10.21, 10.20*f*
   testing and, 10.3, 10.7–10.8
   time-delay ratings of, 10.15, 11.16
   voltage ratings of, 10.14

G

Gains
  bending conduit and, 4.10, 4.10f, 4.11–4.12, 4.12f
  defined, 4.43
Galvanic corrosion, aluminum, 8.9
General-purpose motors, 2.19, 2.49
General purpose (NEMA Type 1) enclosures, 2.45, 5.3
Generators, 1.2–1.3, 1.3f, 1.4f, 1.9, 9.26–9.27, 9.26f
Generator-type separately derived systems, grounding, 9.8, 9.26–9.27, 9.26f
Geometry, bending conduit, 4.6–4.12, 4.7f, 4.8f, 4.9f, 4.10f, 4.12f
GFCI (ground fault circuit interrupters), 10.11–10.13, 10.13f, 10.22
Glass cloth electrical tapes, 8.28
Go and no-go mandrels, 6.8, 6.10f
Going green
  compact fluorescent lamps, 3.12
  fluorescent lamps, 3.12
  lamp efficiency, 3.7
  lamp efficiency ratings, 3.15
  LED lighting systems, 3.51
  motor as generator and brake, 2.43
  power factor, 1.35, 3.19
  retrofitting lighting, 3.6
  synchronous motors, 2.26
  variable-speed drives, 2.26
Gradients, angular representation and, 4.9
Grooming, 8.20
  defined, 8.34
Ground, 9.2
  defined, 9.41
Ground current, 9.28
  defined, 9.41
Grounded, 9.2
  defined, 9.41
Grounded conductors, 9.2, 9.3f, 9.24–9.25, 9.25f
  defined, 9.41
Grounded service conductors, 9.20, 9.22
Ground electrodes, 9.2
Ground fault circuit interrupters (GFCI), 10.11–10.13, 10.13f, 10.22
Ground faults
  clearing, 9.24
  in grounded system, 9.17, 9.17f
  short circuits distinguished from, 9.4, 9.4f, 9.21
Ground grids, 9.28
  defined, 9.41
Grounding clips, 9.19, 9.19f
  defined, 9.41
Grounding conductors, 9.2, 9.3f, 9.18, 9.18f
  defined, 9.41
Grounding connections, 9.3
  defined, 9.41
Grounding electrode conductors
  connection for, 9.12, 9.13f
  defined, 9.41
  grounded conductor distinguished from, 9.2
  installation of, 9.13, 9.14f
  NEC requirements for, 9.4, 9.12–9.13, 9.14
  panelboard showing, 9.3f
  sizing of, 9.12–9.13, 9.14
  splicing of, 9.14, 9.14f
Grounding electrodes
  defined, 9.41
  earth resistance and, 9.28
  grounding methods and, 9.9–9.10, 9.12, 9.15
  grounding systems and, 9.2
  NEC requirements for, 9.10f
  safety and, 9.6
  underground gas piping systems unsuitable for, 9.10, 9.11f, 9.15
  underground tanks as, 9.14, 9.15f
Grounding enclosures, 9.18–9.19, 9.19f, 9.20f
Grounding high-voltage systems, NEC requirements for, 9.28
Grounding insulated bushings, 5.14, 5.15f
Grounding multiple buildings, 9.27, 9.27f
Grounding outlet boxes and devices, 9.19, 9.20f
Grounding resistance curves, 9.33f, 9.34–9.36, 9.34f, 9.36f
Grounding systems
  bonding service equipment, 9.22–9.23, 9.23f, 9.24f
  effective grounding path, 9.24
  for electric services, 9.5, 9.9, 9.14
  equipment grounding, 9.2, 9.5, 9.16–9.21, 9.16f, 9.17f, 9.18f, 9.19f, 9.20f
  grounded conductors, 9.2, 9.3f, 9.24–9.25, 9.25f
  ground faults and, 9.4, 9.4f, 9.17, 9.17f, 9.21, 9.24
  grounding multiple buildings, 9.27, 9.27f
  introduction to, 9.2–9.3, 9.3f
  measuring earth resistance, 9.28, 9.29f, 9.30–9.32, 9.30f, 9.31f
  metal boxes and, 3.45
  NEC requirements for, 9.3–9.4, 9.5, 9.9–9.10, 9.10f, 9.11f, 9.12–9.16, 9.12f, 9.13f, 9.14f, 9.15f
  over 1000 volts, 9.28
  purpose of, 9.3
  separately derived systems, 9.2, 9.7–9.9, 9.8f, 9.9f, 9.25–9.27, 9.25f, 9.26f
  short circuits and, 9.4, 9.4f, 9.21
  testing for effective grounds, 9.28
  three-point testing procedures, 9.33–9.36, 9.33f, 9.34f, 9.35f, 9.36f
  types of, 9.5–9.9, 9.5f, 9.6f, 9.7f, 9.8f, 9.9f
Ground mats, 9.28
  defined, 9.41
Ground resistance, 9.28
  defined, 9.41
Ground rings, 9.9, 9.10, 9.10f
Ground rod depth, 9.32
Ground rods, 9.2, 9.12f, 9.13
  defined, 9.41
Ground testers
  current supply and, 9.31, 9.31f
  safety and, 9.28
  testing methods, 9.30–9.31, 9.30f
  three-point testing and, 9.33–9.36, 9.33f, 9.34f, 9.35f, 9.36f
  voltmeter circuits in, 9.31f, 9.32
Guarded motor enclosures, 2.45
Guards, for motors, 2.63
Guiding conductors, cable pulling, 6.19
Gutters, 5.5, 8.27f

H

HACR circuit breakers, 10.12
Hand bending PVC conduit, 4.37, 4.37f
Handholes
  defined, 5.21
  pull and junction boxes and, 5.10–5.12, 5.11f, 5.12f
Handling lamps, 3.16–3.17
Handling lighting fixtures, 3.26–3.28
Hand-operated cable pullers, 6.17, 6.19, 6.20f

Hand-operated crimping tools, 8.13, 8.15, 8.15f, 8.16, 8.16f, 8.17
Hand-to-body resistance, 10.12
Hangers, for suspended lighting fixtures, 3.39, 3.40f, 3.44
Hard-wired installation, lighting fixtures, 3.45–3.46, 3.46f
Hazardous location Class II (NEMA Type 9) enclosures, 5.3
Hazardous location Class I (NEMA Type 7) enclosures, 5.3
Hazardous locations, pull and junction boxes and, 5.2, 5.3
H bars, 7.11
Heat, amplitude and, 1.6
Heating blankets, bending PVC conduit and, 4.36
Heat-shrinkable motor connection kit insulators, 8.30, 8.30f
Heat-shrink insulators, 8.10–8.11, 8.10f
Henry (H), 1.12
Hertz (Hz), 1.4
    defined, 1.48
Hickeys, 4.12
High-efficiency ballasts, 3.18–3.19, 3.20
High-force cable pulling, 6.22–6.24, 6.24f, 6.25f, 6.26
High-intensity discharge (HID) lamps, 3.13–3.15, 3.14f
High-intensity discharge (HID) lighting fixtures
    ballasts, 3.21–3.22, 3.21f, 3.22f, 3.23f, 3.24
    installation, 3.37, 3.37f
High interrupting capacity, 10.8–10.9, 10.8f
High-output (HO) fluorescent lamps, 3.12
High-pressure sodium lamps, 3.14
High-pressure sodium service checklist, 3.62
High-temperature silicone rubber tapes, 4.28
High-voltage, grounding systems, 9.28
HO (high-output) fluorescent lamps, 3.12
Horizontal adjustment splice plates, 7.11
Horsepower, 2.12, 2.51
    defined, 2.68
Hot air blowers, bending PVC conduit, 4.36
Hotboxes, 4.36
Hours
    defined, 2.68
    duty cycles and, 2.51
Human eyes, structure of, 3.2, 3.2f
Human vision, and electrical lighting, 3.2, 3.3f
Hybrid ballasts, 3.18
Hydraulic cable benders, 8.24, 8.25f
Hydraulic conduit benders, 4.18–4.23, 4.19f, 4.20f, 4.21f, 4.22f
Hydraulic crimping tools, 8.15, 8.16, 8.16f, 8.17
Hydraulic knockout kits, 5.16, 5.16f
Hydraulic-type cable cutters, 8.6
Hypotenuse, 4.6–4.7, 4.7f
Hysteresis synchronous motors, 2.38
Hz (hertz), 1.4, 1.48

I

ID (inside diameters), 4.31, 4.43
Ignitors
    HID lighting fixture ballasts and, 3.21, 3.21f
    testing procedure and checklist, 3.59
IMC conduit fittings, 5.13–5.14, 5.14f
Impedance
    defined, 1.48
    formulas for, 1.21, 1.27, 1.29
    frequency and, 1.21
    grounding and, 9.4, 9.24
Incandescence, 3.6
    defined, 3.55
Incandescent lamps, 3.6–3.7, 3.6f, 3.8f, 3.9–3.10, 3.9f, 3.10f, 3.37, 3.37f, 11.4
Incident light, 3.3
    defined, 3.55
Incoming line connections, 8.14

Indent position, 8.20–8.21, 8.20f
Indicators, bending conduit and, 4.22
Inductance
    in AC circuits, 1.12–1.15, 1.13f, 1.14f
    defined, 1.48
Induction
    in AC motors, 2.18
    three-phase induction motors, 2.18–2.24, 2.18f, 2.19f, 2.20f, 2.21f, 2.23f, 2.24f, 2.25f
Inductive loads, 11.4, 11.15
    defined, 11.35
Inductive reactance, 1.14–1.15
Inductors
    reactive power and, 1.32–1.33
    voltage/current relationship and, 1.12–1.14, 1.14f, 1.20, 1.21, 1.22, 1.23
Industrial use (NEMA Type 12) enclosures, 5.4
Infrared detectors, 3.47, 3.47f
Ingress protection (IP) classification system, 5.3, 5.4
In-line splice connections, 8.30, 8.30f
Inside diameters (ID), 4.31
    defined, 4.43
Inspecting cable, 8.20
Instantaneous
    defined, 11.35
    overcurrent protection, 11.2
Instant start fluorescent lamps, 3.11–3.12, 3.18, 3.18f
Instrumentation cable, 8.12, 8.12f
Instrument transformers, 1.40
Insulated bushings, 5.14, 5.15f
Insulating materials, lighting fixtures, 3.32
Insulating tapes, 8.28, 8.28f, 8.29, 8.29f
    defined, 8.34
Insulating wedges, 6.26, 6.26f
Insulation class, motor nameplate information, 2.52–2.53, 2.52f
Intensity threshold, 3.2
Interlocked armor cables
    cable trays and, 7.24
    defined, 7.28
    minimum bending radii for, 8.24
Intermittent duty, 2.49
    defined, 2.68
Internal shrouds, metal halide lamps, 3.14, 3.14f
Interpoles, 2.9, 2.9f
Interrupting capacity ratings, 10.5–10.10, 10.6f, 10.7f, 10.8f, 10.14
Interrupting pulls, 6.28, 6.32
Inverter systems, grounding, 9.8
IP (ingress protection) classification system, 5.3, 5.4
Iris, 3.2, 3.2f
Iron cores, 1.36, 1.36f
Isolation transformers, 1.40–1.41, 1.41f, 1.42

J

Jigs, bending PVC conduit, 4.37
Joining PVC conduit, 4.35–4.36
Junction boxes, 5.2, 5.2f. *See also* Pull and junction boxes
    defined, 5.21

K

Kelvins (K), 3.15
Kicks, 4.3, 4.5, 4.6f
    defined, 4.43
Knockout punches, 5.16, 5.16f, 5.17f
Knockouts, 5.16, 5.16f, 5.17f
Kynar heat-shrink tubing, 8.11
Kynar tubing, 8.11

L

Ladder trays
    cable trays, 7.2, 7.5
    defined, 7.28
Lambda, 1.5
Laminated steel cores, 1.36, 1.37f
Lamps
    color rendering index, 3.15, 3.55
    color rendition, 3.11
    color temperature and, 3.11, 3.15
    comparison charts, 3.15, 3.16
    fluorescent lamps, 3.10–3.13, 3.10f, 3.11f, 3.13f, 3.17
    high-intensity discharge (HID) lamps, 3.13–3.15, 3.14f
    incandescent lamps, 3.6–3.7, 3.6f, 3.8f, 3.9–3.10, 3.9f, 3.10f, 3.37, 3.37f, 11.4
    installation of, 3.16–3.17
    testing procedure and checklist, 3.59
    tungsten halogen incandescent lamps, 3.9–3.10, 3.10f
Large conductors, stripping, 8.3, 8.4, 8.4f, 8.5, 8.5f
Latch coils, 11.9
    defined, 11.35
LC circuits, 1.27–1.29, 1.28f, 1.29f
Leak resistance, capacitors, 1.18
LED (light emitting diode) lighting systems, 3.51
Left-hand rule for generators, 1.3, 1.7
Leg length, 4.19–4.20
    defined, 4.43
Length of coils, inductance, 1.12, 1.13f
Leveraged crimping tools, 8.15, 8.15f
Life expectancy, of lamps, 3.6, 3.7, 3.9f, 3.12
Lifting motors, 2.63
Light
    absorption of, 3.3–3.4
    characteristics of, 3.3–3.6, 3.4f, 3.5f
    colors and, 3.4–3.6, 3.5f
    reflected light, 3.2, 3.3–3.4, 3.4f, 3.55
    refraction of, 3.3–3.4, 3.4f, 3.5, 3.5f
    spectrum of visible light, 3.4–3.5, 3.5f
Light emitting diode (LED) lighting systems, 3.51
Lighting contactors, 11.4–11.5, 11.4f, 11.5f, 11.6–11.7, 11.8, 11.8f
Lighting fixture installation
    electrical connection of lighting fixtures, 3.45–3.46, 3.46f
    manufacturer's instructions, 3.29, 3.30f
    mechanical installation of recessed lighting fixtures, 3.36–3.39, 3.36f, 3.37f, 3.38f, 3.39f
    mechanical installation of surface-mounted fixtures, 3.29, 3.31–3.35, 3.31f, 3.32f, 3.33f, 3.34f, 3.35f
    mechanical installation of suspended lighting fixtures, 3.39, 3.40, 3.40f, 3.41, 3.41f, 3.42f, 3.44
    mechanical installation of track lighting fixtures, 3.41, 3.43–3.45, 3.43f, 3.45f
    NEC requirements for, 3.29, 3.38
    opening lighting fixture cartons, 3.28–3.29
    schedules, 3.28, 3.28f
Lighting fixtures
    fluorescent lighting fixture ballasts, 3.17–3.21, 3.17f, 3.18f, 3.20f
    recessed lighting fixtures, 3.24, 3.25, 3.26f
    storage, handling, and security, 3.26–3.28
    surface-mounted lighting fixtures, 3.24, 3.24f, 3.25f
    suspended lighting fixtures, 3.25, 3.27f
    symbols for, 3.28, 3.57–3.58
    track-mounted lighting fixtures, 3.25–3.26, 3.27f
Lighting floor plans, 3.28, 3.28f
Lighting poles, handholes, 5.12
Linear nonregulation circuit ballasts, 3.22, 3.23f
Lines of flux. *See also* Magnetic fields
    magnetism and, 1.3, 1.3f, 2.2, 2.2f

Liquidtight flexible metal conduit, as equipment grounding conductor (EGC), 9.17, 9.18f
lm (lumens), 3.6, 3.55
Loaded torque, 2.22–2.23, 2.23f
Load factors, cable trays, 7.7
Loading, of cable trays, 7.7–7.12, 7.7f, 7.8f, 7.9f, 7.10f, 7.11f, 7.12f, 7.13f
Locked-rotor current, motor nameplate data, 2.53, 2.54
Locking clips, fluorescent troffers, 3.39, 3.39f
Locknuts and bushings, 5.14–5.17, 5.14f, 5.15f, 5.17f
Longitudinal cuts, cable stripping, 8.5
Longitudinal side rails, 7.2, 7.5f
Long shunts, 2.14, 2.14f
Low-impedance conductors, grounding, 9.4, 9.24
Low-pressure sodium lamps, 3.15
Low-voltage cables, cable trays, 7.21
Low-voltage connectors and terminals, 8.13–8.14, 8.13f, 8.14f
Low-voltage halogen lamps, 3.9–3.10, 3.10f
Low-voltage remote control switching
    advantages of, 11.19–11.20, 11.20f
    applications for, 11.18–11.19, 11.19f
    design of, 11.27, 11.28f, 11.29f
    installing, 11.27, 11.29–11.30
    operation of, 11.24, 11.24f
    planning systems, 11.25
    symbols for low-voltage drawings, 11.25, 11.26f
    system components, 11.20, 11.21f, 11.22, 11.22f, 11.23f
LPW (lumens per watt), 3.6, 3.55
Lubrication, for installing conductors, 6.2, 6.18, 6.19, 6.32
Lugs, 8.2, 8.8
    defined, 8.34
Lumen maintenance, 3.9
    defined, 3.55
Lumens (lm), 3.6
    defined, 3.55
Lumens per watt (LPW), 3.6
    defined, 3.55
Luminaires (fixtures), 3.6
    defined, 3.55
Luminance, 3.2
    defined, 3.55

M

Magnetic angle finders, 4.23–4.24, 4.26
Magnetic circuit breakers, 10.8
Magnetic contactors
    components of, 11.2–11.3, 11.2f
    lighting contactors, 11.4–11.5, 11.4f, 11.5f, 11.6–11.7, 11.8, 11.8f
    mechanically held contacts, 11.9
    NEMA ratings and, 11.2, 11.4, 11.6–11.7
    reversing motor starters, 11.2, 11.8–11.9, 11.8f
    solenoid principle, 11.3–11.4, 11.3f
Magnetic fields
    in AC motors, 2.16–2.18, 2.21, 2.26, 2.29, 2.31
    in DC motors, 2.2, 2.2f, 2.9–2.10
    electrical generation and, 1.2–1.3, 1.7
    inductance and, 1.13–1.14
    principles of, 2.2, 2.2f
    transformers and, 1.36, 1.37, 1.42
    voltage and, 1.2–1.3
Magnetic overload relays, 11.17, 11.18
Magnetic torpedo levels, 4.36
Magnetism, 1.2, 2.2, 2.2f
Main bonding jumpers
    defined, 9.41
    NEC requirements for, 9.20–9.21, 9.20f
Main disconnects, 8.26, 8.27

INDEX ◆ I.9

Main lugs only (MLO), 8.26
Make, 11.2
    defined, 11.35
MA (mechanical advantage), 8.15, 8.34
Mandrels, 6.8, 6.10f
Manhole feed-in setups, 6.24
Manufacturer's catalog data, 3.44, 6.27
Manufacturer's installation instructions, 2.63, 3.10, 3.29, 3.30f, 8.16
Master-selector switches, 11.24, 11.25f
Master sequencers, low-voltage remote control switching, 11.22, 11.23f
Mastic, 8.30
Maximum pulling tension, 6.29–6.30
MCC (motor control centers), 8.26, 8.27f
Measuring earth resistance, 9.28, 9.29f, 9.30–9.32, 9.30f, 9.31f
Mechanical advantage (MA), 8.15
    defined, 8.34
Mechanical benders, 4.12–4.15, 4.12f, 4.13f, 4.14f, 4.15f
Mechanical compression-type connectors, 8.8, 8.9f
Mechanical compression-type terminators, 8.8
Mechanically held contacts, 11.9
Mechanically held relays, 11.9
    defined, 11.35
Mechanical offsets, 4.16, 4.16f, 4.17
Mechanical strength, crimp connectors, 8.14, 8.15f
Mechanical stub-ups, 4.15–4.16
Mechanical terminal and connector installation, 8.18
Medium machine frame designations, 2.47–2.49, 2.47f
Melting alloy (eutectic) thermal overload relays, 11.17, 11.17f
Mercury vapor lamps, 3.14
Mercury vapor service checklist, 3.61
Metal braid tapes, 8.28
Metal building frames, grounding, 9.9, 9.10f
Metal-clad (Type MC) cable, connecting, 8.18–8.19, 8.19f
Metal conduit couplings, 5.14f
Metal halide lamps, 3.14, 3.14f
Metal halide service checklist, 3.61
Metal water pipes, grounding, 9.9, 9.10, 9.10f, 9.11f
Metering equipment, 1.44
Metric conversions
    chart for, 7.30
    international differences and, 4.4
Micro, 1.13
    defined, 1.48
Minimum cable bending radii, 8.24
Minimum radius requirements, bending conduit, 4.2
Minimum starting voltage, motor nameplate information, 2.53
Missiles. *See* Conduit pistons
MLO (main lugs only), 8.26
Modular wiring systems, lighting fixture installation, 3.46
Moguls, 5.9, 5.10f
    defined, 5.21
Moisture. *See also* Damp and wet locations
    motors and, 2.63
Molded-case circuit breakers, 10.3
    defined, 10.25
Monitored cable pullers, 6.23
Monorail cable tray systems. *See* Center rail cable tray systems
Morse, Samuel, 11.2
Motor action, 2.3, 2.3f
Motor connection kits, 8.30, 8.30f
Motor connections, dual-voltage and/or multi-speed motors, 2.62
Motor control centers (MCC), 8.26, 8.27f

Motor enclosures, 2.45–2.46
Motor heating, variable-speed drives, 2.42
Motor lugs, taping, 8.28, 8.28f
Motor protection, 2.54, 2.55, 2.56–2.57
Motor ratings, 2.49
Motors. *See also* AC motors; DC motors
    braking and, 2.62–2.63
    early electric motors, 2.3
    installing, 2.63, 2.64
    motor enclosures, 2.45–2.46
    multiple-speed induction motors, 2.37–2.39, 2.37f, 2.38f, 2.39f
    nameplate data, 2.49–2.54, 2.50f, 2.52f, 2.54f
    NEMA frame designations, 2.46–2.49, 2.47f
    overload protection, 2.54, 2.55, 2.56–2.57, 11.16–11.18, 11.17f
    ratings, 2.49
    variable-speed drives, 2.26, 2.40–2.45, 2.40f, 2.41f, 2.42f, 2.43f
Mouse. *See* Conduit pistons
Multi-cable pulls, 6.4, 6.4f
Multiple contacts, relays, 11.9
Multiple load centers, 9.22
Multiple-speed induction motors, 2.37–2.39, 2.37f, 2.38f, 2.39f
Multiple-winding motors, 2.37, 2.37f
Multiplex polyester pulling lines, 6.17
Multipoint switching, low-voltage remote control switching, 11.24, 11.25f
Myers hubs, 5.14–5.16, 5.16f

N

Nameplate data, 2.49–2.54, 2.50f, 2.52f, 2.54f, 2.55, 2.56–2.57
Nanometers, 3.5, 3.5f
National Electrical Code (NEC)
    Occupational Safety and Health Administration (OSHA) and, 9.9
    Underwriters Laboratories, Inc. (UL) and, 9.9
National Electrical Code (NEC) requirements
    aluminum connections, 8.10
    barrier strips, 7.12
    bending conduit, 4.3–4.4, 4.3f
    bending spaces, 8.24, 8.25f, 8.26, 8.26f
    bonding electric services, 9.3–9.4, 9.5, 9.22–9.23, 9.23f
    bonding multiple raceway systems, 9.23
    bonding multiple service disconnecting means, 9.22
    bonding of enclosures and equipment, 5.12
    bonding of services over 250 volts, 9.23, 9.23f
    bonding of services over 1000 volts, 9.28
    bonding service equipment, 9.22–9.23
    bushings and locknuts, 5.14
    cable bending radii, 8.24
    cable exits, 7.9
    cable tray grounding, 7.19f
    cable trays, 7.2, 7.7, 7.10, 7.17, 7.17f, 7.18f, 7.19, 7.19f
    ceiling fans/fixtures, 3.34
    circuit breaker ratings, 10.4, 10.10–10.11
    circuit breakers and fuses, 10.10–10.11, 10.15, 10.17, 10.19–10.21
    circuits that cannot be grounded, 9.3, 9.9, 9.9f
    for communication circuits, 8.12
    for conductors entering or leaving enclosures, 8.24, 8.25, 8.26f
    conductors in cable trays, 7.19–7.22
    conductor support, 6.26
    conductor terminations and splices, 8.8, 8.26–8.27, 8.27f
    conduit bodies, 5.7, 5.8, 5.9, 5.10
    control and signal cables, 8.12

current interrupting ratings, 10.5
damp and wet locations, 5.2
electrical connection of lighting fixtures, 3.45–3.46
electrical connections, 8.8
equipment grounding, 9.3–9.4
equipment grounding conductors (EGC), 9.17
exothermic welding, 9.14
fire alarm systems, 8.12
fluorescent fixture ballasts, 3.19–3.20
ground faults, 9.4
grounding conductors, 9.3
grounding electrode conductor installation, 9.9–9.10, 9.10*f*, 9.11*f*, 9.12, 9.12*f*, 9.13, 9.14
grounding electrode conductors, 9.4, 9.12–9.13, 9.14
grounding high-voltage systems, 9.28
grounding methods, 9.3, 9.9–9.10, 9.10*f*, 9.11*f*, 9.12, 9.12*f*
grounding multiple buildings, 9.27
grounding outlet boxes and devices, 9.19
grounding systems, 9.3–9.4, 9.5, 9.9–9.10, 9.10*f*, 9.11*f*, 9.12–9.16, 9.12*f*, 9.13*f*, 9.14*f*, 9.15*f*
grounding systems 50 volts to 1000 volts, 9.6, 9.8*f*
grounding systems less than 50 volts, 9.6, 9.7
ground path, 9.4
handholes, 5.10, 5.11, 5.12
hazardous locations, 4.35
instrumentation tray cables, 8.12
interlocked armor cable bending radii, 8.24
intermixing dissimilar metal conductors, 8.10
labeling circuit breakers and fuses, 10.10
lighting fixture installation, 3.29, 3.38
locked-rotor current, 2.53, 2.54
low-voltage conductors, 11.27
main bonding jumpers, 9.20–9.21, 9.20*f*
maximum full-load motor current, 2.56
metal-clad (Type MC) cable, 8.18
metal halide lamps, 3.14, 3.14*f*
moguls, 5.9
motor branch circuits, 2.56–2.57
motor installations, 2.60–2.61, 2.61*f*, 2.63
motor overload protection, 2.54, 2.55, 2.56
optical fiber cables and raceways, 8.12
outlet boxes, 3.31
overcurrent protection devices, 8.26, 8.27*f*, 10.7, 10.10–10.11
pendants and hanging-type fixtures, 3.34
pull and junction boxes, 5.2, 5.5, 5.6, 5.7
PVC conduit installations, 4.35
recessed fixtures, 3.25, 3.36, 3.36*f*, 3.38, 3.39
remote control, signaling, and power-limited circuits (Class 1, 2, and 3 circuits), 8.12
remote control wiring, 11.19
separately derived systems, 9.7–9.9, 9.8*f*, 9.9*f*, 9.25, 9.26
service disconnects, 9.22, 10.19
sizing fuses, 10.19–10.21
sizing neutral conductors, 9.2
sizing pull and junction boxes, 5.5, 5.6, 5.7
supplemental electrodes, 9.10, 9.12, 9.12*f*
switched neutral breakers, 10.22
switchgear, 8.26
systems permitted but not required to be grounded, 9.8
track lighting, 3.41, 3.43, 3.44
vending machines, 9.30
National Electrical Manufacturers Association (NEMA)
 frame designations for motors, 2.46–2.49, 2.47*f*
 motor code letters, 2.53
 motor design letters, 2.51, 2.52
National Electrical Manufacturers Association (NEMA) ratings, magnetic contactors, 11.2, 11.4, 11.6–11.7

National Electrical Manufacturers Association (NEMA) requirements
 cable trays, 7.2
 classifying and rating control and signal cable, 8.12
 classifying motor enclosures, 2.45–2.46
 classifying motor frame designations, 2.46–2.49, 2.47*f*
 enclosures, 5.3–5.4, 8.27*f*, 11.18
 FS and FD boxes, 5.8
 interrupting rating, 10.8, 10.14
 knockouts, 5.3–5.4
 motor nameplate information, 2.49, 2.50
 protective enclosures, 11.18
National Energy Policy Act (EPACT), 3.6
National Institute for Occupational Safety and Health (NIOSH), 1.43
NC (normally closed) contacts, 11.3, 11.4–11.5, 11.4*f*, 11.5*f*, 11.35
NEC. *See* *National Electrical Code (NEC)*
NEMA. *See* National Electrical Manufacturers Association (NEMA)
Neoprene heat-shrink tubing, 8.11
Net reactance, 1.21, 1.27, 1.30
Neutral, 9.2. *See also* Grounding systems
 defined, 9.41
Neutral conductors, *NEC* requirements for, 9.2
Neutral plane, 2.6–2.7, 2.6*f*, 2.7*f*
90 degree bends, 4.10–4.12, 4.10*f*
 defined, 4.43
90 degree segment bends, 4.20–4.23, 4.21*f*, 4.22*f*, 4.23*f*
NIOSH (National Institute for Occupational Safety and Health), 1.43
No-Dogs, 4.17
Noise, shielding, 8.12
No-load conditions, transformers, 1.37
Nominal rated voltage, motor nameplate information, 2.53
Non-automatic breakers, 10.22
Nonconductive fish tapes, 6.3
Non-IC fixtures, 3.29, 3.36
NO (normally open) contacts, 11.3, 11.4–11.5, 11.4*f*, 11.5*f*, 11.35
Nonrenewable fuses, 10.16
 defined, 10.25
Nonsinusoidal waveforms, 1.9–1.11, 1.10*f*
Nontime-delay fuses, 10.19–10.21
Nonventilated motor enclosures, 2.46
Normal current ratings, 10.5
Normally closed (NC) contacts, 11.3, 11.4–11.5, 11.4*f*, 11.5*f*
 defined, 11.35
Normally open (NO) contacts, 11.3, 11.4–11.5, 11.4*f*, 11.5*f*
 defined, 11.35
Null balance metering systems, 9.31
Number of pumps method, segment bending techniques, 4.23, 4.24
Nylon pulling lines, 6.17

O

Occupancy sensors, 3.47, 3.47*f*
Occupational Safety and Health Administration (OSHA)
 ground fault circuit interrupters (GFCI) and, 10.12, 10.13
 grounding and, 9.9
 NEC and, 9.9
Octagon outlet boxes, 3.29, 5.2
OD (outside diameter), 4.27, 4.43
Offset bends, 4.5, 4.5*f*, 4.28–4.31, 4.28*f*, 4.30*f*, 4.31*f*
Offsets
 cable trays and, 7.11
 concentric offsets, 4.31–4.33, 4.31*f*, 4.32*f*, 4.33*f*
 conduit and, 4.2, 4.5, 4.5*f*, 4.28–4.31, 4.28*f*, 4.30*f*, 4.31*f*
 defined, 4.43

Ohmmeters, ground testing and, 9.28
Ohm's law, 1.11
Oil-tight, dust-tight (NEMA Type 13) enclosures, 5.4
One-loop rotary generators, 1.2, 1.2f
One-shot bending, 4.19–4.20, 4.20f
One-shot shoes, 4.18–4.19
   defined, 4.43
Open motors, 2.45–2.46
Operating characteristics
   of circuit breakers, 10.2–10.3
   of DC motors, 2.15–2.16, 2.16f
Operating modes, variable-speed drives and, 2.40–2.41, 2.41f
OSHA. *See* Occupational Safety and Health Administration (OSHA)
Outdoor boxes, 3.35, 5.3
Outdoor lighting fixtures, installing, 3.34–3.35, 3.35f
Outlet boxes. *See also* Pull and junction boxes
   electrical connection of lighting fixtures and, 3.45
   for surface-mounted fixtures, 3.29, 3.31, 3.31f, 3.32, 3.33, 3.33f, 3.34
Outside diameter (OD)
   bending and, 4.27
   defined, 4.43
Overcurrent protection
   defined, 10.25
   motor ratings and, 2.50
   NEC requirements for, 8.26–8.27, 8.27f
   protective relays, 11.2
   purpose of, 10.2
   safety and, 10.21
Overcurrent protection devices
   circuit breakers and fuses as, 10.2
   coordination of, 10.21–10.22
   NEC requirements for, 8.26, 8.27f, 10.7, 10.10–10.11
   single-line schematic diagram showing, 10.20f
Overcurrents
   circuit breakers and fuses and, 10.2, 10.17–10.18, 10.18f
   defined, 2.68
Overexciting, AC motors, 2.28
Overhead cable tray feed-in setups, 6.24
Overload conditions
   relays and, 11.15–11.18, 11.16f, 11.17f
   three-phase induction motors and, 2.23–2.24, 2.24f
Overload protection
   fuses and, 2.54, 10.2, 11.16
   motors and, 2.54, 2.55, 2.56–2.57, 11.16–11.18, 11.17f
Overload relays, 11.15–11.18, 11.16f, 11.17f
Overloads
   circuit breakers and fuses and, 10.17–10.18, 11.16
   defined, 2.68
   motor ratings and, 2.49
Oxide film, aluminum, 8.9–8.10
Oxide-inhibiting joint compound, 8.18

P

Paddle fan boxes, 3.34
Panelboards, 9.6, 9.20
Parallel connected conductors, cable trays, 7.19–7.20
Parallel LC circuits, 1.28–1.29, 1.29f
Parallel RC circuits, 1.26–1.27, 1.27f
Parallel RLC circuits, 1.30, 1.31f
Parallel RL circuits, 1.24–1.25, 1.25f
Parallel-tap connectors, 8.8, 8.9f
PCBs, ballasts, 3.20
Peak-to-peak (p-p) values, 1.5, 1.5f
Peak values, 1.5, 1.5f
Peak voltages, 1.5, 1.5f
   defined, 1.48

Pendant lighting fixtures, installing, 3.33, 3.35
Periodic duty, 2.49
   defined, 2.68
Periods, of waveforms, 1.4–1.5, 1.4f
Permanent magnet DC motors, 2.15
Permeability, 2.2
Phase angle diagrams, 1.7–1.8, 1.8f
Phase angles, 1.6, 1.8f
Phase-indicating dots, 1.38, 1.38f
Phase lags, 1.6
Phase leads, 1.6
Phase relationships
   alternating current and, 1.6–1.8, 1.8f, 1.21, 1.22f
   transformers and, 1.37–1.38, 1.38f
Photoelectric sensors, 3.47, 3.47f
Photosensors, 3.47, 3.47f
Physical limitations of cable, 6.28–6.32, 6.31f
Pi, 4.8
Pipe electrodes, 9.9, 9.10, 9.10f, 9.11f, 9.14, 9.15, 9.16
Pipe rack mounting, 7.13, 7.15
Pipe racks, 7.15
   defined, 7.28
Pipe-ventilated motor enclosures, 2.45, 2.46
Planning
   conductor installations, 6.3–6.5, 6.4f, 6.6f
   low-voltage remote control switching systems, 11.25
Plaster frame kits, recessed fixtures, 3.38, 3.38f
Plate area, of capacitors, 1.17
Plate distance, capacitance, 1.17
Plate electrodes, grounding, 9.9, 9.10f, 9.12, 9.14, 9.15, 9.15f, 9.16
Plug fuses, 10.5, 10.15–10.16, 10.15f
   defined, 10.25
Plug-in relays, 11.11, 11.12f, 11.13f, 11.14
Plywood sheets, using to calculate added length, 4.35, 4.35f
Pole-mounted transformers, 9.5, 9.5f
Poles
   circuit breakers and, 10.9
   defined, 10.25
   field poles, 2.9–2.10, 2.68
   fish poles, 6.12
   lighting poles, 5.12
   magnetism and, 1.3, 2.2, 2.2f
Polygon curvatures, feed-in setups, 6.26
Polyolefin heat-shrink tubing, 8.10, 8.11
Polyphase motor theory, 2.16–2.18, 2.17f
Polypropylene pulling lines, 6.17
Poorly grounded systems, 9.28, 9.29f
Poor soil conductivity, 9.36
Potential transformers, 1.42–1.43, 1.43f, 1.44
Power, in AC circuits, 1.32–1.34, 1.33f, 1.34f, 1.35f
Power cables
   cutting, 8.6
   stripping, 8.3, 8.4, 8.4f, 8.5, 8.5f
Power dissipation, 1.11
Power factor
   in AC circuits, 1.33–1.34, 1.34f, 1.35
   ballasts and, 3.19
   motor nameplate data and, 2.54
   three-phase induction motors and, 2.24, 2.25f
Power-operated cable pullers, 6.17, 6.18f, 6.19, 6.20, 6.20f, 6.23
Power supplies, low-voltage remote control switching, 11.22, 11.22f, 11.27
Power transistors, for DC loads, 11.14–11.15
Power triangle, 1.34, 1.35f
Preheat fluorescent lamps, 3.11–3.12, 3.17, 3.18f
Pressure connectors, 8.2
   defined, 8.34

Pre-U-frame motors, 2.46
Primary conductors, 10.20
Primary windings, 1.35–1.37, 1.36f
Prisms, 3.5, 3.5f
Protecting grounding electrode conductors, 9.13, 9.15f
Protective enclosures, for contactors and relays, 11.18
Pull and junction boxes
    conduit bodies, 4.2, 5.7–5.10, 5.8f, 5.9f, 5.10f
    for damp and wet locations, 5.2–5.3
    fittings, 5.12–5.16, 5.13f, 5.14f, 5.15f, 5.16f, 5.17f
    handholes, 5.10–5.12, 5.11f, 5.12f
    in hazardous locations, 5.2, 5.3
    NEMA and IP enclosure classifications, 5.3–5.4
    sizing in over 600V systems, 5.6–5.7
    sizing in under 600V systems, 5.5–5.6, 5.6f
Pull boxes, 5.2, 5.2f. *See also* Pull and junction boxes
    defined, 5.21
Pulleys, 7.9, 7.9f
Pulling blocks, 6.11
Pulling cable, in cable trays, 6.27–6.28, 6.28f, 6.29, 7.24
Pulling elbows, 5.9, 5.10f
Pulling eyes, 6.16
Pulling force, 6.18
Pulling grips, 6.11, 6.15, 6.15f, 6.16
Pulling lines, 6.17
Pulling locations, selecting, 6.4
Pulling rope ratings, 6.17
Pulling ropes, 6.11, 6.17, 6.22
Pulling tension
    calculating, 6.31–6.32, 6.31f
    in horizontal pulls, 6.30–6.31
    maximum, 6.29–6.30
    sidewall loading and, 6.31, 6.31f
Pull-in wires, 6.11
Pull lines, installing, 6.11, 6.12, 6.13, 6.13f, 6.14f
Pullout, synchronous motors, 2.28–2.29
Pushbutton control circuits, relays, 11.4, 11.4f, 11.11
PVC conduit
    bending, 4.36–4.38, 4.36f, 4.37f, 4.38f, 4.39
    installing, 4.35–4.36
PVC conduit heating units, 4.36, 4.36f
PVC heat-shrink tubing, 8.10, 8.11
Pythagorean theorem, 1.34

R

Raceway inspection devices, 6.8, 6.10f
Raceways
    measuring for conductor installations, 6.8, 6.10, 6.11, 6.11f
    preparing for conductors, 6.8, 6.10, 6.10f
Radians
    angular representation and, 4.9
    defined, 1.48
Radiant energy, 3.4
Radius, 4.3
    defined, 4.43
Rainproof, sleet resistant (NEMA Type 3R) enclosures, 5.3
Raintight, 5.3
    defined, 5.21
Raintight boxes, 5.3
Raised floors, cable trays beneath, 7.16
Rapid start fluorescent lamps, 3.11–3.12, 3.17–3.18, 3.18f
Ratchet cable bending tools, 8.24, 8.24f
Ratchet-type cable cutters, 8.3, 8.4f
Rated amperage, motor nameplate information, 2.53, 2.54f
Rated full-load speed, motor nameplate data, 2.50–2.51
Rated horsepower, motor nameplate data, 2.53, 2.54f
Rated voltage, motor nameplate data, 2.49, 2.50, 2.50f
RC circuits, 1.25–1.27, 1.26f, 1.27f

Reactance, 1.14–1.15, 1.20–1.21, 1.27, 1.30
    defined, 1.48
Reactive power, 1.32–1.33, 1.33f
Reactor ballasts, 3.22, 3.23f
Recessed lighting fixtures
    installation, 3.36–3.39, 3.36f, 3.37f, 3.38f, 3.39f
    types of, 3.24, 3.25, 3.26f
Rectangular waveforms, 1.10, 1.10f
Rectifier systems, grounding, 9.8
Reducing connectors, 8.8
    defined, 8.34
Reed-controlled AC relays, 11.15
Reel jacks, 6.7, 6.10f
Reel stands, 6.7, 6.10f
Reflected, 3.2
    defined, 3.55
Reflected light, 3.2, 3.3–3.4, 3.4f
    defined, 3.55
Refracted, 3.3
    defined, 3.55
Refraction of light, 3.3–3.4, 3.4f, 3.5, 3.5f
Relay controlled lighting circuits, 3.49, 3.50f
Relay operation, low-voltage remote control switching, 11.24, 11.24f
Relays
    control relays, 11.9–11.11, 11.10f, 11.11f, 11.12f, 11.13f
    introduction to, 11.2
    low-voltage remote control switching and, 11.24, 11.24f
    overload relays, 11.15–11.18, 11.16f, 11.17f
    solid-state relays, 11.13–11.15, 11.13f, 11.14f, 11.15
    timers and timing relays, 11.11, 11.15
Reluctance, 2.2
Removing tape from splices and joints, 8.29
Repulsion-start motors, 2.31
Resistance
    in AC circuits, 1.11, 1.11f, 1.12f
    body resistance, 9.30, 10.12
    earth electrode resistance, 9.32
    earth ground resistance testers, 9.28, 9.28f
    effective resistance area, 9.32
    fall-of-potential resistance testing, 9.30–9.31, 9.30f
    grounding resistance, 9.28, 9.41
    grounding resistance curves, 9.33f, 9.34–9.36, 9.34f, 9.36f
    hand-to-body resistance, 10.12
    leak resistance, 1.18
    measuring earth resistance, 9.28, 9.29f, 9.30–9.32, 9.30f, 9.31f
    starting resistance, 2.10–2.12
Resistivity, 9.28
    defined, 9.41
Retinas, 3.2, 3.2f
Retrofitting, lighting and, 3.6
Reversing motor starters, 11.2, 11.8–11.9, 11.8f
Reversing rotation, three-phase induction motors, 2.24, 2.25f, 11.8–11.9, 11.8f
Revolutions-per-minute (rpm), 2.13
    defined, 2.68
Right triangles, bending conduit, 4.6–4.7, 4.7f, 4.8f
Rigid, aluminum, and IMC conduit, 4.2
Rigid aluminum, bending, 4.19
Rigid conduit fittings, 5.13–5.14, 5.14f
Rise
    bending conduit and, 4.6
    defined, 4.43
RLC circuits, 1.29–1.30, 1.30f, 1.31f
RL circuits, 1.21, 1.23–1.25, 1.23f, 1.24f, 1.25f
rms (root-mean-square) values, 1.6, 1.9, 1.48
Rod electrodes, 9.9, 9.10f, 9.14, 9.15, 9.16

Rods and cones, 3.2, 3.3
Rollers, 6.27, 6.29
Root-mean-square (rms) values, 1.6, 1.9
   defined, 1.48
Rotating fields, AC motors, 2.17–2.18, 2.17f
Rotation, 2.4
   defined, 2.68
Rotor behavior in a rotating field, 2.17f, 2.18
Rotor pole assemblies, 2.27, 2.28f
Rotors. *See* Armatures
Routing cabling, 8.22–8.23, 8.23f
Runaways, in vertical runs, 6.27
Running protection. *See* Overload protection

S
Saddle bends, 4.5, 4.6f, 4.31–4.33, 4.31f, 4.32f, 4.33f
Safety
   aluminum connections and, 8.10
   bending conduit and, 4.2
   cable pullers and, 6.19, 6.33
   cable trays and, 7.25
   circuit breakers and fuses and, 10.21
   current transformers and, 1.42
   cutters and, 6.16, 8.5
   de-energizing circuits and, 3.40
   disconnecting grounds and, 3.21
   electrical connection of lighting fixtures and, 3.40, 3.45
   exothermic welding and, 9.14
   ground fault circuit interrupters (GFCI) and, 9.30
   grounding electrodes and, 9.6
   ground testers and, 9.28
   handholes and, 5.12
   hydraulic conduit benders and, 4.18, 4.19
   incoming lines and, 8.27
   installing conductors and, 6.27
   isolation transformers and, 1.40–1.41, 1.41f, 1.42
   lamps and, 3.17
   lighting fixture installation and, 3.40, 3.45
   maintenance and, 9.17
   mechanical grounds and, 9.17
   metal halide lamps and, 3.14
   motor installation and, 2.60, 2.63
   overcurrent protection and, 10.21
   pulling cable and, 6.7, 6.17, 7.24
   PVC conduit and, 4.36
   sheaves and, 6.28
   steel fish tape and, 6.2
   stripping and, 6.16, 8.5
   testing station grounds and, 9.36
   wire-pulling lubricant and, 6.18
Safety sleeves, fluorescent lamps, 3.17
Safety switches, 10.5
Safety valves, circuit breakers and fuses as, 10.2
Sawtooth waveforms, 1.9, 1.10f
Scissor-action cutting tools, 8.5
Sconce lighting fixtures, installing, 3.32–3.33
Screw-type terminal block connectors, 8.21, 8.22, 8.22f
Sealed windings motor enclosures, 2.46
Seasonal variations, ground resistance testing, 9.36
Secants, 4.6–4.7
Secondary conductors, 8.26
Secondary windings, 1.35–1.37, 1.36f
Security, of lamps and lighting fixtures, 3.26–3.28
Segment bending techniques, 4.23–4.33, 4.25f, 4.26f, 4.27f, 4.28f, 4.30f, 4.31f, 4.32f, 4.33f
Segment bends, 4.19, 4.20–4.23, 4.21f, 4.22f, 4.23f
   defined, 4.43

Segmented bending shoes, 4.18–4.19
   defined, 4.43
Segment spacing, segment bends, 4.22–4.23
Selective absorption, 3.5
Selective coordination, 10.21–10.22
Self-cooling motors, 2.45
Self-inductance, 1.12
   defined, 1.48
Self-starting motors, 2.5–2.6, 2.5f
Semi-guarded motor enclosures, 2.45
Separately derived systems
   defined, 9.41
   equipment grounding conductor and, 9.2, 9.25–9.26
   generator-type, 9.26–9.27, 9.26f
   grounding of, 9.25–9.26, 9.25f, 9.26f
   NEC grounding requirements, 9.7–9.9, 9.8f, 9.9f, 9.25, 9.26
Series LC circuits, 1.28, 1.28f
Series motors, 2.12, 2.13–2.14, 2.13f, 2.15–2.16, 2.16f
Series-rated circuit breakers, 10.9, 10.10, 10.10f
Series RC circuits, 1.26, 1.26f
Series RLC circuits, 1.29–1.30, 1.30f
Series RL circuits, 1.21, 1.23–1.24, 1.23f, 1.24f
Service conductors, 9.13, 9.22, 9.25f
Service disconnecting means, 10.19
Service drops, 9.5
   in buildings, 9.27
Service entrance conductors, 9.2, 9.12, 9.14, 9.22
Service entrances, 9.5–9.6, 9.7f
Service equipment, 9.7f, 9.17, 9.27
Service factors, motor nameplate information, 2.53
Service raceways, 9.22
Setscrew cable grips, 6.11, 6.15, 6.15f, 6.16
   defined, 6.37
Setscrew couplings, 5.13
Setscrew fittings, 5.13
Shaded-pole induction motors, 2.31, 2.34, 2.35, 2.35f, 2.36f, 2.37
Shading coils, 11.2f, 11.3–11.4, 11.3f
   defined, 11.35
Shapes
   fluorescent lamps and, 3.10–3.11, 3.11f
   incandescent lamps and, 3.7, 3.8f
Sheaves, 6.23, 6.24, 6.25f, 6.26, 6.27
   defined, 6.37
Sheet metal screws, grounding conductors, 9.19
Shielded conductors, minimum bending radii for, 8.24
Shielding, 8.12
   defined, 8.34
Shipping reels, planning installations, 6.3–6.4
Short circuit bracing, incoming lines, 8.27
Short circuit calculations, 9.30
Short circuits
   defined, 9.42, 10.25
   grounding systems and, 9.4, 9.4f, 9.21
   interrupting rating and, 10.5, 10.18
   overcurrents, 10.17, 10.18f
Short shunts, 2.14, 2.14f
Shunt motors, 2.11, 2.11f, 2.12–2.13, 2.12f, 2.15, 2.15f
Shunt trip circuit breakers, 10.22
Sidewall loading, pulling tension, 6.31, 6.31f
Sines, 1.3, 4.6–4.7
Sine wave generation, 1.2–1.3, 1.4f
Sine waves, compared with nonsinusoidal waveforms, 1.10–1.11, 1.10f
Sine wave terminology, 1.4–1.6
Single-conductor cables, cable trays, 7.22
Single-element cartridge fuses, 10.16–10.17

Single-loop armature DC motors, 2.5–2.7
Single-phase AC motors
  capacitor-type induction motors, 2.33–2.34, 2.33f
  induction motors, 2.30–2.31, 2.31f
  shaded-pole induction motors, 2.31, 2.34, 2.35, 2.35f, 2.36f, 2.37
  single-phase synchronous motors, 2.37
  split-phase induction motors, 2.31, 2.31f, 2.32, 2.33
Single-phase three-wire 120/240V electric service, 9.5
Single-switch control, low-voltage remote control switching, 11.24, 11.24f
Sizing dual-element time delay fuses, 10.19
Sizing equipment grounding conductors (EGC), 9.17–9.18, 9.18f
Sizing fuses, 10.19–10.21
Sizing grounding electrode conductors, 9.12–9.13, 9.14
Sizing neutral conductors, *NEC* requirements for, 9.2
Sizing nontime-delay fuses, 10.19–10.21
Sizing pull and junction boxes, 5.5–5.7, 5.6f
SLBs (entrance ells), 5.9, 5.10f
Slimline fluorescent lamps, 3.12
Slip, 2.21–2.22
Slip rings, 2.27
Small conductors, stripping, 8.2–8.3, 8.3f
Small machine frame designations, 2.46–2.47, 2.47f
Smooth sheath cable, minimum bending radii for, 8.24
Snap-in spindle and cable rollers, 6.29
Soap, 6.18. *See also* Lubrication
  defined, 6.37
Software programs, for calculating pulling tension, 6.30
Soil differences, earth resistance testing, 9.28
Soil resistivity, 9.28
Solid bottom cable trays, 7.2, 7.4f
  cable fill and, 7.22
Solid-state relays, 11.13–11.15, 11.13f, 11.14f, 11.15
  defined, 11.35
Sound ratings, ballasts, 3.20
Special-purpose motors, 2.49
Specifications, 11.25, 11.27
Spectrum, of visible light, 3.4–3.5, 3.5f
Speed benders, 4.18
Speed control
  AC motors and, 2.44–2.45
  compound motors and, 2.15
  DC motors and, 2.13, 2.14, 2.15, 2.42–2.44
  series motors and, 2.14
  shunt motors and, 2.13, 2.43–2.44, 2.43f
  three-phase induction motors and, 2.20–2.21, 2.24
  variable-speed drives and, 2.40, 2.42–2.45, 2.43f
  wound rotor motors and, 2.20–2.21
Spindles, reel stands, 6.7
Splash-proof motor enclosures, 2.45
Splice plates, cable trays, 7.6, 7.11–7.12, 7.12f, 7.13f
Splices. *See also* Conductor terminations and splices
  basic requirements of, 8.2
  defined, 8.34
  grounding electrode conductors and, 9.14, 9.14f
Splicing, cable trays, 7.8–7.9, 7.8f
Split-phase induction motors, 2.31, 2.31f, 2.32, 2.33
Springback, 4.17
  defined, 4.43
Spring-loaded terminal block connectors, 8.21, 8.22, 8.22f
Square waveforms, 1.9, 1.10f
Squirrel cage induction motors, 2.18–2.19, 2.19f, 2.20f
Stainless steel gasketed hubs, 5.15
Standard ballasts, 3.18
Standard interrupting capacity, 10.8

Starters, 3.17
  defined, 3.55
Starting current
  motor nameplate data and, 2.51, 2.52, 2.54
  three-phase induction motors and, 2.22, 2.23f
Starting resistance, 2.10–2.12
Step bits, 5.16
Step-down transformers, 1.39, 11.19
Step-up transformers, 1.39
Step voltage, 9.28, 9.29
  defined, 9.42
Stopping power cable pulls, 6.28, 6.32
Storage, of lighting fixtures, 3.26–3.28
Strands, 8.2
  defined, 8.34
Stripping and cleaning conductors
  cutting large power cable, 8.6
  stripping control and signal cable/conductors, 8.5, 8.6f, 8.7, 8.7f
  stripping power cables and large conductors, 8.3, 8.4, 8.4f, 8.5, 8.5f
  stripping small conductors, 8.2–8.3, 8.3f
Stripping cable ends for pulling, 6.2, 6.11, 6.15
Stripping length, 8.6–8.7, 8.7f
Stripping tools, 8.2–8.3, 8.3f, 8.4, 8.4f, 8.5, 8.5f, 8.6f
Stub (butt splice) connections, 8.30, 8.30f
Stub-ups
  bending conduit and, 4.6, 4.10–4.12, 4.10f, 4.15–4.16
  defined, 4.43
Submersion, overcurrent and ground fault protective devices, 10.14
Supplemental electrodes, 9.10, 9.12, 9.12f
Supporting cable trays, 7.7, 7.7f, 7.11, 7.11f, 7.13, 7.14, 7.14f, 7.15, 7.15f
Supporting conductors, 6.26–6.27, 6.26f
Supporting PVC conduit, 4.38
Surface-mounted lighting fixtures
  installation, 3.29, 3.31–3.35, 3.31f, 3.32f, 3.33f, 3.34f, 3.35f
  types of, 3.24, 3.24f, 3.25f
Surface-mounted wall fixtures, 3.24, 3.25f
Suspended ceilings, troffers, 3.25, 3.26f, 3.38–3.39, 3.39f
Suspended lighting fixtures
  installation, 3.39, 3.40, 3.40f, 3.41, 3.41f, 3.42f, 3.44
  types of, 3.25, 3.27f
Sweep, pull boxes, 6.10, 6.11f
Sweep bends, 4.4, 4.4f
  defined, 4.43
Switch contacts, 2.31, 2.32
Switched neutral breakers, 10.22
Switches, low-voltage remote control switching, 11.22, 11.23f
Switching functions, multiplying with relays, 11.2
Switch points, low-voltage remote control switching and, 11.27, 11.29
Swivel plates, 7.11, 7.12f
  defined, 7.28
Swivel rope clevises, 6.11, 6.15f
Swivels, 6.16
Symbols
  AC voltage sources, 1.11, 1.11f
  lighting fixtures, 3.28, 3.57–3.58
  low-voltage drawings, 11.25, 11.26f
Synchronous motors
  characteristics and construction of, 2.24, 2.26–2.27, 2.27f
  energy efficiency and, 2.26
  hysteresis synchronous motors, 2.38
  operating principles of, 2.27

Synchronous motors, *continued*
  as polyphase motor, 2.16
  pullout and, 2.28–2.29
  rotor field excitation and, 2.28, 2.29f
  torque angle and, 2.27, 2.29–2.30, 2.29f
Synchronous speed, 2.16, 2.22
  defined, 2.68
System expansion, conduit bending, 4.29
System grounding, 9.5
  defined, 9.42
Systems permitted but not required to be grounded, 9.8
Systems that must be grounded, 2.60, 8.12, 9.6, 9.7, 9.8f, 9.9, 9.16
Systems that must not be grounded, 9.9, 9.9f

T
Table corners, using to calculate added length, 4.35, 4.35f
Take-up (comeback), 4.10
  defined, 4.43
Tangents, 4.6–4.7
Taping electrical joints, 8.28, 8.28f, 8.29, 8.29f
Taping split-bolt connectors, 8.28, 8.29f
Tapped transformers, 1.40, 1.40f
Tees, 7.2
  defined, 7.28
Teflon heat-shrink tubing, 8.11
Temperature rise, motor nameplate data, 2.54
Templates
  bending conduit and, 4.5, 4.37, 4.37f
  installing surface mounted fixtures and, 3.33
Terminal block connections, 8.21–8.23, 8.22f
Terminals. *See also* Conductor terminations and splices
  crimp-type terminals and connectors, 8.2, 8.13, 8.13f, 8.14f
  defined, 8.34
  of delta-connected motors, 2.58–2.59, 2.59f
  low-voltage connectors and terminals, 8.13–8.14, 8.13f, 8.14f
  mechanical terminals and connectors, 8.18
  screw-type terminal block connectors, 8.21, 8.22, 8.22f
  of wye-connected motors, 2.58, 2.58f
Termination inspections, 8.20–8.21, 8.20f, 8.21f
Terminations, 8.2. *See also* Conductor terminations and splices
  defined, 8.34
Testing. *See also* Ground testers
  ballasts, 3.59
  earth ground resistance testers, 9.28, 9.28f
  for effective grounds, 9.28
  fall-of-potential resistance testing, 9.30–9.31, 9.30f
  fuses, 10.3, 10.7–10.8
  for interrupting rating, 10.7–10.8
  three-point testing procedures, 9.31f, 9.33–9.36, 9.33f, 9.34f, 9.35f, 9.36f
Testing station grounds, 9.36
Test pulls, 6.8, 6.10f
T-frame motors, 2.46
Thermal circuit breakers, 10.8
Thermal expansion coefficient, aluminum, 8.9
Thermal overload relays, 11.17, 11.17f, 11.18
Thermal protectors, 2.56, 2.63
  defined, 2.68
Thermomagnetic circuit breakers, 10.8
Theta, 1.3
Third-order harmonics, 1.32
Threaded couplings, 5.13, 5.14f
Three-coil ballasts, 3.22, 3.23f, 3.24
Three-phase induction motors
  overload conditions and, 2.23–2.24, 2.24f
  power factor and, 2.24, 2.25f
  reversing rotation and, 2.24, 2.25f, 11.8–11.9, 11.8f
  slip, 2.21–2.22
  speed control and, 2.20–2.21, 2.24
  squirrel cage induction motors, 2.18–2.19, 2.19f, 2.20f
  starting current and, 2.22, 2.23f
  torque and, 2.18, 2.18f, 2.21, 2.22, 2.23f
  wound rotor induction motors, 2.19–2.20, 2.20f, 2.21, 2.21f
Three-phase service, 9.2
Three-point testing procedures, 9.31f, 9.33–9.36, 9.33f, 9.34f, 9.35f, 9.36f
Time-delay ratings of fuses, 10.15, 11.16
Time rating, motor nameplate data, 2.51
Timers, lighting control, 3.48, 3.48f, 11.15
Timers and timing relays, 11.11, 11.15
Time threshold, 3.2
Torque
  compound motors and, 2.15
  DC motors and, 2.3, 2.4, 2.4f, 2.13, 2.14
  leverage and, 2.3–2.4, 2.7
  mechanical compression-type connectors and, 8.8, 8.18
  neutral plane and, 2.6–2.7, 2.6f, 2.7f
  series motors and, 2.14
  shunt motors and, 2.13
  terminal bolts and, 8.17, 8.18
  three-phase induction motors and, 2.18, 2.18f, 2.21, 2.22, 2.23f
Torque angle, 2.27, 2.29–2.30, 2.29f
Torque slip curves, 2.33, 2.34f
Torque-speed curves, 2.22, 2.23f, 2.41–2.42, 2.41f, 2.42f
Touch voltage, 9.28, 9.30
  defined, 9.42
Track-mounted lighting fixtures
  installation, 3.41, 3.43–3.45, 3.43f, 3.45f
  types of, 3.25–3.26, 3.27f
Training conductors, 8.24–8.25, 8.24f, 8.25f, 8.26f
Transformer action, 1.35–1.36, 1.36f
Transformers
  AC versus DC generation and, 1.2
  construction of, 1.35–1.37, 1.36f, 1.37f
  grounding and, 9.5, 9.5f
  operating characteristics of, 1.37–1.38
  turns and voltage ratios, 1.38–1.39, 1.39f
  types of, 1.39–1.44, 1.40f, 1.41f, 1.42f, 1.43f
Transformer-type separately derived systems, grounding, 9.8, 9.25, 9.25f
Transporting cable reels, 6.7, 6.8f, 6.9f, 6.10f
Trapeze hangers, 4.34
Trapeze mounting
  for cable trays, 7.13, 7.14f, 7.15, 7.15f
  defined, 7.28
Tray covers
  cable trays, 7.9
  defined, 7.28
Triac-controlled AC relays, 11.15
Trigger-start fluorescent ballasts, 3.18
Trigonometry fundamentals, bending conduit, 4.6–4.7, 4.7f
Trip indications, circuit breakers, 10.3
Triple pulleys, 6.29
Tripping transformers, 1.40
Troffers, 3.25, 3.26f
  defined, 3.55
Troubleshooting
  air gaps, 11.30, 11.31
  contactors and relays, 11.3, 11.30, 11.31–11.32
Troughs, 7.2
  defined, 7.28
Trough type cable trays, 7.2, 7.4f
True power, 1.32

Tubing selector guide, 8.11
Tungsten halogen incandescent lamps, 3.9–3.10, 3.10*f*
Turns, number of and inductance, 1.12, 1.13*f*
Turns ratio, transformers, 1.38–1.39, 1.39*f*, 1.40
Two-loop DC motors, 2.8, 2.8*f*
Type C conduit bodies, 5.7, 5.8*f*
Type L conduit bodies, 5.7, 5.8*f*
Type LL conduit bodies, 5.7, 5.8*f*
Type LR conduit bodies, 5.7, 5.8*f*
Type S fuses, 10.15, 10.15*f*, 10.16
Type T conduit bodies, 5.8, 5.8*f*
Type T fuses, 10.16
Type TL fuses, 10.16
Type W fuses, 10.16
Type X conduit bodies, 5.8, 5.9*f*

U

U-frame motors, 2.46
Ultrasonic sensors, 3.47
Uncrating motors, 2.63
Underfloor duct feed-in setups, 6.24, 6.25*f*
Underground gas piping systems, as unsuitable for grounding electrodes, 9.10, 9.11*f*, 9.15
Underground tanks, as grounding electrodes, 9.14, 9.15*f*
Underwriters Laboratories, Inc. (UL)
 circuit breaker ratings and, 10.3–10.4, 10.7–10.8
 fuse classes and, 10.17, 10.18
 ground fault circuit interrupters (GFCI) and, 10.13
 handholes and, 5.12
 NEC and, 9.9
 requirements for protective enclosures, 11.18
 tests for interrupting rating, 10.7–10.8
 time-delay fuses and, 10.15
Ungrounded conductors, 9.5
 defined, 9.42
Unistrut, 7.13
 defined, 7.28
Unit circles, bending conduit, 4.8, 4.9*f*
Universal crimping tools, 8.17, 8.17*f*
Unlatch coils, 11.9
 defined, 11.35
Unused wire taps, ballasts, 3.22
Up cable pulls, 6.21, 6.22*f*

V

Vacuum fish tape systems, 6.2
Variable-speed drives, 2.26, 2.40–2.45, 2.40*f*, 2.41*f*, 2.42*f*, 2.43*f*
Varnish treatments, 1.36
VARs (volt-amperes-reactive), 1.34
Varying duty, 2.49
 defined, 2.68
Vector analysis, 1.21, 1.23–1.24
Vector diagrams, 1.7–1.8, 1.8*f*
Vectors, 1.7
Vertical adjustment splice plates, 7.11, 7.12*f*
Vertical pulls, runaways in, 6.27
Vertical raceways, supporting conductors in, 6.26–6.27, 6.26*f*
Very high-output (VHO) fluorescent lamps, 3.12
Visual angle, 3.2, 3.3*f*
Vitreous humor, 3.2, 3.2*f*
Voltage
 in inductive AC circuits, 1.12–1.14, 1.14*f*, 1.15, 1.20, 1.21, 1.22, 1.23
 magnetic fields and, 1.2–1.3
Voltage/current relationship
 capacitive circuits and, 1.18–1.20, 1.19*f*
 inductive circuits and, 1.12–1.14, 1.14*f*, 1.15, 1.20, 1.21, 1.22, 1.23
 resistive circuits and, 1.11, 1.12*f*
Voltage ratings
 for capacitors, 1.18
 of fuses, 10.14
Volt-amperes-reactive (VARs), 1.34
Voltmeter circuits, in ground testers, 9.31*f*, 9.32

W

Wall mounting
 cable trays and, 7.13, 7.15*f*
 defined, 7.28
Water-cooled motor enclosures, 2.46
Waterproof, 5.3
 defined, 5.21
Waterproof outlet boxes, 10.12
Waterproof polyester measuring tapes, 6.5
Watertight, 5.3
 defined, 5.21
Watertight, corrosion-resistant (NEMA Type 4X) enclosures, 5.3
Watertight (NEMA Type 4) enclosures, 5.3
Water-to-air cooled motor enclosures, 2.46
Waveforms, 1.2
Wavelengths, 1.5, 3.5
Weatherproof, 5.3
 defined, 5.21
Weatherproof connectors, 8.18–8.19
Weather-protected motor enclosures, 2.45
Wet locations
 boxes for, 5.2–5.3
 outdoor lighting and, 3.34–3.35, 3.35*f*
Winches, 6.17, 7.25
Winding coils, inductance, 1.12, 1.13*f*
Windings. *See also* Coils
 damper windings, 2.27
 encapsulated windings motor enclosures, 2.45
 primary windings, 1.35–1.37, 1.36*f*
 sealed windings motor enclosures, 2.46
 secondary windings, 1.35–1.37, 1.36*f*
 transformers and, 1.35, 1.36, 1.37, 1.38, 1.38*f*
Window cuts, cable stripping, 8.3, 8.4*f*, 8.5
Wire caddies, 6.2, 6.3*f*
Wire connections under 600 volts, 8.8–8.11, 8.8*f*, 8.9*f*, 8.10*f*
Wire dispensers, 6.2, 6.3*f*
Wire nuts, 3.46
Wire pulling setup
 cable reels, 6.7, 6.8*f*, 6.9*f*, 6.10*f*
 installing pull lines, 6.11, 6.12, 6.13, 6.13*f*, 6.14*f*
 preparing cable ends, 6.2, 6.11, 6.15–6.17
 preparing for conductors, 6.8, 6.10, 6.10*f*
 pulling grips, 6.11, 6.15, 6.15*f*, 6.16
 types of pulling lines, 6.17
Wire sorters, 6.33
Wire strippers, 8.2–8.3, 8.3*f*, 8.5, 8.6*f*
Working load rating, cable-pulling equipment, 6.17
Wound rotor induction motors, 2.19–2.20, 2.20*f*, 2.21, 2.21*f*
Wraps on puller drums, pulling force, 6.18
Wye-connected motors, 2.57, 2.57*f*, 2.58, 2.58*f*
Wye connections, 2.57, 2.57*f*, 2.58, 2.58*f*
Wyes
 cable trays and, 7.2, 7.4*f*
 defined, 7.28

Z

Zone lighting control circuit, 3.49, 3.49*f*